Theoretical Foundation Engineering

J.ROSS PUBLISHING *Classics*

TITLES IN THE SERIES

Architectural Acoustics
M. David Egan
ISBN 13: 978-1-932159-78-3, ISBN 10: 1-932159-78-9, 448 pages

Earth Anchors
By Braja M. Das
ISBN 13: 978-1-932159-72-1, ISBN 10: 1-932159-72-X, 242 pages

Limit Analysis and Soil Plasticity
By Wai-Fah Chen
ISBN 13: 978-1-932159-73-8, ISBN 10: 1-932159-73-8, 638 pages

Plasticity in Reinforced Concrete
By Wai-Fah Chen
ISBN 13: 978-1-932159-74-5, ISBN 10: 1-932159-74-6, 474 pages

Plasticity for Structural Engineers
By Wai-Fah Chen & Da-Jian Han
ISBN 13: 978-1-932159-75-2, ISBN 10: 1-932159-75-4, 606 pages

Theoretical Foundation Engineering
By Braja M. Das
ISBN 13: 978-1-932159-71-4, ISBN 10: 1-932159-71-1, 440 pages

Theory of Beam-Columns, Volume 1: Space Behavior and Design
By Wai-Fah Chen & Toshio Atsuta
ISBN 13: 978-1-932159-77-6, ISBN 10: 1-932159-77-0, 732 pages

Theory of Beam-Columns, Volume 2: In-Plane Behavior and Design
By Wai-Fah Chen & Toshio Atsuta
ISBN 13: 978-1-932159-76-2, ISBN 10: 1-932159-76-9, 513 pages

Theoretical Foundation Engineering

by Braja M. Das, Ph.D.
Henderson, Nevada

J.ROSS
PUBLISHING

ISBN-10: 1-932159-71-1
ISBN-13: 978-1-932159-71-4

Printed and bound in the U.S.A. Printed on acid-free paper
10 9 8 7 6 5 4 3 2 1

This J. Ross Publishing edition, first published in 2007, is an unabridged republication of the work originally published by Elsevier Science Publishers B.V., in 1987.

Library of Congress Cataloging-in-Publication Data

Das, Braja M., 1941-
Theoretical foundation engineering / by Braja M. Das.
p. cm.
Reprint. Originally published: Amsterdam : Elsevier Science, c1987.
Includes bibliographical references and index.
ISBN-10: 1-932159-71-1 (pbk : alk. paper)
ISBN-13: 978-1-932159-71-4 (pbk : alk. paper)
1. Foundations. 2. Soil mechanics. I. Title.
TA775.D228 2007
624.1'5—dc22 2006101132

Direct all inquiries to J. Ross Publishing, Inc., 5765 N. Andrews Way, Fort Lauderdale, FL 33309.

Phone: (954) 727-9333
Fax: (561) 892-0700
Web: www.jrosspub.com

To

My wife Janice

and

daughter Valerie

PREFACE

Theoretical Foundation Engineering is divided into five major chapters: Lateral Earth Pressure, Sheet Pile Walls, Bearing Capacity of Anchor Slabs and Helical Anchors, Ultimate Bearing Capacity of Shallow Foundations and Slope Stability. Many foundation engineering books now in print cover the above-mentioned topics which are directed toward providing a hands-on type of approach to design engineers. The goal of this book is somewhat different, in that it provides a state-of-the-art review of the advances made in understanding the behavior of earth-retaining and earth-supported structures. Most of the material is drawn from open literature, and a list of references is given at the end of each chapter. The text is liberally illustrated. To demonstrate the application of theoretical derivations, several worked-out problems are given in each chapter. In some instances, the final solutions to the problems under consideration are given, and detailed mathematical treatment has been deleted to keep the text within its scope and space limitations. It is my hope that this text will be useful to students, researchers, and practicing geotechnical engineers.

I am grateful to my wife Janice for typing the entire manuscript in camera-ready form. Several others have been extremely helpful during the preparation of the manuscript. Dr. Samuel P. Clemence, Professor and Chairman of the Department of Civil Engineering at Syracuse

University (New York), and Dr. Ronald B. McPherson, Associate Professor of Civil Engineering, New Mexico State University, Las Cruces, were very helpful in providing several reference papers. Dr. Donald E. Bobbitt and Mr. Richard S. Erdel of the A, B. Chance Company, Centralia, Missouri, have provided the photographs of helical anchors given in Appendix C. Mr. Robert L. Goodman, Director of Editorial Research-North America, for Elsevier Scientific Publishing Company, was most instrumentalin the initiation of this manuscript. Thanks are also due to

Dr. Haskell Monroe, President of The University of Texas at El Paso, for his continuous encouragement and assistance in my professional development.

— **Braja M. Das**
El Paso, Texas
March, 1987

CONTENTS

CHAPTER 1
Lateral Earth Pressure

1.1 INTRODUCTION

In the construction and design of many structures, a thorough knowledge of the earth pressure to which these structures will be subjected is required. Important among these structures are retaining walls, sheet pile bulkheads, temporary sheathings for supporting vertical or near-vertical cuts in soils, and earth anchors. The earth pressure to which the above types of retaining structures are subjected is commonly referred to as *lateral earth pressure*. The lateral earth pressure on an earth-retaining structure can generally be divided into three major categories. They are

1. Earth pressure at rest
2. Active earth pressure
3. Passive earth pressure

The above-stated earth pressure conditions can be explained by means of Fig. 1.1, in which *AB* is a retaining wall supporting soil at its left. The backfill soil has a horizontal ground surface. If the retaining wall does not yield at all from its original position, the horizontal lateral earth pressure at any depth to which the wall will be subjected is called the *lateral earth pressure at-rest*. The total force per unit length of the wall is equal to P_o. This type of condition is shown in Fig. 1.1a. However, if the wall tends to *yield away* from the soil sufficiently to create a plastic state of equilibrium in the soil mass located immediately behind it, the lateral earth pressure at any depth to which it will be subjected is called the *active earth pressure*. The conditions for development of active earth pressure are shown in Fig. 1.1b, c, and d. This can happen by *rotation of the wall* about its bottom or top, and also by lateral *translation* of the wall away from the

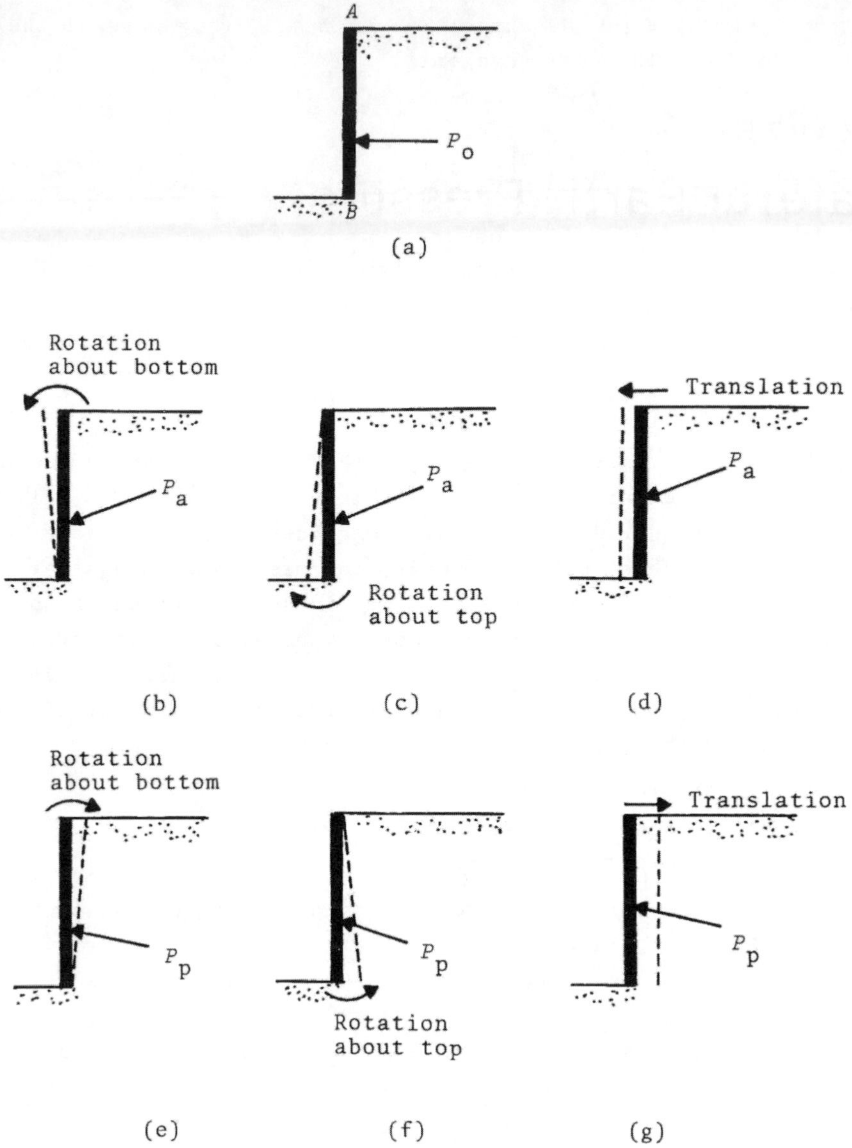

Figure 1.1. Definition of (a) Earth pressure at rest; (b), (c), and (d) Active earth pressure; (e), (f), and (g) Passive earth pressure.

backfill. The magnitude, direction, and the location of the resultant of the *active force* P_a per unit length of the wall will depend on several factors, such as the shear strength parameters of the backfill, profile of the top of the backfill, roughness of the wall and the mode of the wall movement.

Figure 1.1e, f, and g shows the condition where the wall yields into the soil mass located behind it by *rotation* about its bottom and top, and also by lateral *translation*. If the movement of the wall is sufficient to create a plastic state in the soil mass near it, failure will occur. This is referred to as the *passive state*. The lateral earth pressure on the wall at any given depth is called as the *passive earth pressure*. The magnitude, direction, and the location of the resultant of *passive force* will primarily depend on the same factors as listed above for the *active case*.

This chapter has been devoted to analyzing various lateral earth pressure theories presently available in literature. In all cases, it has been assumed that the shear strength of the soil can be defined by Mohr-Coulomb's failure criteria, or

$$s = p'\tan\phi + c \tag{1.1}$$

where s = shear strength

p' = effective normal stress

ϕ = drained angle of friction

c = cohesion

AT-REST EARTH PRESSURE

1.2 AT-REST EARTH PRESSURE

As explained in the preceding section, the *at-rest* condition exists if a wall does not yield at all (that is, either away from the soil mass or toward the soil mass). Figure 1.2 shows a retaining wall with a horizontal backfill. At a depth z below the ground surface, the vertical and horizontal effective stresses are p'_v and p'_h, respectively. The ratio of p'_h to p'_v is usually referred to as the *lateral earth pressure coefficient at-rest*, or

$$K_o = p'_h/p'_v \tag{1.2}$$

where K_o = earth pressure coefficient at-rest

The most common and widely used relationship of K_o for design is that given by Jaky (1944), which is of the form

$$K_o = 1 \quad \sin\phi \tag{1.3}$$

where ϕ = drained friction angle

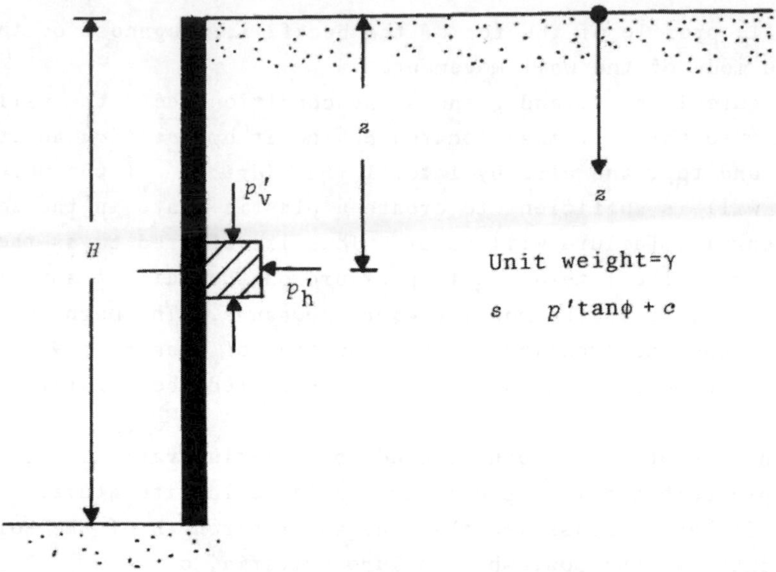

Figure 1.2. Definition of lateral earth pressure at-rest.

In recent times, it has been shown by several investigators that Eq. (1.3) is applicable only to normally consolidated loose soils. This fact can be explained by referring to Fig. 1.3, which shows a plot of p'_h against p'_v. Along the branch *OA* (initial loading), the soil is normally consolidated. Mayne and Kulhawy (1982) have analyzed the published results of about 121 soils. Based on their analysis they found that, for both sand and clayey soils, the best fit line from linear regressions analysis can be given as

$$K_{o(nc)} = 1 - 1.003\sin\phi \tag{1.4}$$

where $K_{o(nc)}$ = at-rest earth pressure coefficient for normally consolidated soils

So it appears that the original Jaky equation is generally valid for all normally consolidated sands and clays. Brooker and Ireland (1965) have given the following relationships for normally consolidated clays

$$K_{o(nc)} \quad 0.4 + 0.007(PI) \quad \text{(for } PI \text{ between 0 and 40)} \tag{1.5}$$

and

$$K_{o(nc)} \quad 0.64 + 0.001(PI) \quad \text{(for } PI \text{ between 40 and 80)} \tag{1.6}$$

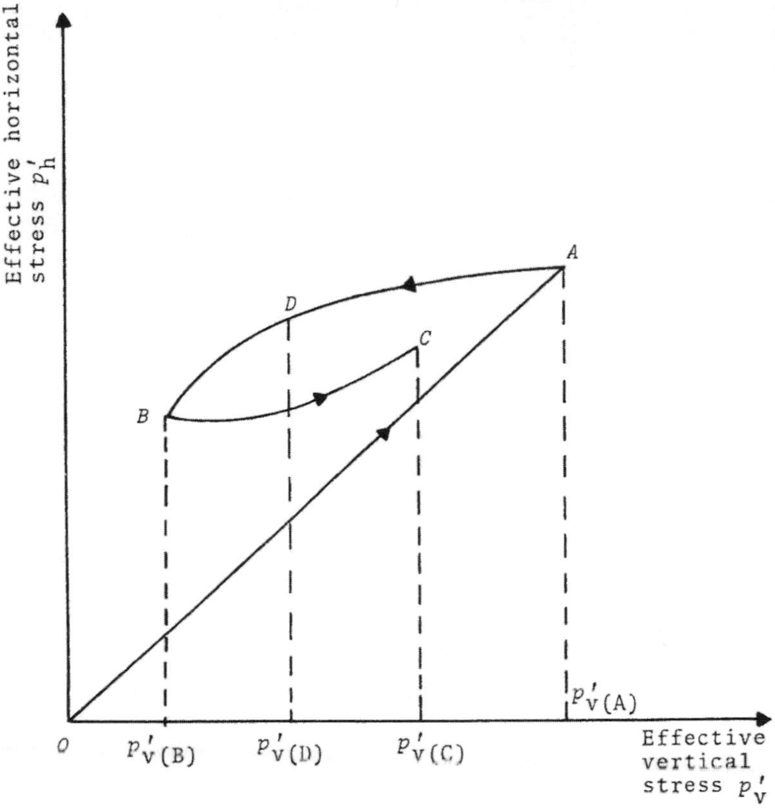

Figure 1.3. Variation of p'_h against p'_v for loading, unloading, and reloading cycles. (*NOTE:* *OA*--initial loading; *ADB*--initial unloading; and *BC*--reloading branch.)

It can be seen however from Fig. 1.3 that at point *D*, which is on the *initial unloading path*, the magnitude of K_o is higher than that observed along *OA* because the soil is overconsolidated. Schmidt (1966) has suggested that

$$K_{o(u)} \quad K_{o(nc)}(OCR)^{A} \tag{1.7}$$

where $K_{o(u)}$ = at-rest earth pressure coefficient during initial unloading

OCR = overconsolidated ratio = $p'_{v(A)}/p'_{v(D)}$

A   a coefficient

Mayne and Kulhawy (1982) have suggested that

$$A \approx \sin\phi \tag{1.8}$$

Thus

$$K_{o(u)} = (1 - \sin\phi)(OCR)^{\sin\phi} \tag{1.9}$$

The vertical effective stress p_v' and thus the lateral effective stress p_h' can change due to several factors, such as change of ground water table, cut and fill operation, and so forth. In Fig. 1.3, BC is a *reloading* path. For the first reloading path, Mayne and Kulhawy (1982) have presented the following relationships

$$K_{o(r)} = K_{o(nc)}\left[\frac{OCR}{OCR_{(max)}^{1-A}}\right] + \alpha\left[1 - \frac{OCR}{OCR_{(max)}}\right] \tag{1.10}$$

where $K_{o(r)}$ = at-rest earth pressure coefficient for first reloading path

$$OCR_{(max)} = \frac{p'_{v(A)}}{p'_{v(B)}} \tag{1.11}$$

$$\alpha = (\tfrac{3}{4})(1-\sin\phi) = (\tfrac{3}{4})[K_{o(nc)}] \tag{1.12}$$

Sherif, Fang, and Sherif (1984) have conducted several well-instrumented laboratory model tests to determine the magnitude of K_o. Figure 1.4 shows a summary of the results obtained from these

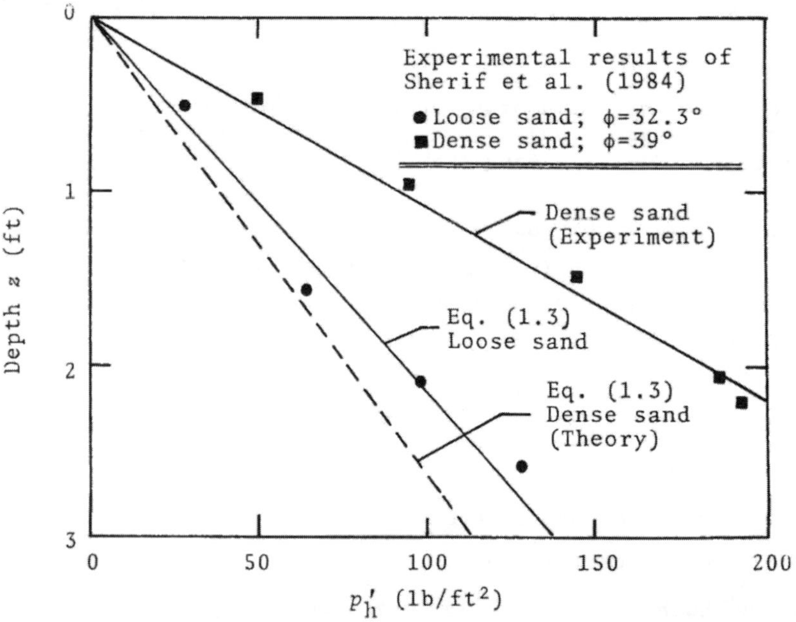

Figure 1.4. Laboratory tests for horizontal effective stress p_h' in loose and dense sand.

tests for loose and dense sands. From this figure, it can be seen that Jaky's original equation [Eq. (1.3)] is only in good agreement for loose state of sand compaction (normally consolidated). However for dense state of sand compaction, Eq. (1.3) grossly underestimates the at-rest earth pressure and hence K_o. This is probably due to the overstressing induced by compaction of sand. Based on their laboratory test results, Sherif et al. (1984) have proposed the following relationship for the at-rest earth pressure coefficient in sand

$$K_{o(d)} = (1-\sin\phi) + \left[\frac{\gamma_d}{\gamma_{d(min)}} - 1\right]5.5 \qquad (1.13)$$

where $K_{o(d)}$ at-rest earth pressure coefficient for design
$\gamma_{d(min)}$ = dry unit weight of sand in loosest state
γ_d = compacted dry unit weight of sand behind the retaining wall

Readers are also referred to the study of Feda (1984) for further review of K_o in sand. Kavazanjian and Mitchell (1984) have studied the variation of K_o in San Francisco Bay mud during consolidation. The study shows that for normally consolidated clay the magnitude of K_o increases with time. Behavioral reasoning indicates that for all soils K_o should ultimately approach one. For overconsolidated clays, the magnitude of K_o is greater than 1.0, however this should decrease with time to reach a value of one.

1.3 AT-REST FORCE AND RESULTANT

According to Eq. (1.2)

$$p'_h = K_o p'_v$$

The preceding equation is valid for effective stress conditions. If ground water is present, the total horizontal stress can be expressed as

$$p_h = K_o p'_v + u \qquad (1.14)$$

Figure 1.5a shows a retaining wall of height H with the ground water level at a depth $z=z_1$. A surcharge q (per unit area) is also applied at the ground surface. The *at-rest force* per unit length of the retaining wall can be determined by plotting the variation of p_h with depth z. Figure 1.5b, c, and d shows three cases which are described below.

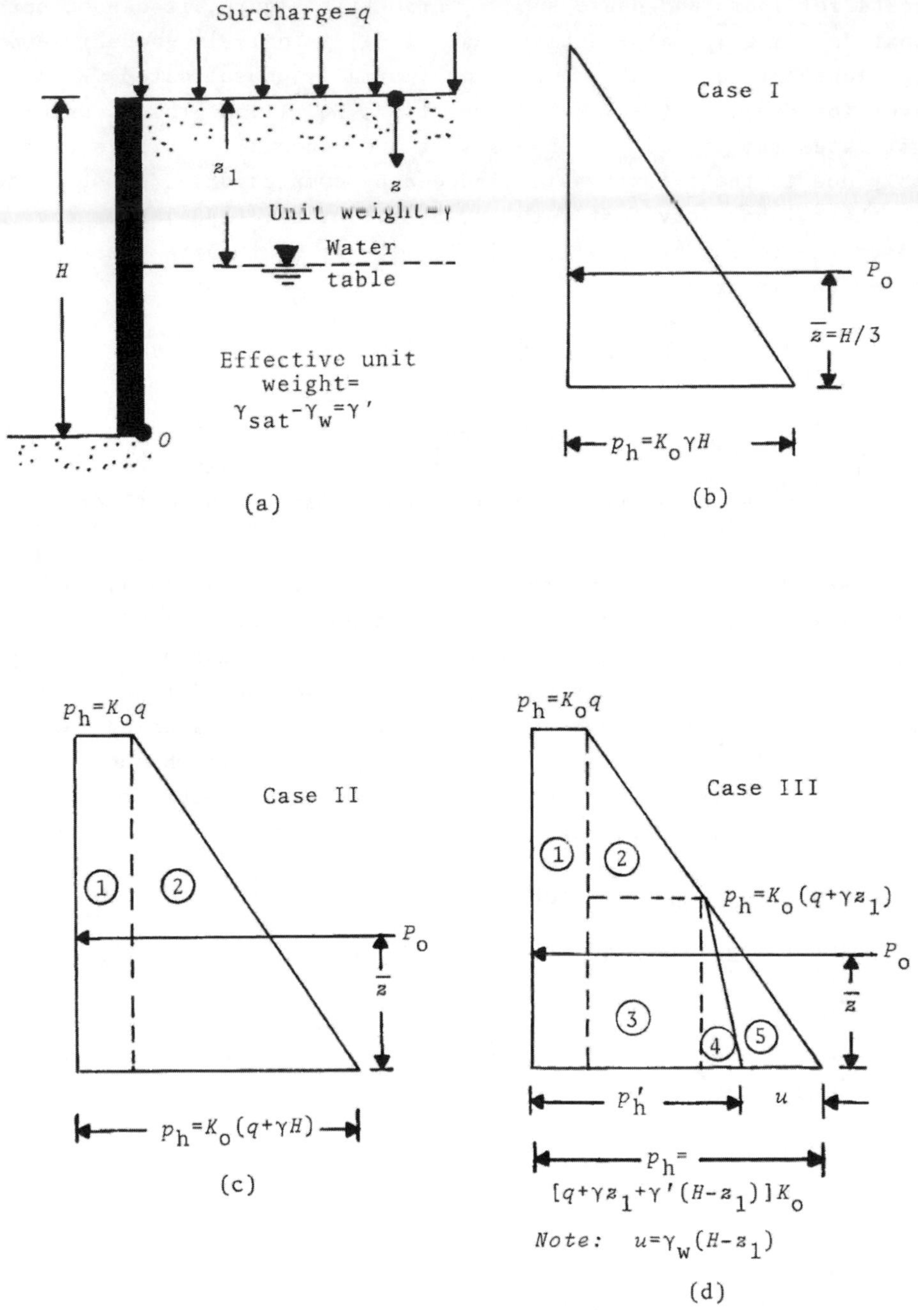

Figure 1.5. At-rest force and resultant.

Case I--$q=0$; $z_1 \geq H$ (Fig. 1.5b)

At $z=0$, $p_h'=0$

At $z=H$, $p_h = K_o \gamma H$; $u=0$

where γ = unit weight of soil

So, force per unit length

$$P_o \quad \text{area of the pressure diagram} = \tfrac{1}{2} K_o \gamma H^2$$

Case II--$q \neq 0$; $z_1 \geq H$ (Fig. 1.5c)

$$P_o \quad \text{Area 1 + Area 2}$$

$$= K_o q H + \tfrac{1}{2} K_o \gamma H^2$$

Case III--$q \neq 0$; $z_1 < H$ (Fig. 1.5d)

$$P_o = \Sigma \text{ Area 1 + Area 2 . . . + Area 5}$$

$$= K_o q H + \tfrac{1}{2} K_o \gamma z_1^2 + K_o \gamma z_1 (H - z_1) + \tfrac{1}{2} K_o \gamma' (H - z_1)^2 + \tfrac{1}{2} \gamma_w (H - z_1)^2$$

The location of the line of action of the resultant $\bar{z}$ can be determined by taking the moment of the areas about the bottom of the retaining wall, or

$$\bar{z} \quad \frac{\Sigma \text{ Moment of the areas about } O}{P_o} \tag{1.15}$$

In certain circumstances, it may be required to determine the at-rest force per unit length of a retaining wall which has an inclined backface as shown in Fig. 1.6. For this case, the at-rest force on the vertical face BC, $P_{o(BC)}$, should be determined first, or

$$P_{o(BC)} = \tfrac{1}{2} K_o \gamma H^2$$

The at-rest force on the back of the wall, P_o, can be given as

$$\vec{P}_o \quad \vec{P}_{o(BC)} + \vec{W}_s$$

where W_s weight of the soil wedge ABC per unit length of the wall

However

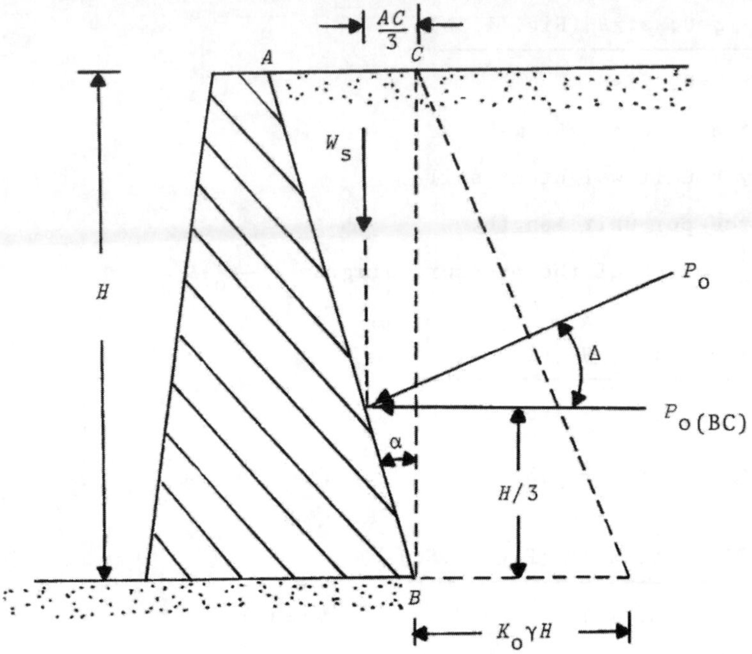

Figure 1.6. At-rest force for a retaining wall with an inclined backface.

$$W_s = (\gamma)[(\tfrac{1}{2})(H)(H)(\tan\alpha)] = \tfrac{1}{2}\gamma H^2 \tan\alpha$$

Hence, the magnitude of the force

$$P_o = \sqrt{P^2_{o(BC)} + W^2_s} = \frac{\gamma H^2}{2}\sqrt{K^2_o + (\tan\alpha)^2}$$

From Fig. 1.6, it is evident that $\bar{z}=H/3$. Also, the force P_o will be inclined at an angle $\Delta=\tan^{-1}(\tan\alpha/K_o)$ with the horizontal.

Example 1.1. A retaining wall is shown in Fig. 1.7a. Assume that both soil layers are overconsolidated and the K_o relationships follow Eq. (1.10). Given:

$OCR_{(max)}$ for sand = 1.5 $\quad$ $OCR_{(max)}$ for clay = 2.0
OCR for sand = 1.2 $\quad$ OCR for clay = 1.7

Determine the lateral force P_o per unit length of the wall assuming that the wall does not yield. Also determine the location of the line of action of the resultant force P_o.

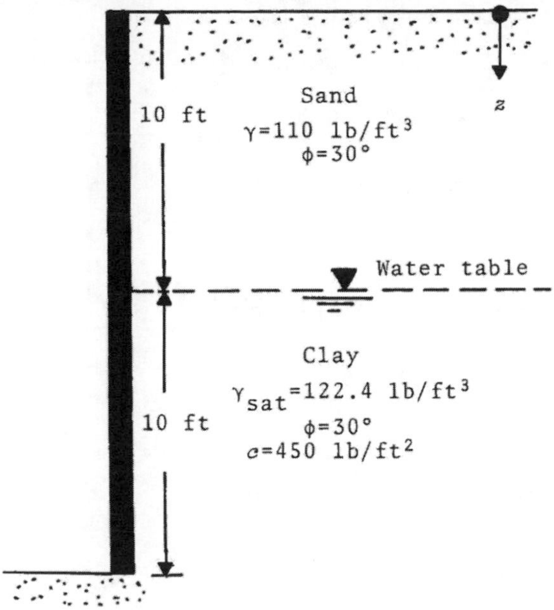

10 ft
Sand
γ=110 lb/ft³
φ=30°
z
Water table
Clay
γsat=122.4 lb/ft³
10 ft
φ=30°
c=450 lb/ft²
(a)

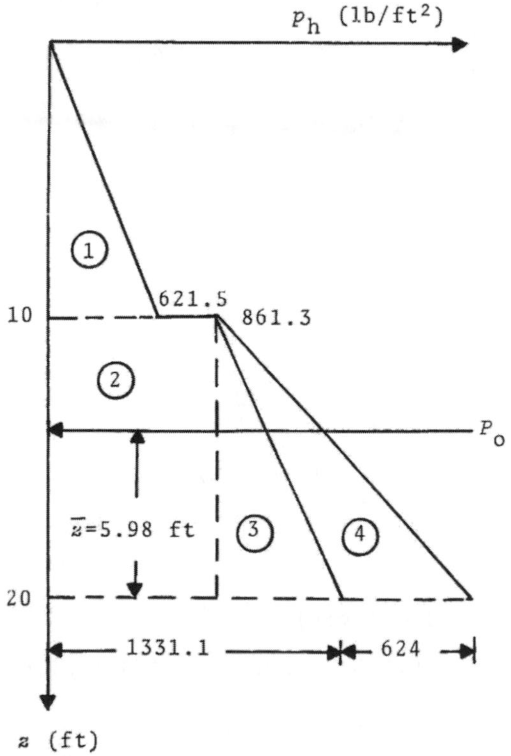

ph (lb/ft²)
1
621.5
10
861.3
2
(b)
Po
z̄=5.98 ft
3
4
20
1331.1
624
z (ft)

Figure 1.7.

Solution

For sand layer:

$$K_{o(nc)} = 1\text{-}\sin\phi \quad 1\text{-}\sin 30 = 0.5$$

$$\alpha \quad \tfrac{3}{4}[K_{o(nc)}] \quad (0.75)(0.5) = 0.375$$

$$A = \sin\phi = 0.5$$

$$K_{o(r)} = 0.5\left[\frac{1.2}{1.5^{1-0.5}}\right] + 0.375\left[1 - \frac{1.2}{1.5}\right]$$

$$= 0.49 + 0.075 \quad 0.565$$

For clay layer:

$$K_{o(nc)} \quad 1\text{-}\sin 20 = 1\text{-}0.342 = 0.658$$

$$\alpha \quad \tfrac{3}{4}(0.658) = 0.494$$

$$A \quad 0.342$$

$$K_{o(r)} = 0.658\left[\frac{1.7}{2.0^{1-0.342}}\right] + 0.494\left[1 - \frac{1.7}{2.0}\right]$$

$$= 0.709 + 0.074 = 0.783$$

Determination of at-rest pressure:

At $z=0$:

$$p'_v = 0$$
$$p'_h = 0$$
$$u = 0$$

At $z=10$ ft (inside sand layer):

$$p'_v = 10\gamma \quad (10)(110) = 1{,}100 \text{ lb/ft}^2$$
$$p'_h = K_{o(r)}p'_v = (0.565)(1{,}100) \quad 621.5 \text{ lb/ft}^2$$
$$u = 0$$

At z=10 ft (inside clay layer):

$$p'_h \quad K_{o(r)}p'_v \quad (0.783)(1{,}100) = 861.3 \text{ lb/ft}^2$$
$$u \quad 0$$

At $z=20$ ft:

$$p'_v \quad 10\gamma_{sand} + 10\gamma'_{clay} = (10)(110) + (10)(122.4\text{-}62.4)$$

$= 1{,}100 + 600 = 1{,}700\ \text{lb/ft}^2$

$p_h' = K_{o(r)} p_v' \quad (0.783)(1{,}700) \quad 1{,}331.1\ \text{lb/ft}^2$

$u = 10\gamma_w = (10)(62.4) \quad 624\ \text{lb/ft}^2$

Figure 1.7b shows the distribution of p_h with depth.

Determination of force P_o:

$P_o = \Sigma$ Areas of the pressure diagram

$A_1 = (\frac{1}{2})(10)(621.5)$	3,107.5
$A_2 \quad (10)(861.3)$	8,613
$A_3 \quad (\frac{1}{2})(10)(1{,}331.1-861.3)$	2,349
$A_4 = (\frac{1}{2})(10)(624)$	3,120
P_o	17,189.5 lb/ft

Determination of the line of action of the resultant force:

Taking the moment about the bottom of the wall

$$P_o\bar{z} \quad (3{,}107.5)(10 + \frac{10}{3}) + (8{,}613)(\frac{10}{2}) + (2{,}349)(\frac{10}{3}) + (3{,}120)(\frac{10}{3})$$

$$= 41{,}423 + 43{,}065 + 7{,}830 + 10{,}400 = 102{,}718$$

So, $\bar{z}$ = 5.98 ft

ACTIVE PRESSURE

1.4 RANKINE ACTIVE PRESSURE (ROTATION ABOUT BOTTOM OF WALL)

Consider a vertical wall of height H with a horizontal backfill as shown in Fig. 1.8a. The wall is assumed to be *frictionless*. The unit weight of the backfill is equal to γ. Initially, if the wall has not moved at all, the *vertical effective stress* on a soil element near the wall is

$$p_v' = \gamma z$$

The *horizontal effective stress* on the same soil element is equal to

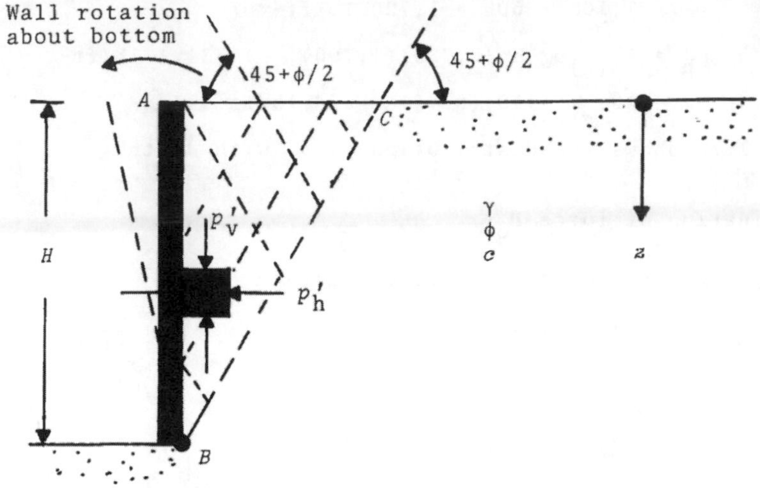

(a)

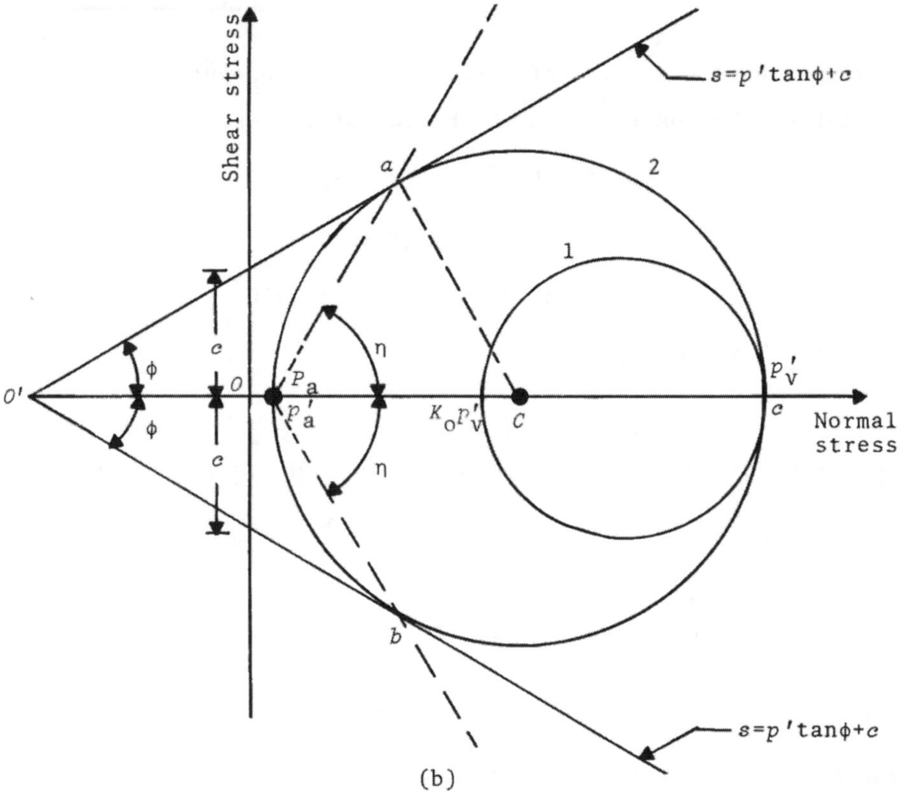

(b)

Figure 1.8. Rankine active earth pressure.

$$p'_h = K_o \gamma z \quad K_o p'_v$$

where K_o = at-rest earth pressure coefficient

There will be no shear stress on the faces of the soil element since the wall is frictionless. The corresponding Mohr's circle for the state of stress on the soil element is shown in Fig. 1.8b. (circle No. 1). If the wall *AB* moves away from the soil (that is, it moves to the left), the magnitude of p'_v will remain the same, however p'_h will gradually decrease. Ultimately it will reach a condition at which the horizontal stress $p'_h = p'_a$, and the correspon-ding Mohr's circle will be represented by circle No. 2. Note that this circle touches the Mohr-Coulomb failure envelope at points *a* and *b* which may be defined by the equation

$$s \quad p'\tan\phi + c \tag{1.1}$$

Mohr's circle No. 2 represents the failure condition, that is, the soil behind the wall (soil wedge *ABC*) has reached a plastic condition. The effective soil pressure p'_a is called the *Rankine active pressure*. Note that, for the Rankine active state, p'_v is the major principal stress and p'_a is the minor principal stress. At this point P_A on Mohr's circle No. 2 represents the *pole*. The reader is referred to Appendix A for definition and discussion of pole. The potential failure planes in soil at the Rankine active state can be represented by lines $P_A a$ and $P_A b$. Since $P_A c$ repre-sents the horizontal plane, the *failure planes* ($P_A a$ and $P_A b$) make an angle $\pm\eta$ with the horizontal. Also, from geometry it can be seen that

$$\eta \quad \pm(45+\phi/2) \tag{1.17}$$

Figure 1.8a shows the potential failure wedge with the above-men-tioned failure planes inclined at angles of $\pm(45+\phi/2)$. These lines are referred to as the *slip lines*.

The magnitude of the active pressure p'_a can be obtained from the geometry of Mohr's circle No. 2 shown in Fig. 1.8b. This can be done as follows

$$\overline{OC} = \frac{p'_v + p'_a}{2}$$

$$\overline{aC} \quad \frac{p'_v - p'_a}{2}$$

$$\overline{O'O} = c \cot\phi$$

So

$$\sin\phi = \frac{\overline{aC}}{\overline{O'O}+\overline{OC}} \quad \frac{\left(\frac{p'_v - p'_a}{2}\right)}{c\cot\phi + \left(\frac{p'_v + p'_a}{2}\right)}$$

or

$$p'_a \quad p'_v\left(\frac{1-\sin\phi}{1+\sin\phi}\right) \quad 2c\left(\frac{\cos\phi}{1+\sin\phi}\right)$$

Since

$$\frac{1-\sin\phi}{1+\sin\phi} = \tan^2(45-\phi/2)$$

and

$$\frac{\cos\phi}{1+\sin\phi} = \tan(45-\phi/2)$$

$$p'_a = p'_v\tan^2(45-\phi/2) \quad 2c\tan(45-\phi/2) = K_a p'_v - 2c\sqrt{K_a} \tag{1.18}$$

where K_a = Rankine active earth pressure coefficient

$$= \tan^2(45-\phi/2) \tag{1.19}$$

For granular soil backfill, $c=0$. So Eq. (1.18) becomes

$$p'_a \quad K_a p'_v \tag{1.20}$$

Figure 1.9a shows the distribution of p'_a with depth z for a horizontal granular soil backfill behind a retaining wall of height H. The active force P_a per unit length of the wall can be given as

$$P_a = \int_0^H p'_a dz = \int_0^H (K_a \gamma z) dz \quad \tfrac{1}{2}K_a\gamma H^2 \tag{1.21}$$

Similarly, the active force per unit length of the wall with *horizontal c-φ soil backfill* can be obtained as

$$P_a \quad \int_0^H (K_a p'_v - 2c\sqrt{K_a}) dz \quad \int_0^H (K_a\gamma z - 2c\sqrt{K_a}) dz$$

$$\tfrac{1}{2}K_a\gamma H^2 \quad 2cH\sqrt{K_a} \tag{1.22}$$

The corresponding pressure distribution diagram is shown in Fig. 1.9b. However, one thing should be kept in mind. From Eq. (1.18)

$$p'_a = K_a\gamma z - 2c\sqrt{K_a}$$

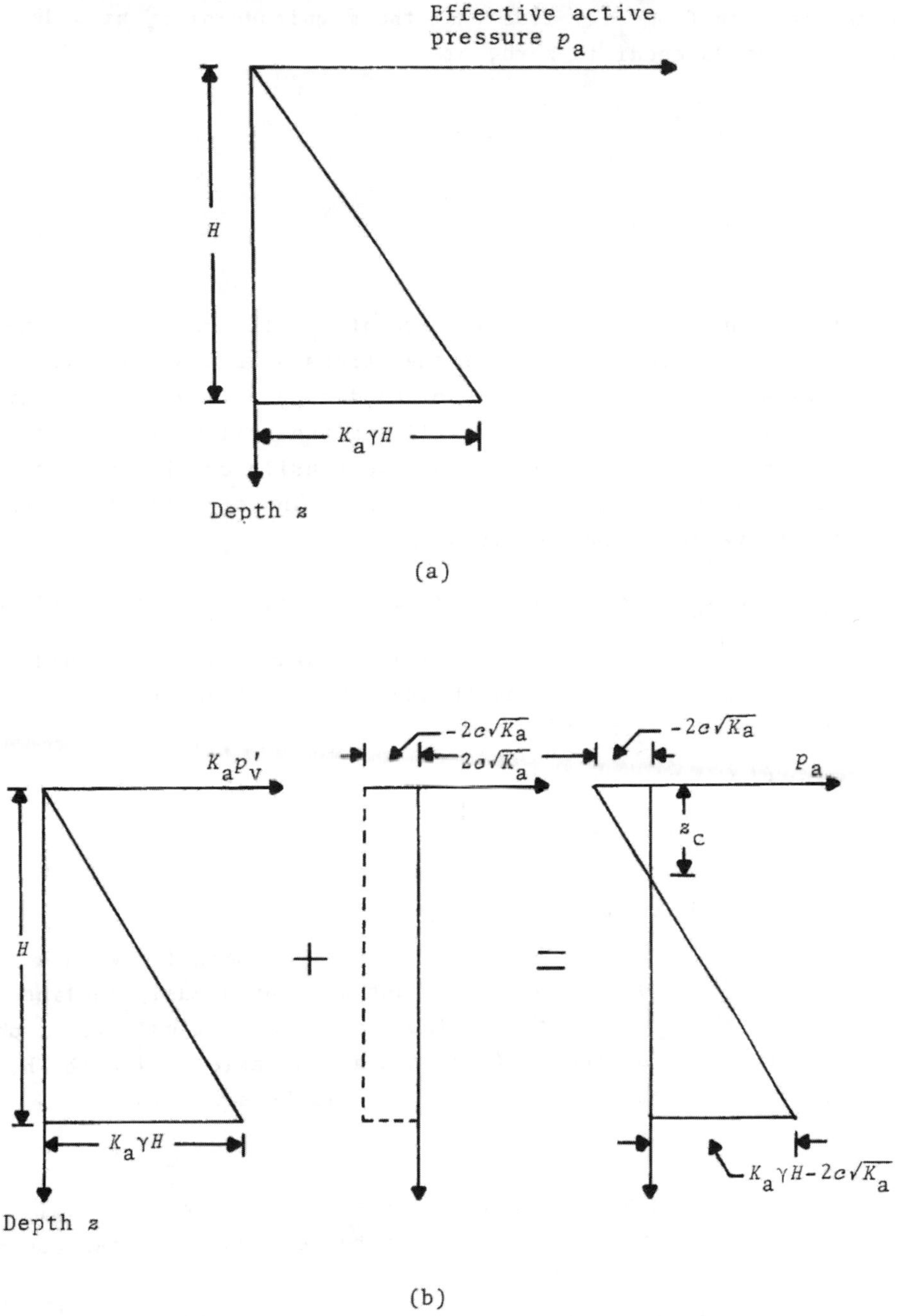

Figure 1.9. Distribution of Rankine active pressure behind a vertical wall with a horizontal backfill: (a) Granular soil; (b) Cohesive soil.

It can be seen from Fig. 1.9b that the magnitude of p_a' at a depth of $z = z_c$ becomes equal to zero, or

$$K_a \gamma z_c - 2c\sqrt{K_a} \quad 0$$

So

$$z_c \quad \frac{2c\sqrt{K_a}}{\gamma K_a} = \frac{2c}{\gamma\sqrt{K_a}} \tag{1.23a}$$

From the ground surface up to a depth of z_c, the magnitude of net p_a' is negative. This means that the stress acts away from the wall and towards the soil. This type of condition will eventually create a tensile crack between the wall and the soil up to a depth of z_c (*depth of tensile crack*). After the tensile crack develops, since there is no contact between the soil and the wall the magnitude of active force can be given as

$$P_a \quad \tfrac{1}{2}(H - z_c)(\gamma H K_a - 2c\sqrt{K_a}) = \tfrac{1}{2}K_a \gamma H^2 - 2cH\sqrt{K_a} + \frac{2c^2}{\gamma} \tag{1.23b}$$

As in the case of at-rest lateral pressure, if the ground water table is located at a depth $z < H$, the total active pressure at any depth can be given as

$$p_a = p_a' + u \tag{1.24}$$

where u = pore water pressure

Inclined Granular Backfill

Figure 1.10a shows a *frictionless wall* of height H with a granular backfill ($s = p'\tan\phi$; $c = 0$) which is continuously inclined at an angle i with the horizontal. The effective vertical stress on the inclined faces *ab* and *cd* (inclined at an angle i with the horizontal) of a soil element at a depth z will be equal to

$$p_v' = \gamma z \cos i$$

An enlarged drawing of this soil element *abcd* is shown in Fig. 1.10b. The stress p_v' on *ab* and *cd* can be resolved into two components

$$\sigma' \quad \text{normal stress on faces } ab \text{ and } cd = \gamma z \cos^2 i \tag{1.25}$$

and

$$\tau = \text{tangential (shear) stress on faces } ab \text{ and } cd$$
$$\gamma z \cos i \sin i \tag{1.26}$$

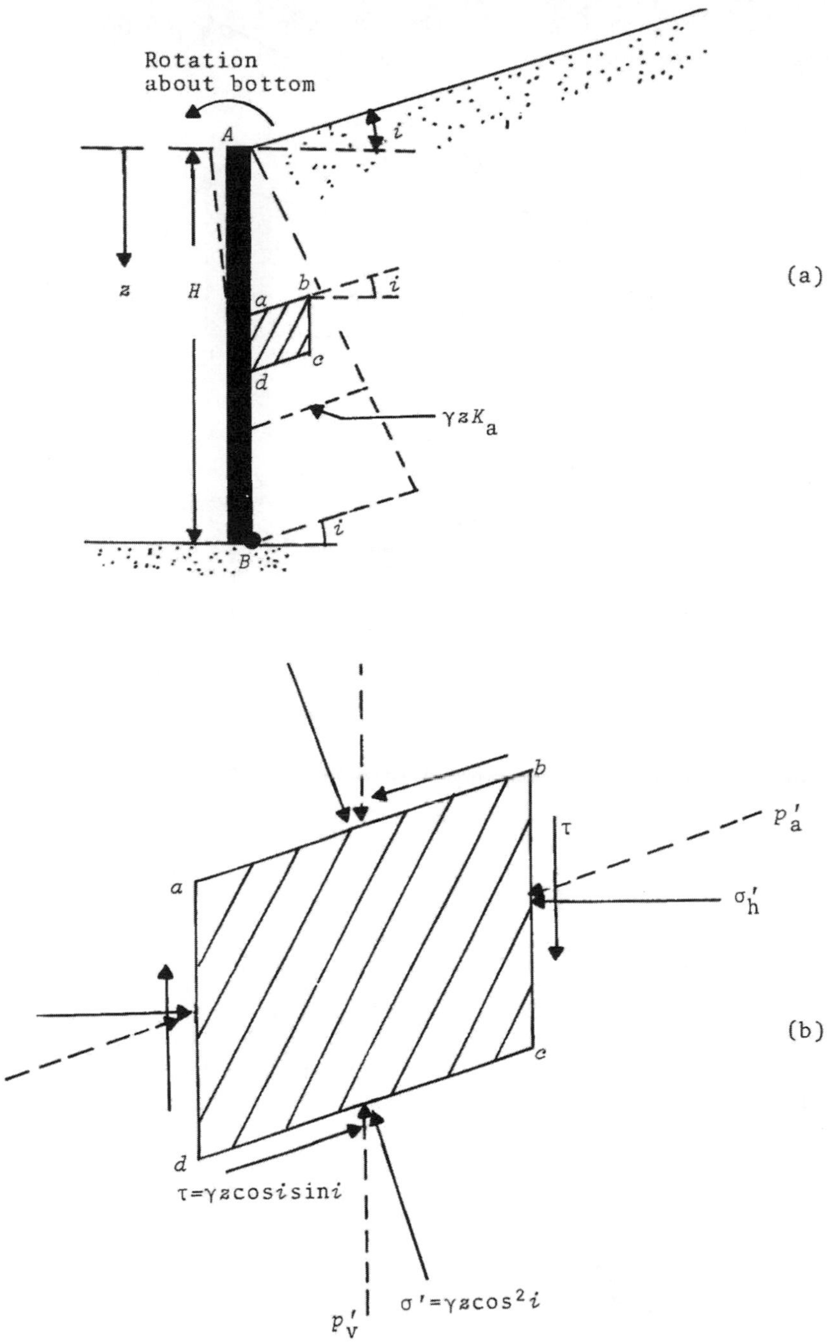

Figure 1.10. Rankine active pressure for a vertical wall with sloping granular backfill (c=0).

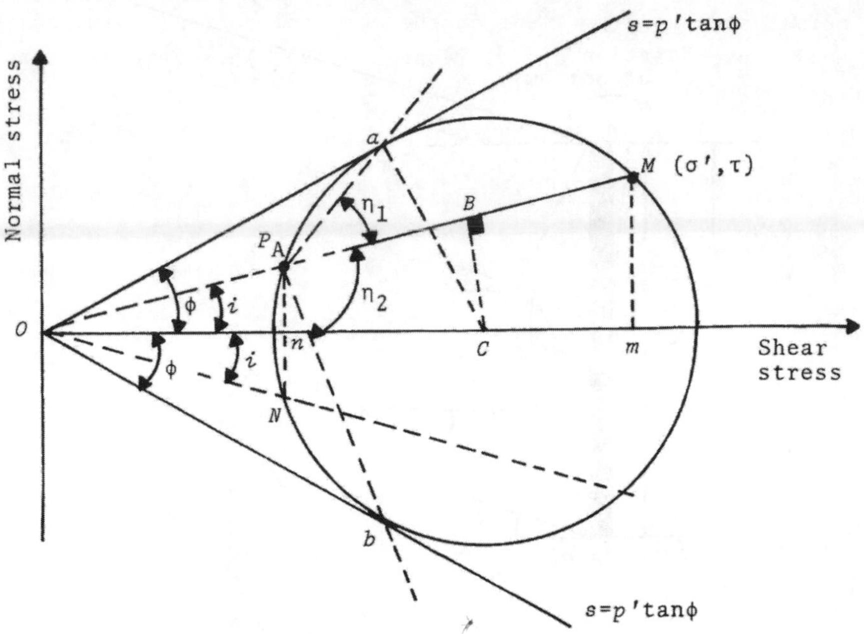

(c)

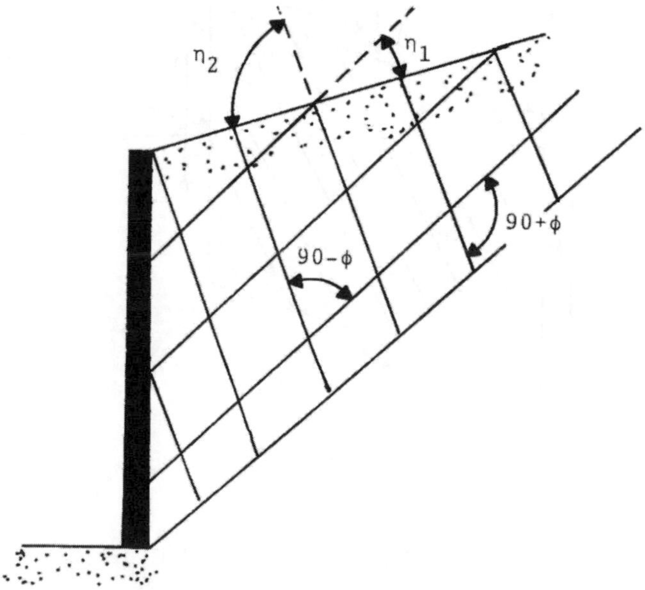

(d)

Figure 1.10. (Continued).

The normal and shear stresses on the vertical faces (*ad* and *bc*) of the soil element are σ_h' and τ, respectively. These are also shown in Fig. 1.10b. When the wall rotates sufficiently to create a plastic state (Rankine active state), the corresponding Mohr's circle will be as shown in Fig. 1.10c. Point *M* on the Mohr's circle represents the stress conditions on the plane *cd*. Note that the Mohr's circle is touching the Mohr-Coulomb failure envelopes since this is the state of failure. The pole for this state of stress is point P_a. Point *N* on the Mohr's circle represents the state of stress on the vertical plane *ad*. Thus, the Rankine active pressure on the vertical plane *ad* is

$$p_a' = \sqrt{\sigma_h'^2+\tau^2} = \sqrt{(\overline{On})^2+(\overline{nN})^2} = \overline{ON} \quad \overline{OP}_A$$

Also

$$\overline{Om} = \sigma' = \gamma z\cos^2 i$$

and

$$\overline{mM} \quad \gamma z\cos i\sin i$$

So

$$OM \quad \sqrt{(\overline{Om})^2+(\overline{mM})^2} \quad \sqrt{(\gamma z\cos^2 i)^2+(\gamma z\cos i\sin i)^2} \quad \gamma z\cos i$$

Now

$$\frac{\overline{OP}_A}{\overline{OM}} = \frac{\overline{OB}-\overline{BP}_A}{\overline{OB}+BM} \tag{1.27}$$

But

$$\overline{OB} = \overline{OC}(\cos i)$$

$$\overline{BP}_A \quad \overline{BM} = \sqrt{r^2-\overline{BC}^2}$$

where r = radius of the Mohr's circle = $OC\sin\phi$

$$\overline{BC} = \overline{OC}(\sin i)$$

So

$$\overline{BP}_A = \overline{BM} = \sqrt{(\overline{OC}\sin\phi)^2-(\overline{OC}\sin i)^2} = \overline{OC}\sqrt{\sin^2\phi-\sin^2 i}$$

Hence, from Eq. (1.27)

$$\frac{\overline{OP}_A}{\overline{OM}} \quad \frac{\overline{OC}(\cos i - \sqrt{\sin^2\phi-\sin^2 i})}{\overline{OC}(\cos i + \sqrt{\sin^2\phi-\sin^2 i})} = \frac{\cos i - \sqrt{\cos^2 i-\cos^2\phi}}{\cos i + \sqrt{\cos^2 i-\cos^2\phi}}$$

However, it has been shown that $\overline{OP}_A = p_a'$ and $\overline{OM} = \gamma z\cos i$. So

$$p_a' = \gamma z\left[\cos i\left(\frac{\cos i-\sqrt{\cos^2 i-\cos^2\phi}}{\cos i+\sqrt{\cos^2 i+\cos^2\phi}}\right)\right] = \gamma z K_a \tag{1.28}$$

where K_a Rankine active pressure coefficient (for inclined backfill)

The variations of K_a with i and ϕ are given in Fig. 1.11. The active force per unit length of the wall can be given as

$$P_a \quad \int_0^H p_a\,dz \quad \int_0^H (\gamma z K_a)\,dz = \frac{1}{2}K_a\gamma H^2 \tag{1.29}$$

Several important facts need to be kept in mind, and they are

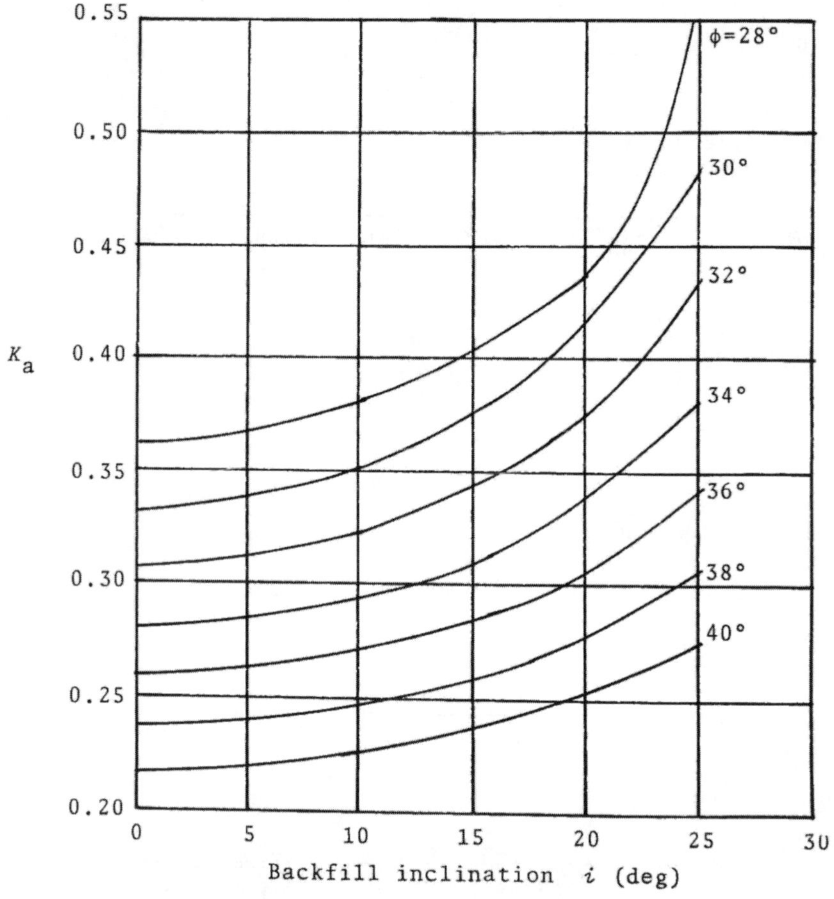

Figure 1.11. Variation of K_a with i and ϕ [Eq. (1.28)].

1. The active pressure p_a' is not the minor principal stress as is evident from the Mohr's circle given in Fig. 1.10c.

2. The active earth pressure p_a' at any given depth z makes an angle α with the horizontal.

3. From Fig. 1.10c, it is evident that slip lines in the failure zone make angles of η_1 and η_2 with the ground surface. The slip planes correspond to the lines P_Aa and P_Ab Figure 1.10d shows the directions of the slip lines. Note that the angle between the two sets of slip lines are $90+\phi$ and $90-\phi$.

1.5 ACTIVE PRESSURE WITH WALL FRICTION--COULOMB'S ACTIVE PRESSURE THEORY

Rankine active pressure refers to the condition where *the friction angle at the soil-wall interface is equal to zero.* However, in all practical cases, walls are not frictionless. Figure 1.12 shows the nature of the failure surface in soil at active state behind a retaining wall with friction. The actual failure surface is *curved*; however, calculations have shows that, for active condition, the approximation of a plane failure surface in soil yields fairly good results. The first analysis for determination of active earth pressure behind a retaining wall with a plane

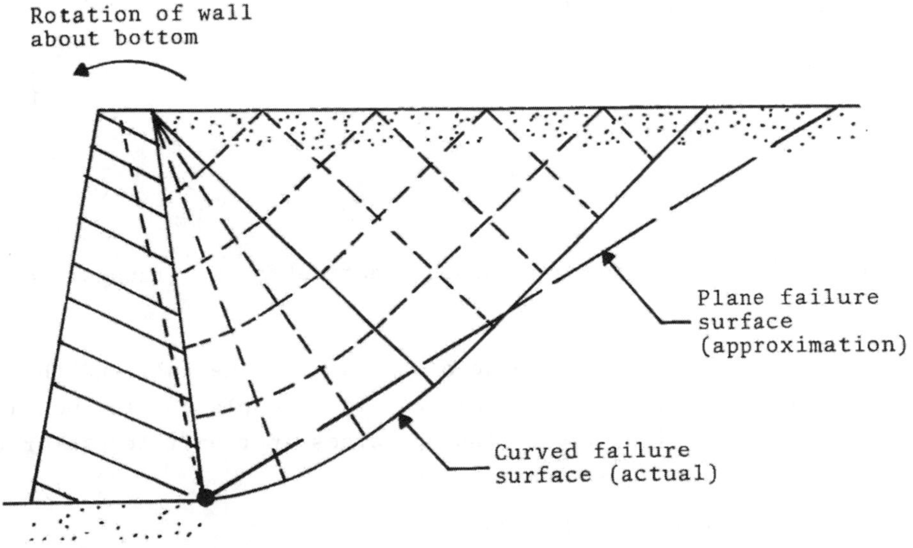

Figure 1.12. Failure surface in soil for rotation of wall (with friction) about bottom.

failure assumption is generally credited to Coulomb (1776). Coulomb's active earth pressure theory can be explained by referring to Fig. 1.13. The retaining wall has a *sloping granular backfill*, and the shear strength of the soil can be given by the equation $s = p' \tan\phi$.

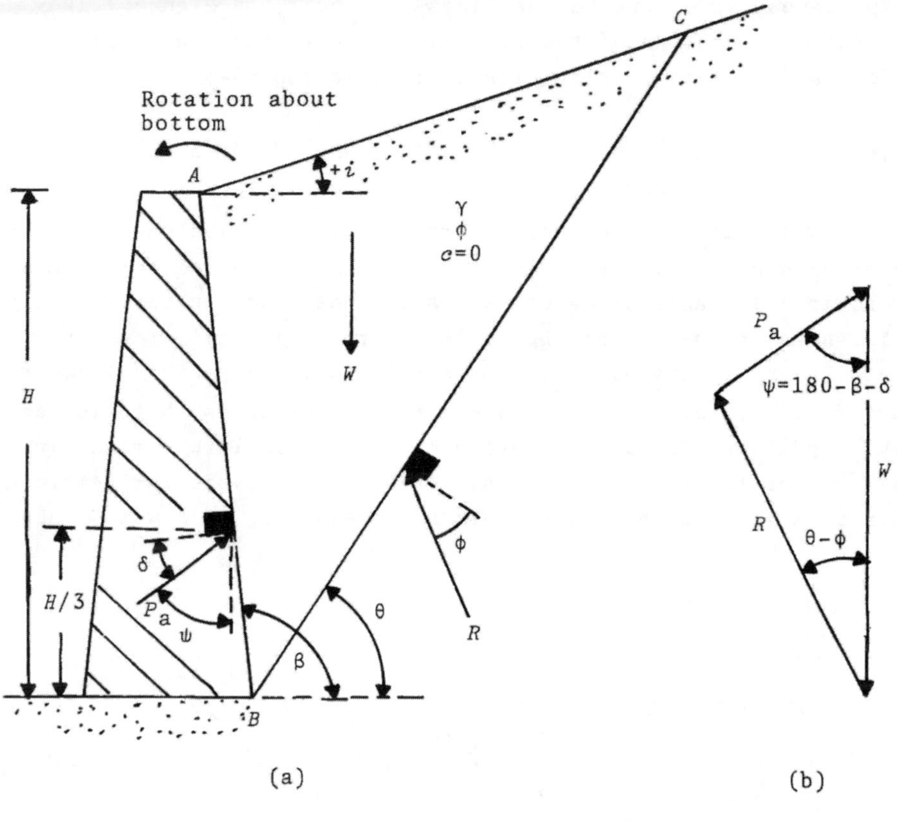

Figure 1.13. Derivation of Coulomb's active force (granular backfill; $c=0$).

The angle of friction between the back of the wall and the soil is equal to δ. Let *BC* be a trial failure plane. In such a case, for equilibrium, the following forces (per unit length of the wall) will act on the soil wedge *ABC*:

1. W--Weight of the wedge
2. R--Resultant of the normal and shear forces along the trial failure surface *BC*
3. P_a--Active force inclined at an angle δ to the normal

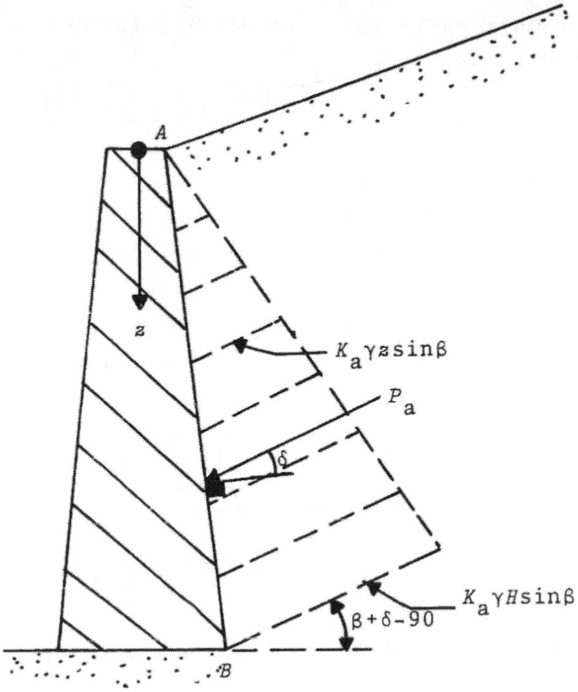

(c)

Figure 1.13. (Continued).

drawn to the back side of the wall.

The force triangle for the wedge equilibrium system is also shown in Fig. 1.13b. From the force triangle, using the law of sines

$$P_a = (W)\frac{\sin(\theta-\phi)}{\sin[(180°)-(\theta-\phi)-(180-\beta-\delta)]} = (W)\frac{\sin(\theta-\phi)}{\sin(\beta+\delta-\theta+\phi)} \quad (1.30)$$

However, the wedge *ABC* is a trial failure wedge. The maximum active force can be obtained by using the theorem of maxima and minima, or

$$\frac{\partial p_A}{\partial \theta} = 0 \quad (1.31)$$

The solution of the preceding equation results in

$$p_A = \frac{1}{2}K_a \gamma H^2 \quad (1.32)$$

where H = height of the wall

K_a = Coulomb's earth pressure coefficient

$$= \left[\frac{\sin(\beta-\phi)}{\sin\beta \left\{ \sqrt{\sin(\beta+\delta)} + \sqrt{\frac{\sin(\phi+\delta).\sin(\phi-i)}{\sin(\beta-i)}} \right\}} \right]^2 \quad (1.33)$$

The nature of variation of the Coulomb's active earth pressure against the wall is shown in Fig. 1.13b.

In using Eq. (1.33) for K_a, the proper sign convention for the angle i (Fig. 1.13c) should be used. Figure 1.14 shows a plot of K_a for various values of ϕ, δ, and i for a retaining wall with a vertical back (β=90°). In most practical design works, the angle

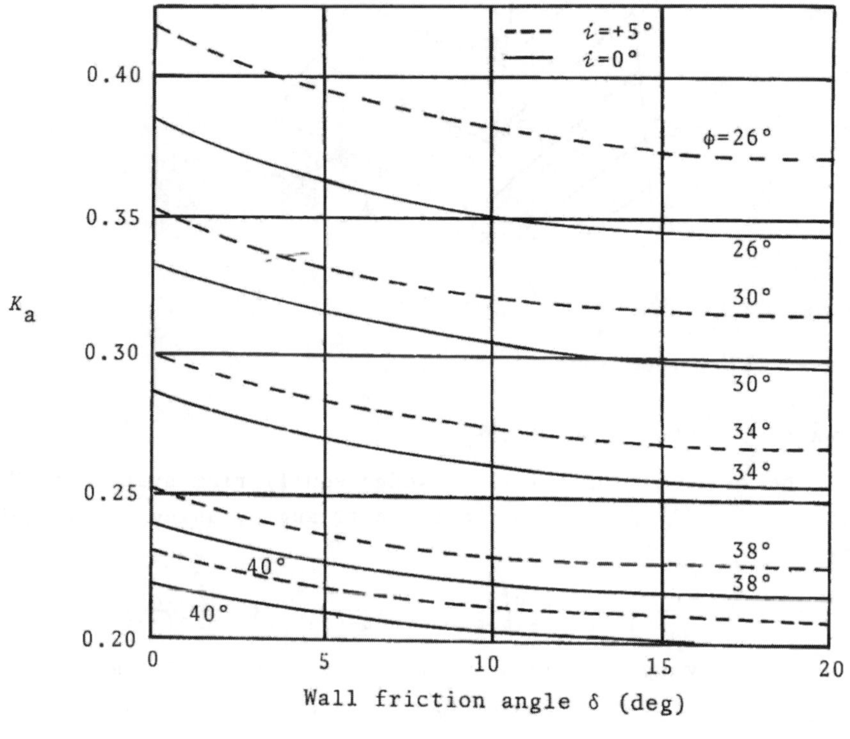

Figure 1.14. Plot of K_a [Eq. (1.33)] for vertical backface of the wall (β=90°).

of wall friction δ is assumed to be between $\phi/2$ and $2\phi/3$. Table 1.1 provides the values of K_a for various values of ϕ, β, and i with $\delta=2\phi/3$.

In special cases, for a vertical wall (β=90°) with a horizontal granular backfill (i=0°) Eq. (1.32) simplifies to the form

Table 1.1. Values of K_a [Eq. (1.33)] For Various Values of ϕ, β, and i With $\delta=2\phi/3$

β (deg)	i (deg)	ϕ (deg) 28	30	32	34	36	38	40
90	0	0.321	0.297	0.275	0.254	0.235	0.217	0.200
	+5	0.343	0.317	0.292	0.269	0.248	0.228	0.210
	+10	0.370	0.340	0.312	0.287	0.263	0.242	0.221
95	0	0.359	0.335	0.313	0.292	0.272	0.254	0.236
	+5	0.383	0.358	0.333	0.310	0.288	0.268	0.249
	+10	0.416	0.386	0.358	0.331	0.307	0.285	0.264
100	0	0.400	0.377	0.355	0.334	0.314	0.295	0.277
	+5	0.431	0.404	0.379	0.356	0.334	0.313	0.294
	+10	0.469	0.438	0.409	0.382	0.357	0.334	0.313
105	0	0.448	0.425	0.402	0.381	0.362	0.343	0.325
	+5	0.484	0.458	0.432	0.409	0.387	0.366	0.346
	+10	0.529	0.497	0.468	0.441	0.416	0.392	0.370
110	0	0.503	0.479	0.457	0.437	0.417	0.398	0.381
	+5	0.546	0.519	0.494	0.471	0.448	0.427	0.407
	+10	0.599	0.568	0.538	0.511	0.485	0.461	0.438

$$P_a = \frac{\gamma H^2}{2\cos\delta}\left[\frac{1}{(1/\cos\phi) + \sqrt{\tan^2\phi+\tan\phi\tan\delta}}\right]^2 \tag{1.34}$$

Equation (1.34) is for the estimation of active force P_a against a retaining wall with granular backfill having an inclination of i with the horizontal. This equation can be modified somewhat to include the effect of surcharge (q per unit area) on the surface of the backfill as shown in Fig. 1.15. Let ABC be the failure wedge. The weight of this soil wedge per unit length is equal to

$$W_{(ABC)} = \tfrac{1}{2}\gamma HL\left[\frac{\sin(\beta-i)}{\sin\beta}\right] \tag{1.35}$$

The total weight within the wedge ABC (per unit length) which includes the surcharge is equal to

$$W_{(total)} = W_{(ABC)} + q \quad \tfrac{1}{2}\gamma HL\left[\frac{\sin(\beta-i)}{\sin\beta}\right] + qL$$

or

$$W_{(total)} = \tfrac{1}{2}\gamma_{eq}HL\left[\frac{\sin(\beta-i)}{\sin\beta}\right] = \tfrac{1}{2}\gamma HL\left[\frac{\sin(\beta-i)}{\sin\beta}\right] + qL$$

where γ_{eq} = equivalent unit weight of the backfill material which will include the effect of surcharge

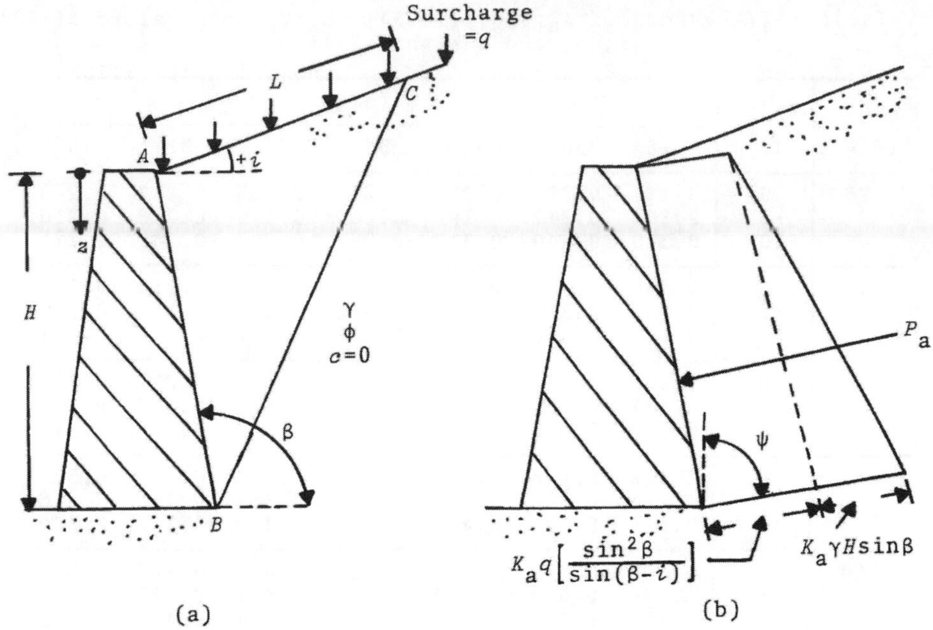

Figure 1.15. Coulomb's active pressure with a surcharge on the backfill. (*Note:* ψ=180°-β-δ.)

So

$$\gamma_{eq} \quad \gamma + \left[\frac{\sin\beta}{\sin(\beta - i)}\right] \frac{2q}{H} \tag{1.36}$$

Now, the equivalent unit weight can be used in Eq. (1.32 in place of γ to determine the active force on the wall as

$$P_a = \tfrac{1}{2} K_a \gamma_{eq} H^2 \tag{1.37}$$

The distribution of the active pressure along the wall back-face for such a condition can be obtained by replacing H with z in Eq. (1.37). So

$$P_a = \tfrac{1}{2} K_a \gamma_{eq} z^2$$

Hence, pressure

$$p_a = \frac{\partial P_a}{\partial (z/\sin\beta)} = \frac{\partial P_a}{\partial z}(\sin\beta) = \tfrac{1}{2} K_a \sin\beta \left[\frac{\partial}{\partial z}(\gamma_{eq} z^2)\right]$$

$$= \tfrac{1}{2} K_a \sin\beta \left\{\frac{\partial}{\partial z}\left[\gamma z^2 + \frac{\sin\beta}{\sin(\beta - i)}\left(\frac{2qz^2}{z}\right)\right]\right\}$$

$$= K_a \gamma z \sin\beta + K_a\left[\frac{\sin^2\beta}{\sin(\beta-i)}\right]q \tag{1.38}$$

The variation of p_a along the wall face is shown in Fig. 1.15b.

Example 1.2. Refer to Fig. 1.15a. Given: H=20 ft, ϕ=30°, δ=20°, i=5°, β=95°, q=2,000 lb/ft^2, and γ=115 lb/ft^3. Determine Coulomb's active force and the location of the line of action of the resultant P_a.

Solution

For β=95°, i=5°, δ=20°, and ϕ=30°, K_a=0.358 [Eq. (1.33)].

$$P_a = \tfrac{1}{2}K_a\gamma_{eq}H^2 = \tfrac{1}{2}K_a\left[\gamma + \frac{2q}{H}\,\frac{\sin\beta}{\sin(\beta-i)}\right]H^2$$

$$= \underbrace{\tfrac{1}{2}K_a\gamma H^2}_{P_{a(1)}} + \underbrace{K_a Hq\left[\frac{\sin\beta}{\sin(\beta-i)}\right]}_{P_{a(2)}}$$

$$(0.5)(0.358)(115)(20)^2 + (0.358)(20)(2,000)\left[\frac{\sin 95}{\sin(95-5)}\right]$$

$$= 8,234 + 14,265.5 = 22,499.5 \text{ lb/ft}$$

Location of the line of action of the resultant:

$$P_a\bar{z} = P_{a(1)}(H/3) + P_{a(2)}(H/2)$$

or

$$\bar{z} = \frac{(8,234)(20/3) + (14,265.5)(20/2)}{22,499.5} = \frac{54,893.3 + 142,655}{22,499.5}$$

$$= \underline{8.78 \text{ ft}} \text{ (measured vertically from the bottom of the wall)}$$

1.6 GRAPHICAL SOLUTIONS FOR COULOMB'S ACTIVE FORCE

There are several graphical procedures for determination of Coulomb's active force behind a retaining wall. Two of these procedures will be discussed in this section. They are Poncelet's procedure and Culmann's procedure.

Poncelet's Procedure

In order to use this procedure, the following steps need to be followed (Fig. 1.16).

1. Draw the retaining wall and the constantly sloping backfill to a certain scale.

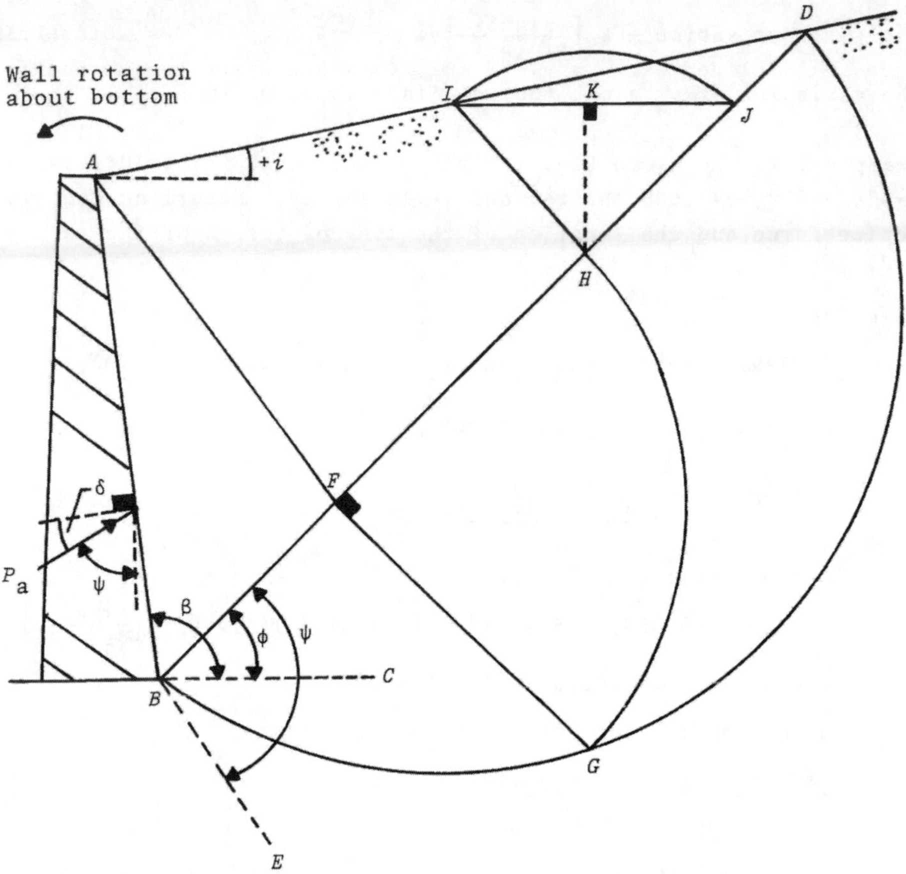

Figure 1.16. Poncelet's graphical solution of Coulomb's active force. (*Note:* Height of wall=*H*; granular backfill--*c*=0,)

2. Draw line *BD* which makes an angle ϕ with the horizontal (that is $\angle CBD=\phi$).
3. Draw line *BE* which makes an angle ψ=180-β-δ with *BD*.
4. Draw a semicircle with *BD* as the diameter.
5. Draw *AF* parallel to *BE*.
6. Draw *FG perpendicular* to *BD*.
7. Draw the arc *GH* with *FG* as the radius.
8. Draw *HI* parallel to *AF* (or *BE*).
9. Draw arc *IJ* with *HI* as the radius.
10. Draw line *IJ*.
11. Calculate the force as

$$P_a = (\text{Area of triangle } IHJ)\gamma = \frac{1}{2}(\overline{IJ})(\overline{HK})\gamma$$

The resultant force P_a acts at a distance $H/3$ above the bottom of the wall inclined at an angle δ to the normal drawn to the backface of the wall.

Culmann's Procedure

This procedure is a versatile one and can take into account any type of irregular backfill and surcharges such as line loads and strip loads. The graphical procedure is explained below in a step-by-step manner.

1. Draw the retaining wall and the backfill to a given scale as shown in Fig. 1.17.

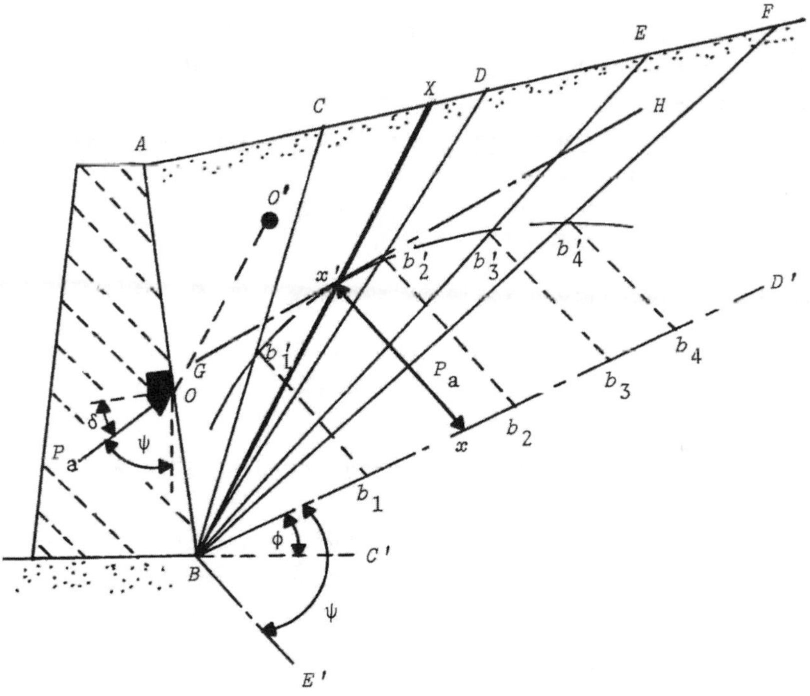

Figure 1.17. Culmann's procedure for graphical determination of Coulomb's active force with granular backfill. (*Note:* $\angle ABC' = \beta$.)

2. Draw several trial failure wedges such as *ABC*, *ABD*, *ABE*, and so forth.

3. Determine the areas of the wedges and, thus, their weights (per unit length of the wall) as

$$W_1 = (\text{Area of } ABC)(\gamma)$$

$$W_2 = (\text{Area of ABD})(\gamma)$$

.

.

.

4. Draw line BD' which makes an angle ϕ with the horizontal (line BC').

5. Determine $\psi = 180° - \beta - \delta$.

6. Draw line BE' which makes an angle ψ with BD'.

7. Adopt a load scale.

8. With the load scale adopted in Step 7, draw $Bb_1 = W_1$, $Bb_2 = W_2$, . . . on line BD'.

9. Draw b_1b_1', b_2b_2', . . . parallel to BE'.

10. Join points b_1', b_2', . . . by a smooth curve. This is called the *Culmann's line*.

11. Draw line GH parallel to BD' which will be tangent to the Culmann line at point x'.

12. Draw line $x'x$ parallel to bine BE'.

13. The active force is given by

$$P_a = (\overline{x'x})(\text{load scale})$$

The failure wedge can now be determined to be ABX. The location of the line of action of the resultant can be determined as follows.

1. Determine the centroid of the failure wedge ABX. Let this be O'.

2. Draw a line $O'O$ parallel to BD'. This line intersects the back of the wall at O. The line of action of the resultant will pass through point O (Terzaghi, 1943).

The above procedure can be modified to take into account a *line load* located on the surface of the backfill, such as q as shown in Fig. 1.18. The entire procedure for construction of the Culmann's line without a line load will be the same except for Step 3, in which the weights of the wedges are obtained. As shown in Fig. 1.18, let the line load be located at C. In that case

$$W_1 \quad (\text{Area of } ABC)(\gamma) + q$$

$$W_2 \quad (\text{Area of } ABD)(\gamma) + q$$

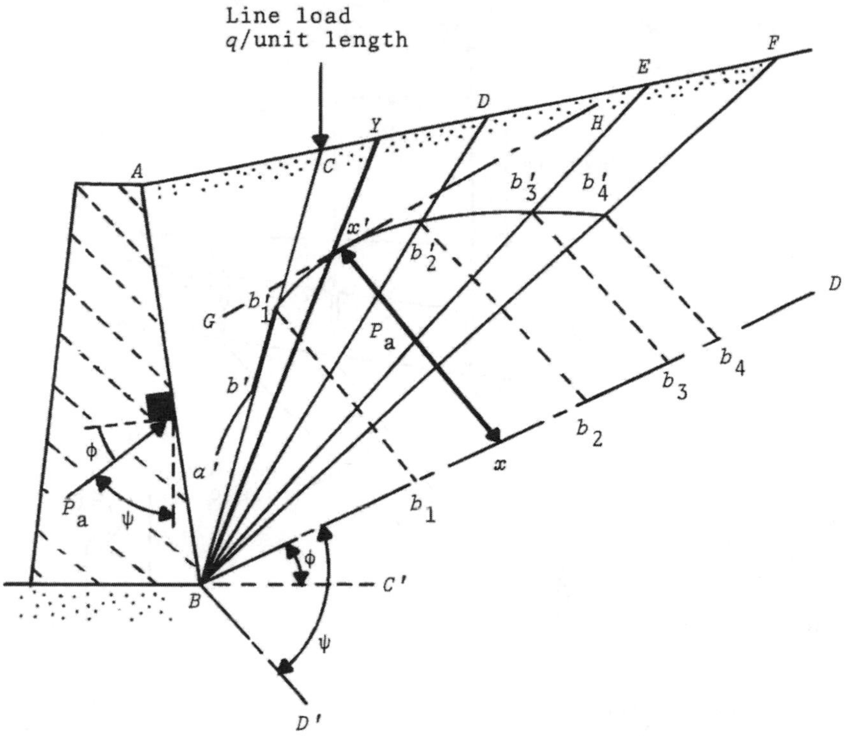

Figure 1.18. Culmann's graphical procedure for determination of Coulomb's active pressure with a line load as surcharge. (*Note:* Backfill is granular.)

.

.

.

Based on the above weights of the wedges, the Culmann's line can be darwn and P_a can be determined. This is shown in Fig. 1.18. Note that the Culmann's line is now $a'b'b_1'b_2'b_3'b_4'$, and the failure wedge with the line load is *ABY*.

1.7 GRAPHICAL PROCEDURE FOR COULOMB'S ACTIVE FORCE FOR A c-ϕ SOIL BACKFILL

The graphical procedures described in Section 1.6 are for estimation of the active force behind a retaining wall with a granular backfill. A similar procedure can also be adopted to estimate the active force for a c-ϕ type of soil as a backfill material. This can be explained with reference to Fig. 1.19. Figure 1.19a

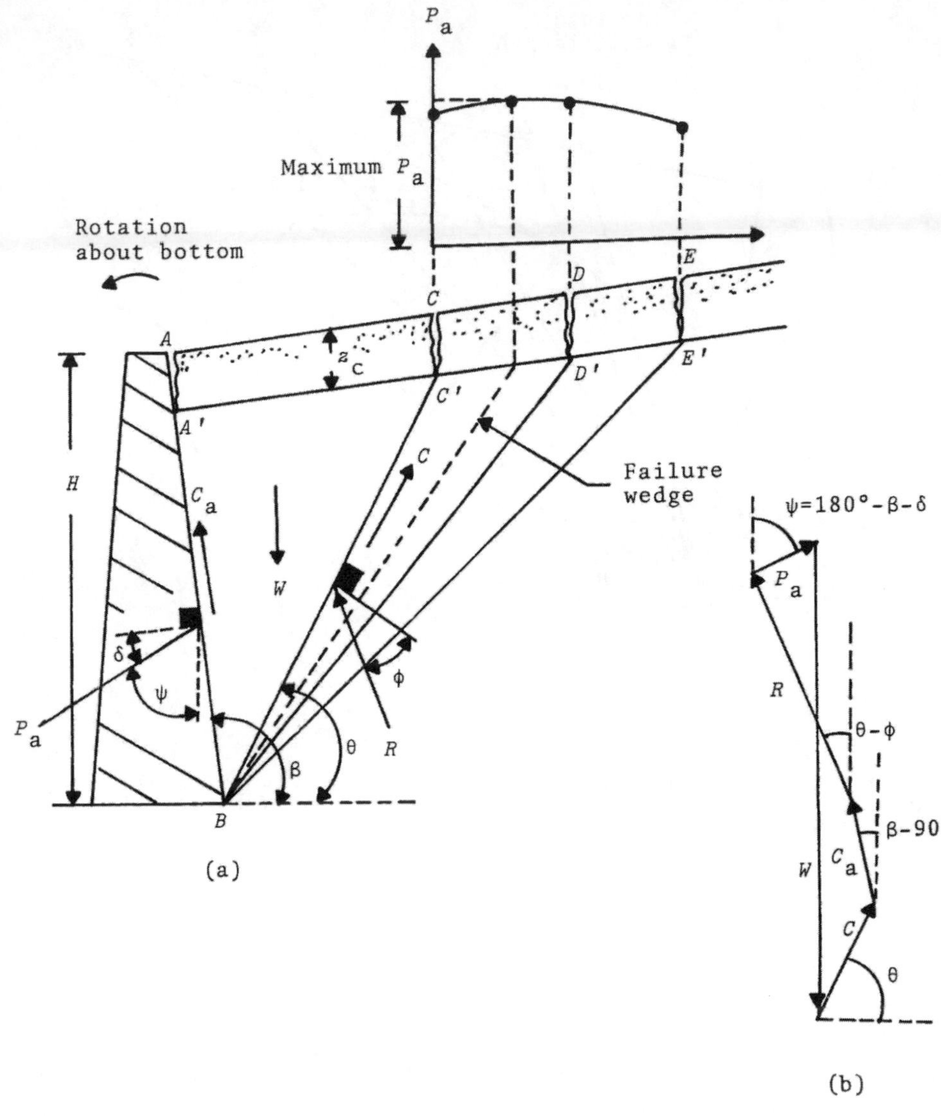

Figure 1.19. Coulomb's active force for c-ϕ soil as backfill material.

shows a retaining wall of height H. The shear strength of the backfill can be given by the relationship $s=p'\tan\phi+c$. For cohesive soil, it is possible that with the passage of time tensile cracks in the backfill will develop. The maximum depth of the tensile cracks may be given as

$$z_c = \frac{2c}{\gamma\sqrt{K_a}} \tag{1.23}$$

Now, the following step-by-step procedure may be adopted for determination of P_a.

1. Draw the retaining wall and the backfill to a given scale (Fig. 1.19a).

2. Draw the line $A'C'D'$ for the backfill to define the depth of tensile crack. The vertical distance between lines ACD and $A'C'D'$ is equal to z_c [Eq. (1.23)].

3. Draw several trial wedges such as $ABC'C$, $ABD'D$, ...

4. For a given trial wedge (for example, wedge $ABC'C$), the forces per unit length of the wall are as follows.

   a. Weight of the wedge--W. [= (Area of $ABC'C$)(γ)]

   b. Adhesive force along soil-wall interface--C_a (= $c_a A'B$; where c_a=adhesion between the wall and the soil)

   c. Cohesive force along the trial failure surface of the wedge--C. [= (c)(BC')]

   d. Resultant of the normal and shear forces (frictional) along the trial failure surface--R.

   e. Active force--P_a.

5. Draw the force polygons for the failure wedges. Figure 1.19b shows the force polygon for the wedge $ABC'C$.

6. Determine the magnitudes of P_a for all failure wedges.

7. Plot the values of P_a as shown in the upper part of Fig. 1.19a. Join all of the points with a smooth curve.

8. The maximum value of P_a obtained from the curve drawn in Step 7 is the desired active force.

1.8 ACTIVE EARTH PRESSURE FOR BACKFILL WITH VERTICAL DRAIN

Figure 1.20a shows a retaining wall with a vertical backface. Let the base of the retaining wall be located on an *impermeable surface*. The backfill is of *granular material* with a horizontal surface. During rainstorms, water will seep from the ground surface into the backfill. This will tend to increase the lateral active force (P_a) against the wall. For that reason, in many cases vertical drains along the backface of the wall are provided to drain the seepage water. Figure 1.20a also shows the vertical drain and the flow net for seepage during rainstorms. The lateral active force against the wall during the rainstorm can be calcu-

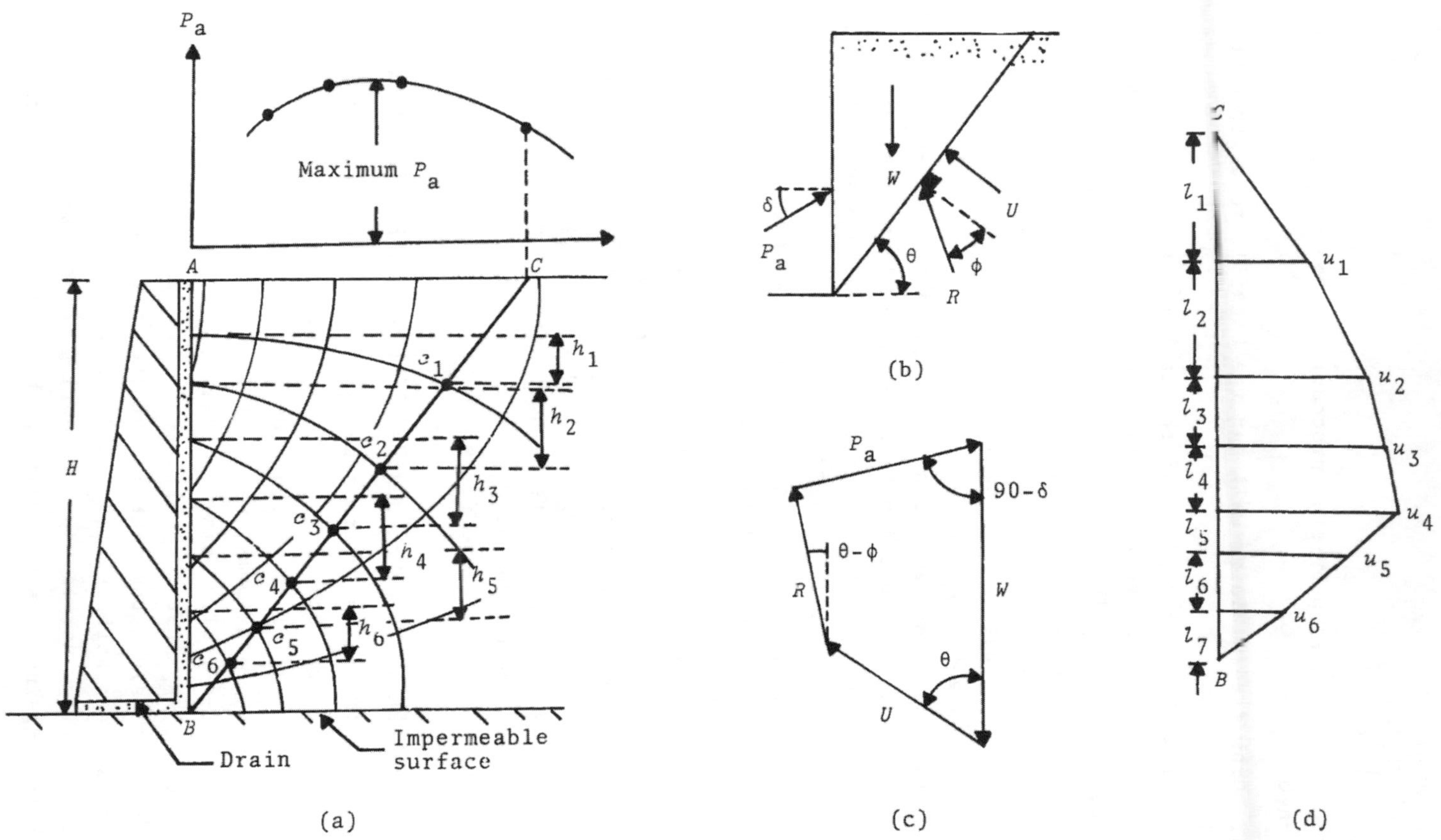

Figure 1.20. Active force P_a for granular backfill with vertical drain. (*Note:* For backfill, unit weight=γ_{sat}, angle of friction=ϕ, and cohesion c=0.)

lated by using the principles of the trial wedge solution. This can be explained by referring to Fig. 1.20a in which *ABC* is a trial wedge. The forces (per unit length of the wall) on the trial wedge are as follows (Fig. 1.20b).

1. Saturated weight of the wedge--W. [= (Area of *ABC*)(γ_{sat})]
2. Active force--P_a.
3. Resultant of the normal and shearing resistance along *BC*--R.
4. Force due to pore water pressure on the trial failure surface *BC*--U.

The force polygon for these forces is shown in Fig. 1.20c. Based on this force polygon

$$P_a = \frac{(W - U\cos\theta)\tan(\theta-\phi) + U\sin\theta}{\cos\delta + \sin\delta\tan(\theta-\phi)} \tag{1.39}$$

The magnitude of U can be calculated by referring to Fig. 1.20a. The trial wedge line intersects the equipotential lines at $c_1, c_2, \ldots, c_6$. At these points, the pore water pressures (u) are as follows.

At c_1, $u_1 = \gamma_w h_1$

c_2, $u_2 = \gamma_w h_2$

.

.

.

c_6, $u_6 = \gamma_w h_6$

These pore water pressures are plotted in Fig. 1.20d. So

U Area of the pore water pressure diagram

$$\frac{u_1}{2}(l_1) + \frac{u_1+u_2}{2}(l_2) + \frac{u_2+u_3}{2}(l_3) + \ldots + \frac{u_6}{2}(l_7) \tag{1.40}$$

Once the magnitude of U is calculated, P_a can be determined from Eq. (1.39). This procedure can be repeated for several trial wedges and the maximum value of P_a can be determined (see upper part of Fig. 1.20a).

1.9 WALL ROTATION REQUIRED FOR ACTIVE EARTH PRESSURE CONDITION

It has been previously mentioned that the wall must rotate

sufficiently about its bottom to develop Rankine's or Coulomb's active earth pressure condition. Table 1.2 shows the approximate rotational limits for various types of backfills. Sherif, Fang, and Sherif (1984) have provided some laboratory model test results

Table 1.2. Wall Rotation About the Bottom For Development of Active State

Soil type	ω (rad)
Loose sand	10×10^{-4} to 20×10^{-4}
Dense sand	6×10^{-4} to 12×10^{-4}
Soft clay	200×10^{-4}
Stiff clay	100×10^{-4}

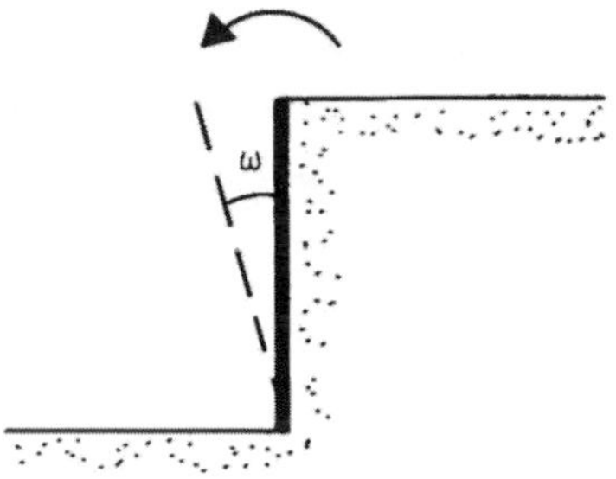

for active pressure behind a retaining wall with a granular backfill. According to their model test results, the wall rotation required to develop active state is approximately the same in dense and loose sand.

1.10 ACTIVE FORCE ON WALL WITH ROTATION ABOUT TOP

The Coulomb and Rankine active forces described in the preceding section are valid for the case where the wall rotates about its bottom. However in some cases, such as in construction of braced cuts (Fig. 1.21a), the rotation of the wall is about the top. In such cases, the distribution of active earth pressure with depth is different than that obtained for wall rotation about its top. This can be explained with reference to Fig. 1.21b. For simplicity, let us consider that the soil considered is sand and the sheet pile walls are frictionless. If the wall does not yield at all, it will be the case of at-rest earth pressure. The lateral pressure at any

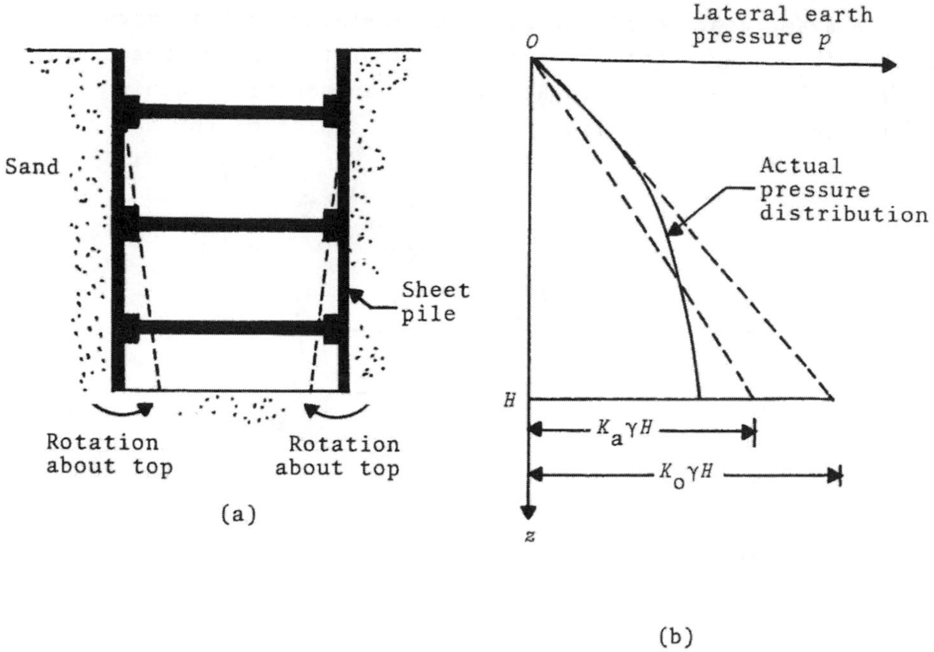

Figure 1.21. Active force for braced cut (rotation of wall about top).

depth can be given as

$$p = \gamma z K_o$$

However, if the wall yields sufficiently at all depths with rotation about its bottom, then the Rankine active state will prevail. So, the pressure at any depth can be expressed as

$$p = \gamma z K_a$$

For rotation about its top, the wall movement gradually increases from $z=0$ (that is, ground surface) downward. At smaller depths, the wall movement may not be sufficient to initiate Rankine active state. Hence the actual magnitude of the lateral earth pressure p will be between $p_{(\text{at-rest})}$ and $p_{(\text{active-Rankine})}$. However, at greater depths, the wall movement may be substantially larger than that required to reach the Rankine active state. For that case, the lateral pressure p may be smaller than $p_{(\text{active-Rankine})}$. This nature of pressure distribution thus derived will not by hydrostatic. This is shown in Fig. 1.21b.

An example of this type of nonhydrostatic distribution of lateral earth pressure can be seen from laboratory model tests (Fig. 1.22) of Sherif and Fang (1984) with a sand backfill behind the model retaining wall. Figure 1.22 shows a plot of the horizontal

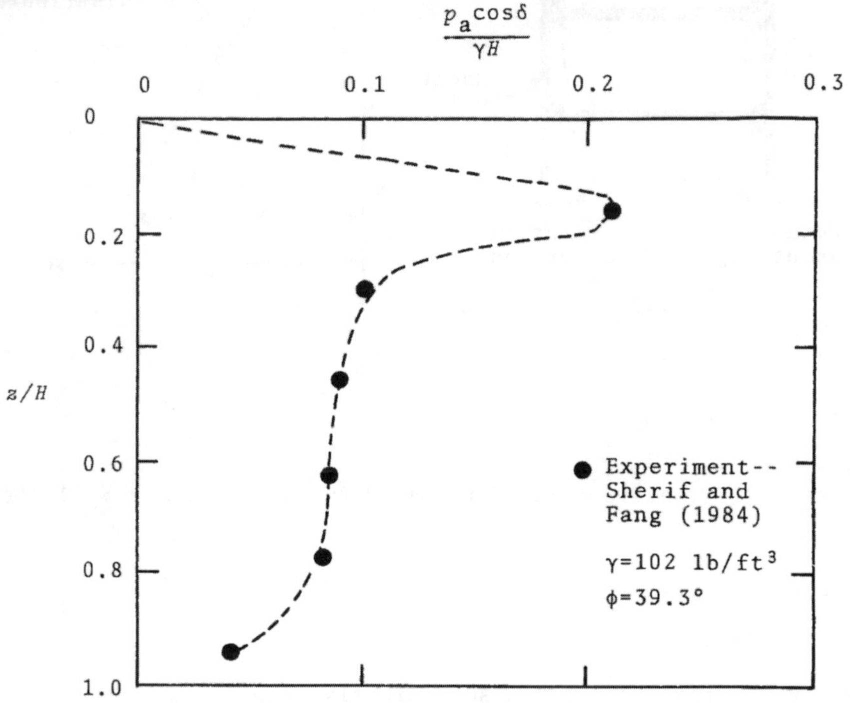

Figure 1.22. Experimental measurement of horizontal component of lateral earth pressure for wall rotation about its top.

component of the lateral earth pressure, that is, $p_{(\text{horizontal})}/\gamma H = (p_a \cos\delta)/\gamma H$ vs. z/H. At the present time, sufficient research results are not available to establish the nature of pressure distribution behind a wall with rotation about its top. In the following two subsections, the existing theories will be briefly described.

1.10.1 Dubrova's Solution

Dubrova (1963) has proposed a method by which the distribution of active earth pressure behind a wall rotating about its top can be determined (see also Harr, 1966). According to this method, Coulomb's active pressure solution for wall rotation about its bottom is accepted as valid. This method can be explained by considering a vertical wall with a horizontal granular backfill as shown in Fig. 1.23. Let BC be the failure plane as per Coulomb's active

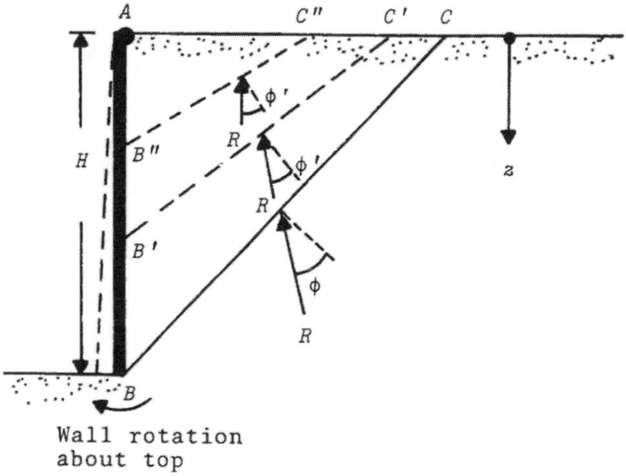

Figure 1.23. Dubrova's solution (1963) of active earth pressure for wall rotation about its top.

earth pressure theory. However, there exists an infinite number of *quasi-rupture* lines in the failure wedge between $z=0$ to $z=H$. The resultant reaction along BC is equal to R which is inclined at an angle ϕ to the normal drawn to BC. This indicates full shear strength mobilization for the rupture line BC. At a depth $z<H$, the movement of the wall is less, and so the degree of shear strength mobilization will be less. Hence quasi-rupture lines like $B'C'$ and $B''C''$ will have their resultant reaction R inclined at an angle ϕ' which is less than ϕ. Dubrova (1963) assumed that the magnitude of ϕ' varies as

$$\phi' = \frac{\phi z}{H} \tag{1.41}$$

where z = depth of intersection of the quasi-rupture line with wall AB

With this assumption, the active force against the wall for any depth z can be given as [modification of Eq. (1.34)]

$$P_a = \frac{\gamma z^2}{2\cos\delta}\left[\frac{1}{(1/\cos\phi') + \sqrt{\tan^2\phi' + \tan\phi'\tan\delta}}\right]^2 \tag{1.42}$$

The distribution of active pressure against the wall at any depth z can be obtained by substituting Eq. (1.41) into Eq. (1.42) and differentiating with respect to z, or

$$\frac{dP_a}{dz} = p_a = \frac{\gamma z}{\cos\delta}\left[\frac{\cos^2\phi'}{(1+n\sin\phi')^2} \quad \frac{z\phi\cos\phi'}{H(1+n\sin\phi')^2}\left(\sin\phi' + \frac{1+n^2}{2n}\right)\right] \tag{1.43}$$

where p_a = active pressure at any depth z

$$n \quad \left(1 + \frac{\tan\delta}{\tan\phi'}\right)^2 \tag{1.44a}$$

However, without loss of much accuracy, one can assume that

$$\frac{1+n^2}{2n} \approx n \tag{1.44b}$$

With the above assumption, Eq. (1.42) simplifies to the form

$$p_a = \frac{\gamma z}{\cos\delta}\left\{\left(\frac{\cos\phi'}{1+n\sin\phi'}\right)^2 - \left[\frac{z\phi\cos\phi'}{H(1+n\sin\phi')^2}\right](\sin\phi' + n)\right\} \tag{1.45a}$$

For a frictionless wall $\delta=0$, so $n=1$. Hence

$$p_a = \gamma z\tan^2(45-\phi'/2)\left(1 \quad \frac{z\phi}{H\cos\phi'}\right) \tag{1.45b}$$

Example 1.3. A retaining wall is 5 m high. The backfill is horizontal sand with $\phi=30°$. Use Dubrova's method and determine the variation of active pressure p_a at $z=0$ (ground surface), 1 m, 2 m, 3 m, 4 m, and 5 m for wall rotation about the top. Assume $\delta=0$ and γ - 17.5 kN/m³. Also determine the active force P_a per unit length of the wall.

Solution

Determination of variation of active pressure:

For this case, we will use Eq. (1.45b) ($\delta=0$).

$$p_a = \gamma z\tan^2(45-\phi'/2)\left(1 - \frac{z\phi}{H\cos\phi'}\right)$$

Now the following table can be prepared.

z (m)	ϕ'[a] (deg)	$\tan^2(45-\phi'/2)$	$1 \quad \frac{z\phi}{H\cos\phi'}$[b]	p_a (kN/m²)
0	0	1.0	0	0
1	6	0.81	0.895	12.69
2	12	0.656	0.786	18.05
3	18	0.528	0.670	18.57
4	24	0.422	0.541	15.98
5	30	0.333	0.395	11.51

[a]Eq. (1.41)
[b]ϕ is in radians

Determination of active force P_a:

The approximate value of the active force per unit length can be determined by using a trapezoidal rule as

$$P_a = \left[\frac{p_{(0)}+p_{(5)}}{2} + p_{(1)} + p_{(2)} + p_{(3)} + p_{(4)}\right]\Delta z$$

$$\left(\frac{0+11.51}{2} + 12.69 + 18.05 + 18.57 + 15.98\right)(1)$$

$$\underline{71.05 \text{ kN/m}}$$

NOTE: The line of action of the resultant force will be horizontal since $\delta=0$.

For rotation about the bottom of the wall

$$P_a = \tfrac{1}{2}\gamma H^2\tan^2(45-\phi/2) = (\tfrac{1}{2})(17.5)(5)^2\tan^2(45-30/2) - 72.91$$

$$= 72.91 \text{ kN/m}$$

There is not a substantial difference in the value of P_a for rotation about the top or rotation about the bottom of the wall. However, the nature of the pressure distribution is different.

1.10.2 Terzaghi's General Wedge Theory

According to the general wedge theory of Terzaghi (1941), the failure surface in soil for wall rotation about its top can be given as shown in Fig. 1.24. The failure surface *BC* is an *arc of a logarithmic spiral*, which can be expressed by the equation

$$r \quad r_o e^{\theta\tan\phi} \tag{1.46}$$

where ϕ = soil friction angle

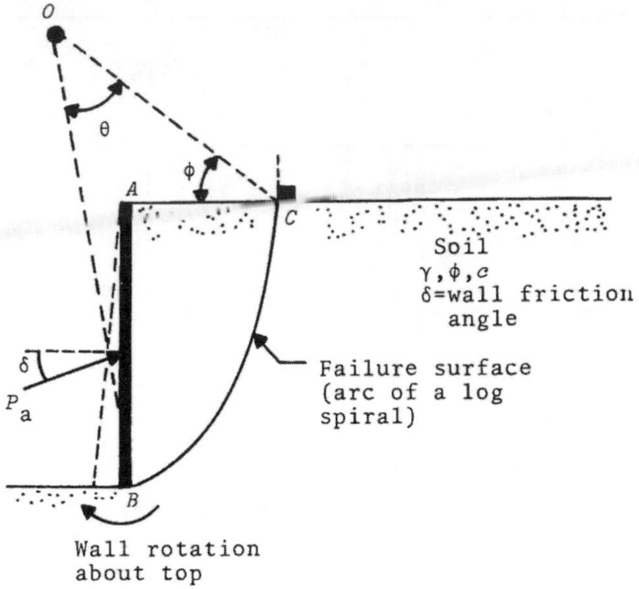

Figure 1.24. Failure wedge according to the general wedge theory (Terzaghi, 1941) for wall rotation about its top. (*Note:* $OB=r_2$, and $OC=r_1$.)

In Fig. 1.24, *O* is the center of the log spiral. The properties of a log spiral have been discussed in detail in Appendix B. The log spiral arc intersects the ground surface (at *C*) at 90 degrees. Note that, according to the properties of a log spiral, the radial line *OC* will make an angle ϕ with the horizontal.

In order to determine the active force P_a, the following procedure needs to be followed.

1. Plot a log spiral with the desired value of soil friction ϕ. This can be done with any desired value of r_0.

2. Plot the profile of the wall to a given scale (Fig. 1.25).

3. Draw several lines such as O_1C_1, O_2C_2, O_3C_3, . . ., as shown in Fig. 1.25. Each of these lines make an angle ϕ with the horizontal.

4. Superimpose the wall profile in Step 3 on the log spiral drawn in Step 1. By trial and error, trace out the arcs of the log spirals BC_1, BC_2, BC_3, . . ., with centers at O_1, O_2, O_3, ... So, ABC_1, ABC_2, ABC_3, . . . are trial failure wedges.

5. The forces acting *per unit length of the wall* on the wedge ABC_1 are as follows.

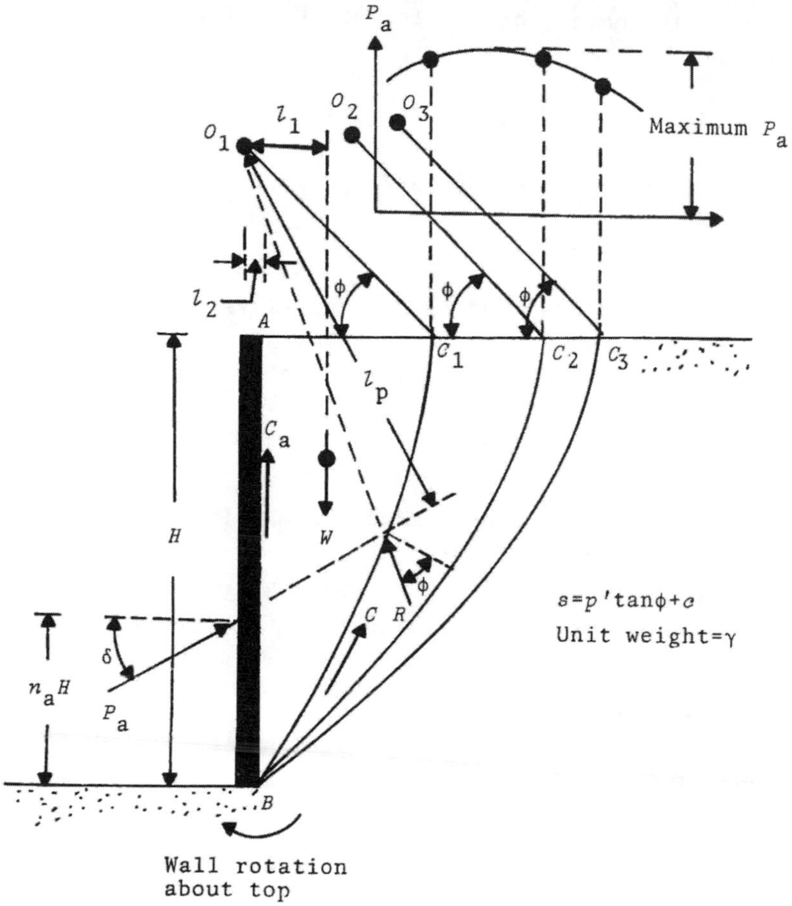

Figure 1.25. Determination of active force P_a for wall rotation about its top (general wedge theory).

a. Weight of the wedge--W. [= (Area of the wedge)(γ)] This force acts through the centroid of the wedge.

b. Active force per unit length--P_a. It acts at a distance of $n_a H$ measured from the bottom of the wall.

c. Adhesion along the soil-wall interface--C_a. Note that $C_a = Hc_a$ (where c_a=adhesion).

d. Cohesive force along the arc BC--C. (Note that the force per unit area is c=cohesion.)

e. Resultant of the frictional resistance along the surface of sliding--R.

6. Take the moment of all forces about the center of the log spiral O_1, or

$$P_a l_p - W l_1 + C_a l_2 + M_c + R(0) = 0$$

where M_c moment of the cohesive force C

$$= \frac{c}{2\tan\phi}(O_1B^2 - O_1C_1^2) \qquad (1.47)$$

Note that, based on the property of the log spiral, the line of action of the force R will pass through the origin of the spiral. So the moment arm for R is equal to zero. So

$$P_a = \frac{1}{l_p}(W l_1 \quad C_a l_2 \quad M_c) \qquad (1.48)$$

7. Using Eq. (1.48), determine the magnitude of P_a for other trial wedges, that is, ABC_2, ABC_3, ...

8. Plot these values of P_a as shown at the top of Fig. 1.25. Plot a smooth curve through these points. The maximum value of P_a obtained from this curve is the desired active pressure.

There are several points that we need to keep in mind. The general wedge theory gives only the magnitude of P_a. It does not provide any idea regarding the distribution of pressure with depth. Also, in order to estimate P_a, one needs to assume a certain value of n_a. In many practical cases, it has been shown that n_a can be as high as 0.5 to 0.6.

The active force thus determined can be expressed as

$$P_a = \frac{1}{2}\gamma H^2 K' \qquad (1.49)$$

where K' = earth pressure coefficient = $f(c, c_a, \delta, \phi, \text{and } n_a)$

Table 1.3 gives a tabulation of $K'/K_{a(\text{Rankine})}$ for various values of δ, ϕ, n_a, and $c/\gamma H$. In the preparation of this table, it has been assumed that

$$c_a = c\left(\frac{\tan\delta}{\tan\phi}\right) \qquad (1.50)$$

If the soil behind the retaining wall is *saturated clay* (c=0; total stress analysis), the failure surfaces BC_1, BC_2, BC_3, . . . shown in Fig. 1.25 become arcs of circles [when ϕ=0 is substituted into Eq. (1.46), r becomes equal to r_0, which is the equation of a *circle*]. For such a case, the forces on a trial wedge ABC_1 will be

Table 1.3. Variation of $\frac{K'}{K_{a(Rankine)}}$[a]

δ (deg)	$n_a=0.3$		$n_a=0.4$		$n_a=0.5$	
	$c/\gamma H$		$c/\gamma H$		$c/\gamma H$	
	0	0.1	0	0.1	0	0.1
φ=15°						
0	0.921	0.431	1.022	0.484	1.155	0.564
5	0.880	0.363	0.977	0.408	1.097	0.459
10	0.858	0.318	0.949	0.357	1.068	0.404
15	0.847	0.287	0.941	0.324	1.058	0.370
φ=20°						
0	1.018	0.390	1.010	0.429	1.124	0.482
5	0.878	0.327	0.965	0.365	1.073	0.408
10	0.855	0.286	0.939	0.318	1.124	0.353
15	0.843	0.249	0.927	0.278	1.029	0.314
φ=25°						
0	0.913	0.340	0.998	0.369	1.101	0.411
5	0.877	0.286	0.958	0.315	1.054	0.347
10	0.855	0.244	0.931	0.271	1.025	0.300
15	0.842	0.209	0.919	0.234	1.010	0.261
φ=30°						
0	0.912	0.279	0.990	0.310	1.083	0.339
5	0.879	0.234	0.954	0.258	1.041	0.282
10	0.858	0.198	0.930	0.219	1.017	0.240
15	0.846	0.168	0.918	0.180	1.002	0.201
20	0.843	0.141	0.915	0.153	0.996	0.168
φ=35°						
0	0.911	0.218	0.985	0.236	1.070	0.255
5	0.882	0.173	0.952	0.192	1.033	0.210
10	0.863	0.140	0.930	0.151	1.007	0.170
15	0.852	0.111	0.919	0.122	0.996	0.129
20	0.852	0.081	0.915	0.092	0.993	0.010
φ=40°						
0	0.912	0.111	0.982	0.147	1.060	0.166
5	0.885	0.077	0.949	0.111	1.028	0.120
10	0.871	0.069	0.931	0.074	1.009	0.083
15	0.862	0.037	0.922	0.046	0.995	0.051
20	0.862	0.014	0.922	0.014	0.995	0.018

[a] $K_{a(Rankine)}$ $\tan^2(45-\phi/2)$

c_a $c\left(\frac{\tan\delta}{\tan\phi}\right)$

as shown in Fig. 1.26. Note that $C_a = c_a H$ and $C = c_u r\theta$. Taking the moment about point O_1

$$P_a \quad \frac{1}{H(1-n_a)}(Wl_1 - c_u r^2\theta - c_a H^2 \cot\theta) \tag{1.51}$$

where c_u = undrained shear strength of clay

The preceding equation can be applied to several trial wedges such as ABC_2, ABC_3, . . ., and the maximum value of P_a can be obtained (see the top portion of Fig. 1.26). Using this procedure, the desired active thrust can be expressed as (Das and Seeley, 1975)

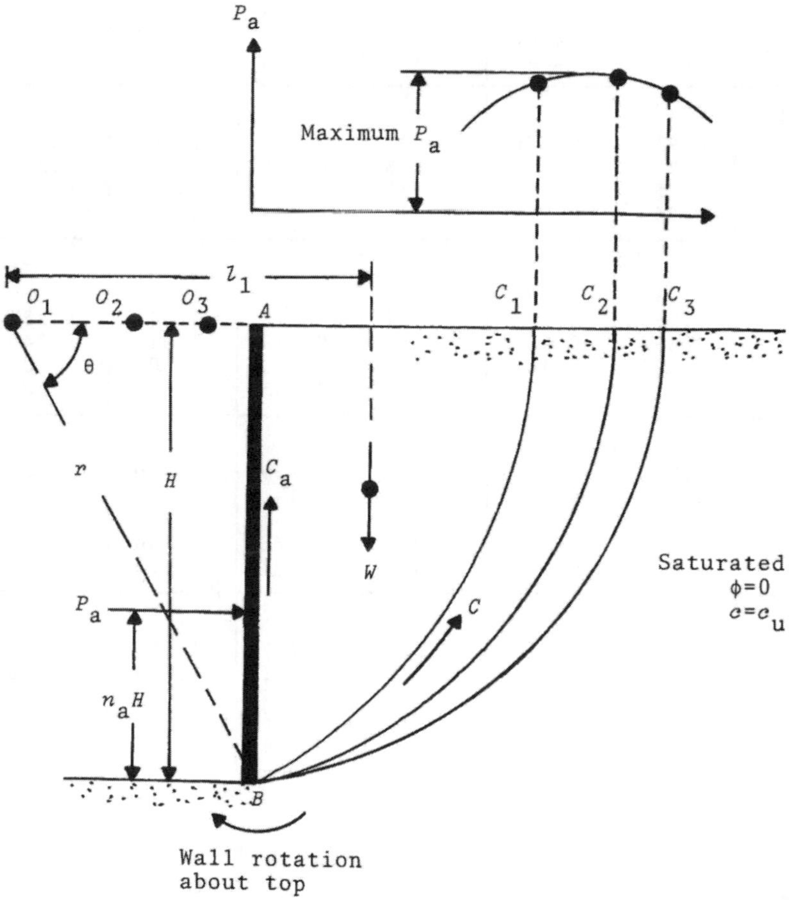

Figure 1.26. Determination of active force P_a for wall rotation about its top (saturated soil; φ=0 concept).

$$P_a = \frac{1}{2(1-n_a)}(0.667-K''N_c)\gamma H^2 \tag{1.52}$$

where $N_c \quad \frac{c_u}{\gamma H}$ (1.53)

γ = unit weight (total) of the clay

$K'' = f\ \frac{c_a}{c_u}$

The values of K'' are as follows.

(c_u/c_a)	K''
0	2.762
0.5	3.056
1.0	3.143

Example 1.4. Consider the retaining wall given in Example 1.3. Use Table 1.3 to calculate P_a for n_a=0.3, 0.4, and 0.5.

Solution

For this problem, ϕ=30°, δ=0°, H=5 m, γ=17.5 kN/m³, and c=0. So

$$\frac{c}{\gamma H} = \frac{0}{(17.5)(5)} = 0; \text{ also } K_{a(\text{Rankine})} \quad \tan^2(45-\phi/2) = \tfrac{1}{3}$$

From Table 1.3, the following can be prepared.

n_a	$\frac{K'}{K_{a(\text{Rankine})}}$	K'	$P_a=\frac{1}{2}\gamma H^2 K'$ (kN/m)
0.3	0.912	0.304	66.5
0.4	0.990	0.33	72.19
0.5	1.083	0.361	78.97

1.10.3 Empirical Pressure Diagrams For Design of Braced Cuts

In practical cases, the lateral active earth pressure (p_a) distribution with depth on sheet pile walls may change with time.

The magnitude of p_a also depends on soil type, construction conditions, and workmanship of the construction crew. For these reasons empirical pressure distribution diagrams are used in all design works. These empirical pressure distribution diagrams are prepared from envelopes of observed pressures from various braced cuts. At the present time, there are two empirical pressure distribution systems used in practice. They have been proposed by Peck (1969) and Tschebotarioff (1973), and are shown in Figs. 1.27 and 1.28.

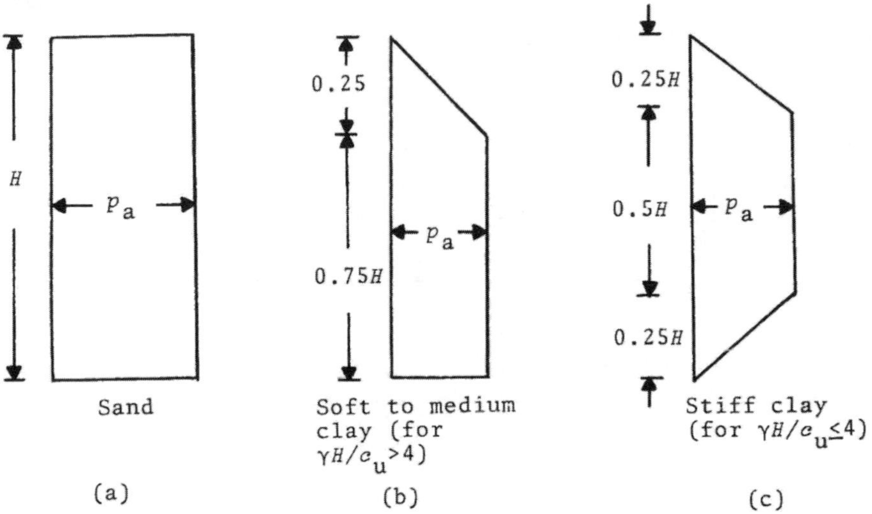

Figure 1.27. Peck's empirical pressure distribution diagrams (1969) for braced cuts: (a) $p_a = K_a \gamma H$ [where $K_a = \tan^2(45-\phi/2)$]; (b) $p_a = \gamma H[1-(4c_u)/(\gamma H)]$ or $0.3\gamma H$, whichever is greater; (c) p_a $0.2\gamma H$ to $0.4\gamma H$.

1.11 ACTIVE EARTH PRESSURE FOR LATERAL TRANSLATION OF WALL

In many earth-retaining structures such as rigid retaining diaphragms, there is a *lateral translation* of the wall. In this type of movement, free formation of a sliding wedge is restricted. The variation of lateral earth pressure with depth is not *hydrostatic* in nature. Very few experimental and theoretical studies relating to active earth pressure for lateral wall movement are available at the present time. In the following subsections, a general review of the available studies will be presented.

Rendulic's Approximation

Rendulic (1938) has given an approximation for lateral earth

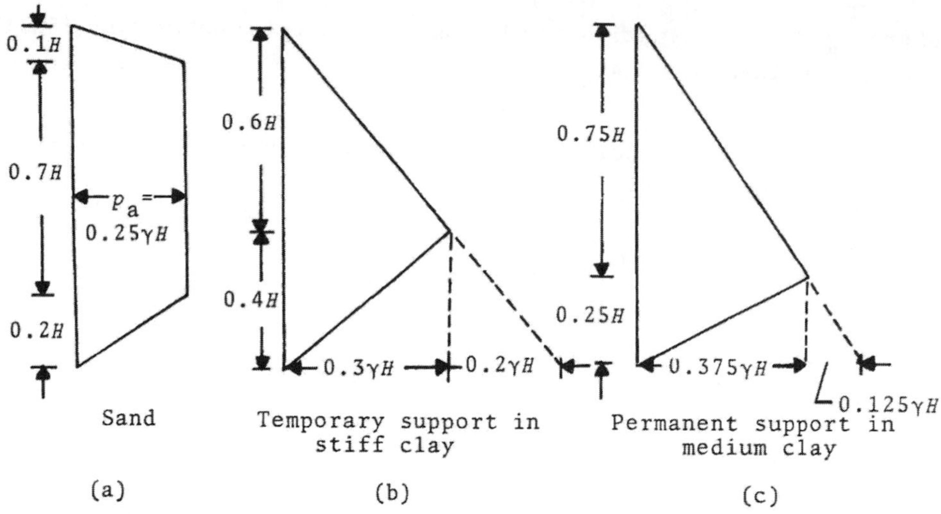

Figure 1.28. Tschebotarioff's (1973) empirical pressure distribution diagrams for braced cuts.

pressure distribution behind a vertical retaining wall (δ=0) with a horizontal *granular backfill* (Fig. 1.29). According to this, the lateral pressure at any depth z can be given as

$$p_a = Kz\gamma - \frac{3}{2H}(K-K_a)z^2\gamma \qquad (1.54)$$

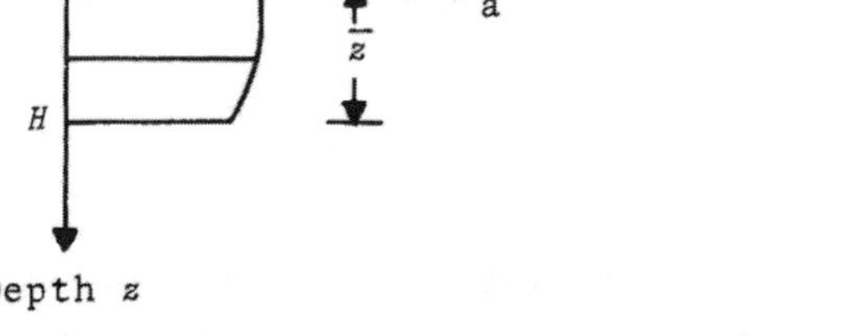

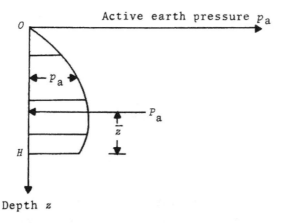

Figure 1.29. Rendulic's approximation (1938) for lateral earth pressure for lateral translation of the wall with granular backfill.

where K_a = Rankine active earth pressure coefficient = $\tan^2(45-\phi/2)$
$K = 1.8K_a$ to $2.4K_a$
γ = unit weight of backfill

The total horizontal active force can be estimated by integrating Eq. (1.53), or

$$P_a = \int p_a dz = K\gamma \int_0^H z\,dz - \left(\frac{3\gamma}{2H}\right)(K-K_a)\int_0^H z^2 dz$$

$$= \frac{K\gamma H^2}{2} - \frac{3\gamma(K-K_a)H^3}{6H} = \frac{K\gamma H^2}{2} \quad \frac{\gamma(K-K_a)H^2}{2}$$

$$= \frac{1}{2}\gamma H^2(K-K+K_a) = \frac{1}{2}K_a\gamma H^2 \tag{1.55}$$

From the preceding equation, it can be seen that the total active force P_a in this case is the same as that obtained from the Rankine active case. However, the pressure distribution is non-hydrostatic.

The location of the line of action of the resultant force P_a has been approximated by Rendulic (1938) as

$$\bar{z} = H\left(\frac{1}{4} + \frac{K}{12K_a}\right) \tag{1.56}$$

Dubrova's Solution

Dubrova (1963) has suggested the following procedure to estimate the variation of active earth pressure p_a with depth (Fig. 1.30) for a vertical retaining wall with a horizontal granular backfill. The procedure is given below in a step-by-step manner.

1. Calculate the variation of active earth pressure against a retaining wall rotating about the bottom by using Coulomb's equation [Eq. (1.34)], or

$$p_{a(\text{rotation about bottom})} = \frac{dP_a}{dz}$$

$$= \left(\frac{\gamma z}{\cos\delta}\right)\left(\frac{1}{\frac{1}{\cos\phi} + \sqrt{\tan^2\phi+\tan\phi\tan\delta}}\right)^2 \tag{1.57}$$

The nature of this pressure distribution is shown in Fig. 1.30b.

2. Calculate the variation of p_a for wall rotation about its top by using Eq. (1.45a). The nature of this pressure distribution is shown in Fig. 1.30b.

3. Calculate the variation of p_a (Fig. 1.30b) for lateral

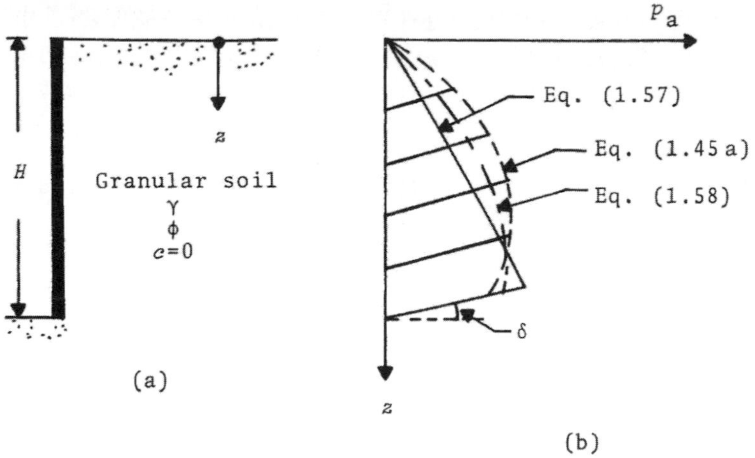

Figure 1.30. Dubrova's method (1963) for calculation of p_a for lateral translation of wall.

translation of the wall as

$$p_{a(\text{lateral translation})} = \frac{p_{a(\text{rotation about bottom})} + p_{a(\text{rotation about top})}}{2} \tag{1.58}$$

Experimental Studies

Sherif, Ishibashi, and Lee (1982) have conducted several laboratory model tests to study the behavior of retaining walls with vertical back and horizontal granular backfill being subjected to lateral translation. Based on their experimental results, they have concluded that the lateral force P_a for lateral translation is approximately the same as that calculated by using Coulomb's equation. However, the line of action of the resultant intersects the wall at a distance $(\bar{z})$ $0.42H$ measured from the bottom and is inclined at an angle δ with the horizontal. Based on their laboratory experiments, they also determined that the lateral translation required to mobilize the active state can be expressed by the equation

$$\frac{S}{H} = (7.0 - 0.13\phi)10^{-4} \tag{1.59}$$

where S = lateral translation for active state and ϕ is in degrees.

Example 1.5. Refer to Example 1.3. Assume the wall to be 4 m high. Use Eq. (1.54) to determine the earth pressure distribution with depth for lateral translation of the wall. Use $K \approx 2$.

Solution

Given: $z=4$ m, $\phi=30°$, $\delta=0$, $\gamma=17.5$ kN/m^3, $K=2.0$, and K_a $\tan^2(45-30/2)=0.33$.

From Eq. (1.54)

$$p_a \quad Kz\gamma - \frac{3}{2H}(K-K_a)z^2\gamma$$

Now the following table can be prepared.

z (m)	$Kz\gamma$ (kN/m^2)	$\frac{3}{2H}(K-K_a)z^2\gamma$ (kN/m^2)	p_a
0	0	0	0
1	35	8.77	26.23
2	70	35.08	34.92
3	105	78.93	26.07
4	140	140.32	-0.32

Example 1.6. Refer to Example 1.3. Use Dubrova's method (Fig. 1.30).

a. Determine the active earth pressure distribution with depth for lateral translation of the wall.
b. Determine the active force per meter of the wall.
c. Determine the location of the line of action of the resultant force.

Solution

Part a. Given: $\delta=0$; $\phi=30°$.

$$K_a = \tan^2(45-30/2) = 0.33$$

For rotation about the bottom

$$p_{a(\text{rotation about bottom})} \quad \gamma z K_a$$

For rotation about the top, the distribution of p_a is given in Example 1.3. So now the following table can be prepared.

z (m)	p_a(rotation about bottom) (kN/m²)	p_a(rotation about top) (kN/m²)	p_a(translation) (kN/m²)
0	0	0	0
1	5.78	12.69	9.24
2	11.55	18.05	14.8
3	17.33	18.57	17.95
4	23.1	15.98	19.54
5	28.88	11.51	20.2

Part b. The total active force per unit length can be obtained as (trapezoidal rule)

$$P_a = \Delta z\left[\frac{p_{(0)}+p_{(5)}}{2} + p_{(1)} + p_{(2)} + p_{(3)} + p_{(4)}\right]$$

$$(1)\left(\frac{0+20.2}{2} + 9.24 + 14.8 + 17.95 + 19.54\right)$$

$$= \underline{71.63 \text{ kN/m}^3}$$

Part c. The location of the line of action of the resultant can be determined by taking the moment of the pressure diagram about the bottom of the wall, or

$$\bar{z} = \frac{1}{P_a}\left\{4.5\left[\frac{p_{(0)}+p_{(1)}}{2}\right]\Delta z + 3.5\left[\frac{p_{(1)}+p_{(2)}}{2}\right]\Delta z + 2.5\left[\frac{p_{(2)}+p_{(3)}}{2}\right]\Delta z + 1.5\left[\frac{p_{(3)}+p_{(4)}}{2}\right]\Delta z + 0.5\left[\frac{p_{(4)}+p_{(5)}}{2}\right]\Delta z\right\}$$

$$= \frac{1}{71.63}\left[4.5\left(\frac{0+9.24}{2}\right)(1) + 3.5\left(\frac{9.24+14.8}{2}\right)(1) + 2.5\left(\frac{14.8+17.95}{2}\right)(1) + 1.5\left(\frac{17.95+19.54}{2}\right)(1) + 0.5\left(\frac{19.54+20.2}{2}\right)(1)\right]$$

$$= \frac{1}{71.63}(20.79 + 42.07 + 40.94 + 28.12 + 9.94) = \underline{1.98 \text{ m}}$$

NOTE: According to the experimental observations of Sherif et al. (1982)

$$\bar{z} = 0.42H \quad 0.42(5) = \underline{2.08 \text{ m}}$$

1.12 RANKINE PASSIVE PRESSURE (ROTATION ABOUT THE BOTTOM OF THE WALL)

Figure 1.31a shows a frictionless retaining wall with a vertical back and a horizontal backfill whose shear strength can be

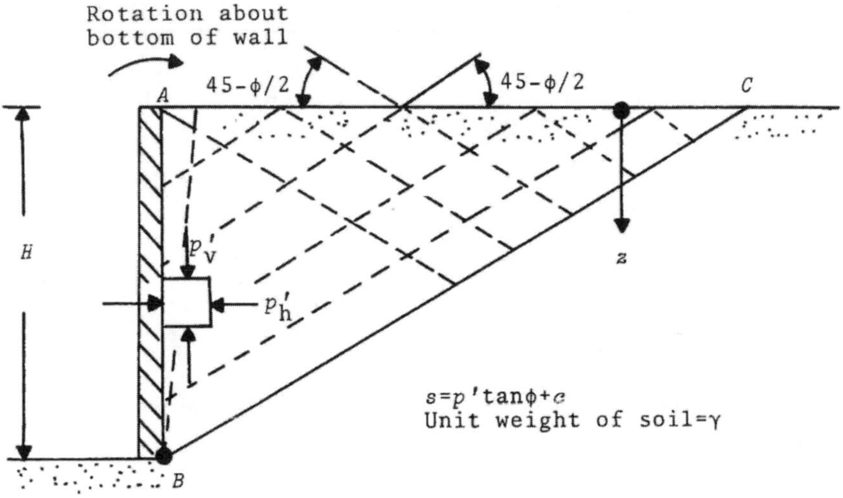

(a)

Figure 1.31. Rankine passive pressure.

given by the Mohr-Coulomb equation as

$$s \quad p'\tan\phi + c$$

When the wall is at its original position *without yielding*, the vertical and horizontal effective stresses on a soil element near the wall at a depth z can be given as

$$p_v' = \gamma z$$

and

$$p_h' \quad K_o \gamma z$$

where K_o = at-rest earth pressure coefficient

The corresponding Mohr's circle for the state of stress on the soil element is shown by circle No. 1 in Fig. 1.31b.

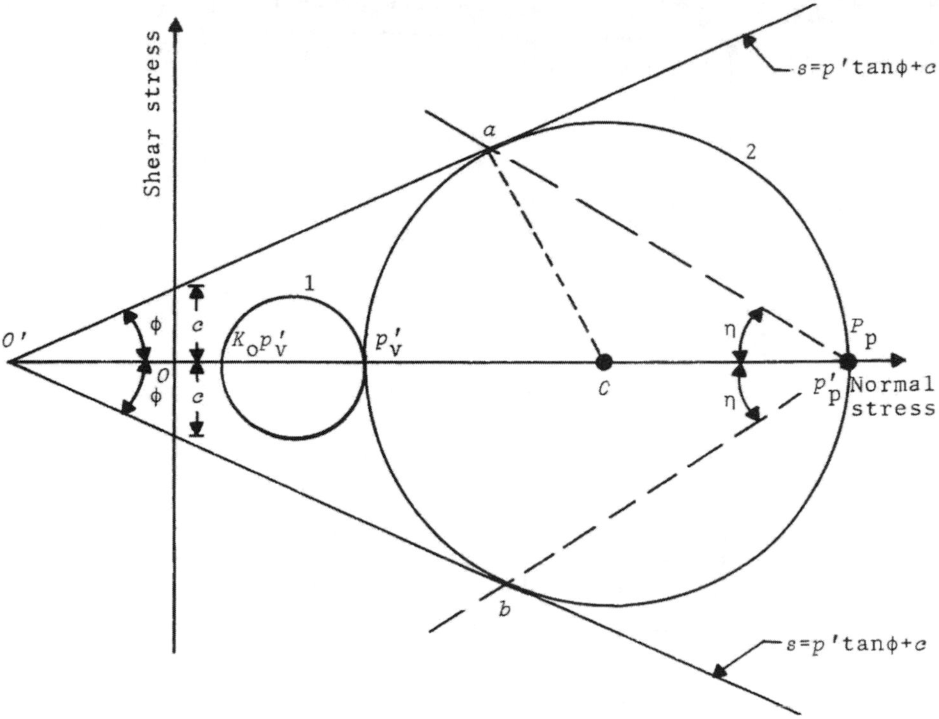

(b)

Figure 1.31. (Continued).

If the retaining wall is pushed into the soil and if wall movement takes place by rotation about its bottom (point *B*) the horizontal stress on the soil element will continue to increase. When the effective horizontal stress p_h' becomes equal to p_p', the stress conditions on the soil element can be represented by Mohr's circle No. 2 in Fig. 1.31b. Note that Mohr's circle No. 2 touches the Mohr-Coulomb failure envelope signifying failure condition in the soil wedge *ABC* (Fig. 1.31a) adjacent to the wall. The lateral effective stress p_p' is referred to as the *Rankine passive pressure*. It also needs to be pointed out that at Rankine passive state the vertical effective stress p_v' is the *minor principal stress*, and the horizontal effective stress p_h' is the *major principal stress*.

Referring to Fig. 1.31b, it is obvious that P_p is the *passive pole*. The potential failure surfaces (slip lines) in the soil wedge *ABC* (Fig. 1.31a) are represented by the lines $P_p a$ and $P_p b$

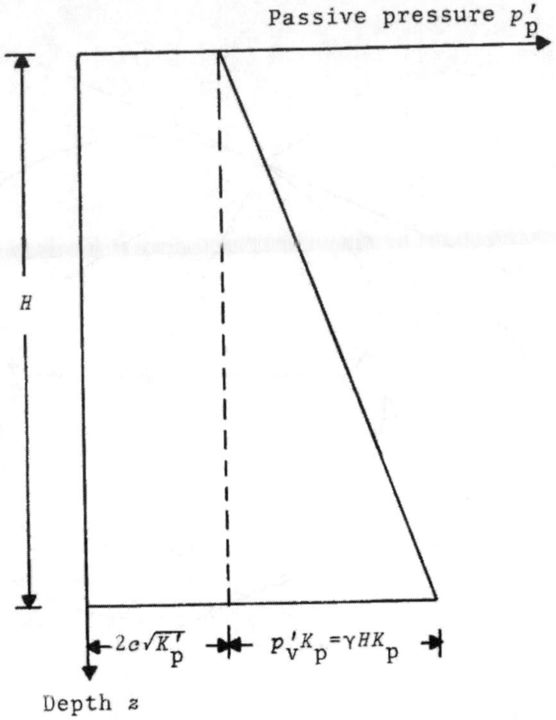

(c)

Figure 1.31. (Continued).

which are inclined at angles of $\pm\eta$ to the horizontal. For Rankine passive case

$$\eta = \pm(45-\phi/2) \tag{1.60}$$

Proceeding in a similar manner as that in Section 1.4, for Rankine active state it can be seen that

$$p'_p = p'_v \tan^2(45+\phi/2) + 2c\tan(45+\phi/2) = \gamma z K_p + 2c\sqrt{K_p} \tag{1.61}$$

where p'_p = Rankine passive pressure

K_p = Rankine passive earth pressure coefficient

$= \tan^2(45+\phi/2)$

The directions of the slip lines in the passive failure wedge *ABC* behind the retaining wall are shown in Fig.11.31a. The nature of distribution p'_p behind the retaining wall is shown in Fig. 1.31c. The passive force per unit length of the wall (P_p) can thus be given as

$$P_p = \int_0^H p_p' dz = \int_0^H (\gamma z K_p + 2c\sqrt{K_p}) dz = \tfrac{1}{2}K_p \gamma H^2 + 2cH\sqrt{K_p} \qquad (1.62)$$

In case of presence of a ground water table behind the retaining wall, the pore water pressure must be accounted for, or

$$p_p = K_p p_v' + u \qquad (1.63)$$

where p_p = passive pressure (total) at any depth z
u = pore water pressure

Inclined Granular Backfill

If the frictionless retaining wall discussed above has a granular backfill which is continuously inclined at an angle i with the horizontal, the effective passive pressure at any depth z developed due to sufficient wall rotation about its bottom (point B) can be derived in a similar manner as that for active case discussed in Section 1.4. Figure 1.32a shows a frictionless retaining wall. If

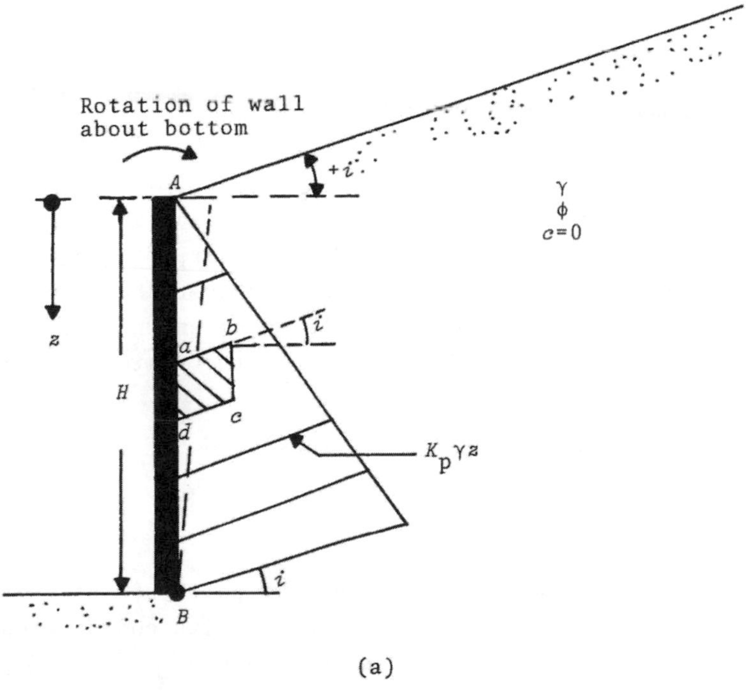

Figure 1.32. Rankine passive pressure for a frictionless retaining wall with an inclined granular backfill.

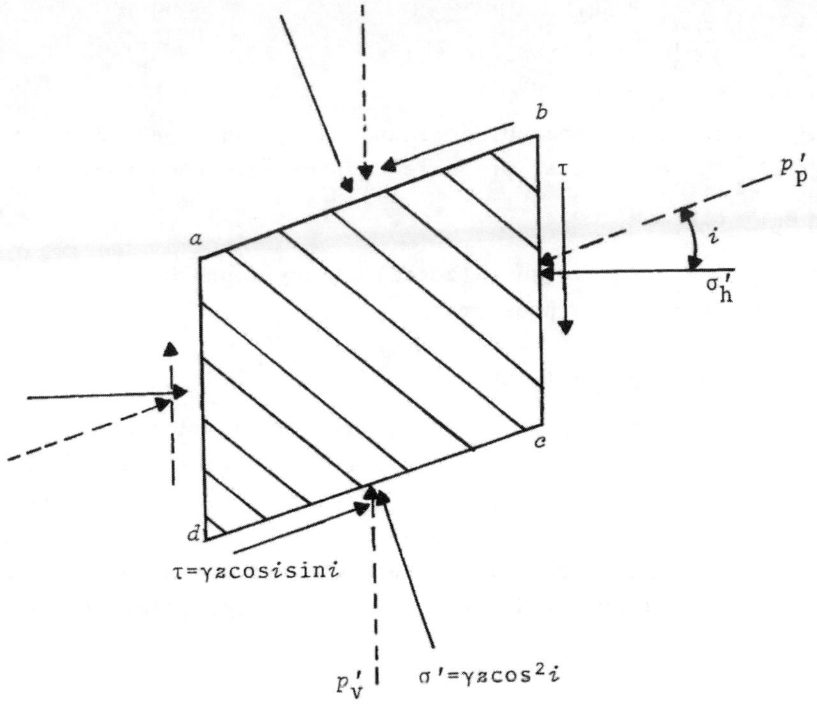

(b)

Figure 1.32. (Continued).

the wall is pushed into the soil mass by rotation about point *B*, the stresses on the soil element *abcd* at failure will be the same as shown in Fig. 1.32b. This is similar to that shown in Fig. 1.10b. The corresponding Mohr's circle is shown in Fig. 1.32c. Note that point *M* in Fig. 1.32c represents the state of stress on the inclined plane *cd*, or

Effective normal stress, $\sigma' = \gamma z \cos^2 i$

and

Shear stress, $\tau = \gamma z \cos i \sin i$

For this case, the point P_p is the passive pole. Point *N* represents the stresses on the vertical plane *ad* with coordinates of σ'_h and τ. However, the effective passive pressure

$$p'_p = \sqrt{(\sigma'_h)^2 + (\tau)^2} = \overline{ON} = \overline{OP_p} \tag{1.64}$$

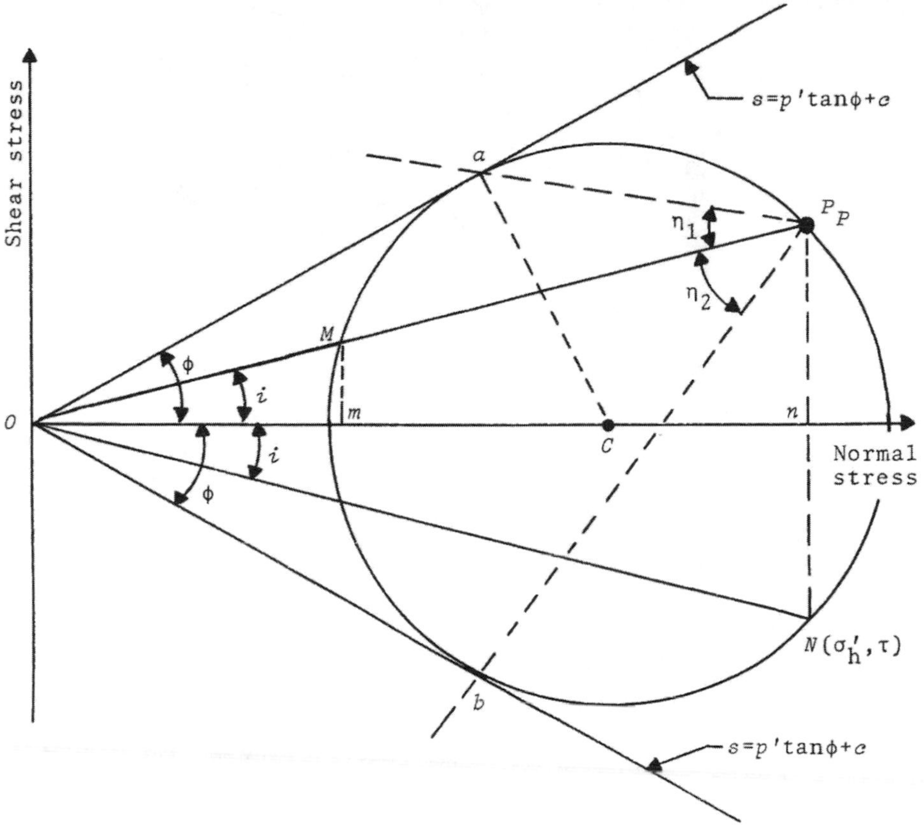

(c)

Figure 1.32. (Continued).

As for the active pressure case discussed in Section 1.4, it can be shown that

$$p_p' = \gamma z\left[\cos i\left(\frac{\cos i + \sqrt{\cos^2 i - \cos^2\phi}}{\cos i \quad \sqrt{\cos^2 i - \cos^2\phi}}\right)\right] = \gamma z K_p \tag{1.65}$$

where K_p $\quad \cos i\left(\frac{\cos i + \sqrt{\cos^2 i - \cos^2\phi}}{\cos i - \sqrt{\cos^2 i - \cos^2\phi}}\right)$ (1.66)

The variation of K_p with i and ϕ are given in Table 1.4.

The total passive force per unit length of the wall can be given as

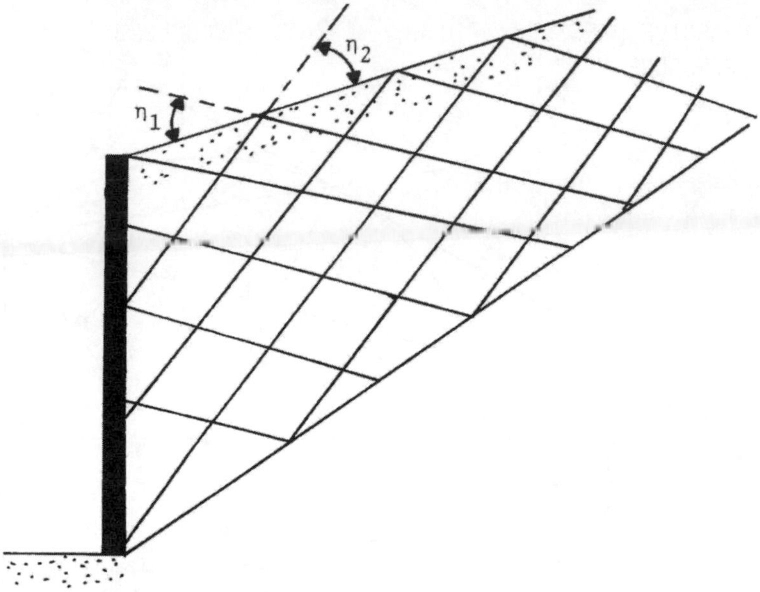

(d)

Figure 1.32. (Continued).

Table 1.4. Passive Earth Pressure Coefficient K_p [Eq. (1.66)]

i	K_p						
	ϕ (deg)						
(deg)	28	30	32	34	36	38	40
0	2.77	3.0	3.25	3.54	3.85	4.2	4.60
5	2.71	2.94	3.20	3.48	3.79	4.14	4.53
10	2.55	2.77	3.02	3.29	3.60	3.94	4.32
15	2.28	2.50	2.74	3.00	3.29	3.62	3.98
20	1.92	2.13	2.36	2.61	2.89	3.19	3.53
25	1.43	1.66	1.89	2.14	2.39	2.68	2.99

$$P_p = \int_0^H p_p' dz \quad \tfrac{1}{2}K_p \gamma H^2 \tag{1.67}$$

The direction of the slip lines in the failure wedge behind the retaining wall will be parallel to the lines P_pa and P_pb which are inclined at angles of $-\eta_1$ and $+\eta_2$ to the sloping backfill.

1.13 PASSIVE PRESSURE WITH WALL FRICTION (COULOMB'S PASSIVE PRESSURE

As discussed in Section 1.5, Coulomb's earth pressure theory assumes that the friction between the backface of the wall and the soil does exist, and the failure plane in the soil is a plane. Figure 1.33 shows a wall with a cohesionless soil backfill.

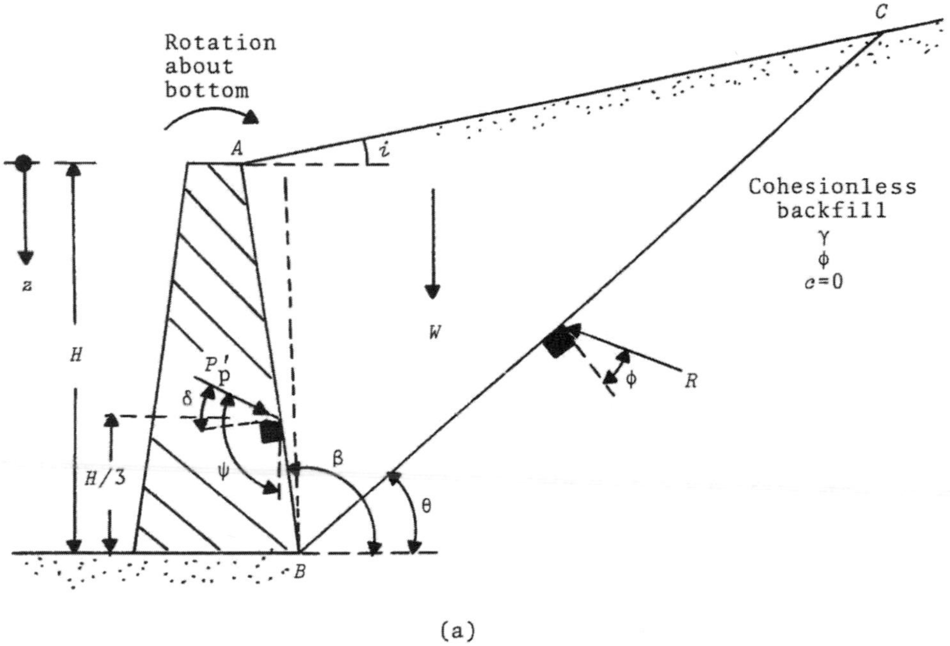

Figure 1.33. Coulomb's passive pressure.

If the wall rotates about its bottom into the soil mass, the failure wedge in the soil will be a triangular wedge such as *ABC* as shown in Fig. 1.33a. The forces acing on the wedge per unit length of the wall are

1. Weight of the wedge--*W*.
2. Passive force--P_p.
3. The resultant of the shear and normal forces along the potential failure surface *BC*--*R*.

For equilibrium of the wedge *ABC*, the force triangle is shown in Fig. 1.33b. Using the law of sines

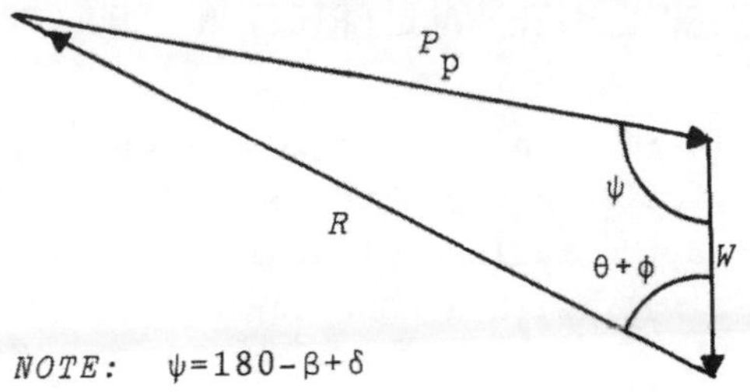

(b)

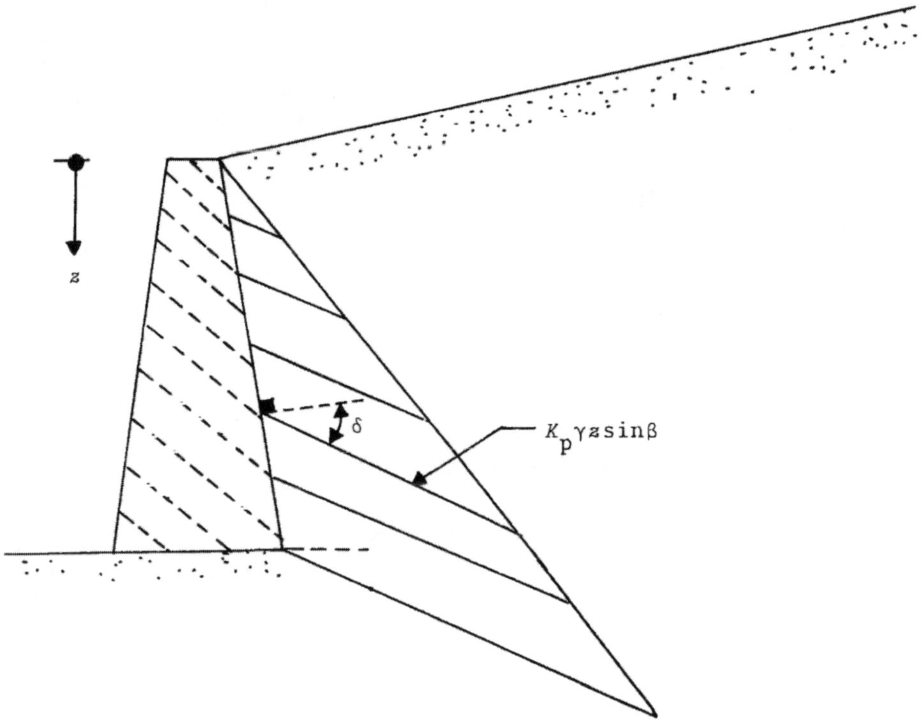

(c)

Figure 1.33. (Continued).

$$\frac{W}{\sin[180-(\theta+\phi)-(180-\beta+\delta)]} = \frac{P_p}{\sin(\theta+\phi)}$$

or

$$P_p = W\left[\frac{\sin(\theta+\phi)}{\sin(\beta-\delta-\theta-\phi)}\right] \quad (1.68)$$

For maximum value of P_p

$$\frac{\partial P_p}{\partial \theta} = 0 \quad (1.69)$$

Combining Eqs. (1.68) and (1.69), we obtain

$$P_p = \frac{1}{2}\gamma H^2 K_p \quad (1.70)$$

where K_p = Coulomb's passive pressure coefficient

$$\left[\frac{\sin(\beta+\phi)}{\sin\beta\left\{\sqrt{\sin(\beta+\delta)} - \sqrt{\dfrac{\sin(\phi+\delta)\sin(\phi+i)}{\sin(\beta-i)}}\right\}}\right]^2 \quad (1.71)$$

For a retaining wall with a vertical back and a horizontal granular backfill (β=90° and i=0), Eq. (1.70) simplifies to

$$P_p \quad \frac{\gamma H^2}{2\cos\delta}\left[\frac{1}{(1/\cos\phi) \quad \sqrt{\tan^2\phi+\tan\phi\tan\delta}}\right]^2 \quad (1.72)$$

The variations of Coulomb's earth pressure coefficient [Eq. (1.71)] with β=90°) (vertical backface) are given in Table 1.5. The variation of the passive pressure p_p with z can be obtained by substituting z for H in Eq. (1.70). So

$$P_p = \frac{1}{2}\gamma z^2 K_p$$

So

$$p_p = \frac{\partial P_p}{\partial(z/\sin\beta)} \quad \gamma z K_p \sin\beta \quad (1.73)$$

Figure 1.33c shows the nature of variation of p_p along the backface of the wall.

1.14 PASSIVE PRESSURE BY USING CURVED FAILURE SURFACE

Experience has shown that Coulomb's passive pressure (Section 1.13) derived by assuming plane failure surface in soil gives large values of P_p which are on the *unsafe side of design*. Several laboratory model tests have shown that the actual failure surface in

Table 1.5. Variation of K_p--Coulomb's Case For Vertical Backface of Wall (β=90°) [Eq. (1.71)][a]

φ (deg)	δ (deg)	*i* (deg) 0	5	10
26	0	2.561	2.943	3.385
	16	4.195	5.250	6.652
	17	4.346	5.475	6.992
	20	4.857	6.249	8.186
	22	5.253	6.864	9.164
28	0	2.770	3.203	3.713
	16	4.652	5.878	7.545
	17	4.830	6.146	7.956
	20	5.436	7.074	9.414
	22	5.910	7.820	10.625
30	0	3.000	3.492	4.080
	16	5.174	6.609	8.605
	17	5.385	6.929	9.105
	20	6.105	8.049	10.903
	22	6.675	8.960	12.421
32	0	3.255	3.815	4.496
	16	5.775	7.464	9.876
	17	6.025	7.850	10.492
	20	6.886	9.212	12.733
	22	7.574	10.334	14.659
34	0	3.537	4.177	4.968
	16	6.469	8.474	11.417
	17	6.767	8.942	12.183
	20	7.804	10.613	15.014
	22	8.641	12.011	17.497
36	0	3.852	4.585	5.507
	16	7.279	9.678	13.309
	17	7.636	10.251	14.274
	20	8.892	12.321	17.903
	22	9.919	14.083	21.164
38	0	4.204	5.046	6.125
	16	8.230	11.128	15.665
	17	8.662	11.836	16.899
	20	10.194	14.433	21.636
	22	11.466	16.685	26.013
40	0	4.599	5.572	6.841
	16	9.356	12.894	18.647
	17	9.882	13.781	20.254
	20	11.771	17.083	26.569
	22	13.364	20.011	32.602

[a]After Bowles (1982)

soil at passive state is curved as shown in Fig. 1.34. In Fig. 1.34, *BCD* is the failure surface in which *BC* is curved and *CD* is a straight line. The zone *ACD* is a *Rankine passive zone* in which the

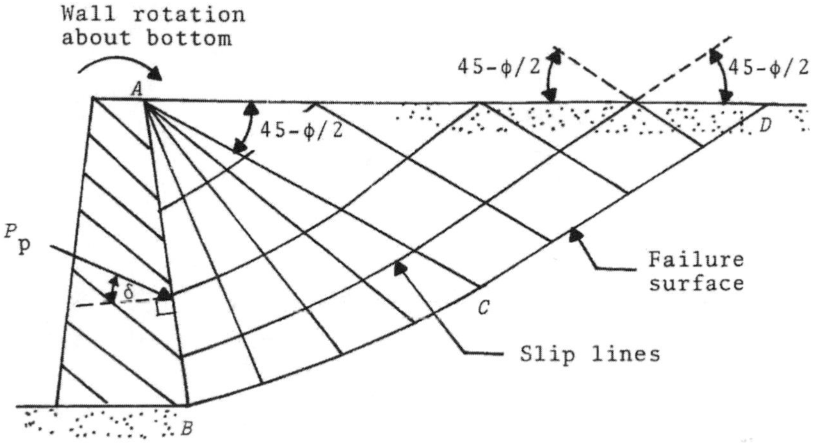

Figure 1.34. Nature of failure surface in soil (at passive state) behind retaining wall with friction.

slip lines intersect the horizontal ground surface at an angle of $\pm(45-\phi/2)$. Several investiagors have made different assumptions about the nature of the curved surface *BC* to obtain the magnitude of the passive pressure coefficient. A summary of these is given below.

Investigator	Year	Assumption on *BC* (Fig. 1.34)
Caquot and Kerisel	1948	Arc of an ellipse
Terzaghi and Peck	1967	Arc of a log spiral
Janbu	1957	Arc of a log spiral
Shields and Tolunay	1973	Arc of a log spiral

In this section, Terzaghi and Peck's method (1967) for determination of the passive pressure using a curved failure surface will be discussed in detail. Figure 1.35 shows a retaining wall with an inclined back face and a horizontal backfill which is a c-ϕ soil. If the wall rotates about its bottom, the nature of the failure surface in soil will be like *BCD* in which *BC* is an arc of a log spiral. *ACD* is a Rankine passive zone. The center of the

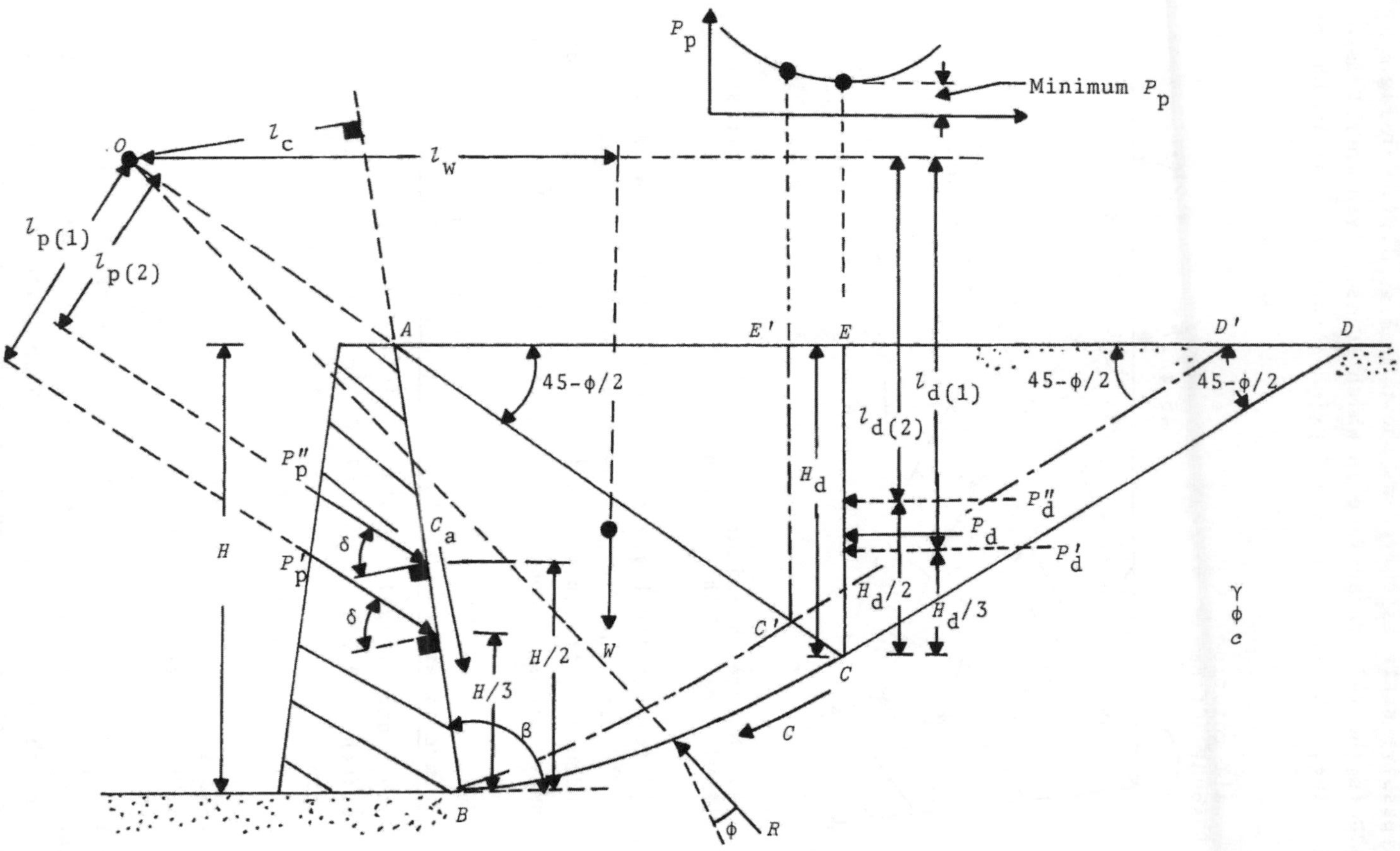

Figure 1.35. Passive pressure with the curved failure surface assumption. (*Note:* *BC* and *BC″* are arcs of log spirals.)

log spiral will be along the line *OA*. The point *O* is the center of the log spiral. Now, let the failure wedge shown in Fig. 1.35 be a trial wedge. The forces acting on the soil mass (per unit length of the wall), *ABCE*, are as follows.

1. Weight of the wedge *ABCE*--*W*. The line of action of *W* passes through the centroid of the wedge *ABCE*.

2. Rankine passive force on the face *CE*--P_d. Or

$$P_d = P'_d + P''_d$$

where $P'_d = \frac{1}{2}\gamma H_d^2 K_{p(Rankine)}$

and

$$P''_d \quad 2cH_d\sqrt{K_{p(Rankine)}}$$

$$H_d = \overline{CE}$$

3. The resultant of the frictional shear and normal forces along *BC*--*R*. Note that, based on the property of the log sprial, the force *R* will be directed along a radial line of the log spiral.

4. The resisting cohesive force along arc *BC*--*C*.

5. The resisting adhesive force along the wall face *AB*--C_a.

6. Passive force--$P_p = P'_p + P''_p$.

The force P'_p is the contribution of the weight and friction of the soil, and P''_p is the contribution of cohesion of the soil. The above-mentioned forces are shown in Fig. 1.35. Note that the forces due to the weight and friction are P'_p, W, R, and P'_d.

Now, taking the moment of these forces about the center of the log spiral *O*

$$P'_p l_{p(1)} - Wl_w \quad \frac{1}{2}\gamma H_d^2 K_{p(Rankine)} l_{d(1)} + R(O) = 0 \tag{1.74}$$

or

$$P'_p = \frac{1}{l_{p(1)}}[Wl_w + \frac{1}{2}\gamma H_d^2 K_{p(Rankine)} l_{d(1)}] \tag{1.75}$$

In a similar manner, the forces due to cohesion on the soil wedge *ABCE* are C_a, C, $P''_d = 2cH_d\sqrt{K_{p(Rankine)}}$, and P''_p.

Taking the moment about the center of the log spiral *O*

$$P''_p l_{p(2)} \quad C_d l_c \quad 2cH_d\sqrt{K_{p(Rankine)}} l_{d(2)} - M_c = 0 \tag{1.76}$$

where M_c = moment due to the cohesive force *C*

$$= \frac{c}{2\tan\phi}(\overline{OC}^2-\overline{OB}^2) \tag{1.77}$$

or

$$P''_p = \frac{1}{l_{p(2)}} = [M_c + C_d l_c + 2cH_d\sqrt{K_{p(Rankine)}}\, l_{d(2)}] \tag{1.78}$$

$$\text{where } C_d = c_d AB \quad c_d \frac{H}{\sin(180-\beta)} = c_c(\frac{H}{\sin\beta}) \tag{1.79}$$

c_d = adhesion per unit area at the soil-wall interface

So, combining Eqs. (1.75) and (1.78)

$$P_p = P'_p + P''_p = \frac{1}{l_{p(1)}}[Wl_w + \tfrac{1}{2}\gamma H_d^2 K_{p(Rankine)}\, l_{d(1)}]$$

$$+ \frac{1}{l_{p(2)}}[M_c + C_d l_c + 2cH_d\sqrt{K_{p(Rankine)}}\, l_{d(2)}] \tag{1.80}$$

Several trial wedges like *ABCD* (such as *ABC'D'*) can now be taken. For all trials, the centers of the log spiral will lie along line *OA*. The magnitude of P_p calculated for each trial wedge can be plotted as shown in the upper part of Fig. 1.35. A smooth curve is plotted through these points. The minimum value of P_p obtained from the curve is the desired passive force.

It needs to be pointed out that, for granular backfills, $c=0$ and hence $P''_p=0$. So

$$P_p \quad P'_p = \tfrac{1}{2}\gamma H^2 K_p \tag{1.81}$$

where K_p = passive earth pressure coefficient

The passive earth pressure coefficient determined in this manner for horizontal granular backfill and vertical backface of the wall ($\beta=90°$) is shown in Fig. 1.36.

The analysis of Caquot and Kerisel (1948) who assumed *BC* (Fig. 1.34) to be an arc of an ellipse will not be discussed here in detail. However, the variation of K_p for vertical backface of the wall with horizontal granular backfill ($c=0$) as obtained by using Caquot and Kerisel's method (1948) is shown in Fig. 1.37.

1.15 OTHER SOLUTIONS FOR PASSIVE PRESSURE (ROTATION ABOUT THE BOTTOM)

Use of Theory of Plasticity

Rosenfarb and Chen (1972) have developed a closed form solu-

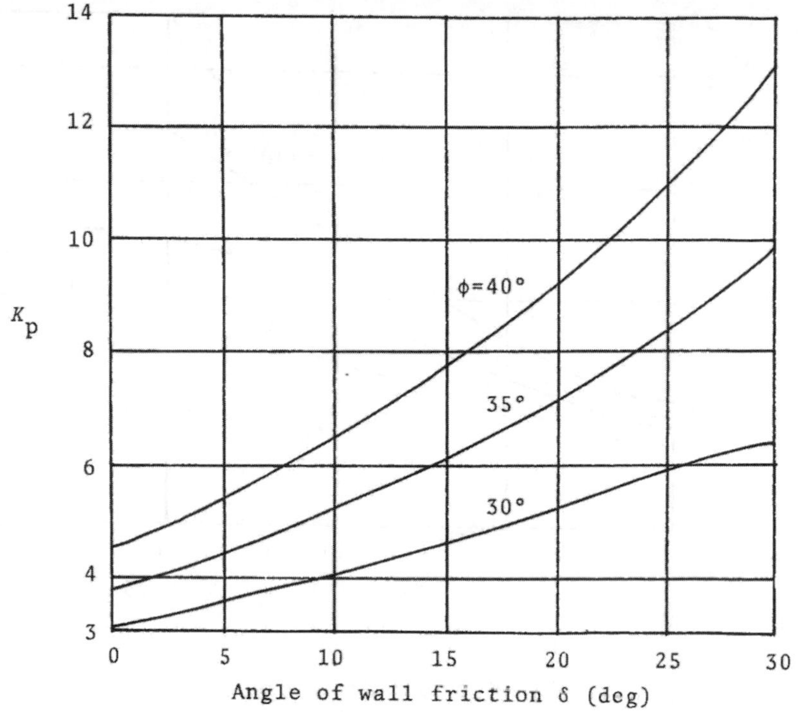

Figure 1.36. Variation of K_p obtained from Terzaghi and Peck's method [vertical backface of wall (β=90°) and horizontal granular backfill].

tion for determination of P_p by using the theory of plasticity. The *passive log-sandwich mechanism* of failure wedge developed by them is shown in Fig. 1.38, in which the backfill is a granular soil with c=0. Along the failure surface, *BC* and *DE* are straight lines, and *CD* is an arc of a log spiral. According to this theory

$$P_p \quad \tfrac{1}{2}K_p\gamma H^2$$

The values of K_p obtained from Rosenfarb and Chen's theory (1972) are given in Table 1.6. It needs to be pointed out that when i=0 the values of K_p coincide with those obtained from Coulomb's theory.

Use of Method of Slices

Shields and Tolunay (1973) have proposed the method of slices to determine the passive force P_p with a curved failure surface.

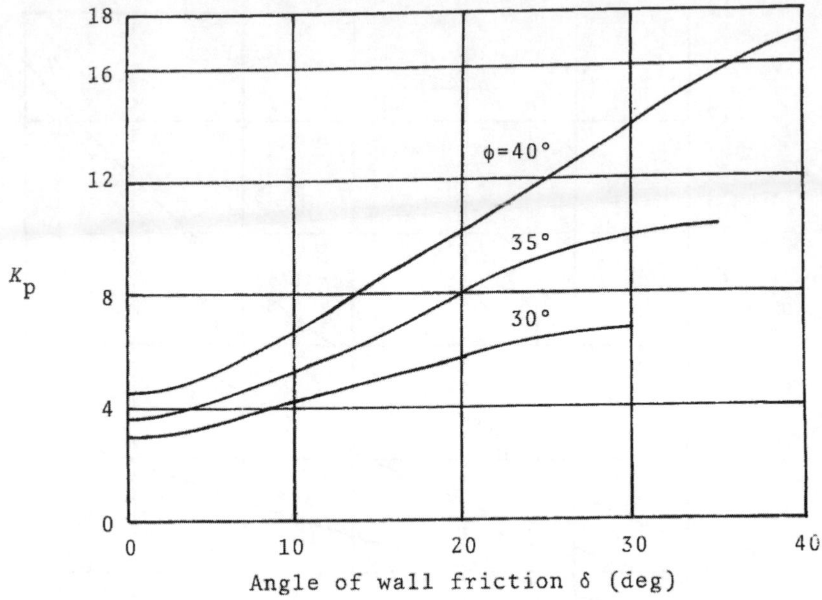

Figure 1.37. Variation of K_p obtained from Caquot and Kerisel's method (1948) [vertical backface of wall (β=90°) and horizontal granular backfill].

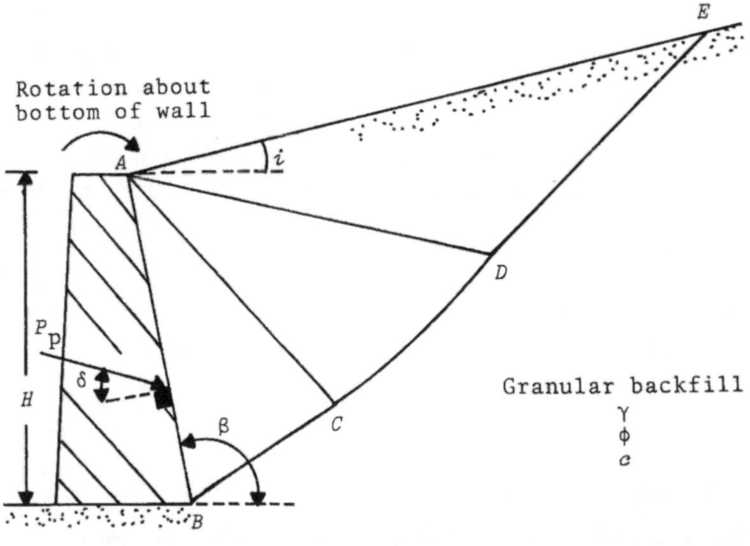

Figure 1.38. Passive *log-sandwich* mechanism for determination of P_p.

Table 1.6. Values of K_p Obtained From Theory of Plasticity--Log-Sandwich Failure Mechanism[a]

i (deg)	δ (deg)	ϕ (deg) 30	35	40	45
10	0	4.01	5.2	6.68	8.93
20	0	5.25	7.03	9.68	13.8
30	0	6.74	9.50	14.0	21.5
10	10	5.7	7.61	10.4	14.9
20	10	7.79	10.9	15.9	24.4
30	10	10.3	14.7	23.6	39.6
10	20	7.94	11.2	16.3	24.9
20	20	11.2	16.5	25.6	42.4
30	20	15.1	23.2	41.0	70.2
10	30	10.6	15.8	24.6	40.7
20	30	15.2	23.2	39.5	70.3
30	30	20.8	34.8	62.0	--

[a]After Bowles, 1982

This can be explained by means of Fig. 1.39. Figure 1.39a shows a vertical retaining wall with a horizontal granular backfill. At passive state *ABCD* is the failure surface in soil. *BC* is an arc of a log spiral. *O* is the center of the log spiral. The zone *ACD* is the Rankine passive zone. By trigometric procedure, it can be shown that the angle $B'BB''' = \Delta$, or

$$\Delta = 90° - \phi - \alpha_w \tag{1.82}$$

(*NOTE:* *BB'* is a horizontal line.)

$$\alpha_w = \frac{1}{2}\left\{\text{arc cos}\left[\cos(\phi-\delta) - \frac{\sin(\phi-\delta)}{\tan\phi}\right] - \phi - \delta\right\} \tag{1.83}$$

The angle α_w is considered *positive* which it is *above the horizontal line* and negative when it is below the horizontal line. Hence

$$\theta = 180° - (90-\Delta) - (90+45-\phi/2) = \Delta - 45° + \phi/2 \tag{1.84}$$

Using the law of sines for the triangle *OAB*

$$\frac{H}{\sin\theta} \quad \frac{\overline{OB}}{\sin(135-\phi/2)}$$

or

$$\overline{OB} = \left[\frac{\sin(135-\phi/2)}{\sin\theta}\right]H \tag{1.85}$$

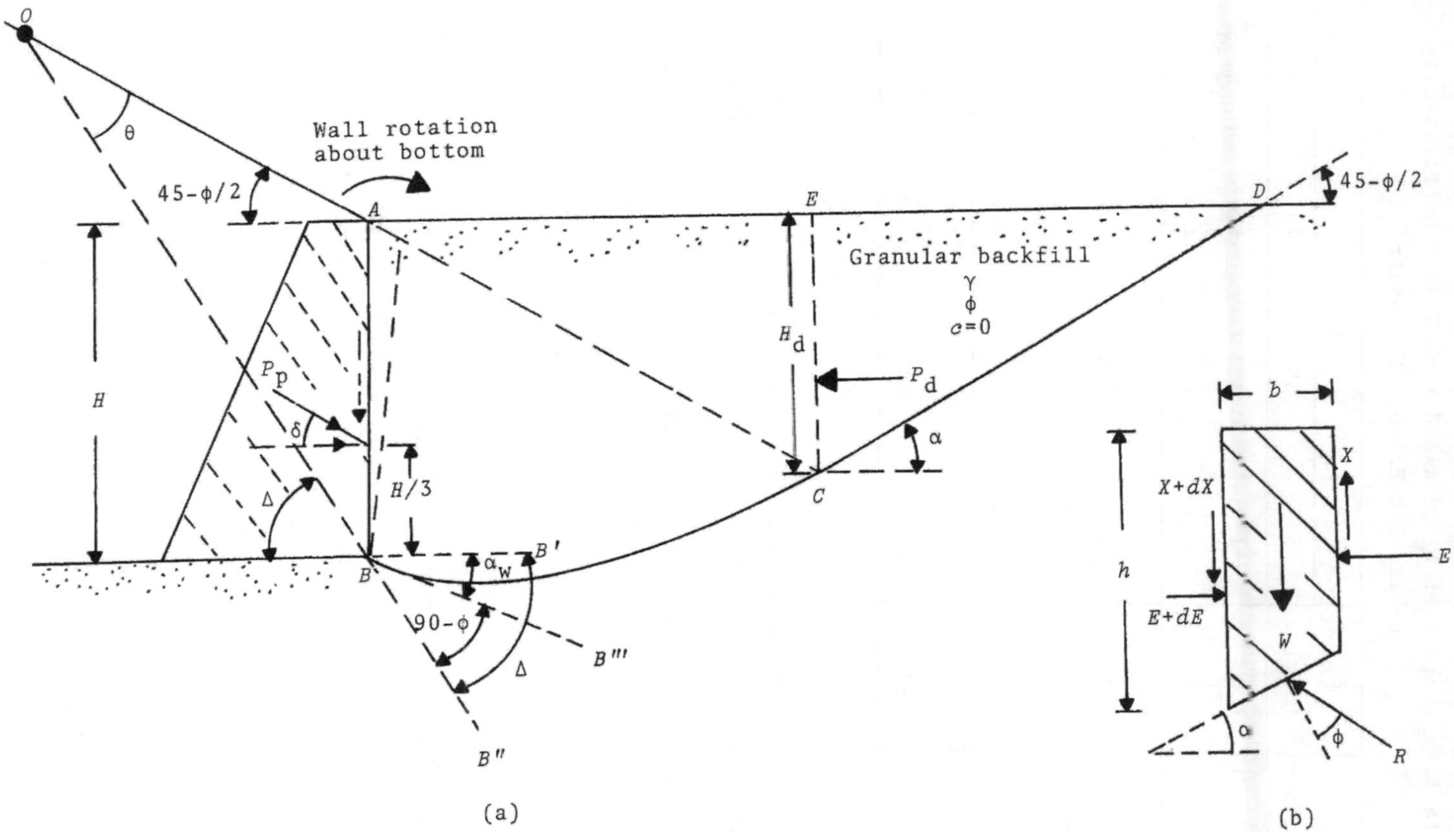

Figure 1.39. Determination of P_p by the method of slices. (*Note:* In part a, *BB'''* is tangent to the log spiral at *B*.)

Similarly

$$\overline{OA} = \left[\frac{\sin(90-\Delta)}{\sin\theta}\right]H \tag{1.86}$$

Also, from the properties of a log spiral

$$\overline{OC} = \overline{OB}(e^{\theta\tan\phi}) \tag{1.87}$$

Knowing $\overline{OA}$ and $\overline{OC}$, the height of $\overline{CE}$ can be determined as

$$H_d = \overline{CE} \quad (\overline{OC}-\overline{OA})\sin(45-\phi/2) \tag{1.88}$$

As in Section 1.14, the Rankine passive force on the soil wall CE is equal to

$$P_d = \tfrac{1}{2}\gamma H_d^2\tan^2(45+\phi/2) \tag{1.89}$$

Now the wedge $ABCE$ can be divided into several vertical slices slices. Figure 1.39b shows one of the slices with the forces (per unit length of the wall) acting on it. Summing the horizontal forces

$$E + dE - E \quad R\sin(\alpha+\phi) = 0$$

or

$$dE = R\sin(\alpha+\phi) \tag{1.90}$$

Again, summing the horizontal forces

$$W + X + dX - X - R\cos(\alpha+\phi) \quad 0$$

or

$$W + dX = R\cos(\alpha+\phi) \tag{1.91}$$

Combining Eqs. (1.90) and (1.91)

$$dE = (W+dX)\tan(\alpha+\phi) \tag{1.92}$$

So

$$P_p\cos\delta = (\sum_{BC} dE) + P_d = [\sum_{BC} (W+dX)\tan(\alpha+\phi)] + P_d$$

Without losing too much accuracy, we can assume that for all slices except for the one closest to the wall, $dX \to 0$. At the wall, $\alpha=\alpha_w$ and $dX=P_p\sin\delta$. So

$$P_p\cos\delta = \left\{[\sum_{BC} W\tan(\alpha+\phi)] + P_p\sin\delta\tan(\alpha_w+\phi)\right\} + P_d \tag{1.93}$$

$$P_p[\cos\delta-\sin\delta\tan(\alpha_w+\phi)] = P_d + \sum_{BC} [W\tan(\alpha+\phi)]$$

or

$$P_p \quad \frac{P_d + \Sigma\,[W\tan(\alpha+\phi)]}{\frac{BC}{\cos\delta - \sin\delta\tan(\alpha_w+\phi)}} = \tfrac{1}{2}\gamma H^2 K_p \tag{1.94}$$

where K_p = passive earth pressure coefficient

$$= \left(\frac{2}{\gamma H^2}\right) \frac{P_d + \Sigma\,[W\tan(\alpha+\phi)]}{\frac{DC}{\cos\delta - \sin\delta\tan(\alpha_w+\phi)}} \tag{1.95}$$

The values of K_p determined in this manner are given in Table 1.7.

Table 1.7. Values of K_p For Vertical Retaining Wall (β=0) With Horizontal Granular Backfill (c=0) As Determined From The Method of Slices

δ	ϕ (deg)					
(deg)	20	25	30	35	40	45
0	2.04	2.46	3.00	3.69	4.60	5.83
5	2.27	2.78	3.44	4.31	5.46	4.09
10	2.47	3.08	3.86	4.92	6.36	8.43
15	2.64	3.34	4.28	5.53	7.30	9.89
20	2.87	3.61	4.68	6.17	8.3	11.49
25		4.0	5.12	6.85	9.39	13.20
30			5.81	7.61	10.60	15.31
35				8.85	12.00	17.65
40					14.40	20.48
45						25.47

1.16 DUBROVA'S METHOD FOR CALCULATION OF PASSIVE PRESSURE DISTRIBUTION

Dubrova's method (1963) for calculation of the active pressure distribution behind a rough wall with a vertical back (β=90°) and having a horizontal granular backfill has been discussed in Section 1.10.1 for wall rotation about its top and in Section 1.11 for lateral displacement of the wall. A similar procedures can be adopted for determination of passive pressure distribution behind a rough vertical wall with a horizontal granular backfill. Figure 1.40a shows a vertical retaining wall with *quasi-rupture* lines in the failure wedge in the soil. If the rotation of the wall is about the bottom (point B), then the angle which the resultant of the shear and normal forces along a quasi-rupture line will make with the normal is equal to ϕ', or

$$\phi' = \frac{\phi z}{H} - \phi$$

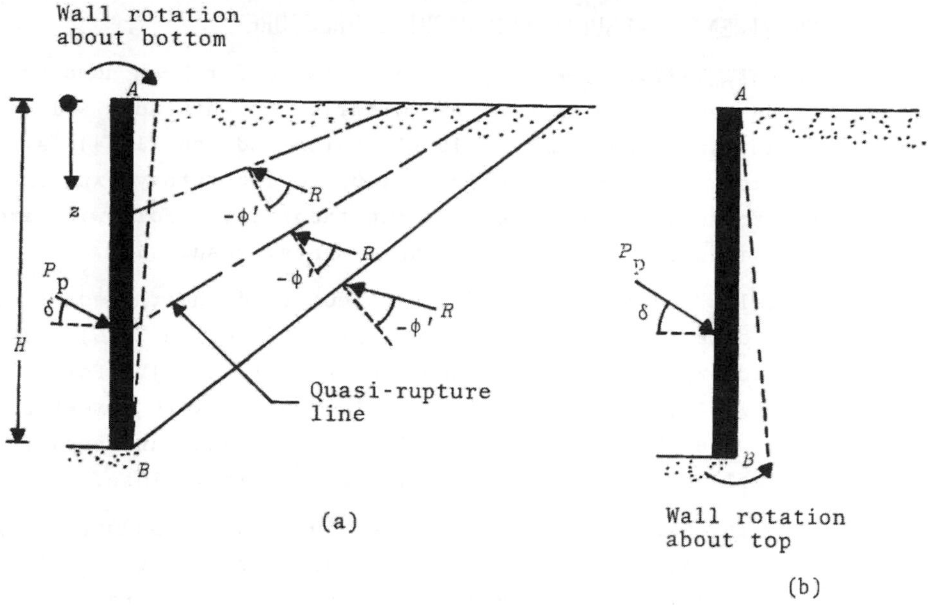

Figure 1.40. Dubrova's method for calculation of passive pressure distribution--granular backfill.

The resulting passive pressure distribution as given by Dubrova (1963) is as follows

$$p_p = \left[\frac{\gamma\cos^2\phi'}{(1+n\sin\phi')^2\cos\delta}\right]\left[z - \frac{\phi z^2(\sin\phi'+n)}{H(1+n\sin\phi')\cos\phi'}\right] \quad (1.96)$$

where $n \quad \sqrt{1 + (\frac{\tan\delta}{\tan\phi})}$ (1.97)

Similarly, for wall rotation about the top (Fig. 1.40b)

$$\phi' \quad - \frac{\phi z}{H} \quad (1.98)$$

$$p_p = \left[\frac{\gamma\cos^2\phi'}{(1+n\sin\phi')^2\cos\delta}\right]\left[z + \frac{\phi z^2(\sin\phi'+n)}{H(1+n\sin\phi')\cos\phi'}\right] \quad (1.99)$$

as before, where $n = \sqrt{1 + (\frac{\tan\delta}{\tan\phi})}$

The passive pressure at a depth z below the ground surface for horizontal wall translation can be obtained as

$$p_{p(\text{translation})} \quad \frac{p_{p(\text{rotation about bottom})} + p_{p(\text{rotation about top})}}{2} \quad (1.100)$$

1.17 EXPERIMENTAL RESULTS FOR PASSIVE PRESSURE

Relatively little experimental work has so far been done to determine the passive earth pressure distribution behind a wall. Narain, Saran, and Nandakumaran (1969), Saran and Deo (1974), and Rowe and Peaker (1965) have provided valuable laboratory experimental results to compare with the existing theories. Following are brief discussions on the available experimental results.

1. Shields and Tolunay (1973) have compared their theoretical values of K_p obtained by the method of slices (Section 1.15) with those obtained experimentally by Rowe and Peaker (1965). For loose sands (ϕ=34°), the experimental values of K_p were somewhat higher than those obtained theoretically. For dense sand backfill, the theoretical and experimental results were fairly close.

2. Narain, Saran, and Nandakumaran (1969) have conducted several laboratory model tests with a rough wall having a vertical back and horizontal granular backfill (loose and dense states of compaction). Passive pressure at various depths of the wall was measured when the wall was subjected to rotation about its bottom, rotation about its top, and horizontal translation. The properties of the backfill sand were as follows

Loose state of compaction

ϕ=38.5°

δ=23.5°

Relative density of compaction=31.5%

Dense state of compaction

ϕ=42°

δ=23.5°

Relative density of compaction=70.25%

The distribution of the horizontal components of passive pressure $p_p\cos\delta$ [$=(\partial P_p/\partial z)(\cos\delta)$] at various depths of the wall for loose and dense sand backfill showed that the passive force P_p is highest when the wall is rotated about the bottom, and minimum when it is rotated about the top. A comparison of $K_p\cos\delta$ values obtained from these tests is as follows.

Loose sand backfill

Experimental $K_p\cos\delta$	Rotation about bottom: 9.10
	Rotation about top: 6.70
	Translation: 8.40

Theoretical $K_p\cos\delta$	Coulomb: 13.6 Rankine: 4.0 Caquot and Kerisel (1948): 11.8 Terzaghi (1943): 7.85 Shields and Tolunay (1973): 7.5

Dense sand backfill

Experimental $K_p\cos\delta$	Rotation about bottom: 10.10 Rotation about top: 6.90 Translation: 8.80
Theoretical $K_p\cos\delta$	Coulomb: 19.94 Rankine: 6.05 Caquot and Kerisel (1948): 18.0 Terzaghi (1943): 16.6 Shields and Tolunay (1973): 9.43

As can be seen from the above comparison, the experimental values of $K_p\cos\delta$ for wall rotation about the bottom with dense sand backfill compares reasonably well with the theory of Shields and Tolunay (1973).

3. Saran and Deo (1974) have also presented some laboratory experimental results for $K_p\cos\delta$ for a vertical retaining wall with a horizontal sand backfill (dense and loose states of compaction) with and without a surcharge q (load per unit area) on the surface of the backfill. According to the fundamental theory of passive pressure

$$P_p\cos\delta \text{ (without surcharge)} = \tfrac{1}{2}\gamma H^2 K_p\cos\delta \qquad (1.101a)$$

and

$$P_p\cos\delta \text{ (with surcharge } q \text{ on the surface of the backfill)}$$

$$= (\tfrac{1}{2}\gamma H^2 + qH)K_p\cos\delta \qquad (1.101b)$$

Following is a summary of the experimental results of the $K_p\cos\delta$ values as obtained by Saran and Deo (1974). From this table, it can be seen that, for all modes of wall rotation, the magnitudes of $K_p\cos\delta$ with surcharge are lower than those obtained without a surcharge. This means that the values of K_p as related to the surcharge q may be somewhat smaller than those related to the unit weight of soil, γ.

Wall movement	$K_p \cos\delta$ without surcharge	$K_p \cos\delta$ with surcharge (q=2.02 kN/m^2)
	Loose sand backfill	
Rotation about bottom	10.3	9.0
Rotation about top	8.0	7.2
Translation	8.7	8.1
	Dense sand backfill	
Rotation about bottom	11.0	9.8
Rotation about top	8.9	8.1
Translation	9.9	9.1

1.18 WALL ROTATION REQUIRED FOR PASSIVE EARTH PRESSURE CONDITION

The approximate magnitudes of the wall movement (rotation about the bottom) necessary to create a passive state in the failure wedge of the soil are given in Table 1.8.

Table 1.8. Approximate Magnitude of the Wall Movement (Rotation About the Bottom) For Development of Passive State

Type of backfill	ω (radians)
Dense sand	50×10^{-4}
Loose sand	100×10^{-4}
Stiff clay	100×10^{-4}
Soft clay	500×10^{-4}

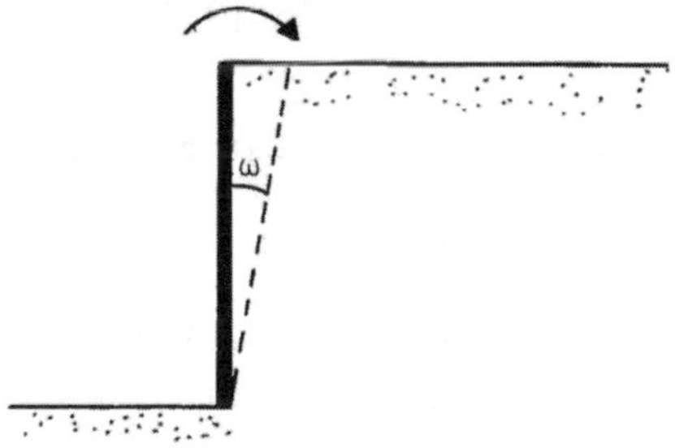

The experimental works conducted by Narain et al. (1969) have shown that the wall rotation about the bottom (ω) required for the development of the passive state is about 85×10^{-4} radians for loose sand, and it is about 65×10^{-4} radians for dense sand backfill. For wall rotation about the top, the magnitudes of ω were 65×10^{-4}

radians for loose sand and 40×10^{-4} radians for dense sand. The same experimental results also showed that the translation of the wall required for development of the passive state in granular soil backfill are as follows:

State of compaction of backfill	S/H^a (%)
Dense	6.4
Loose	8.55

[a] S=lateral translation of the wall toward the soil; H=height of the wall

STABILITY OF CUTS IN CLAY AND PRESSURE ON SHAFT LINING

1.19 STABILITY OF SLURRY TRENCH CUTS IN SATURATED CLAY

Open cuts in clayey soils are quite common in the construction industry. Sometimes these cuts are filled with a clay slurry of negligible shear resistance to increase the depth of the cut without failure. In this section, the short-term stability of open cuts in saturated clay (ϕ=0; total stress concept) will be discussed. The following concept was derived by Meyerhof (1972).

Figure 1.41 shows an open cut in saturated clay having an undrained shear strength of $s=c_u$. The open cut is filled with a slurry having a unit weight of γ_s. For a long cut, according to the Rankine active pressure theory [Eq. (1.18)], with ϕ=0 (K_a=0)

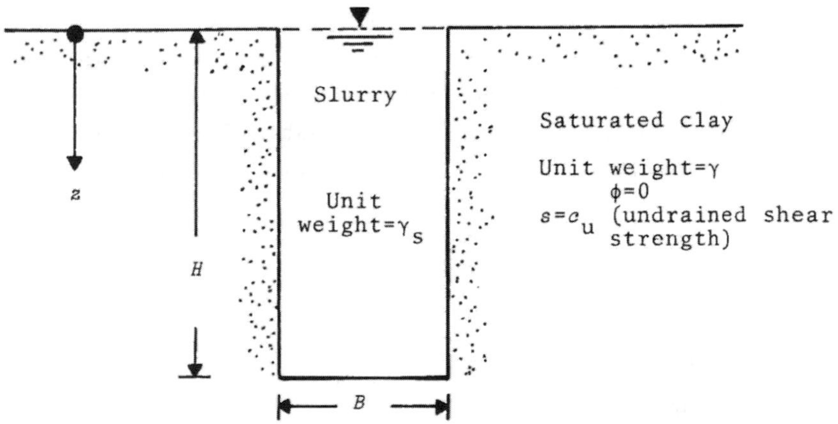

Figure 1.41. Open cut in saturated clay filled with a slurry.

$$p_a = (\gamma - \gamma_s)z - 2c_u \tag{1.102}$$

where p_a = Rankine active earth pressure at a depth z
γ = unit weight of saturated clay
γ_s = unit weight of slurry

The total active force up to a depth z is

$$P_a \quad \int p_a dz \quad \int [\gamma - \gamma_s)z \quad 2c_u]dz$$

$$(\gamma - \gamma_s)\frac{z^2}{2} - 2c_u z \tag{1.103}$$

The critical depth $z = H_c$ at which P_a is equal to zero (for $z > H_c$ failure will occur) can be determined as

$$P_a = (\gamma - \gamma_s)\frac{H_c^2}{2} - 2c_u H_c = 0$$

or

$$H_c = \frac{4c_u}{\gamma - \gamma_s} \tag{1.104}$$

For circular trenches (of diameter B) or square trenches (of side B) Eq. (1.102) can be modified (Hencky, 1923)

$$p_a = (\gamma - \gamma_s)z \quad 2\left[ln\left(\frac{2z}{B} + 1\right) + 1\right]c_u \tag{1.105}$$

The above modification is due to the fact that, for cylindrical cuts, plastic flow of the soil occurs in both planes (that is, horizontal and radial). Hoop stresses act normal to the radial planes. The hoop stresses are approximately equal to the vertical stress near the face of the cut. At greater distances, the hoop and radial stresses both approach the earth pressure at rest.

Meyerhof (1972) has used several simplified approximations and has provided the following relationship for estimation of the critical depth of cut (H_c) for square, circular, and rectangular cuts.

$$H_c \quad \frac{Nc_u}{K_o\gamma' - \gamma_s'} \tag{1.106}$$

where N stability factor $= 4(1 + \frac{B}{L})$ (1.107)
L length of the cut
γ' effective unit weight of clay $= \gamma - \gamma_w$
γ_s' effective unit weight of the slurry $= \gamma_s - \gamma_w$

γ_w = unit weight of water

K_o coefficient of at-rest earth pressure (Section 1.2)

The factor of safety against failure of trenches may be defined as

$$F_s = \frac{Nc_u}{p_{o(H)}-p_{i(H)}} \tag{1.108}$$

where F_s = factor of safety

$p_{o(H)}$ = maximum horizontal earth pressure (that is, at a depth H) from the soil side

$p_{i(H)}$ = maximum horizontal pressure from the slurry side

Meyerhof (1972) has recommended a minimum value of F_s=1.3 for use in construction.

Example 1.8. Consider an open cut in saturated clay with B=5 ft and L=20 ft. For the clay, c_u=600 lb/ft^2. The cut is filled with water. Given: saturated unit weight of clay=γ=122 lb/ft^2 and K_o-0.6. Determine the critical depth of the cut at which failure will occur.

Solution

Given B=5 ft; L=20 ft. B/L=5/20=0.25. So, from Eq. (1.107)

$$N = 4(1+\frac{B}{L}) = 4(1+0.25) = 5$$

Hence

$$H_c = \frac{Nc_u}{K_o\gamma'-\gamma_s'} = \frac{(5)(600)}{(0.6)(122-62.4)-(62.4-62.4)} \quad \frac{3000}{35.76} \quad \underline{\underline{83.9\text{ ft}}}$$

Example 1.9. Consider a cut in saturated clay with B=5 ft and L=20 ft. The cut is 40 ft deep. The water table is located 10 ft below the ground surface. Given: K_o=0.6 and c_u=450 lb/ft^2. Determine the factor of safety F_s.

Solution

As in Example 1.8, B/L=0.25, so N=5. At the bottom of the cut

$$p_{o(H)} \quad K_o[\gamma(10)+\gamma'(30)]+\gamma_w(30)$$

$$= 0.6[(120)(10)+(120-62.4)30]+(62.4)(30)$$

$$= 1.756.8 + 1,872 = 3,628.8 \text{ lb/ft}^2$$

$$p_{i(H)} = \gamma_w(30) \quad (62.40(30) = 1,872 \text{ lb/ft}^2$$

So, from Eq. (1.108)

$$F_s = \frac{Nc_u}{p_{0(H)} - p_{i(H)}} \quad \frac{(5)(450)}{3,628.8-1,872} = \underline{1.28}$$

1.20 STABILITY OF UNSUPPORTED AXISYMMETRIC EXCAVATIONS IN SATURATED CLAY

The stability of unsupported axisymmetric excavations in saturated clays has been analyzed by Britto and Kusakabe (1983). According to their analysis, the undrained shear strength of saturated clays can be expressed as

$$c_{u(z)} = c_{u(o)} + kz \tag{1.109}$$

where $c_{u(z)}$ undrained shear strength at a depth z below the ground surface

$c_{u(o)}$ = undrained shear strength of clay at the ground surface (that is, z=0)

k = a constant

This type of variation of the undrained shear strength is shown in Fig. 1.42a, which generally occurs in *normally consolidated clay* deposits. Figure 1.42b shows a circular excavation having a diameter B. The depth of the excavation is equal to H. For this cut, two nondimensional parameters can now be defined, and they are

$$N \quad \frac{\gamma H}{c_{u(o)} + kH} \tag{1.110}$$

and

$$M = \frac{c_{u(o)}}{c_{u(o)} + kH} \tag{1.111}$$

where N = stability factor

γ = unit weight (total) of the soil

Britto and Kusakabe (1983) analyzed the stability of the excavation by using the upper bound theorem of plasticity. The variation of N against M thus determined by them is shown in Fig. 1.43. For a given excavation, if k, H, B, and $c_{u(o)}$ are known, the value

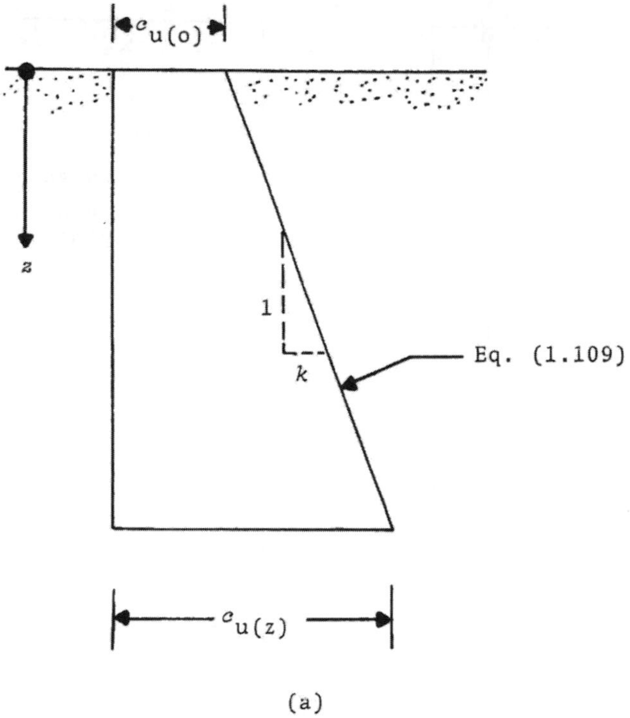

Figure 1.42. (a) Undrained shear strength variation as defined by Eq. (1.109); (b) Circular excavation in saturated clay.

of M [Eq. (1.111)] can be obtained. For known values of M and $2H/B$ Figure 1.43 can be used to obtain the stability factor. The critical depth of the cut can be obtained as

$$H_c = \frac{Nc_{u(o)}}{\gamma M} \tag{1.112}$$

Example 1.10. A cylindrical cut in saturated clay is made. Given: B=3 m and H=2.5 m. For the clay, $c_{u(o)}$=15 kN/m^2, k=1.5 kN/m^3, and γ=17.5 kN/m^3. Determine the critical depth of the cut.

Solution

From Eq. (1.111)

$$M = \frac{c_{u(o)}}{c_{u(o)}+kH} \quad \frac{15}{15+(1.5)(2.5)} = \frac{15}{18.75} \quad 0.8$$

Also

$$2H/B = (2)(2.5)/3 \quad 1.67$$

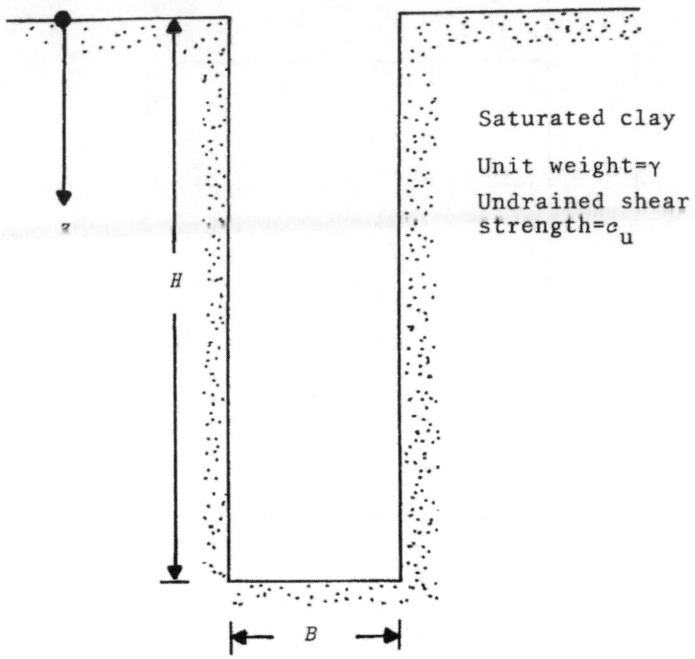

(b)

Figure 1.42. (Continued).

From Fig. 1.43, for M=0.8 and $2H/B$=1.67, the value of N is equal to 5.8. Again, using Eq. (1.112)

$$H_c = \frac{Nc_{u(o)}}{\gamma M} = \frac{(5.8)(15)}{(17.5)(0.8)} = \underline{6.21 \text{ m}}$$

1.21 LATERAL EARTH PRESSURE ON LININGS OF CIRCULAR SHAFTS IN SAND

In many instances in the construction of drilled piers and other substructures, linings are placed in drilled shafts. The problem of lateral earth pressure on drilled shafts has been analyzed by Terzaghi (1936), Karafiath (1953a,b), Schultz (1970), Berezantzev (1958), and Prater (1977). The results of the studies will be presented in this section.

Berezantev's Analysis (1958)

The insert in Fig. 1.44 shows a shaft lining in sand. The diameter of the shaft is equal to B, and the depth of the shaft is

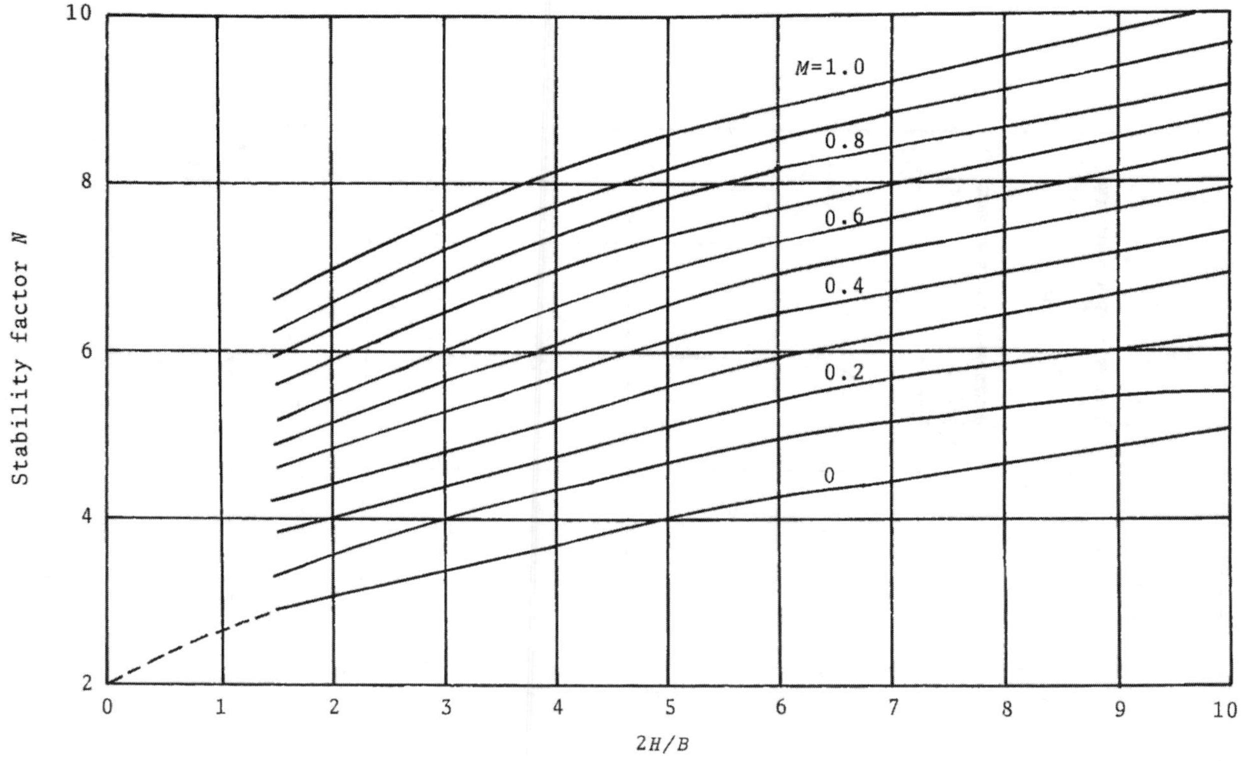

Figure 1.43. Variation of stability factor N with M and $2H/B$ (after Britto and Kusakabe, 1983).

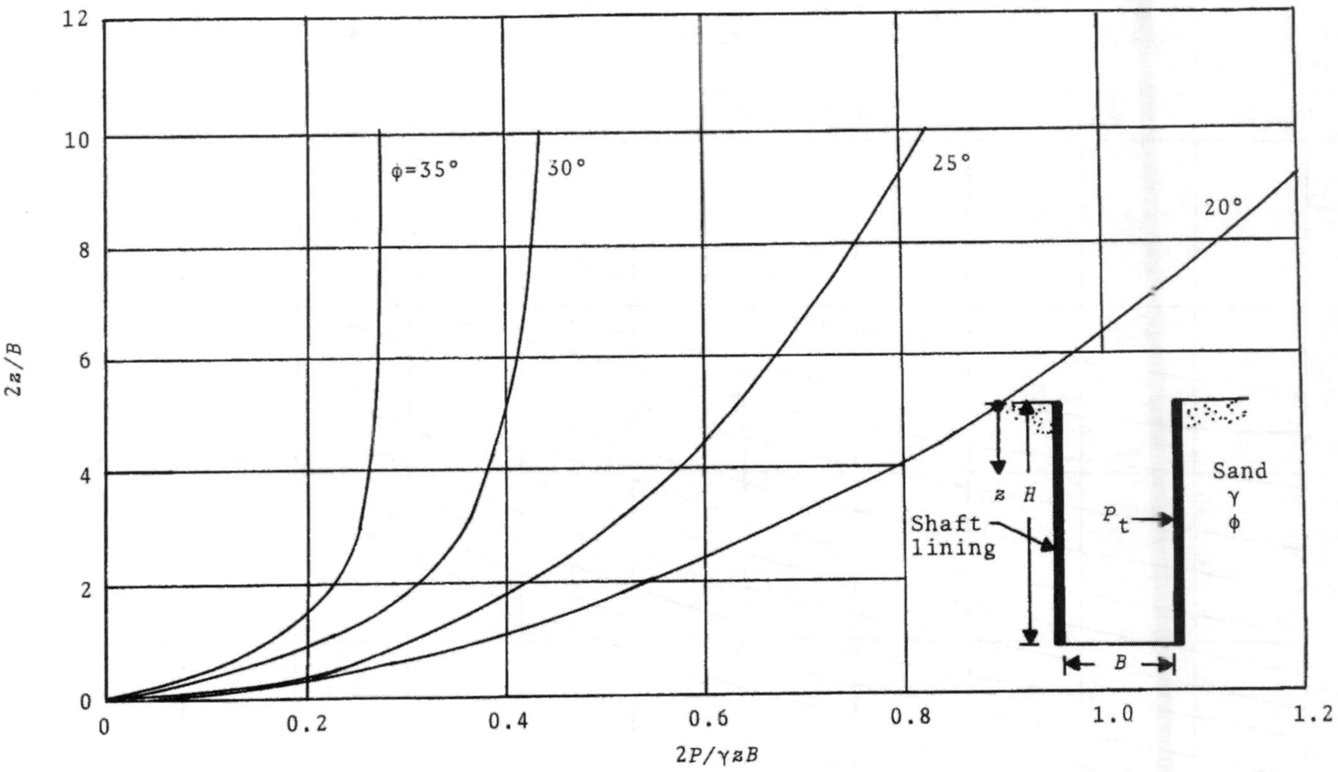

Figure 1.44. Force on shaft lining in sand--based on the analysis of Berezantzev by using the method of limit equilibrium.

equal to H. Berezantzev (1958) applied Sokolovski's step-by-step method of solution of limit equilibrium. According to his study, the total horizontal lateral force on the shaft lining can be given as P_t. The force on the lining per unit circumferential length is equal to P, or

$$P \quad \frac{P_t}{\pi B} \tag{1.113}$$

The magnitude of P can be expressed in a nondimensional form as $2P/\gamma HB$ (or $2P/\gamma zB$). Figure 1.44 shows a plot of $2P/\gamma zB$ vs. $2z/B$ for various values of the friction angle of the sand (ϕ).

Prater's Analysis (1977)

Prater (1977) has extended the theories of Karafiath (1953) and Steinfeld (1958), in which the failure surface around a cylindrical shaft is assumed to be a plane as shown in Fig. 1.45a. The friction angle between the shaft and the soil is assumed to be zero. In order to determine the total force P_t on the shaft lining we need to consider a trial failure wedge. Figure 1.45b shows the elementary sector of a failure wedge. From this, the weight of the soil is

$$\Delta W \quad \frac{(\gamma)(\Delta\alpha')}{2\pi}\left[\frac{\pi H}{3}(R^2+Rr+r^2) + \pi H r^2\right] \tag{1.114}$$

where γ unit weight of soil
r radius of the shaft $B/2$
B = diameter of the shaft
R $H\cot\theta + r$ (Fig. 1.45a)
H = depth of the shaft

Combining Eqs. (1.114) and the above relation for R

$$\Delta W = (\gamma)(\Delta\alpha')\left(\frac{H^3}{6\tan^2\theta} + \frac{H^2 r}{2\tan\theta}\right) \tag{1.115}$$

The tangential force T (Fig. 1.45b) has a radial component which can be given as

$$\Delta F \quad 2T\sin\left(\frac{\Delta\alpha'}{2}\right) \tag{1.116}$$

However

$$\sin\left(\frac{\Delta\alpha'}{2}\right) \approx \frac{\Delta\alpha'}{2} \tag{1.117}$$

So, combining Eqs. (1.116) and (1.117)

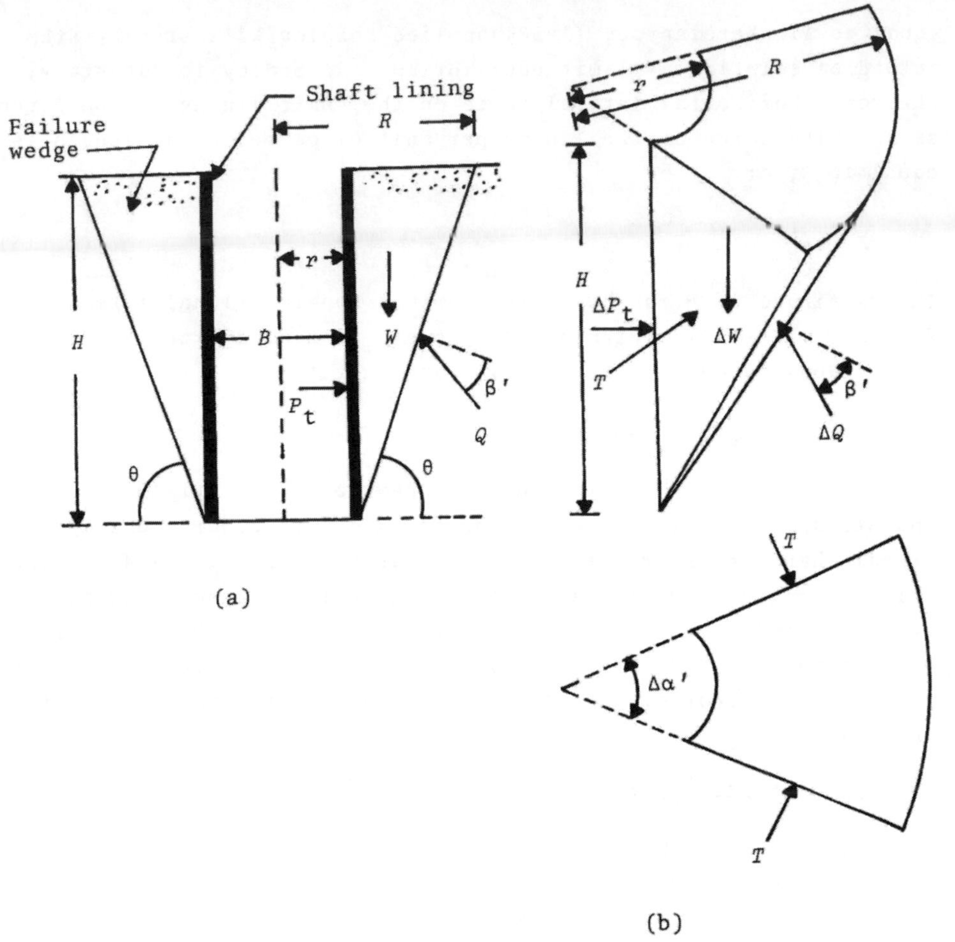

Figure 1.45. Prater's analysis (1977) for lateral force on shaft lining in sand.

$$\Delta F = T(\Delta\alpha') \tag{1.118}$$

The magnitude of T can be given as

$$T = \frac{K\gamma H^3}{6\tan\theta} \tag{1.119}$$

where K = an earth pressure coefficient

Hence

$$\Delta F \quad \frac{K\gamma H^3}{6\tan\theta}(\Delta\alpha') \tag{1.120}$$

In Fig. 1.45a, Q is the resultant of the normal and shearing forces along the trial sliding surface which acts at an angle β'

with the normal (drawn to the trial failure surface). The limits of β' can vary with $-\phi<\beta'<\phi$. The value of $\beta'=-\phi$ represents the active condition, and $\beta'=+\phi$ represents the passive pressure condition. For equilibrium of forces acting on the elementary sector of the wedge shown in Fig. 1.45b

$$\Delta P_t \quad (\Delta W)\tan(\theta+\beta')-\Delta F \tag{1.121}$$

Substituting Eqs. (1.115) and (1.120) into Eq. (1.121)

$$\Delta P_t = \left(\frac{\gamma H^2}{2\tan\theta}\right)(\Delta\alpha')\left\{[\tan(\theta+\beta')]\left(\frac{H}{3\tan\theta}+r\right)-\frac{KH}{3}\right\} \tag{1.122}$$

Hence, the total force on the shaft can be given as

$$P_t = \int \Delta P_t \quad \left(\frac{\pi\gamma H^3}{\tan\theta}\right)\left\{[\tan(\theta+\beta')]\left(\frac{1}{3\tan\theta}+\frac{\gamma}{H}\right)-\frac{K}{3}\right\} \tag{1.123}$$

In order to obtain the value of $\theta=\theta_{cr}$ which will yield the maximum value of P_t, we need to use the theorem of maxima and minima, or

$$\frac{\partial P_t}{\partial\theta} = 0 \tag{1.124}$$

Combining Eqs. (1.123) and (1.124) results in the following

$$\frac{r}{H} = \frac{\frac{1}{3\tan\theta}\left[2 - X' - \frac{K}{3\tan(\theta+\beta')}\right]}{X' - 1} \tag{1.125}$$

where $X' \quad \dfrac{\sin 2\theta}{\sin 2(\theta+\beta')}$ (1.126)

Once Eq. (1.125) is solved, the value of $\theta=\theta_{cr}$ can be substituted into Eq. (1.123) to obtain the maximum value of P_t. Figure 1.46 gives the values of θ_{cr} thus obtained for $\beta'=-\phi$ and $K=1-\sin\phi$. (Note this relationship for $K_{o(n)}$ in Section 1.2.) The total lateral force on the shaft lining can now be expressed as

$$P = \frac{P_t}{2\pi r} = K_r\left(\frac{\gamma H^2}{2}\right) \tag{1.127}$$

where P force per unit length of the circumference of the shaft lining

K_r = coefficient of earth pressure on the cylindrical shaft

$$= \frac{1}{(r/H)\tan\theta_{cr}}\left\{[\tan(\theta_{cr}+\beta')]\left(\frac{1}{3\tan\theta_{cr}}+\frac{r}{H}\right) \quad \frac{K}{3}\right\} \tag{1.128}$$

It needs to be pointed out that the earth pressure coefficient K_r will be equal to zero if $\cot\theta_{cr}=0$, or

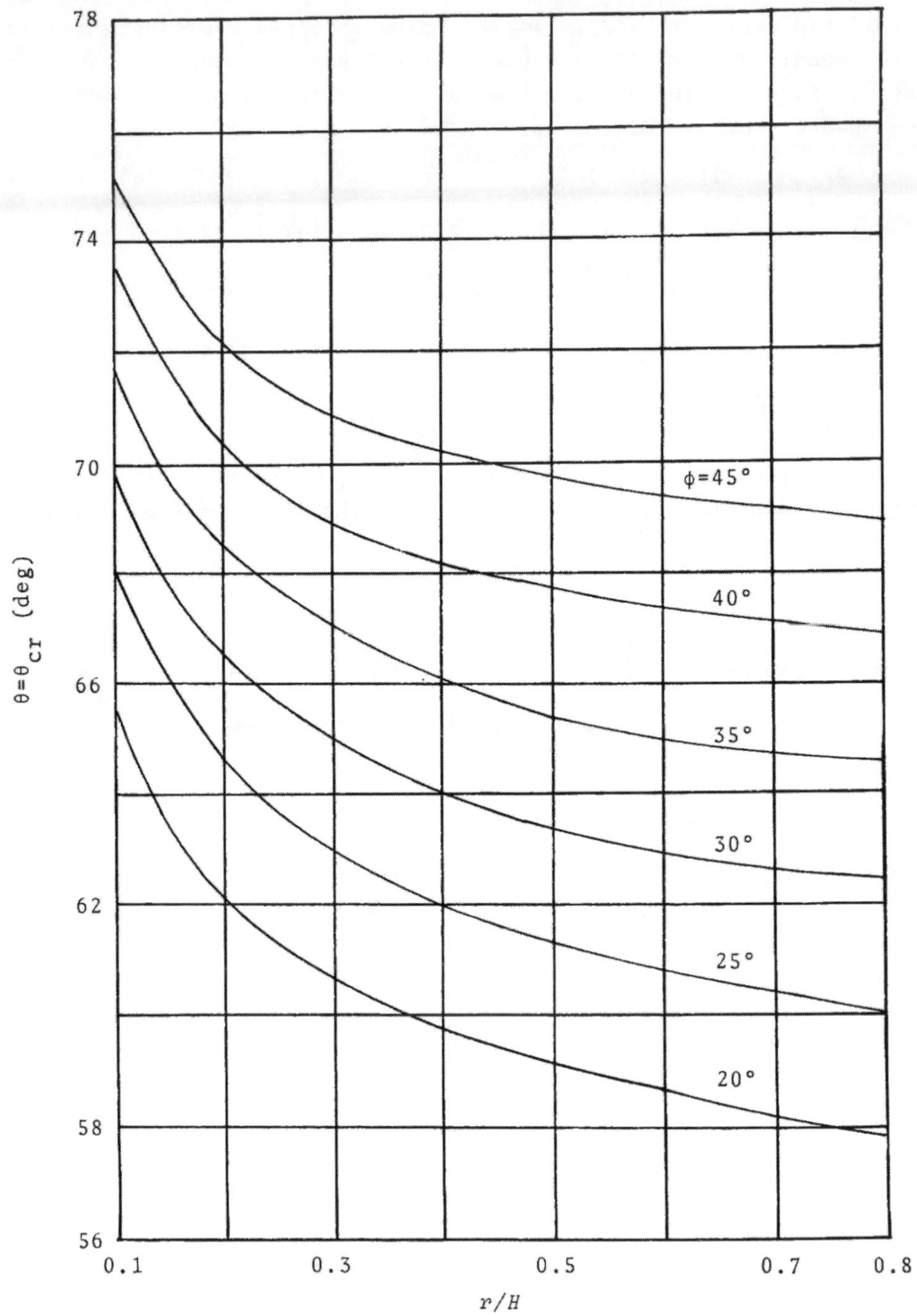

Figure 1.46. Variation of $\theta=\theta_{cr}$ from Eq. (1.125) (for $K=1-\sin\phi$).

$$\sin^2(\theta_{cr}+\beta') \quad K\sin^2\theta \tag{1.129}$$

With proper values of θ_{cr} and β', Eq. (1.128) can be solved to determine the critical values of $r/H=(r/H)_{cr}$. Theoretically, the above exercise means that, if $r/H>(r/H)_{cr}$, no earth pressure is exerted over the shaft lining. Using the proper values of θ_{cr} in Eq. (1.128), the magnitudes of K_r for $\beta'=-\phi$ and $K=1$ and $K=1-\sin\phi$ have been determined. These are given in Figs. 1.47 and 1.48. These graphs can be used to determine the magnitudes of the force per unit length P on the shaft lining. Prater (1977) has suggested

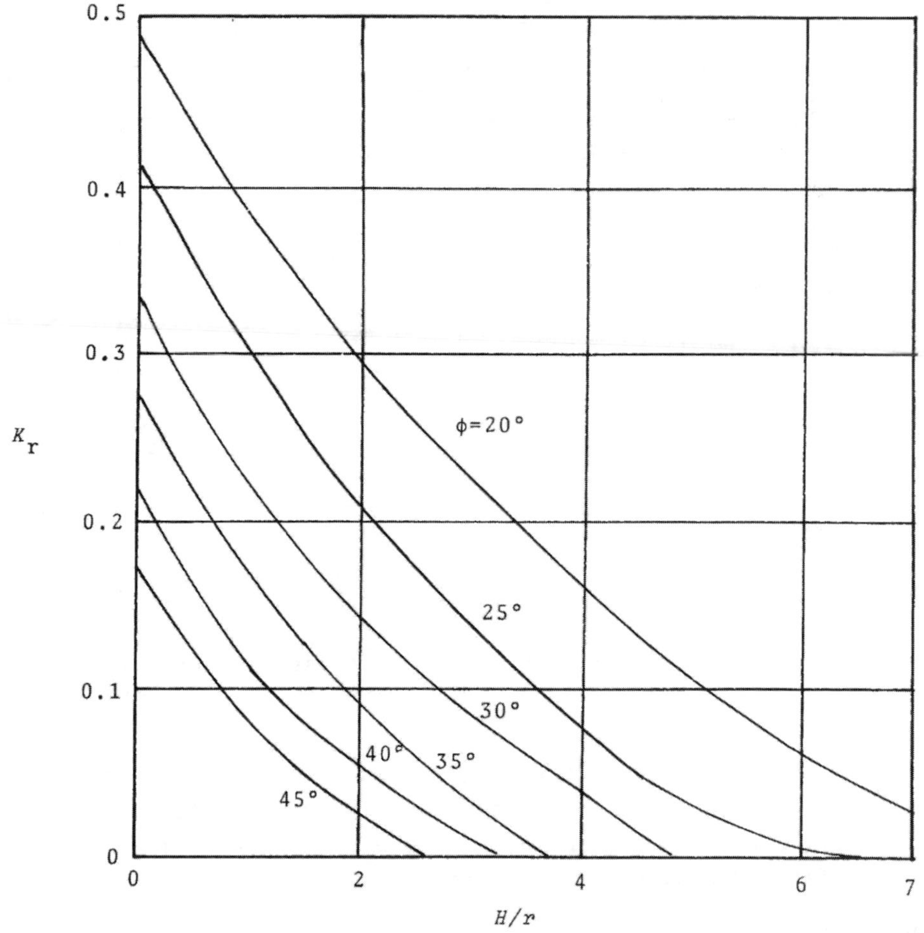

Figure 1.47. Variation of K_r for $\beta'=-\phi$ and $K=1.0$ [Eq. (1.128)].

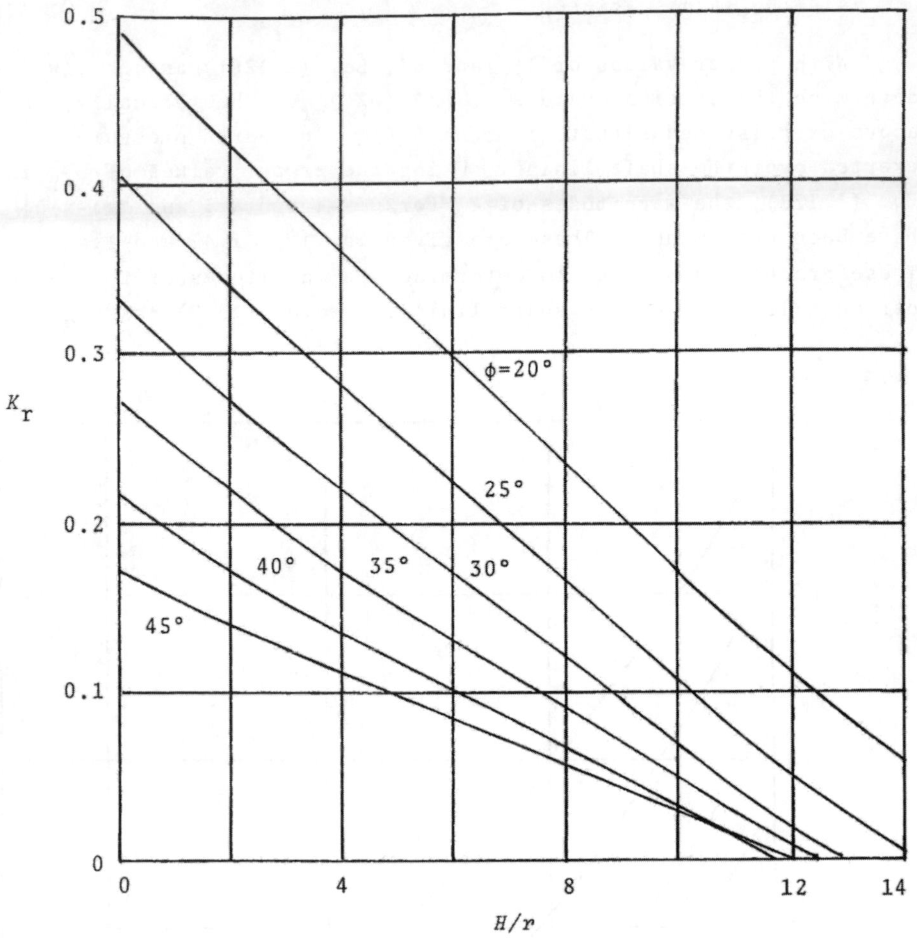

Figure 1.48. Variation of $\beta'=-\phi$ and $K=1-\sin\phi$ [Eq. (1.128)].

that, for realistic values of P, one must assume $\tan^2(45-\phi/2)<K<\tan^2(45+\phi/2)$.

Example 1.11. For a shaft lining in sand which is 6 ft in radius and 24 ft in depth, determine the force P per unit length of the circumference. The soil properties are as follows: $\phi=30°$ and $\gamma=115\ \text{lb/ft}^3$. Use Berezantzev's procedure.

Solution

H/r 24/6 = 4

From Fig. 1.44, for $\phi=30°$ and $\gamma=115\ \text{lb/ft}^3$, $P/\gamma Hr\approx0.39$. So

$$P = (0.39)(115)(24)(6) = \underline{6{,}458.4 \text{ lb/ft}}$$

Example 1.12. Solve Example 1.11 by using Prater's method. Use $\beta=-\phi$ and $K=1-\sin\phi$.

Solution

$$\frac{r}{H} = \frac{6}{24} \quad 0.25$$

$$\phi = 30°$$

From Fig. 1.46, for the above values

$$\theta_{cr} \simeq 65.5°$$

Also, from Fig. 1.48

$$\left(\frac{r}{H}\right)_{cr} \sim \frac{1}{12.75} = 0.0078$$

Also, from the same figure, for $H/r=4$ and $\phi=30°$, the value of $K_r \simeq 0.22$. So

$$P = K_r\left(\frac{\gamma H^2}{2}\right) \quad (0.22)\,\frac{(115)(24)^2}{2} \quad \underline{7{,}286 \text{ lb/ft}}$$

Example 1.13. Consider Example 1.12. Determine the variation of the lateral force per unit area of the shaft lining with depth z.

Solution

We will have to use a step-by-step incremental procedure, that is

$$p_{\left(\frac{z_1+z_2}{2}\right)} \quad \frac{P_{(z_2)}-P_{(z_1)}}{z_2-z_1}$$

where $p_{\left(\frac{z_1+z_2}{2}\right)}$ = force per unit area at a depth of $\frac{z_1+z_2}{2}$

$P_{(z_2)}$ total force per unit length of the circumference for depth z_2

$P_{(z_1)}$ total force per unit length of the circumference for depth z_1

Now the following table can be prepared.

z	$\frac{z_1+z_2}{2}$	$\frac{H}{r}$	K_r[a]	P	$P_{(z_2)}-P_{(z_1)}$	$p\left(\frac{z_1+z_2}{2}\right)$
(ft)	(ft)			(lb/ft)	(lb/ft)	(lb/ft^2)
0		0	0	0		
	2				287.04	72
4		0.67	0.312	287.04		
	6				772.76	193
8		1.33	0.288	1,059.8		
	10				1,192.4	298
12		1	0.272	2,252.2		
	14				1,516.1	379
16		2.67	0.256	3,768.3		
	18				1,659.7	415
20		3.33	0.236	5,428		
	22				1,856	464
24		4	0.22	7,286		

[a]From Fig. 1.48

1.22 LATERAL EARTH PRESSURE ON LININGS OF CIRCULAR SHAFTS IN SATURATED CLAY

Prater (1977) has also extended his theory on earth pressure on shaft lining (Section 1.21) in sand to that in saturated clay (total stress analysis, $\phi=0$ condition). Figure 1.49 shows a shaft of radius r and depth H (similar to that in Fig. 1.45a) along with the forces acting on a trial wedge. The adhesive force between the

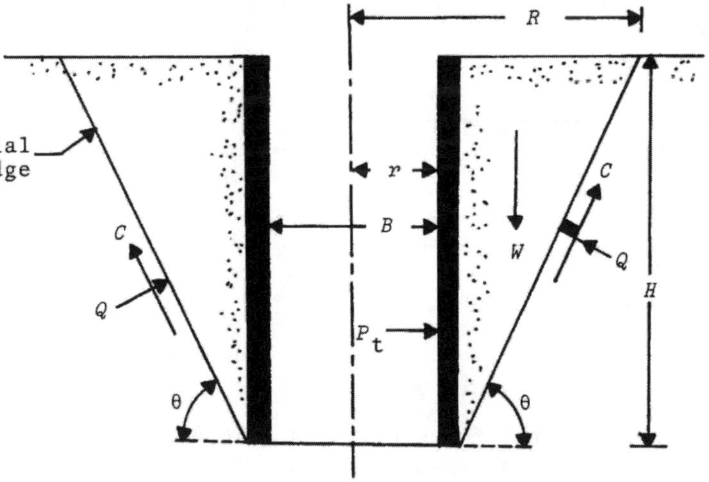

Figure 1.49. Lateral earth pressure on lining of circular shaft in saturated clay.

clay and the shaft lining is neglected. Now, the total weight of the trial failure wedge can be given as

$$W = \gamma\pi\left(\frac{\gamma H^2}{\tan\theta} + \frac{H^3}{3\tan^2\theta}\right) \tag{1.130}$$

The cohesive force C along the failure wedge surface is given by the relation

$$C = \frac{c_u\pi}{\cos\theta}\left(\frac{2rH}{\tan\theta} + \frac{H^2}{\tan^2\theta}\right) \tag{1.131}$$

where c_u = undrained shear strength

As in the case of a shaft lining in sand (Fig. 1.45b)

$$F = \int_0^{2\pi} (T)\,d\alpha' = \int_0^{2\pi} \left(\frac{K\gamma H^3}{6\tan\theta}\right) d\alpha' \tag{1.132}$$

where K an earth pressure coefficient

For equilibrium of the failure wedge

$$C + \cos\theta\left[P_t + \int_0^{2\pi}\left(\frac{K\gamma H^3}{6\tan\theta}\right) d\alpha'\right] - W\sin\theta = 0 \tag{1.133}$$

Substitution of Eqs. (1.130) and (1.131) into Eq. (1.133) yields

$$P_t = \pi r\gamma H^2\left[\frac{H(1-K)}{r(3\tan\theta)} + 1 - \frac{c_u}{\gamma H^2\cos^2\theta}\left(\frac{2}{\tan\theta} + \frac{H}{r\tan^2\theta}\right)\right] \tag{1.134}$$

For maximum value of P_t

$$\frac{\partial P_t}{\partial\theta} = 0$$

Solution of the preceding equation yields

$$\frac{H}{r} \quad \frac{(1-\tan^2\theta)}{\left[\cot\theta - \frac{(1-K)\gamma H}{6c}\right]} \tag{1.135}$$

If the value of K is assumed to be one, Eq. (1.135) takes the form

$$\frac{H}{r} = (\tan\theta)(1-\tan^2\theta) \tag{1.136}$$

The solution of Eq. (1.136) will given $\theta=\theta_{cr}$ for a maximum value of P_t (for K=1). These critical values of $\theta=\theta_{cr}$ are shown in Fig. 1.50. These values of $\theta=\theta_{cr}$, when substituted into Eq. (1.134) will result in the maximum value of P_t. So, the maximum force

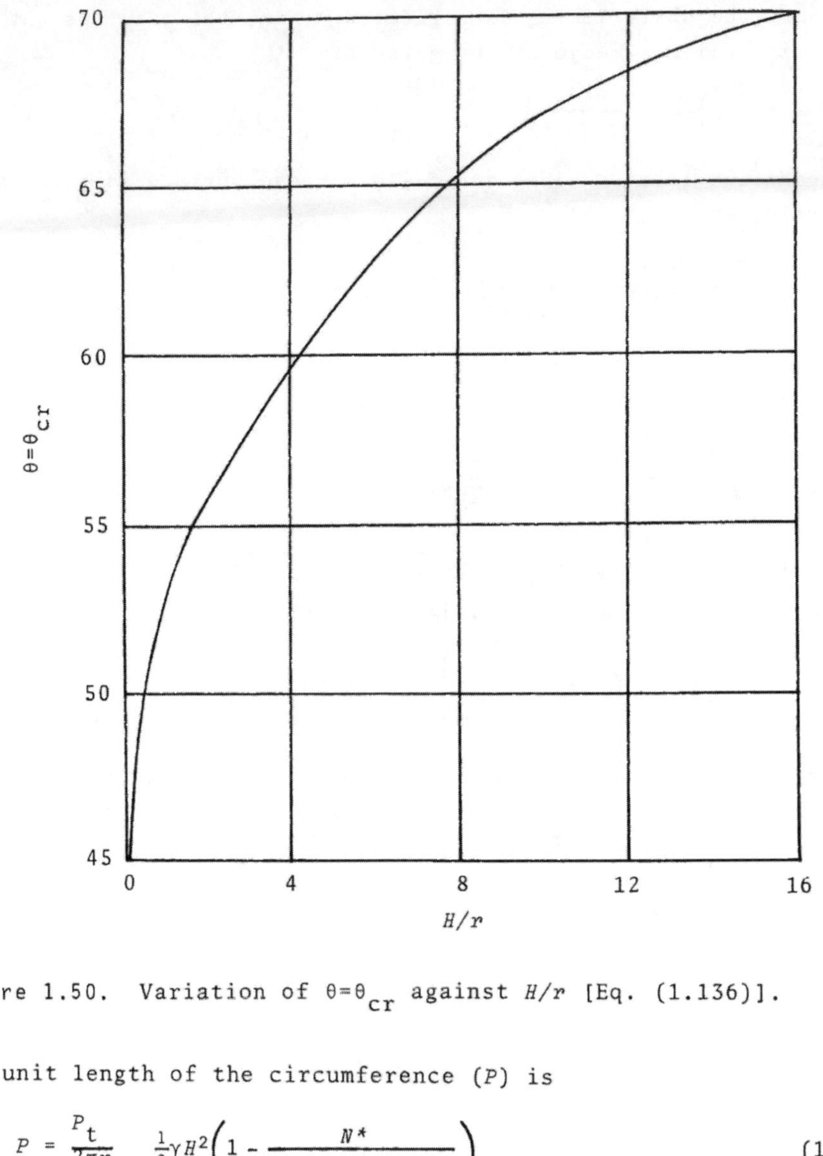

Figure 1.50. Variation of $\theta=\theta_{cr}$ against H/r [Eq. (1.136)].

per unit length of the circumference (P) is

$$P = \frac{P_t}{2\pi r} \quad \frac{1}{2}\gamma H^2\left(1 - \frac{N^*}{\sin\theta_{cr}\cos^3\theta_{cr}}\right) \tag{1.137}$$

where N^* = stability number $\quad \dfrac{c_u}{\gamma H}$ (1.138)

or

$$P = \frac{1}{2}K_r\gamma H^2 \tag{1.139}$$

where K_r = earth pressure coefficient

$$1 - \frac{N^*}{\sin\theta_{cr}\cos^3\theta_{cr}} \qquad (1.140)$$

Figure 1.51 shows the variation of K_r with H/r and N^*.

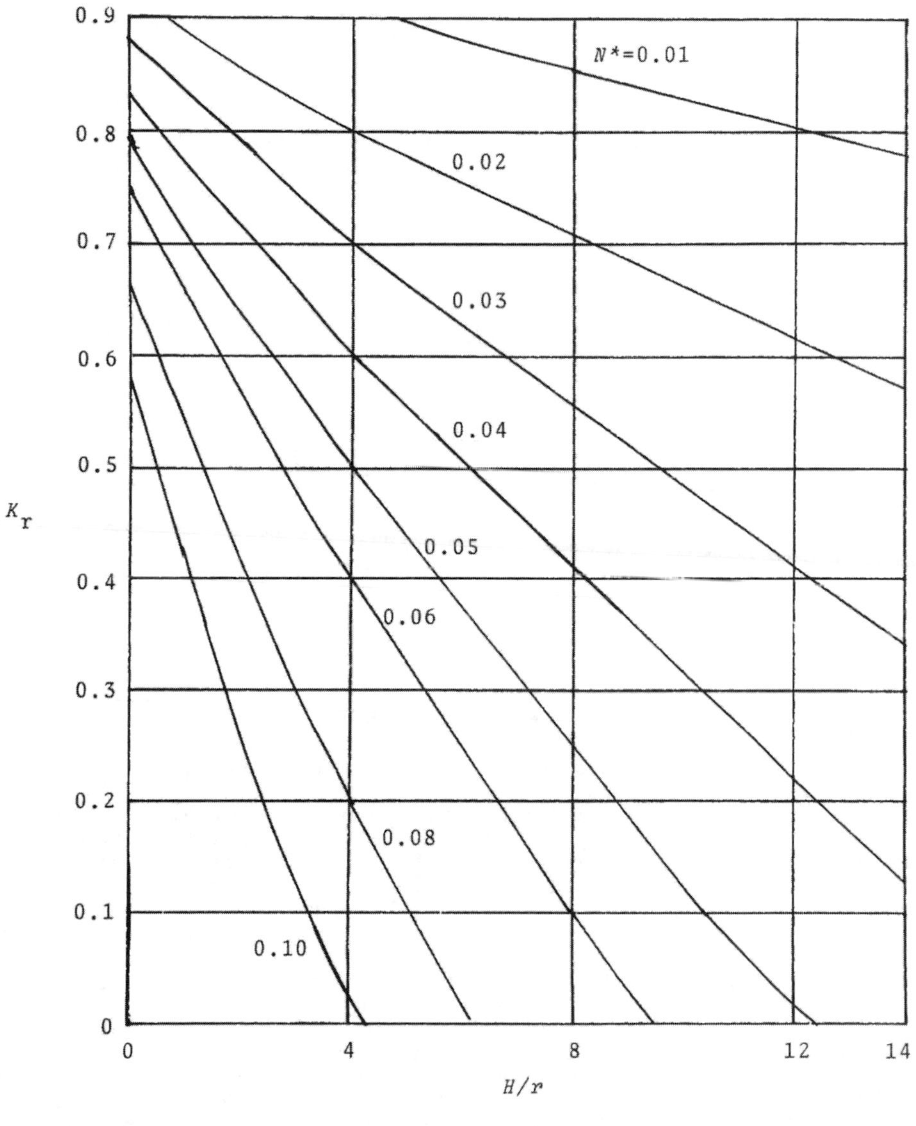

Figure 1.51. Variation of K_r with H/r and N^* [Eq. (1.140)].

From Eq. (1.140), it can be seen that $K_r=0$ when

$$N^* = \sin\theta_{cr}\cos^3\theta_{cr} \tag{1.141}$$

or

$$H_c = \frac{c_u}{\gamma \sin\theta_{cr}\cos^3\theta_{cr}} \tag{1.142}$$

where H_c = critical depth at which the earth pressure on the shaft lining is equal to zero

Example 1.14. Determine the force per unit length of the perimeter for a shaft lining in clay. Given: c_u=24 kN/m², γ=17 kN/m³, H=24 m, and r=3 m.

Solution

$$\frac{H}{r} = \frac{24\text{ m}}{3\text{ m}} = 8$$

$$\frac{c_u}{\gamma H} \quad \frac{24}{(17)(24)} = 0.059$$

From Fig. 1.51, for H/r=8 and $c_u/\gamma H$=0.059, the value of K_r is equal to about 0.12. So

$$P \quad (\tfrac{1}{2})(0.12)(17)(24)^2 \quad \underline{587.5\text{ lb/ft}}$$

References

Berezantzev, V.G., 1958. Earth pressure on the cylindrical retaining wall. Proc., Brussels Conf. on Earth Pressure Problems, 2:21-27.

Bowles, J.E., 1982. Foundation Analysis and Design, 3rd. Ed. McGraw-Hill, New York.

Britto, A.M. and Kusakabe, O., 1983. Stability of axisymmetric excavations in clay. J. Geotech. Eng., ASCE, 109(5):666-681.

Brooker, E.W. and Ireland, H.O., 1965. Earth pressures at rest related to stress history. Canadian Geotech. J., 2(1):1-15.

Caquot, A. and Kerisel, J., 1948. Tables for the calculation of passive pressure, active pressure and bearing capacity of foundations. Gauthier-Villars, Paris, France.

Coulomb, C.A., 1776. Essai sur une application des regles des maximis et minimis a quelques problemes de statique relatifs a l'artitecture. Mem. Acad. Royal Pres. Div. Sav, 7. Paris.

Culmann, C., 1866. Graphische statik. Zurich.

Das, B.M. and Seeley, G.R., 1975. Active thrust on braced cuts in clay. J. Const. Div., ASCE, 101(CO4), 945-949.

Dubrova, G.A., 1963. Interaction of soil and structures. Izd. Rechnoy Transport, Moscow, U.S.S.R.

Feda, J., 1984. K_o coefficient of sand in triaxial apparatus. J. Geotech. Eng., ASCE, 110(4):519-524.

Harr, M.E., 1966. Foundations of theoretical soil mechanics. McGraw-Hill, New York.

Hencky, H., 1923. Uber einige statisch bestimmte falle des gleichgewichts in plastischen korpern. Z. Ang. Math. Mech, 3:241.

Jaky, J., 1944. The coefficient of earth pressure at rest. J. Soc. of Hungarian Arch. Engrs., Budapest, Hungary, pp. 355-358.

Janbu, N., 1957. Earth pressure and bearing capacity calculations by generalized procedure of slices. Proc., IV Intl. Conf. Soil Mech. Found. Eng., 2:207-213.

Karafiath, L., 1953a. Erddruck auf wande mit kreisformigen querschnitt. Bauplanning und Bautechnik, Berlin, 7:319-329.

Karafiath, L, 1953b. On some problems of earth pressure. Acta. Tech. Acad., Budapest, Hungary, pp. 327-337.

Kavazanjian, E. and Mitchell, J.K., 1984. Time dependence of lateral earth pressure. J. Geotech. Eng., ASCE, 110(GT4):530-533.

Mayne, P.W. and Kulhawy, F.H., 1982. K_o-*OCR* relationships in soil. J. Geotech. Eng., ASCE, 108(GT6):851-872.

Meyerhof, G.G., 1972. Stability of slurry trench cuts in saturated clay. Proc., Spec. Conf. on Performance of Earth and Earth Supported Structures, ASCE, I(II):1451-1466.

Narain, J., Saran, S., and Nanadkumaran, P., 1969. Model study of passive pressure in sand. J. Soil Mech. Found. Div., ASCE, 95(SM4):969-983.

Peck, R.B., 1969. Deep excavation and tunneling in soft ground. Proc., VII Intl. Conf. Soil Mech. Found. Eng., State-of-the-Art Volume, 225-290.

Poncelet, V., 1840. Mem. sur la stabilite des revetements et de lear fondations. Mem. de l'officier du genie, 13.

Prater, E.G., 1977. An examination of some theories of earth pressure on shaft linings. Canadian Geotech. J., 14(1):91-106.

Rendulic, L., 1938. Der erddruck im strassenbau und bruckenbau forschungsarbeiten, Volk und Reich Verlag, Berlin.

Rosenfarb, J.L. and Chen, W.F., 1972. Limit analysis solutions of earth pressure problem. Fritz Eng. Lab. Report No. 355.14, Lehigh University, 53 pp.

Rowe, P.W. and Peaker, K., 1965. Passive earth pesssure measurements. Geotechnique, London, 15(1):57-79.

Saran, S. and Deo, P., 1974. Passive pressure in sands with uniform surcharge on backfill. J. Geotech. Eng. Div., ASCE, 100 (GT12):1326-1329.

Schmidt, B., 1966. Discussion of Earth pressure at rest related to stress history. Canadian Geotech. J., 3(4):239-242

Schulz, M., 1970. Berechnung des Raumlichen erddruckes auf die wandung kreiszylindrischer korper. Ph.D. dissertation, University of Stuttgart.

Sherif, M.A. and Fang, Y.S., 1984. Dynamic earth pressure on walls rotating about the top. Soils and Found., 24(4):109-117.

Sherif, M.A., Fang, Y.S., and Sherif, R.I., 1984. K_A and K_O behind rotating and non-yielding walls. J. Geotech. Eng., ASCE, 110 (GT1):41-56.

Sherif, M.A., Ishibashi, I., and Lee, C.D., 1982. Earth pressures against rigid retaining walls. J. Geotech. Eng. Div., ASCE, 108(GT5):679-695.

Shields, D.H. and Tolunay, A.Z., 1973. Passive pressure by method of slices. J. Soil Mech. Found. Div., ASCE, 99(SM12):1043-1053.

Sokolovski, V.V., 1965. Statics of granular media. Pergamon Press, New York.

Steinfeld, K., 1958. Uber den erddruck auf schacht- und brunnenwandungen. Contribution to the Foundation Engineering Meeting, Hamburg, Ger. Soc. Soil Mech. Found. Eng., pp. 111-126.

Terzaghi, K., 1943. Theoretical soil mechanics. John Wiley, New York.

Terzaghi, K., 1941. General wedge theory of earth pressure.

Transactions, ASCE, 106:68-97.
Terzaghi, K., 1936. A fundamental fallacy in earth pressure computations. In Contributions to Soil Mechanics: 1925-1940. Boston Soc. Civ. Eng., 1940, pp. 277-294.
Terzaghi, K. and Peck, R.B., 1967. Soil mechanics in engineering practice, second edition. John Wiley, New York.
Tschebotarioff, G.P., 1973. Foundations, retaining and earth structures, second edition. McGraw-Hill, New York.

CHAPTER 2
Sheet Pile Walls

2.1 INTRODUCTION

Earth retaining walls used for many waterfront structures are generally made from connected or semi-connected sheet piles. An advantage of sheet pile wall construction is that dewatering is not required during construction. Sheet piles can be made from *wood*, *precast concrete*, or *steel*.

Wooden sheet piles are generally used for small or temporary structures. Figure 2.1 shows the sections of some typical wooden

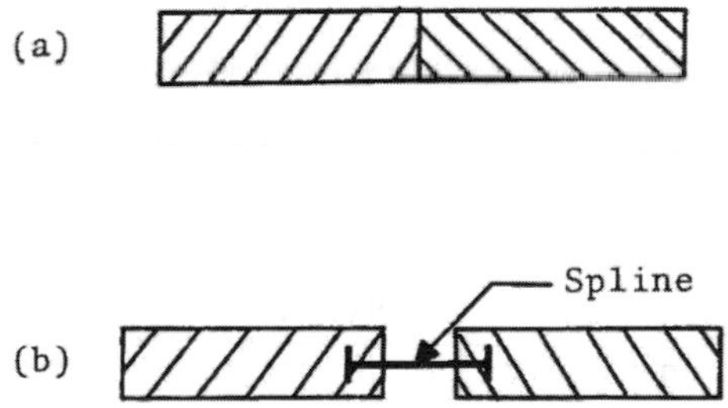

Figure 2.1. Wooden sheet piles: (a) Wooden planks driven side by side; (b) Wooden plank with splines.

sheet piles. Figure 2.1a shows the sections of wooden planks which are driven side by side, and Fig. 2.1b shows the sections of splined wooden sheet piles. For this type of sheet pile wall construction, metal splines are driven into the grooves of adjacent sheetings to hold them together after they are driven. Wooden sheet piles may be 2-3 in. (50-75 mm) thick and 12-16 in. (300-400 mm) wide.

Figure 2.2 shows the section and elevation of a precast concrete sheet pile generally used for construction. These sheet

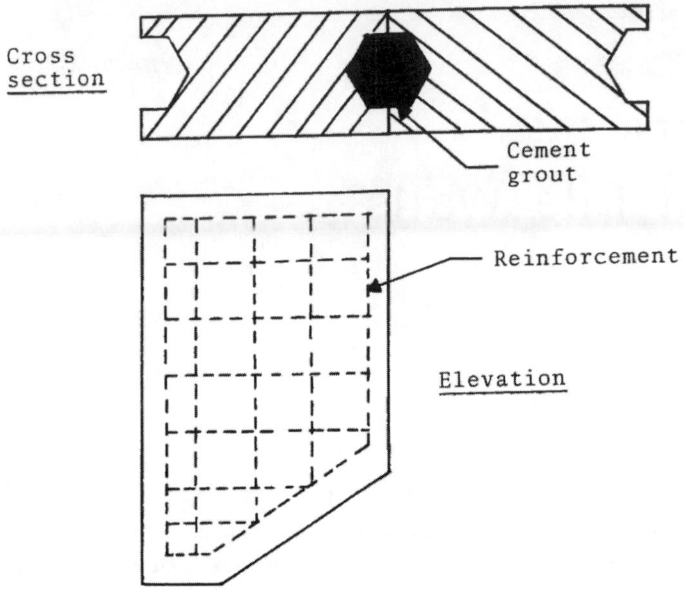

Figure 2.2. Precast concrete sheet pile.

piles are 5-10 in. (125-250 mm) thick and 20-40 in. (500-1000 mm) wide. After the sheet piles are driven, cement grout is placed in the grooves between two adjacent sheet piles.

In most permanent structures, steel sheet piles are used. Steel sheet piles may generally have a Z-type section, a *U*-type section (or *low or high arch* type section), or a *straight web*-type section, as shown in Fig. 2.3. Steel sheet piles are easier to handle, and they can tolerate high driving stresses. The allowable flexural stresses for steel sheet piles are between 30,000-25,000 psi (≈210-170 MPa). Sheet piles generally weigh between 15-65 lb/ft^2 of the wall area. Usual thicknesses of sheet piles vary from 0.25-0.8 in. The sheet pile *interlocking* arrangement may be of two types: *ball-and-socket* type or *thumb-and-finger* type, as shown in Fig. 2.3. Table 2.1 gives the properties of the sections of some of the sheet piles produced by the United States Steel Corporation.

2.2 TYPES OF SHEET PILE WALLS

Sheet pile walls, in general, may be divided into two types: *cantilever* sheet pile walls, and *anchored* sheet pile walls. Figure 2.4 shows a cantilever sheet pile wall with a granular soil back-

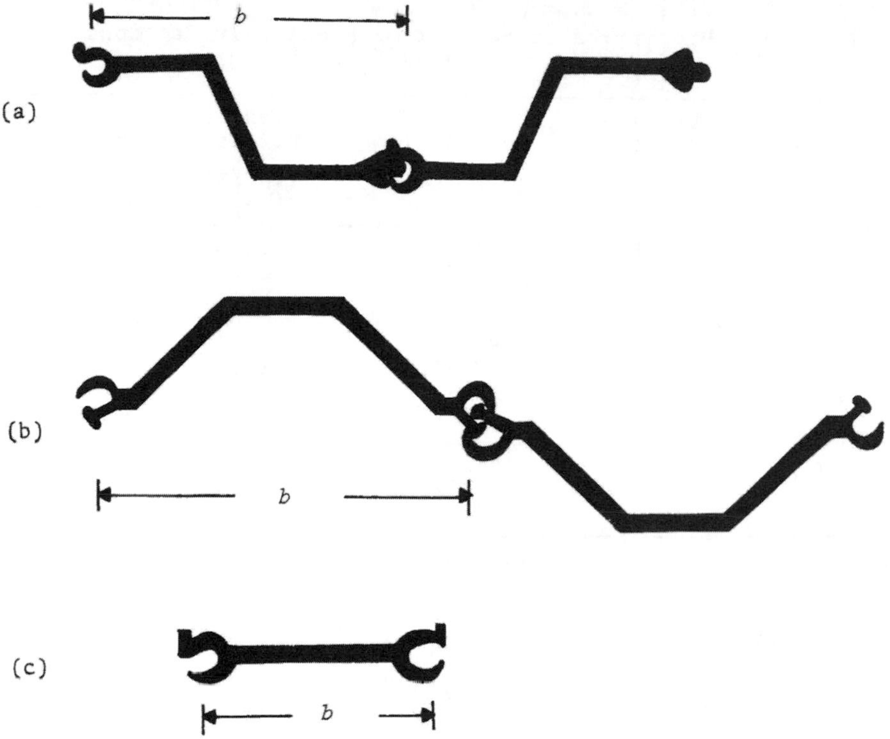

Figure 2.3. Steel sheet pile section: (a) *Z*-type sections with ball-and-socket type of interlocking arrangement; (b) *U*-type section with thumb-and-finger type of interlocking arrangement; (c) *straight web*-type section. (*Note:* b=driving distance; see Table 2.1.)

fill penetrating into a granular soil. It is important to note that this type of wall is usually not recommended for wall heights (that is, height above the dredge line) greater than about 20 ft (≈6 m). These walls are subjected to a *net active pressure* between the levels of *A* and *B* from the land side. Note that the water pressure from the land side and the water side will cancel each other, so only the effective stresses need to be considered. Due to the large active force above the wall, it will rotate about a hinge point *C*. Hence, between the levels of *B* (dredge line) and *C*, there will be active pressure from the land side and passive pressure from the water side. However, between the levels of *C* and *D*, there will be passive pressure from the land side and active pressure from the water side. The probable net pressure diagram along the depth of the sheet pile wall will be as shown by the broken

Table 2.1. Properties of Some Steel Sheet Pile Sections Produced by the United States Steel Corporation

Type	Driving[a] distance b (in.)	Designation	Moment of inertia I (in^4/ft of wall)	Section modulus S (in^3/ft of wall)
Z	18	PZ-38	280.8	46.8
	21	PZ-32	220.4	38.3
	18	PZ-27	184.2	30.2
U	16	PDA-27	39.8	10.7
	19.6	PMA-22	13.7	5.4
	16	PSA-28	4.5	2.5
	16	PSA-23	4.1	2.4
Straight web	16.5	PSX-32	3.7	2.4
	15	PS-32	2.9	1.9
	15	PS-28	2.8	1.9

[a]Refer to Fig. 2.3 for definition

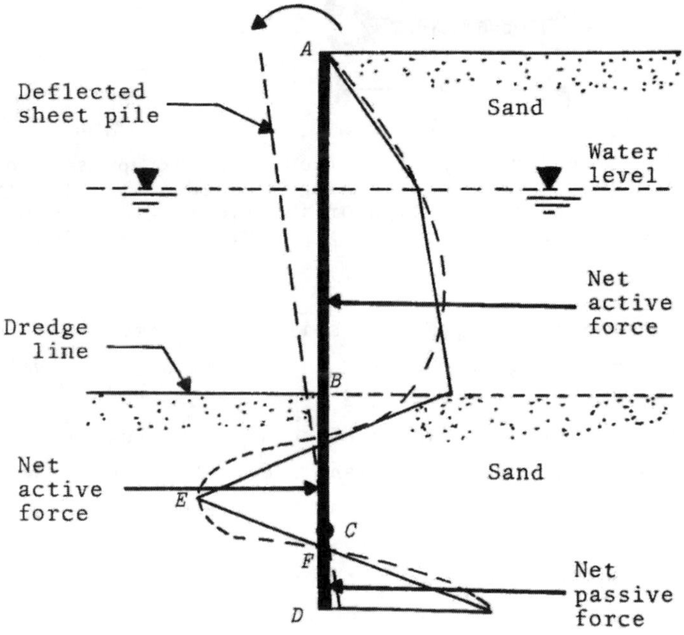

Figure 2.4. Cantilever sheet pile wall.

line. For computation purposes, they are idealized as shown by solid lines in Fig. 2.4. The required depth of penetration is determined considering the *force equilibrium* and the *moment equilibrium conditions.*

Anchored sheet pile walls, or anchored *bulkheads*, can be subdivided into two main categories based on their method of design: *free earth support* method of design and *fixed earth support* method of design. These types of walls can be constructed up to a height of about 50 ft (≈15 m) measured above the dredge line. The free earth support method is usually referred to as the *method of least penetration*. Figure 2.5 demonstrates an anchored sheet pile wall

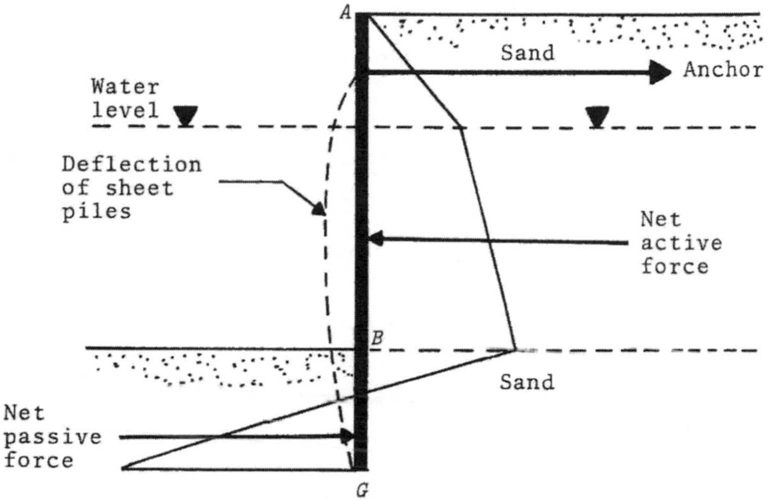

Figure 2.5. Anchored sheet pile wall--free earth support method of design.

designed by the free earth support method. In this method, an anchoring system is provided towards the top of the wall. The depth of penetration below the dredge line is such that it provides some passive force from water side for stability. Compare the lateral earth pressure diagrams shown in Figs. 2.4 and 2.5 and note the level of point *E* in Fig. 2.4. Relatively speaking, the level of point *G* in Fig. 2.5 is between the levels of points *B* and *E* in Fig. 2.4.

A fixed earth support method of design for an anchored sheet pile wall is shown schematically in Fig. 2.6. The required depth of penetration of sheet piles for this type of design is larger than that required by the free earth support method of design. Compare the lateral earth pressure diagrams shown in Figs. 2.4, 2.5 and 2.6. Note that the level of point *H* in Fig. 2.6 is lower than

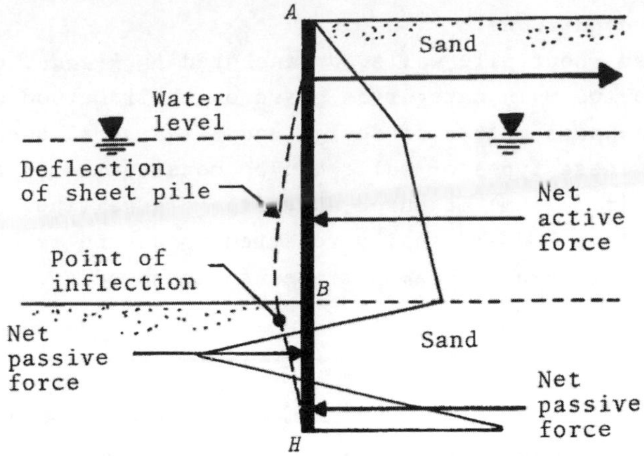

Figure 2.6. Anchored sheet pile wall--fixed earth support method of design.

that of point *G* in Fig. 2.5 and is between the levels of points *F* and *D* in Fig. 2.4.

The free earth support design method is preferable when clay soil is present below the dredge line, or when hard soil is encountered at a limited depth below the dredge line which may cause problems when sheet piles are driven to a greater depth.

2.3 ANCHORS IN SHEET PILE WALL DESIGN

It has been pointed out in the preceding section that anchors are generally used for construction of sheet pile walls. These anchorages can be of various types, such as *vertical plate anchors*, *batter anchor piles*, and *tie backs*. Tie backs are stressed tendons of high-strength steel. Figure 2.7 shows the general nature of placement of these anchors. It is important to note the placement of the anchor (that is, the length of tie rods) in order to avoid interference between the active soil wedge behind the retaining wall and the passive soil wedge in front on the anchor. The general safe locations of the anchors are also shown in Fig. 2.7. The holding capacity of plate anchors (which may consist of prefabricated steel or concrete slabs) in sand and clay is described in more detail in Chapter 3. Note that, in Fig. 2.7, t=two-thirds the design depth of penetration (D) for the fixed earth support method of design, and t=design depth of penetration (D) for the free earth support method of design (Cornfield, 1975).

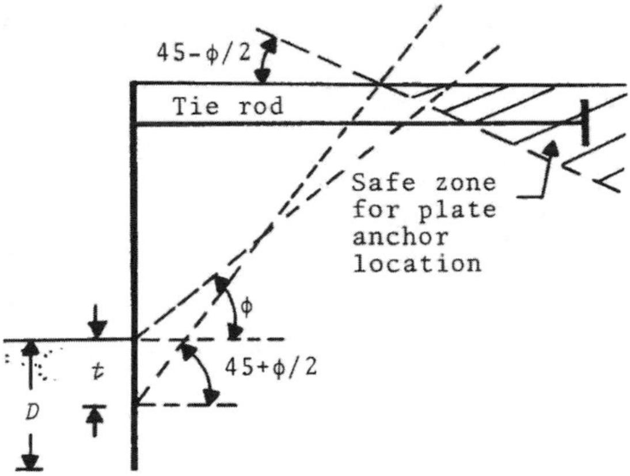

(a) Plate anchor

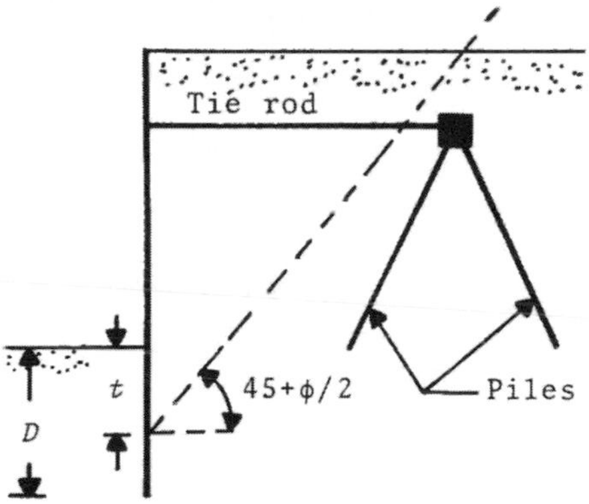

(b) Batter anchor piles

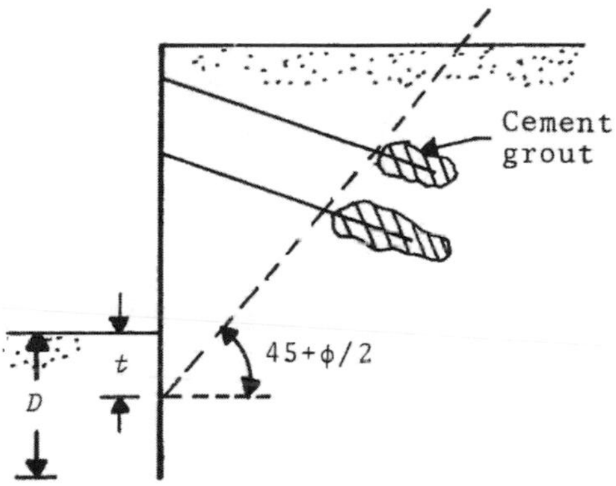

(c) Tie backs

Figure 2.7. Anchors in sheet pile wall design.

2.4 CONSTRUCTION METHODS FOR SHEET PILE WALLS

The construction methods for sheet pile walls can be divided into two basic categories (Tsinker, 1983): *backfilled structures* and *dredged structures*. In the backfilled structure, the general sequence of construction is as follows:

1. Dredging of the *in situ* soil in the front and back of the back of the proposed structure (Fig. 2.8a).
2. Driving of the sheet piles (Fig. 2.8b).
3. Backfilling up to the level of the anchorage (Fig. 2.8c).
4. Backfilling up to the top of the wall (Fig. 2.8d).

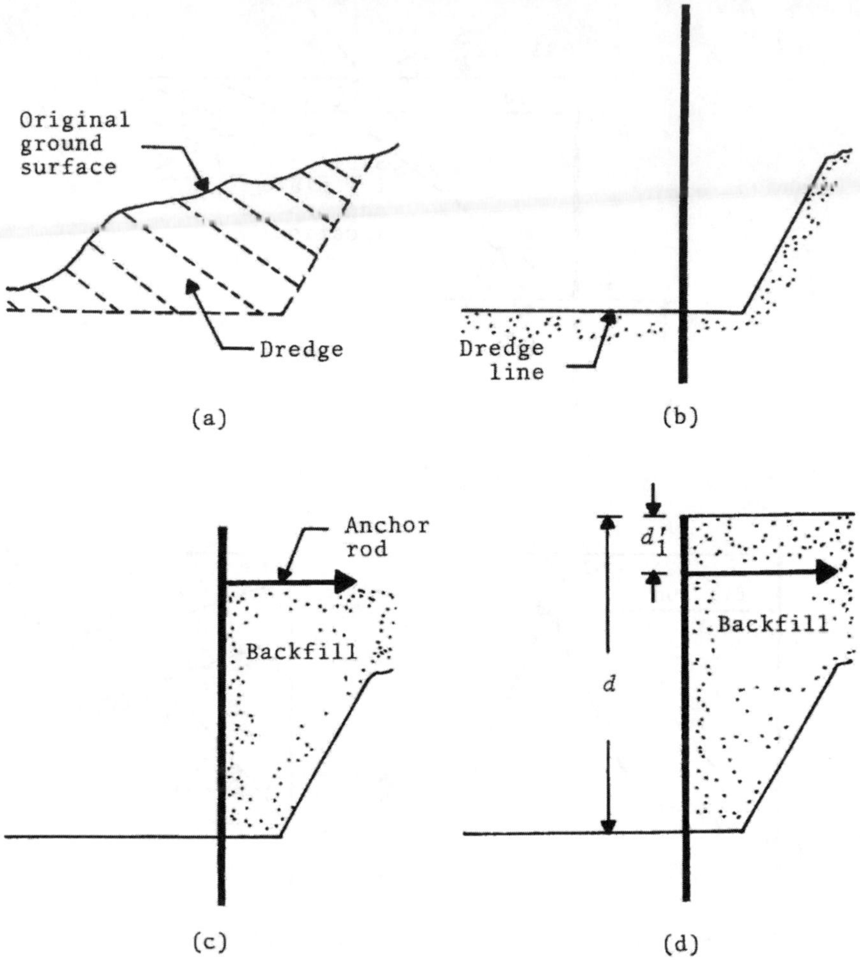

Figure 2.8. Sequence of construction of a backfilled structure.

Steps 3 and 4 are not required for cantilever walls.

The sequence of construction of a dredged structure is as follows:

1. Driving of the sheet piles (Fig. 2.9a); construction of the anchor system.
2. Backfilling up to the anchor level (Fig. 2.9b).
3. Backfilling to the top of the wall (Fig. 2.9c).
4. Dredging the front side of the wall (Fig. 2.9d).

For cantilever sheet pile walls, Step 2 is not required.

The construction sequence has a great impact on the net earth

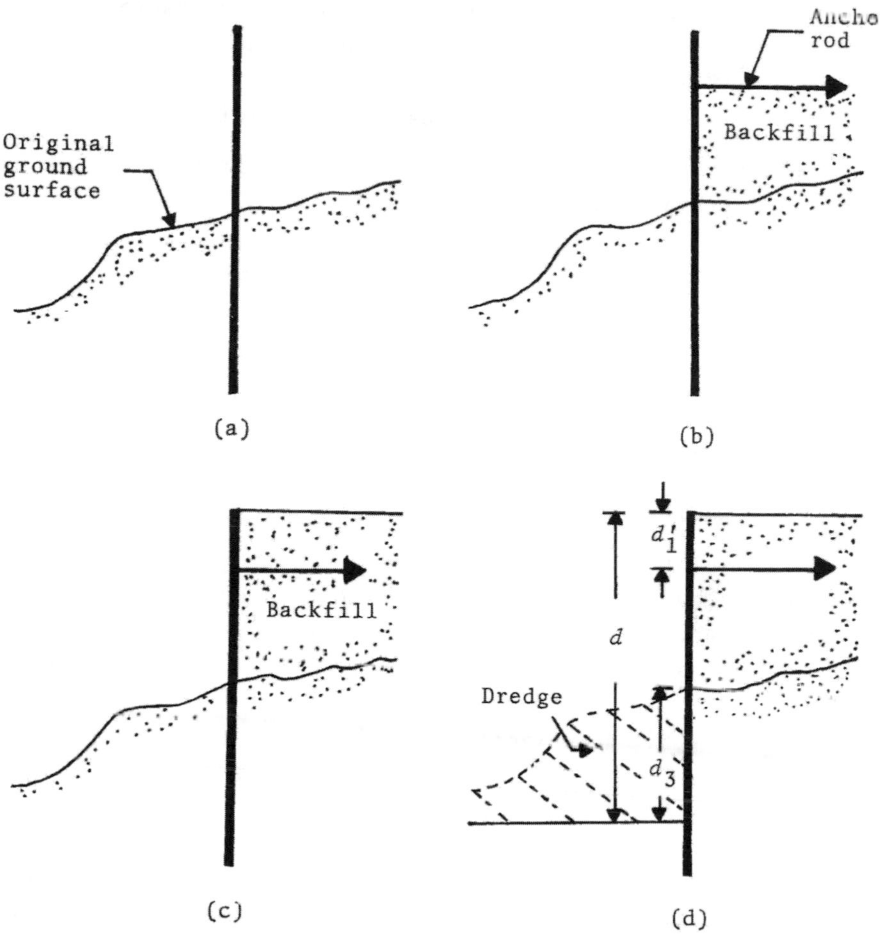

Figure 2.9. Sequence of construction of a dredged structure.

pressure distribution on the sheet pile wall.

2.5 LATERAL EARTH PRESSURE--GENERAL CONSIDERATIONS

For the design and analysis of sheet pile walls, *Coulomb's earth pressures* are considered adequate in most cases. This means that, for active and passive pressures in granular soil

$$p_a = \sigma_v' K_a \tag{2.1}$$

and

$$p_p = \sigma_v' K_p \tag{2.2}$$

where p_a, p_p = intensity of active and passive pressures
σ'_v = vertical effective stress at the depth under consideration
K_a, K_p = Coulomb's earth pressure coefficients

The horizontal components of the lateral pressures (p_{aH} and p_{pH}) at a certain depth can then be given as

$$p_{aH} = p_a \cos\delta = \sigma'_v K_a \cos\delta = \sigma'_v K_{aH} \tag{2.3}$$

and

$$p_{pH} = p_p \cos\delta = \sigma'_v K_p \cos\delta = \sigma'_v K_{pH} \tag{2.4}$$

where $K_{aH} = K_a \cos\delta$

$K_{pH} = K_p \cos\delta$

δ = soil and sheet pile interface friction angle

For calculation of K_a or K_{aH}, the magnitude of δ is often taken as zero, or $\phi/2$. The assumption $\delta=0$ is on the safe side of the calculation. For passive pressure coefficient calculation, $\delta = 2/3(\phi)$ may be assumed. However in many cases, it is not unusual to assume $\delta=0$.

In the following sections, the mathematical derivations for the depth of penetration for sheet pile walls into sand and clay layers will be derived. For these cases, the terms K_{aH} and K_{pH} will be used. For undrained conditions in clay (that is, $\phi=0$ condition), $K_{aH}=1$ and $K_{pH}=1$. Since the stability of sheet pile walls depends primarily on the passive forces developed below the dredge line, some form of factor of safety should be applied. This is conveniently done in the following two ways.

In sand:

1. If the actual K_{pH} value is used to calculate the depth of penetration of a sheet pile into a sand layer, the calculated depth (or theoretical depth) D should be increased by about 20-30% in the actual construction work, or

$$D_{actual} = (1.2 \text{ to } 1.3) D_{theoretical} \tag{2.5}$$

2. In lieu of Step 1, a factor of safety may initially be assigned to K_{pH}, or

$$K_{pF} = K_{pH}/F_s \tag{2.6}$$

where F_s = factor of safety

The term K_{pH} should replace K_{pF} in all derivations for the depth of penetration. The depth of penetration, D, derived this way is the actual depth of penetration (D_{actual}).

In clay:

1. For a sheet pile penetrating a clay layer below the dredge line, if the actual value of the undrained shear strength of clay (c_u) is used, then the calculated (theoretical) D value should be increased by about 35-50%. So

$$D_{actual} = (1.35 \text{ to } 1.5) D_{theoretical} \quad (2.7)$$

2. In lieu of Step 1, a factor of safety may be assigned to c_u, or

$$c_{uF} \quad c_u / F_s \quad (2.8)$$

This value of c_{uF} must replace the c_u terms used in the depth of penetration derivation. The magnitude of D calculated in this manner is the actual depth of penetration.

Although idealized assumptions are made to estimate the distribution of lateral earth pressure on sheet pile walls, the actual pressure distribution is highly complicated. An example of this can be given by the following, which is referred to as the *heredity* effect (Tsinker, 1983; Budin and Demina, 1979). Figure 2.10a shows a flexible sheet pile wall. The active earth pressure distribution behind the wall is hydrostatic. Now, if a surcharge (which is a strip load) is placed on the top of the backfill (Fig. 2.10b), it will contribute an additional pressure Δp which will be a function of depth z. At this time, total pressure at any depth is $p_a + \Delta p$. Due to the additional pressure, the top of the wall would deflect through a distance δ. If the surcharge is now removed (Fig. 2.10c) the magnitudes of Δp and δ do not actually become zero. instead they will be respectively equal to $\Delta p'$ and δ'. The magnitude of δ' may be in the range of about 0.8δ. Also, if $M(z)$ is the bending moment on the sheet pile created by Δp, the residual bending moment due to $\Delta p'$ will be $M'(z)$. The factor $H = M'(z)/M(z)$ will depend on the relative density of the granular material. Approximate values of H are as follows:

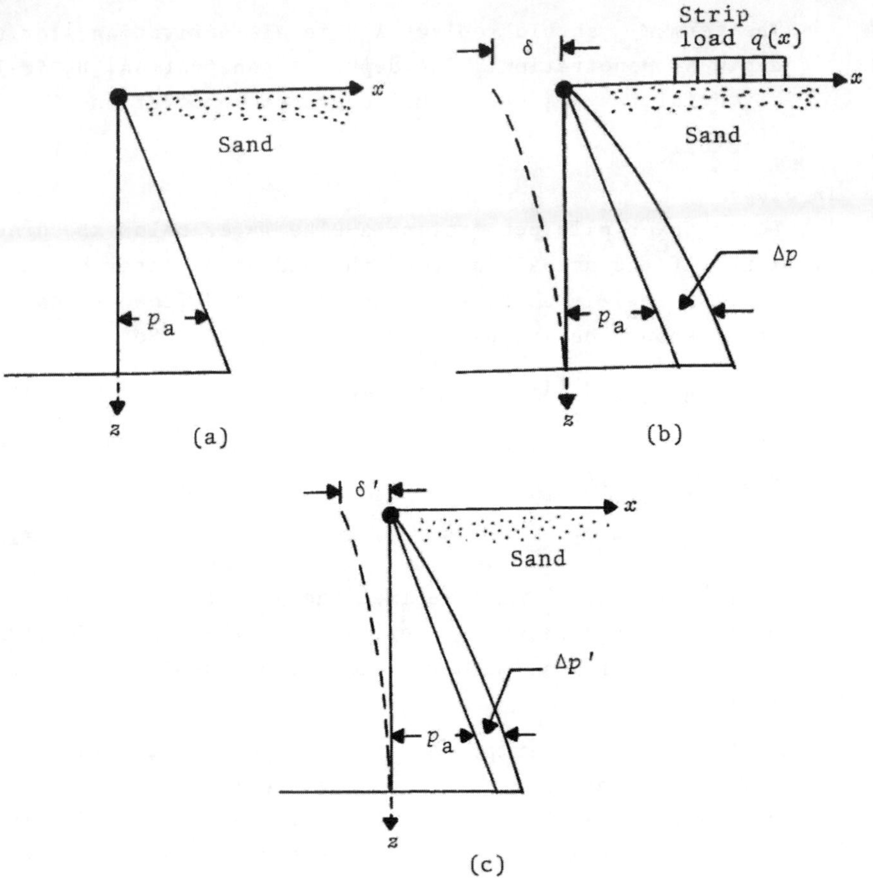

Figure 2.10. Heredity effect.

Relative density D_r (%)	H
≤50%	≈0.7
50-75%	≈0.8
>50%	≈0.85

Also, if a surcharge is again placed on the backfill, the pressure due to the surcharge will not increase until $q(x)$ exceeds a value of $q'(x)$, which is the surcharge corresponding to the residual pressure $\Delta p'$.

For reasons such as those stated above, proper judgment needs to be used in the final design of each structure.

2.6 CANTILEVER SHEET PILE WALL PENETRATING INTO SAND

Cantilever sheet pile walls are usually constructed when the wall height is about 20 ft (≈6 m) or less above the dredge line. Figure 2.11 shows a cantilever wall with sand as backfill above the

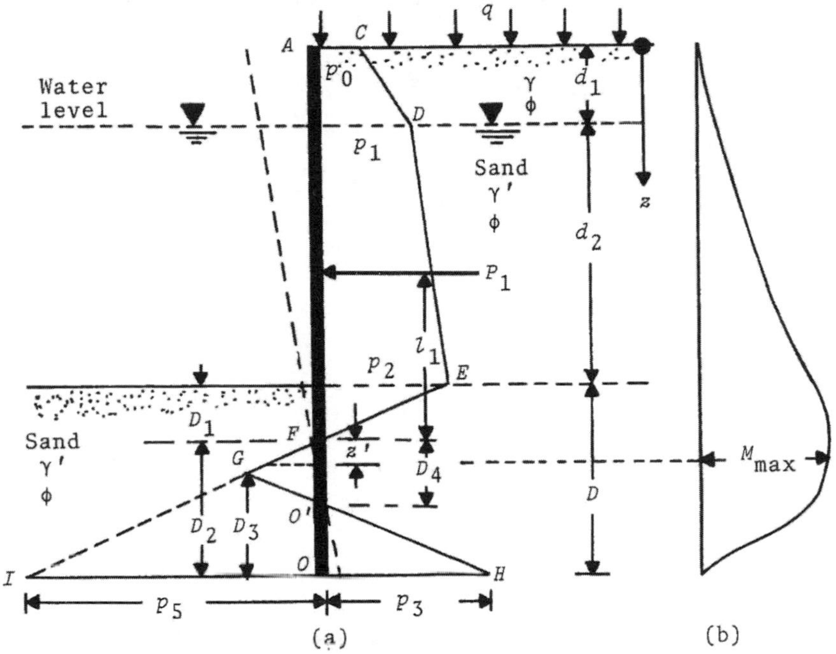

Figure 2.11. Cantilever sheet pile wall penetrating into a sand layer: (a) Net pressure distribution diagram; (b) Moment diagram. (*Note:* $d_1+d_2=d$.)

dredge line. The sheet pile penetrates to a depth D into the sand also. The water level is located at a depth d_1 below the top of the wall. Since the hydrostatic pressures from both sides of the wall will cancel each other, we will consider only the horizontal effective pressure distribution on the wall.

Now, referring to Fig. 2.11a, the net lateral pressures at depths $z=0$, $z=d_1$, and $z=d_1+d_2$ are equal to p_0, p_1, and p_2, respectively, or

$$p_0 = qK_{aH} \tag{2.9}$$

$$p_1 = (q+\gamma d_1)K_{aH} \tag{2.10}$$

$$p_2 = (q+\gamma d_1+\gamma' d_2)K_{aH} \tag{2.11}$$

where q surcharge
γ unit weight of sand

γ' = effective unit weight of sand

At a depth $z>d=d_1+d_2$, the active pressure is acting from right to left, and the passive pressure is acting from left to right. So the net pressure

$$p = [q+\gamma d_1+(z-d_1)\gamma']K_{aH} \quad [(z-d_1-d_2)\gamma']K_{pH}$$

$$p_2-(z-d)(K_{pH}-K_{aH})\gamma' \tag{2.12}$$

The depth D_1 below the dredge line at which p is equal to zero can be determined as

$$p \quad 0 = p_2-(D_1)(K_{pH}-K_{aH})\gamma'$$

or

$$D_1 = \frac{p_2}{\gamma'(K_{pH}-K_{aH})} \tag{2.13}$$

At a depth $z=d+D$, the active pressure acts from left to right, and the passive pressure acts from right to left. So the net pressure

$$p \quad p_3 \quad [\gamma d_1+\gamma'(d_2+D)]K_{pH} -(\gamma' D K_{aH})$$

or

$$p_3 = (\gamma d_1+\gamma' d_2)K_{pH} + \gamma' D(K_{pH}-K_{aH})$$

$$= p_4+\gamma' D_2(K_{pH}-K_{aH}) \tag{2.14}$$

where $p_4 \quad (\gamma d_1+\gamma' d_2)K_{pH} +\gamma' D_1(K_{pH}-K_{aH})$ (2.15)

The net pressure distribution diagram is shown in Fig. 2.11a as *ACDEFGH*. For equilibrium

Σ Horizontal forces per unit length of the sheet pile, $F_H=0$

Σ Moment of forces per unit length of sheet pile about point O, $M_O=0$

$$\Sigma F_H = P_1-P_2+P_3 \tag{2.16}$$

where P_1 area of the pressure diagram *ACDEF*
P_2 area of the triangle *FOI*
P_3 = area of the triangular area *GIH*

Note that $IO=p_5$=(slope of the line EG)(D_2). However, the slope of line EG [from Eq. (2.13)] is 1 vertical to $\gamma'(K_{pH}-K_{aH})$. So

$$\overline{IO} \quad p_5 \quad \gamma'(K_{pH}-K_{aH})D_2 \tag{2.17}$$

So

$$\Sigma \, F_H \quad P_1 - \tfrac{1}{2}p_5D_2 + \tfrac{1}{2}(D_3)(p_3+p_5) \tag{2.18}$$

or

$$D_3 = \frac{p_5D_2-2P_1}{p_3+p_5} \tag{2.19}$$

Again

$$\Sigma \, M_o = P_1(D_2+l_1) \quad (\tfrac{1}{2}p_5D_2)(\tfrac{D_2}{3}) + \tfrac{1}{2}(D_3)(p_3+p_5)(\tfrac{D_3}{3}) \quad 0 \tag{2.20}$$

Combining Eqs. (2.14), (2.15), (2.17), (2.19), and (2.20), we obtain the following relationship

$$D_2^4+AD_2^3-BD_2^2-CD_2-E = 0 \tag{2.21}$$

where

$$A \quad \frac{p_4}{\gamma'(K_{pH}-K_{aH})} \tag{2.22}$$

$$B = \frac{8P_1}{\gamma'(K_{pH}-K_{aH})} \tag{2.23}$$

$$C = \frac{6P_1[2l_1\gamma'(K_{pH}-K_{aH}) + p_4]}{\gamma'^2(K_{pF}-K_{aH})^2} \tag{2.24}$$

$$E \quad \frac{P_1(6l_1p_4+4P_1)}{\gamma'^2(K_{pH}-K_{aH})^2} \tag{2.25}$$

Since the values of A, B, C, and E can be calculated, Eq. (2.21) can be solved by trial and error to determine D_2. Once D_2 is known, the magnitude of p_5 can be calculated by using Eq. (2.17). With known values of p_5 and D_2, the magnitude of D_3 can be obtained from Eqs. (2.14) and (2.19). Hence, the pressure distribution diagram can be plotted with proper magnitudes. The depth of penetration of the sheet pile is given as

$$D = D_1+D_2$$

The maximum moment in the sheet pile will occur at a point where the shear force is equal to zero. For this case, it will be at some level between F and G as shown in Fig. 2.11a. For zero shear force at a distance z' below the level of F

$$P_1 \quad \tfrac{1}{2}(z')[z'\gamma'(K_{pH}-K_{aH})] = 0$$

or

$$z' = \sqrt{\frac{2P_1}{\gamma'(K_{pH}-K_{aH})}} \tag{2.26}$$

The maximum moment (Fig. 2.11b) can then be calculated as

$$M_{max} = P_1(l_1+z') - \frac{z'^3\gamma'(K_{pH}-K_{aH})}{6} \tag{2.27}$$

The unit of M_{max} is ft-lb/ft of the wall. The sheet pile can be chosen by knowing the section modulus as

$$S = \frac{M_{max}}{\sigma_{all}} \tag{2.28}$$

where S = required section modulus per unit length of the wall

Blum (1951, 1955) has suggested a simplified method for obtaining the depth of penetration D. This can be explained by referring to Fig. 2.12, in which the point O' corresponds to point O'

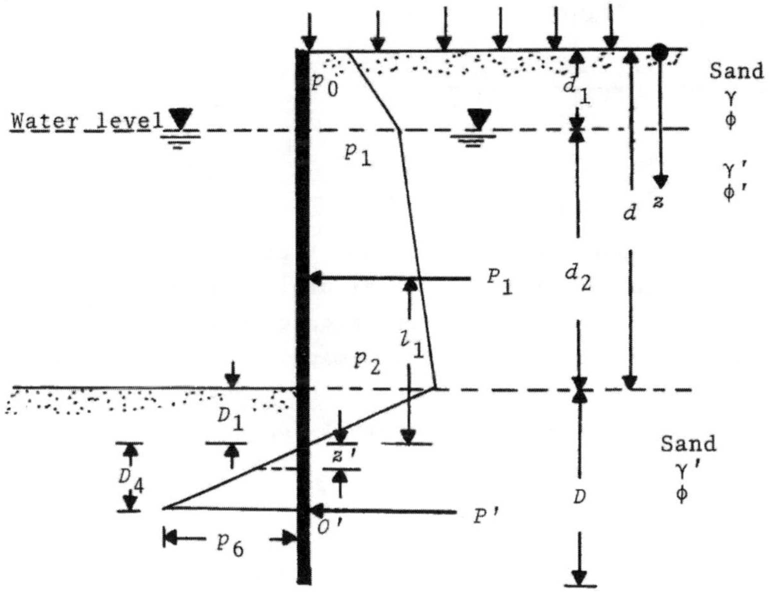

Figure 2.12. Cantilever sheet pile wall--Blum's method.

in Fig. 2.11. According to this method, the pressure distribution diagram will be as shown in Fig. 2.12. Note that a reactive force P' is placed at O'. Also

$$p_6 = \gamma'(K_{pH}-K_{aH})D_4 \tag{2.29}$$

Now, taking the moment at O'

$$\Sigma M_{O'} \quad P_1(l_1+D_4) \quad \tfrac{1}{2}(p_6)(D_4)(\tfrac{D_4}{3})$$

$$= P_1(l_1+D_4) - \tfrac{1}{6}(\gamma')(K_{pH}-K_{aH})D_4^3 = 0$$

or

$$D_4^3 \quad \left[\frac{6P_1}{\gamma'(K_{pH}-K_{aH})}\right]D_4 \quad \left[\frac{6P_1l_1}{\gamma'(K_{pH}-K_{aH})}\right] = 0 \tag{2.30}$$

Once D_4 is determined, the actual depth of penetration can be given as

$$D = 1.2(D_1+D_4) \tag{2.31}$$

The location and the magnitude of the maximum moment can be given by Eqs. (2.26) and (2.27).

Some special cases for cantilever sheet pile walls are shown in Figs. 2.13 and 2.14. Figure 2.13 shows a free cantilever sheet piling penetrating into a sand layer. The sheet piling is subjected to a load of V *per unit length of the pile* at right angles to the cross section shown. So the unit of V is lb/ft or kN/m and so forth. The net pressure distribution diagram ($ABO'CO$) can be drawn in a similar manner as described earlier in relation to Fig. 2.11. For this condition, the following equation needs to be solve solved to obtain the necessary depth of penetration D.

$$D^4 - \left[\frac{8V}{\gamma(K_{pH}-K_{aH})}\right]D^2 \quad \left[\frac{12Vd}{\gamma(K_{pH}-K_{aH})}\right]D - \left[\frac{2V}{\gamma(K_{pH}-K_{aH})}\right]^2 \quad 0 \tag{2.32}$$

where γ unit weight of soil below the ground surface

Once the magnitude of D is determined, the distance D_3 can be calculated as

$$D_3 \quad \frac{\gamma(K_{pH}-K_{aH})D^2-2V}{2D(K_{pH}-K_{aH})\gamma} \tag{2.33}$$

The location z' and magnitude of the maximum moment can be obtained as follows:

For zero shear force:

$$V = (\tfrac{1}{2})(z')[\gamma z'(K_{pH}-K_{aH})] = \frac{\gamma z'^2(K_{pH}-K_{aH})}{2} \tag{2.34}$$

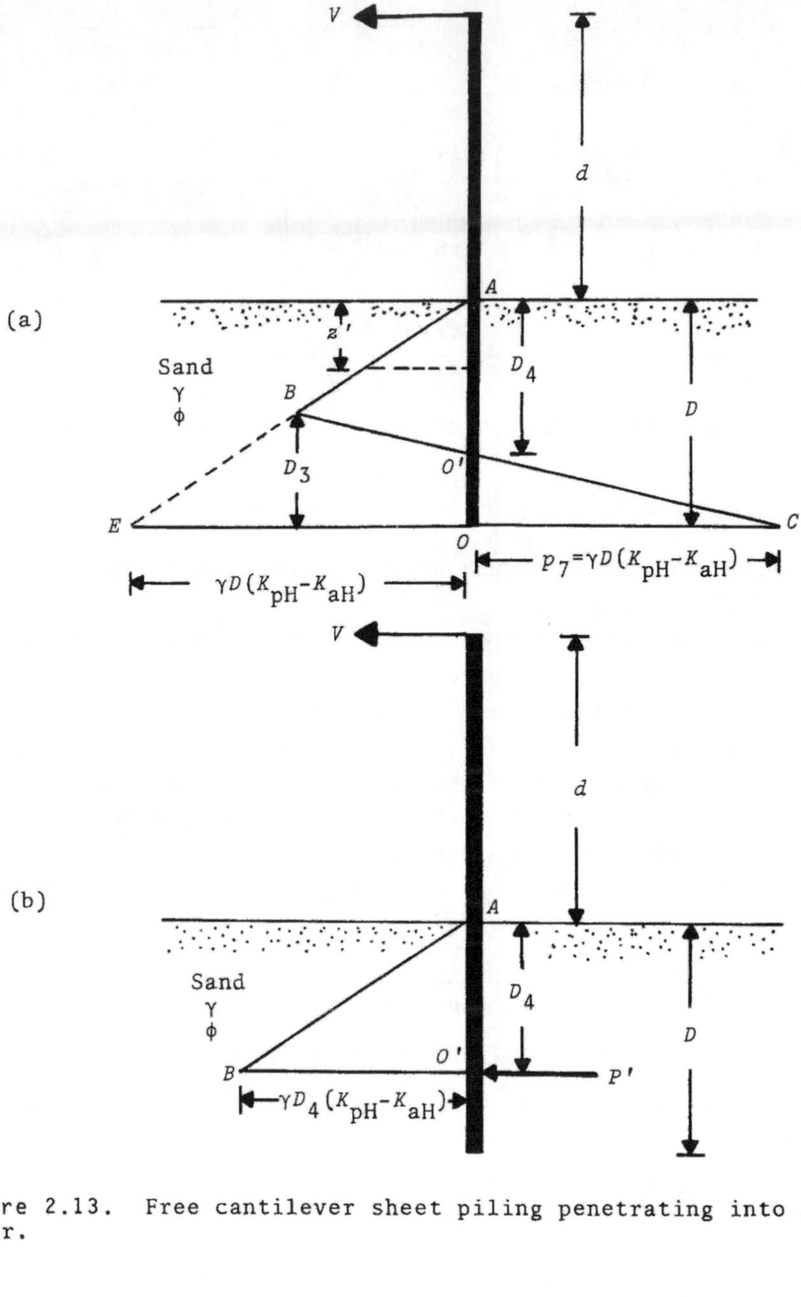

Figure 2.13. Free cantilever sheet piling penetrating into a sand layer.

Hence

$$M_{max}(\text{per unit length of wall}) = H(d+z') - \frac{\gamma z'^3(K_{pH}-K_{aH})}{6} \quad (2.35)$$

Blum's simplified method can also be used in this case to determine

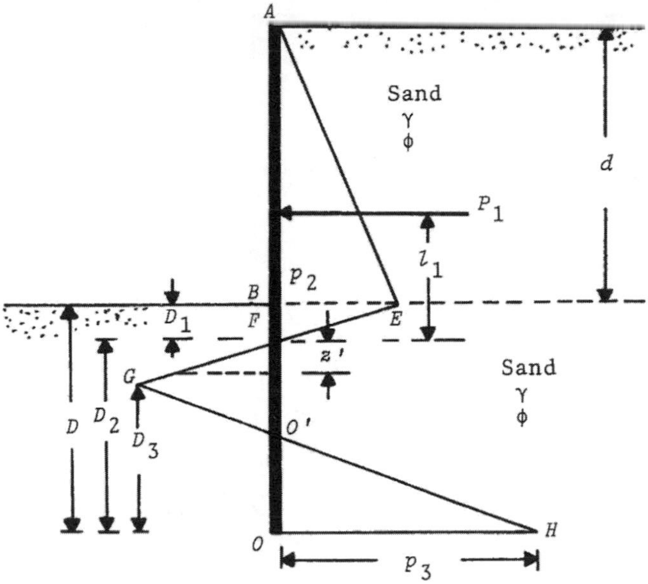

Figure 2.14. Cantilever sheet pile wall with the absence of water table and surcharge.

the depth of penetration. To do this, we introduce a reactive force P' at O' as shown in Fig. 2.13b. Now, summing the moments about O'

$$\Sigma M_{O'} = V(D_4+d) - \tfrac{1}{2}D_4\left[(\gamma D_4)(K_{\mathrm{pH}}-K_{\mathrm{aH}})\right]\frac{D_4}{3} \quad 0 \qquad (2.36)$$

$$D_4V+dV - \gamma(K_{\mathrm{pH}}-K_{\mathrm{aH}})\frac{D_4^3}{6} = 0$$

or

$$D_4^3 - \left[\frac{6V}{\gamma(K_{\mathrm{pH}}-K_{\mathrm{aH}})}\right]D_4 - \frac{6dV}{\gamma(K_{\mathrm{pH}}-K_{\mathrm{aH}})} = 0 \qquad (2.37)$$

The magnitude of D_4 can be calculated from Eq. (2.37), and the required depth of penetration can now be given as

$$D = 1.2D_4 \qquad (2.38)$$

The location and the magnitude of the maximum moment can be calculated by using Eqs. (2.34) and (2.35).

Figure 2.14 shows a cantilever sheet pile wall with no water table and surcharge. The net pressure diagram ($AEFGO'HO$) can be constructed in similar manner as that shown in Fig. 2.11. Note that

$$p_2 = \gamma d K_{aH} \tag{2.39}$$

$$D_1 = \frac{p_2}{\gamma(K_{pH}-K_{aH})} \tag{2.40}$$

The slope of the line EG is 1 vertical to $\gamma(K_{pH}-K_{aH})$ horizontal. Also

$$p_3 = \gamma d K_{pH} + \gamma D K_{pH} - \gamma D K_{aH} = \gamma K_{pH}(d+D) - \gamma D K_{aH} \tag{2.41}$$

The equations for obtaining D and D_3 are as follows (Jumikis, 1971)

$$D^4 + \left[\frac{d(K_{pH}-3K_{aH})}{(K_{pH}-K_{aH})}\right]D^3 - \left[\frac{K_{aH}d^2(7K_{pH}-3K_{aH})}{(K_{pH}-K_{aH})^2}\right]D^2 - \left[\frac{5K_{aH}d^3(K_{pH}-K_{aH})}{(K_{pH}-K_{aH})^2}\right]D - \frac{K_{pH}K_{aH}d^4}{(K_{pH}-K_{aH})^2} = 0 \tag{2.42}$$

and

$$D_3 = \frac{K_{pH}D^2-K_{aH}(d+D)^2}{(K_{pH}-K_{aH})(d+2D)} \tag{2.43}$$

The location z' and the magnitude of the maximum moment M_{max} can be determined as follows:

For zero shear force:

$$P_1 - \tfrac{1}{2}\gamma z'^2(K_{pH}-K_{aH}) = 0 \tag{2.44}$$

where P_1 = area of the pressure diagram $AEFB$

$$= \tfrac{1}{2}p_2 d + \tfrac{1}{2}p_2 D_1 = \tfrac{1}{2}p_2(d+D_1)$$

So

$$\tfrac{1}{2}p_2(d+D_1) = \tfrac{1}{2}\gamma z'^2(K_{pH}-K_{aH})$$

or

$$z' = \sqrt{\frac{p_2(d+D_1)}{\gamma(K_{pH}-K_{aH})}} \tag{2.45}$$

The maximum moment at a point z' below point F can be given as

$$M_{max} = P_1(l_1+z') - \tfrac{1}{6}\gamma z'^3(K_{pH}-K_{aH}) \tag{2.46}$$

Masih (1984) has provided a graphical procedure for determination of the depth of penetration of sheet piles for linear and nonlinear variation of lateral earth pressure.

Example 2.1. Refer to the cantilever sheet pile wall shown in Fig. 2.11. Given q=600 lb/ft², γ=110 lb/ft³, γ'=60 lb/ft³, ϕ=35°, d=15 ft, d_1= 5 ft, and d_2=10 ft. Determine the required depth of penetration D, and also the maximum moment developed in the sheet pile wall. Use δ=0° and a factor of safety of 2 for passive pressure.

Solution

Determination of depth of penetration, D: For δ=0

$$K_{aH} \quad \tan^2(45-\phi/2) = \tan^2(45-35/2) = 0.271$$

$$K_{pH} \quad \tan^2(45+\phi/2) = \tan^2(45+17.5) = 3.69$$

$$K_{pF} = K_{pH}/F_s = 3.69/2 \quad 1.845$$

$$p_0 = qK_{aH} = (600)(0.271) = 162.6 \text{ lb/ft}^2$$

$$p_1 = (q+\gamma d_1)K_{aH} = [600+(5)(110)](0.271) - (1150)(0.271)$$

$$= 311.7 \text{ lb/ft}^2$$

$$p_2 = (q+\gamma d_1+\gamma' d_2)K_{aH} \quad [600+(5)(110)+(10)(60)](0.271)$$

$$(1750)(0.271) \quad 474.3 \text{ lb/ft}^2$$

From Eq. (2.13)

$$D_1 = \frac{p_2}{\gamma'(K_{pF}-K_{aH})} = \frac{474.3}{(60)(1.845-0.271)} = 5.02 \text{ ft}$$

Figure 2.15 shows the plot of the net pressure diagram from the levels of A to F as shown in Fig. 2.11. So

$$P_1 = \Sigma \text{ Areas of pressure diagram} \quad (162.5)(15) + (\tfrac{1}{2})(5)(311.7-$$

$$162.6) + (10)(322.7-162.6) + (\tfrac{1}{2})(10)(474.3-311.7)$$

$$+ (\tfrac{1}{2})(5.02)(474.3)$$

$$= 2{,}439 + 372.75 + 1{,}491 + 813 + 1{,}190.5 = 6{,}306.25 \text{ lb/ft}$$

$$l_1 = \frac{\Sigma M_F}{P_1} = \frac{\begin{bmatrix}(2{,}439)(5.02+7.5) + (372.75)(15.02+1.67) \\ + (1{,}491)(5.02+5) + (813)(5.02+3.33) \\ + (1{,}190.5)(5.02/3)\end{bmatrix}}{6{,}306.25} = \underline{9.59 \text{ ft}}$$

Now, referring to Eqs. (2.21), (2.22), (2.23), (2.24), and (2.25)

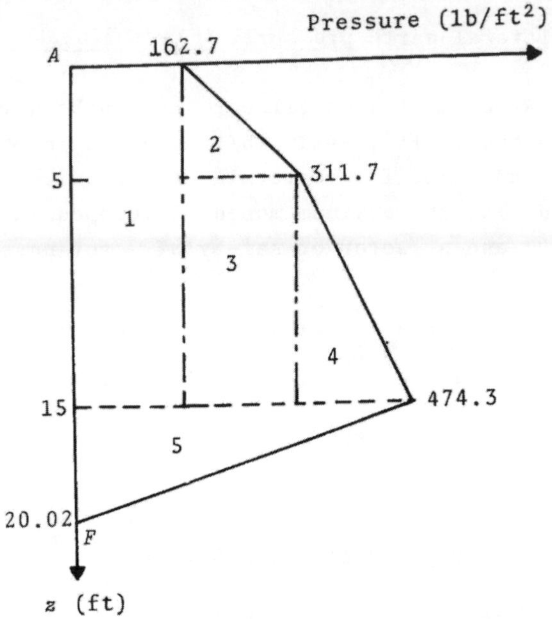

Figure 2.15.

$$A \quad \frac{p_4}{\gamma'(K_{pF}-K_{aH})}$$

$$p_4 = (\gamma d_1+\gamma' d_2)K_{pF} + \gamma' D_1(K_{pF}-K_{aH})$$

$$= [(110)(5)+ (60)(10)](1.845) + (60)(5.02)(1.845-0.271)$$

$$2,121,75 + 474.1 = 2,595.85 \text{ lb/ft}^2$$

$$A \quad \frac{2,595.85}{(60)(1.845-0.271)} = 27.49$$

$$B = \frac{8P_1}{\gamma'(K_{pF}-K_{aH})} = \frac{(8)(6,306.25)}{(60)(1.845-0.271)} = 534.2$$

$$C = \frac{6P_1[2l_1\gamma'(K_{pF}-K_{aH}) + p_4}{\gamma'^2(K_{pF}-K_{aH})^2}$$

$$\frac{(6)(6,306.25)[(2)(9.59)(60)(1.845-0.271)+ 2,595.85]}{(60)^2(1.845-0.271)^2}$$

$$= 18,696.9$$

$$E \quad \frac{P_1(6l_1p_4 + 4P_1)}{\gamma'^2(K_{pF}-K_{aH})^2}$$

$$\frac{(6{,}306.25)[(6)(9.59)(2{,}505.85) + (4)(6{,}306.25)]}{(60)^2(1.845-0.271)^2}$$

$$- 123{,}445.4$$

So

$$D_2^4 + 27.49D_2^3 - 534.2D_2^2 - 18{,}696.9D_2 - 123{,}445.4 \quad 0$$

By trial and error, $D_2 \approx 27$ ft. So, the total actual depth of penetration

$$D \quad D_1 + D_2 = 5.02 + 27 \approx \underline{32\ \text{ft}}$$

Determination of maximum moment, M_{max}: From Eq. (2.26)

$$z' \quad \sqrt{\frac{2P_1}{\gamma'(K_{pF}-K_{aH})}} = \sqrt{\frac{(2)(6{,}306.25)}{(60)(1.845-0.271)}} = 11.56\ \text{ft}$$

So, from Eq. (2.27)

$$M_{max} = (6{,}306.25)(9.59+11.56) - \frac{(11.56)^3(60)(1.845-0.271)}{6}$$

$$= 109{,}062\ \text{ft-lb} \sim \underline{109\ \text{ft-kip}}$$

Example 2.2. Refer to Fig. 2.13b, which shows a free cantilever sheet piling penetrating into a sand layer. Given V=2000 lb/ft, ϕ=35°, γ=115 lb/ft^3, and d=10 ft. Using Blum's method, determine the required depth of penetration D. Use $\delta=\phi/2$ for active pressure coefficient and δ=0° for passive pressure coefficient. Increase the depth of penetration by 25% to obtain the actual depth of penetration.

Solution

From Eq. (1.34)

$$K_a \quad \frac{1}{\cos\delta} \cdot \frac{1}{\left(\frac{1}{\cos\delta}\right) + \sqrt{\tan^2\phi + \tan\phi\tan\delta}}$$

$$K_{aH} = K_a\cos\delta$$

For this problem, ϕ=35° and $\delta=\phi/2$=17.5°. So K_{aH}=0.235.
For passive pressure, with δ=0°

$$K_{pH} = \tan^2(45+\phi/2) = \tan^2(45+17.5) = 3.69$$

From Eq. (2.37)

$$D_4^3 - \left[\frac{6V}{\gamma(K_{pH}-K_{aH})}\right]D_4 - \frac{6dV}{\gamma(K_{pH}-K_{aH})} = 0$$

Substitution of proper values gives

$$D_4^3 \left[\frac{(6)(2,000)}{(115)(3.69-0.235)}\right] D_4 - \frac{(6)(10)(2,000)}{(115)(3.69-0.235)} \quad 0$$

or

$$D_4^3 - 30.2D_4 - 302 = 0$$

So

$$D_4 \approx 8.2 \text{ ft}$$

Hence, $D_{\text{theoretical}} \approx 1.2D_4 = (1.2)(8.2) = 9.84$ ft. The actual depth of penetration is

$$D_{\text{actual}} = (1.25)(D_{\text{theoretical}}) = (1.25)(9.84) = \underline{12.3 \text{ ft}}$$

2.7 CANTILEVER SHEET PILE WALL PENETRATING INTO CLAY LAYER

Figure 2.16 shows a cantilever sheet pile wall with a granular backfill above the dredge line where the sheet pile penetrates a

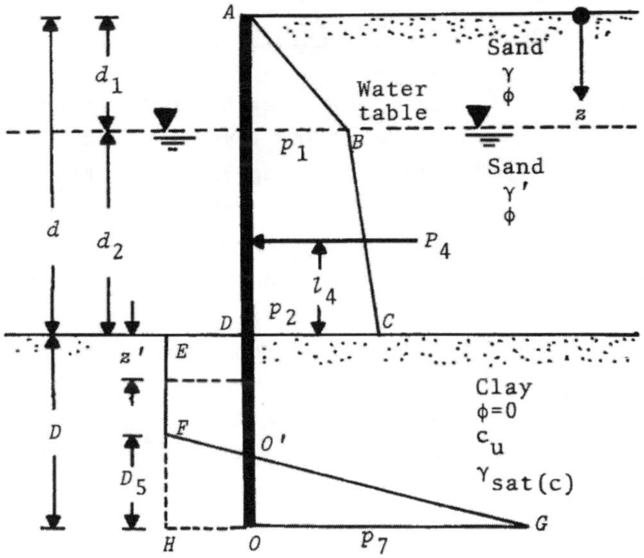

Figure 2.16. Cantilever sheet pile wall penetrating into clay layer.

saturated clay layer. The saturated unit weight of the clay layer is $\gamma_{sat(c)}$, and the undrained shear strength is equal to c_u. For the loading condition shown, the net pressures on the sheet pile are as follows.

$$\text{At } z = 0, \qquad p = p_0 = 0$$

$$\text{At } z = d_1, \qquad p = p_1 = \gamma d_1 K_{aH}$$

$$\text{At } z = d_2, \qquad p = p_2 = (\gamma d_1 + \gamma' d_2) K_{aH}$$

where K_{aH} = horizontal active earth pressure coefficient
$= K_a \cos\delta$

At depth $z>d$ $(=d_1+d_2)$, the horizontal active earth pressure from right to left is p_a, and the horizontal passive earth pressure from left to right is p_p. So

$$\underset{\rightarrow}{p} \quad \underset{\rightarrow}{p_p} - \underset{\leftarrow}{p_a}$$

$$= \{[\gamma_{sat(c)}(z-d)] + 2c_u\} - \{[q' + \gamma_{sat(c)}(z-d)] - 2c_u \tag{2.47}$$

where $q' = \gamma d_1 + \gamma' d_2$ (2.48)

Also note that, for the undrained condition in clay (that is, $\phi=0$), $K_a = K_{aH} = 0$. So

$$p = 4c_u - q' = p_6 \tag{2.49}$$

At $z=d+D$, the passive earth pressure will be from right to left, and the active earth pressure will be from left to right. So

$$p = \underset{\leftarrow}{p_p} \quad \underset{\rightarrow}{p_a} = \{[q' + \gamma_{sat(c)}D] + 2c_u\} - [\gamma_{sat(c)}D - 2c_u]$$

$$= 4c_u + q' \tag{2.50}$$

The net pressure distribution diagram is *ABCDEFO'GO* as shown in Fig. 2.16. Now, let P_4 be the force above the dredge line.

$$P_4 = \frac{1}{2}(p_1 d_1) + p_1 d_2 + \frac{1}{2}(p_2 - p_1) d_2$$

The line of action of the resultant force P_4 is located at a distance l_2 above the dredge line. For equilibrium

$$\Sigma \text{ Horizontal forces} = 0$$

or

$$\underset{\leftarrow}{P_4} + \underset{\leftarrow}{(\text{Area of } FGH)} - \underset{\rightarrow}{(\text{Area of } DEHO)} = 0$$

$$P_4 + \frac{1}{2}[(4c_u - q') + (4c_u + q')]D_5 \quad (4c_u - q')D = 0$$

$$P_4 + 4c_u D_5 \quad 4c_u D + q'D = 0$$

Hence

$$D_5 = \frac{(4c_u - q')D - P_4}{4c_u} \tag{2.51}$$

Also

$$\Sigma \text{ Moment about } O = 0$$

So

$$[P_4(D + l_2)] - [(4c_u D_5)(D_5/3)] - [(4c_u - q')D](D/2) = 0$$

$$D^2(4c_u - q') \quad 2DP_4 \quad \frac{P_4(P_4 + 12c_u l_2)}{q' + 2c_u} = 0 \tag{2.52}$$

The depth of penetration can be calculated by solving Eq. (2.52), after which the magnitude of D_5 can be calculated using Eq. (2.51). Once the magnitudes of D and D_5 are known, the pressure diagram can be plotted.

The maximum moment on the sheet pile will occur at a depth z' below the dredge line as shown in Fig. 2.16. For the maximum moment, the shear force will be zero. So

$$\underset{(\leftarrow)}{P_4} - \underset{(\rightarrow)}{p_6 z'} = 0$$

or

$$z' = P_4/p_6 = P_4/(4c_u - q') \tag{2.53}$$

Hence, the maximum moment

$$M_{max} \quad P_4(l_2 + z') - (4c_u - q')(z'^2/2) \tag{2.54}$$

It needs to be pointed out that if the magnitude of q' is larger than $4c_u$, $4c_u - q' = p_6$ will be negative. In that case, the sheet pile structure will not be stable.

Figure 2.17 shows a free cantilever sheet piling penetrating into a layer. This is a special case of the problem shown in Fig. 2.16. The sheet piling is subjected to a load of V per unit length at right angles to the cross section shown. In this case, the lateral pressure distribution diagram below the dredge line can be given by $BCDO'EO$. For equilibrium of the sheet pile structure

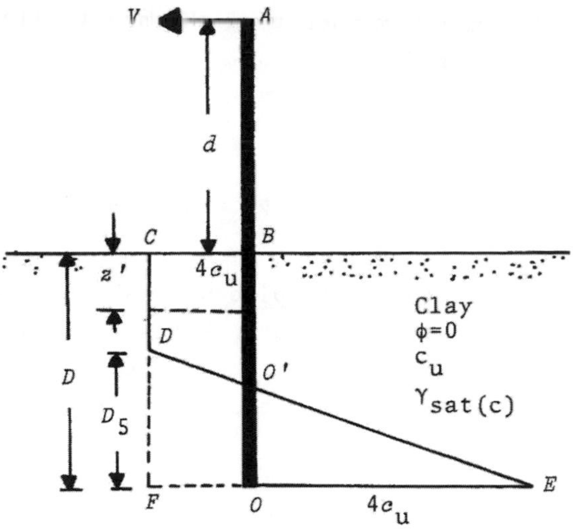

Figure 2.17. Free cantilever sheet piling penetrating into a clay layer.

$$\Sigma \text{ Horizontal forces} = 0 \tag{2.55}$$

and

$$\Sigma \text{ Moment of all forces about } O = 0 \tag{2.56}$$

From Eq. (2.55)

$$\underset{(\leftarrow)}{V} - \underset{(\rightarrow)}{4c_u D} + \underset{(\leftarrow)}{\tfrac{1}{2}(8c_u)D_5} = 0$$

(Area of pressure diagram $BCFO$) (Area of pressure diagram $DO'EOF$)

or

$$D_5 = \frac{4c_u D - V}{4c_u} \tag{2.57}$$

Again, from Eq. (2.56)

$$V(D+d) - (4c_u D)(D/2) + (4c_u D_5)(D_5/3) = 0 \tag{2.58}$$

Combining Eqs. (2.57) and (2.58)

$$4D^2 c_u - 2VD - \frac{V(V+12c_u d)}{2c_u} = 0 \tag{2.59}$$

The preceding equation can be solved to obtain the required depth of penetration D. The maximum moment on the sheet pile will occur at a depth z' below the dredge line, or

$$z' \quad V/4c_u \tag{2.60}$$

Hence

$$M_{max} = V(d+z') \quad 4c_u(z'^2/2) = Vd + Vz' - 2c_u z'^2 \tag{2.61}$$

Example 2.3. Refer to the sheet pile wall shown in Fig. 2.16. Given d_1=1.5 m, d_2=3 m, ϕ=30°, δ=0°. For sand, γ=15.5 kN/m³, γ_{sat}= 18.5 kN/m³, c_u=45 kN/m². Use F_s=1.5 for the undrained shear strength of clay. Determine the required depth of penetration for the sheet pile.

Solution

For $\phi=30°$, $\delta=0$.

$$K_a = K_{aH} = \tfrac{1}{3}$$

$$p_1 = \gamma d_1 K_{aH} = (15.5)(1.5)(\tfrac{1}{3}) = 7.75 \text{ kN/m}^2$$

$$p_2 = (\gamma d_1 + \gamma' d_2)K_{aH} = [(15.5)(1.5) + (18.5-9.81)3](\tfrac{1}{3})$$

$$= (23.25 + 26.07)(\tfrac{1}{3}) \quad 16.44 \text{ kN/m}^2$$

$$P_4 = \tfrac{1}{2}(p_1 d_1) + p_1 d_2 + \tfrac{1}{2}(p_2 - p_1)d_2$$

$$= \tfrac{1}{2}(7.75)(1.5) + (7.75)(3) + \tfrac{1}{2}(16.44-7.75)(3)$$

$$= 5.81 + 23.25 + 13.04 \quad 42.1 \text{ kN/m}$$

Taking the moment about D (Fig. 2.16)

$$l_2 \quad \frac{(5.81)(3+1.5/3) + (23.5)(3/2) + (13.04)(3/3)}{42.1}$$

$$\frac{20.34 + 34.88 + 13.04}{42.1} = 1.62 \text{ m}$$

In order to use Eq. (2.52), we need q' [Eq. (2.48)]. So

$$q' \quad \gamma d_1 + \gamma' d_2 = (15.5)(1.5) + (18.5-9.81)(3)$$

$$23.25 + 26.07 = 49.32 \text{ kN/m}^2$$

Now, from Eq. (2.52)

$$D^2(4c_{uF} - q') - 2DP_4 - \frac{P_4(P_4 + 12c_{uF}l_2)}{q' + 2c_{uF}} = 0$$

$$c_{uF} = c_u/F_s \quad 4.5/1.5 = 30 \text{ kN/m}^2$$

So

$$D^2[(4)(30)-49.32] - (2)(D)(42.1) - \frac{(42.1)[42.1+(12)(30)(1.62)]}{49.32+(2)(30)} = 0$$

or

$$70.68D^2 - 84.2D - 240.81 = 0$$

From the above equation, the actual depth of penetration $D \simeq 2.6$ m

2.8 ANCHORED SHEET PILE WALL PENETRATING INTO SAND--FREE EARTH SUPPORT METHOD

Figure 2.18 shows a sheet pile wall with a sand backfill penetrating into a sand layer. The anchor tie rod is located at a depth d_1' from the ground surface on the land side. Let q be the

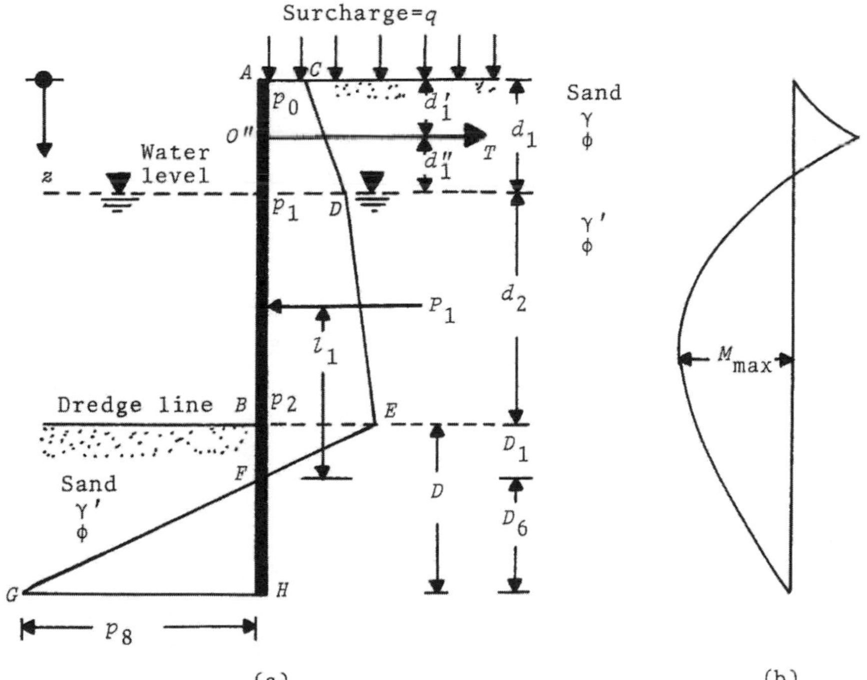

Figure 2.18. Anchored sheet pile wall penetrating into sand--free earth support method: (a) Pressure distribution diagram; (b) Moment diagram. (*Note:* $d=d_1+d_2$; $d_1=d_1'+d_1''$.)

surcharge per unit area. The net pressure distribution diagram between the levels of A and F along the sheet pile wall will be the same as shown in Fig. 2.11a (cantilever sheet pile wall). It was also noted in Section 2.7 (with reference to Fig. 2.11a) that the slope of the line EF is equal to $\gamma'(K_{pH}-K_{aH})$ and

$$D_1 = \frac{p_2}{\gamma'(K_{pH}-K_{aH})} \tag{2.13}$$

Now, let the anchor force per unit length of the sheet pile wall be equal to T, and the area of the pressure diagram $ACDEF$ be P_1. Let the line of action of the resultant force P_1 be located at a distance l_1 from point F. The pressure at the bottom of the sheet pile will be

$$p_8 = \gamma'(K_{pH}-K_{aH})D_6 \tag{2.62}$$

For equilibrium

Σ Horizontal forces = 0

and

Σ Moment of forces about the anchor level (that is, O'') = 0

So

$$P_1 - T - \tfrac{1}{2}p_8 D_6 = 0 \tag{2.63}$$

and

$$P_1(d+D_1-l_1-d_1') - (\tfrac{1}{2}p_8 D_6^2)(d+D_1-d_1' + \tfrac{2}{3}D_6) \quad 0 \tag{2.64}$$

Note that $d=d_1+d_2$, and $d_1=d_1'+d_1''$. Now, from Eq. (2.63)

$$T = P_1 - \tfrac{1}{2}p_8 D_6 \tag{2.65}$$

and, from Eq. (2.64)

$$D_6^3 + \left[\frac{3(d_1''+d_2+D_1)}{2}\right]D_6^2 \quad \frac{3P_1(d_1''+d_2+D_1-l_1)}{\gamma'(K_{pH}-K_{aH})} = 0 \tag{2.66}$$

Equations (2.65) and (2.66) can be used to determine D_6 and T. Hence, the depth of penetration of the sheet pile is equal to

$$D = D_1+D_6 \tag{2.67}$$

As mentioned in Section 2.5, in order to determine the design depth of penetration, the theoretical depth of penetration can be tentatively increased by about 30%, or

$$n_{design} \cong 1.3(n_1 + n_6) \qquad\qquad (2.68)$$

An alternative to the above procedure is to use a factor of safety (F_s) over K_{pH}, or

$$K_{pF} \quad K_{pH}/F_s$$

This value of K_{pF} may be used in place of K_{pH} to determine the magnitude of the design value of the depth of penetration D.

The maximum moment of the sheet pile will be located between the levels of D and E as shown in Fig. 2.18a (that is, $d_1 < z < d$). For the maximum moment, the shear force on the sheet pile should be equal to zero, or referring to Fig. 2.19

$$p_0(d_1 + z') - T + \tfrac{1}{2}d_1(p_1 - p_0) + (p_1 - p_0)z' + \tfrac{1}{2}\gamma'z'^2 K_{aH} \quad 0 \qquad (2.69)$$

Once the magnitude of z' is calculated from Eq. (2.69), the maximum moment (M_{max}) on the sheet pile can be easily obtained.

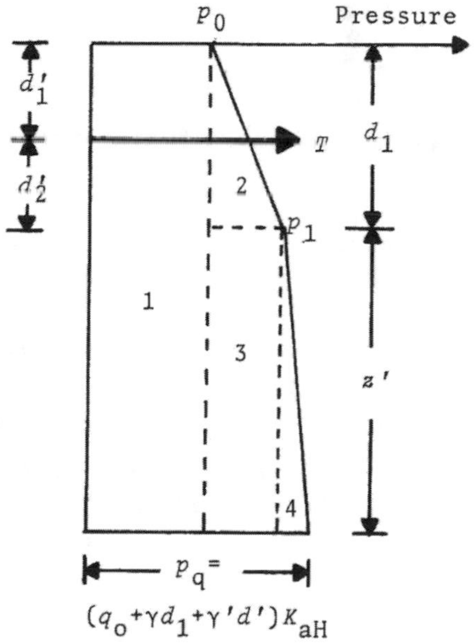

Figure 2.19. Determination of maximum moment.

2.9 LATERAL EARTH PRESSURE DUE TO STRIP LOAD NEAR A WALL

Strip loads of limited width are sometimes encountered near a wall on the land side, as shown in Fig. 2.20. The load per unit area in the strip is q. Steenfelt and Hansen (1984) have used

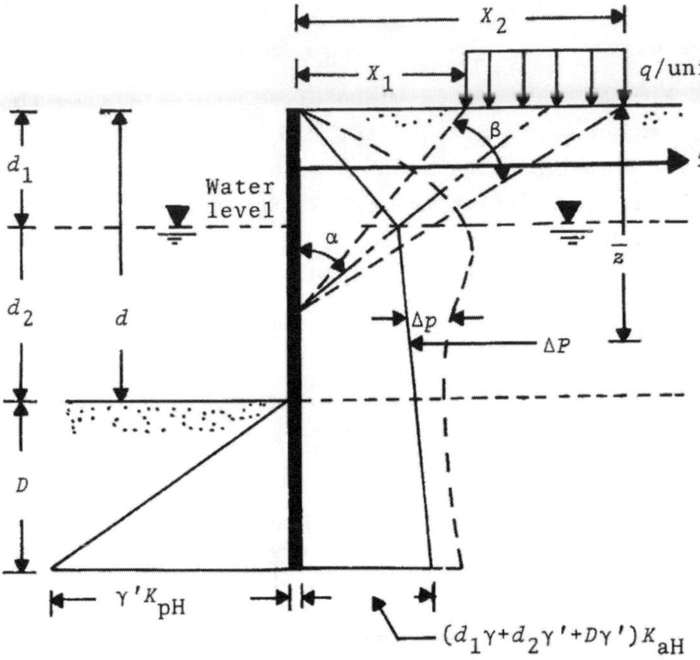

Figure 2.20. Lateral earth pressure due to strip load near the wall. (*Note:* γ=unit weight of soil above the water level and γ'= effective unit weight of soil below the water level.)

Brinch Hansen's theory (1953) to evaluate the additional active lateral force to which a sheet pile wall will be subjected due to the strip load surcharge. For details of this theory, the readers are referred to the original paper. However, a simplified theory for approximate lateral earth pressure distribution can be used by assuming the wall to be rigid. According to the theory of elasticity

$$\Delta p = \frac{q}{d+D}(\beta - \sin\beta\cos 2\alpha) \qquad (2.70)$$

where Δp lateral pressure due to the strip load surcharge only at a depth z measured from the ground surface

The definitions of α and β are shown in Fig. 2.20.

For soils, however, the right side of Eq. (2.70) is doubled due to the yielding of the continuum, or

$$\Delta p = \frac{2q}{d+D}(\beta - \sin\beta\cos 2\alpha) \tag{2.71}$$

Jarquio (1981) has integrated the above equation. So, the increase of lateral force due to the surcharge can be given as

$$\Delta P \int_0^{D+d} (\Delta p)\,dz = q/90[(D+d)(\theta_2-\theta_1)] \tag{2.72}$$

where $\theta_1 = \tan^{-1}[X_1/(d+D)]$ (2.73)

$\theta_2 = \tan^{-1}[X_2/(d+D)]$ (2.74)

In Eq. (2.72), θ_1 and θ_2 are in degrees. The location of the resultant force ΔP can be given as

$$\bar{z} = \frac{(d+D)^2(\theta_2-\theta_1) + (R-Q) - 57.30(X_2-X_1)(d+D)}{2(D+d)(\theta_2-\theta_1)} \tag{2.75}$$

where $R = (X_2)^2(90-\theta_2)$ (2.76)

$Q = (X_1)^2(90-\theta_1)$ (2.77)

2.10 ROWE'S MOMENT REDUCTION METHOD

Observations have shown that, due to the flexibility of the sheet piles, the actual maximum moment on the sheet pile walls is somewhat less than that calculated theoretically by the free earth support method of design. Hence, for the determination of the *actual design moment*, M_{max} should be reduced somewhat. Based on several experimental results, Rowe (1951, 1957),has proposed a semi-empirical procedure for moment reduction which is referred to as *Rowe's moment reduction procedure*. According to this procedure

$$\text{Flexibility number} = \rho \quad \frac{(d+D_{design})^4}{EI} \tag{2.78}$$

where d and D_{design} are in ft

E = modulus of elasticity of sheet piles (lb/in^2)

I = moment of inertia of sheet pile wall (in^4/ft of wall)

Figure 2.21 shows a plot of (M_{design}/M_{max}) vs. $\log_{10}\rho$ for loose sand dense sand and gravel. Note that M_{design}=(allowable flexural stress)(section modulus)=$\sigma_{all}S$. So, any sheet pile section having a section modulus of S per unit length for which the plot of $S/(M_{max}/\sigma_{all})$ vs. $\log_{10}\rho$ falls above the designated curve

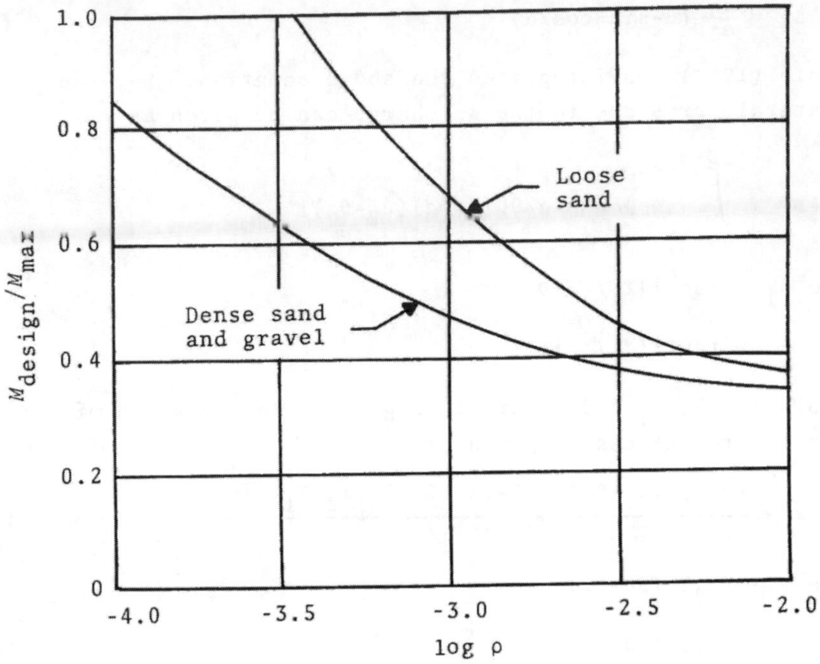

Figure 2.21. Plot of M_{design}/M_{max} against log ρ (after Rowe, 1952).

in Fig. 2.21 will be a safe section for use. In his original work, Rowe suggested that D_{design} be calculated by using wall friction angle $\delta=0$ and $F_s=1.5$ for the *horizontal passive earth pressure* coefficient (K_{pF}).

Example 2.4. Refer to Fig. 2.18. Given $q=0$, $\gamma=115$ lb/ft³, $\gamma'=60$ lb/ft³, $\phi=30°$, $d_1=10$ ft, $d_2=20$ ft, $d_1'=5$ ft, $d_1''=5$ ft, $\delta=0$, and $F_s=1.5$ for passive pressure coefficient calculation. Determine:

a. the design depth of penetration
b. the section modulus of the sheet pile section by using Rowe's moment reduction method

Solution

Part a. Determination of design depth of penetration.

$$K_{aH} = \tan^2(45-\phi/2) = \tfrac{1}{3}$$

$$K_{pF} = \frac{K_{pH}}{F_s} = \frac{\tan^2(45+\phi/2)}{1.5} = 2$$

Referring to Fig. 2.18

$$p_0 \quad 0$$

$$p_1 = \gamma d_1 K_{aH} = (115)(10)\tfrac{1}{3} \quad 383.3\ \text{lb/ft}^2$$

$$p_2 = (\gamma d_1 + \gamma' d_2)K_{aH} = [(115)(10)+(60)(20)]\tfrac{1}{3} \quad 783.3\ \text{lb/ft}^2$$

$$D_1 = \frac{p_2}{\gamma'(K_{pF}-K_{aH})} = \frac{783.3}{(60)(2-0.33)} \quad 7.82\ \text{ft}$$

Now, referring to Fig. 2.22

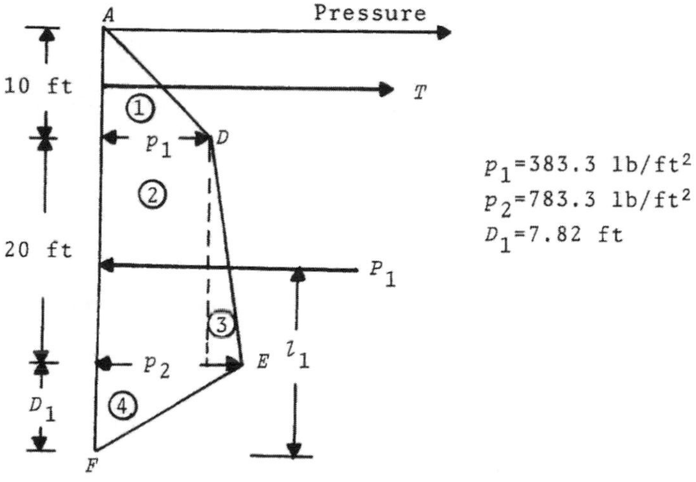

Figure 2.22.

$$P_1 = \tfrac{1}{2}(10)(383.3) + (383.3)(20) + \tfrac{1}{2}(783.3-383.3)(20)$$

$$+ \tfrac{1}{2}(7.82)(783.3)$$

$$= 1{,}916.5 + 7{,}666 + 4{,}000 + 3062.7 = 16{,}645.2\ \text{lb/ft}$$

Taking the moment of the pressure diagram about F

$$l_1 \quad \frac{(1{,}916.5)(27.82+3.33) + (7{,}666)(17.82) + (4{,}000)(7.82+6.67) + (3{,}062.7)(2/3)(7.82)}{16{,}645.2}$$

$$= 16.23\ \text{ft}$$

Referring to Eq. (2.66)

$$D_6^3 + \left[\frac{3(d_1''+d_2+D_1)}{2}\right] D_6^2 \quad \frac{3P_1(d_1''+d_2+D_1-l_1)}{\gamma'(K_{pF}-K_{AH})} = 0$$

$$\frac{3(d_1''+d_2+D_1}{2} = \frac{(3)(5+20+7.82)}{2} = 49.23$$

$$\frac{3P_1(d_1''+d_2+D_1-l_1)}{\gamma'(K_{pF}-K_{aH})} = \frac{(3)(16,645.2)(5+20+7.82-16.23)}{(60)(1.67)} = 8,267.8$$

So

$$D_6^3 + 49.23 D_6^2 \quad 8,267.8 \quad 0$$

By trial and error, $D_6 \sim 12$ ft

Hence

$$D_{\text{design}} \quad D_1 + D_6 = 7.82+12 = 19.82 \text{ ft} \simeq \underline{20 \text{ ft}}$$

Part b. Determination of location of zero shear force.

From Eq. (2.65)

$$T \quad P_1 - \tfrac{1}{2} p_8 D_6 = P_1 - \tfrac{1}{2}(D_6)[(D_6)(\gamma')(K_{pF}-K_{aH})]$$

$$= 16,645.2 - \tfrac{1}{2}(12)[(12)(60)(1.67)] \simeq 9431 \text{ lb/ft}$$

For zero shear force, referring to Fig. 2.23

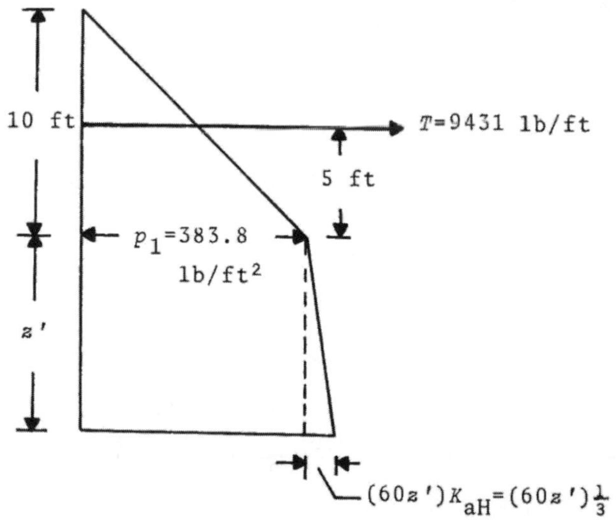

Figure 2.23.

$$\frac{1}{2}(10)(383.3) \quad 9{,}431 + (383.3)z' + [(60)(z')\tfrac{1}{3}]\tfrac{1}{2}(z') = 0$$

or

$$1{,}916.5 \quad 9{,}431 + 383.3z' + 10z'^2 = 0$$

$$z^2 + 383.3z' \quad 751.45 = 0$$

$$z' = 14.3 \text{ ft}$$

So

$$M_{max} = (1{,}916.5)(14.3+3.33) \quad (9{,}431)(5+14.3) + \frac{(383.3)(14.3)^2}{2} + \frac{(10)(14.3)^3}{3}$$

$$= 33{,}787.9 \quad 182{,}018 + 39{,}190.5 + 9{,}747.4 = -99{,}291.2 \text{ ft-lb}$$

So

$$M_{max} \quad (99{,}292.2 \times 12) \text{ in-lb/ft of wall}$$

Now, the following table can be prepared to choose the sheet pile section.

Section[a]	I^b (in^4/ft)	ρ^c	$\log\rho$	S^d (in^3/ft)	$\frac{S^e}{(M_{max}/\sigma_{all})}$
PZ-38	280.8	0.000742	-3.13	46.8	1.18
PMA-22	13.7	0.0152	-1.82	5.4	0.136
PZ-27	184.2	0.00113	-2.947	30.2	0.76

[a]From Table 2.1

[b]From Table 2.1

[c]Eq. (2.78); $\rho = \dfrac{(30+20)^4}{(30\times10^6 \text{ lb/in}^2)(I \text{ in}^4/\text{ft})} \quad \dfrac{0.2083}{I}$ (↑ E)

[d]From Table 2.1

[e] $\dfrac{S}{(M_{max}/\sigma_{all})} = \dfrac{S}{(99{,}292.2)(12)/(30{,}000 \text{ lb/in}^2)} \quad \dfrac{S}{39.72}$

The result for the sections PMA-22 and PZ-27 are plotted in Fig. 2.24. It can be seen that PZ-27 falls above the Rowe's moment reduction curve for loose sand. So this is a safe section. However, PMA-22 is an unsafe section.

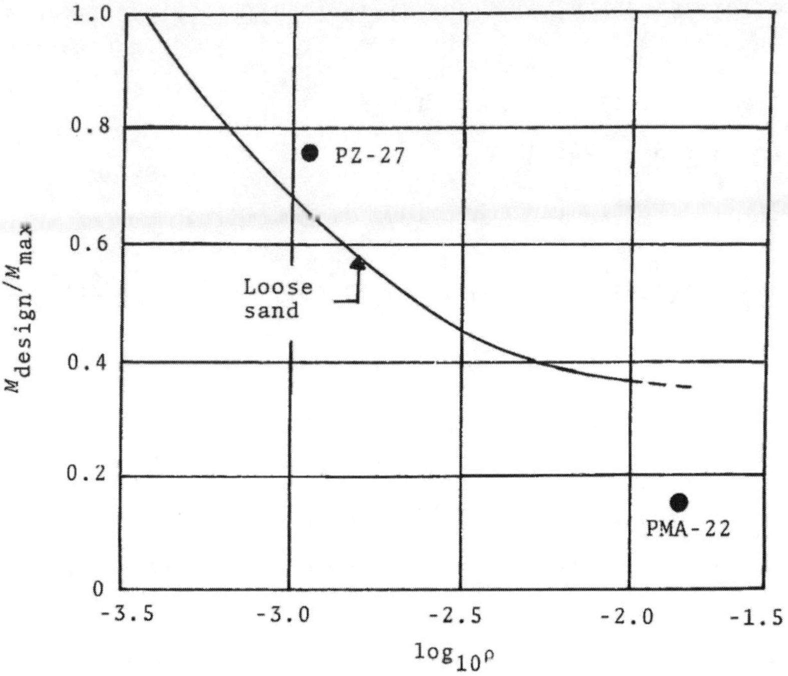

Figure 2.24.

2.11 ANCHORED SHEET PILE WALL WITH SLOPING DREDGE LINE

In construction under certain circumstances, the dredge line in front of an anchored bulkhead is sloping (Fig. 2.25). Theoreti-

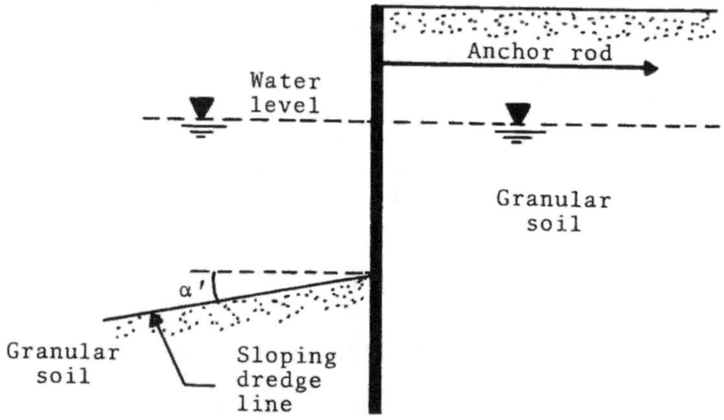

Figure 2.25. Anchored sheet pile wall with sloping dredge line.

cal and experimental results presently available in literature for anchored sheet pile wall design are for the condition where the dredge line is horizontal. Schroeder and Roumillac (1983) have conducted several small-scale laboratory model tests in *granular soil* to develop a design procedure for an anchored sheet pile wall via the *free earth support method*. Based on this study, it appears that reasonable results may be obtained by using the standard procedure described in Section 2.8 and by using the proper K_{pH} values as given by Caquot and Kerisel (1948). These values are:

Soil friction angle ϕ (deg)	δ (deg)	α'	K_{pH}
30	$\phi/2$	0	4.68
30	$2\phi/3$	0	5.22
30	$\phi/2$	10	3.17
30	$2\phi/3$	10	3.54
30	$\phi/2$	20	1.73
30	$2\phi/3$	20	1.93
35	$\phi/2$	0	6.55
35	$2\phi/3$	0	7.56
35	$\phi/2$	10	4.18
35	$2\phi/3$	10	4.82
35	$\phi/2$	20	2.44
35	$2\phi/3$	20	2.82

Schroeder and Roumillac (1983) have suggested that, for the design of the anchored sheet pile wall by the free earth support method, a factor of safety of 2 may be used over the K_{pH} values to obtain the actual depth of penetration D. They have also proposed a plot of M/M_{max} against $\log_{10}\rho$ for the moment reduction process. This is shown in Fig. 2.26 along with the similar plots given by Rowe (1952) for the case of a horizontal dredge line (loose sand).

2.12 COMPUTATIONAL PRESSURE DIAGRAM METHOD

In order to simplify the calculation procedures by the *free earth support method*, Nataraj and Hoadley (1985) have proposed a procedure which they call the *computational pressure diagram method* (CPD). This procedure was developed based upon the experimental observations of Tschebotarioff (1949), Rowe (1952), Packshaw and

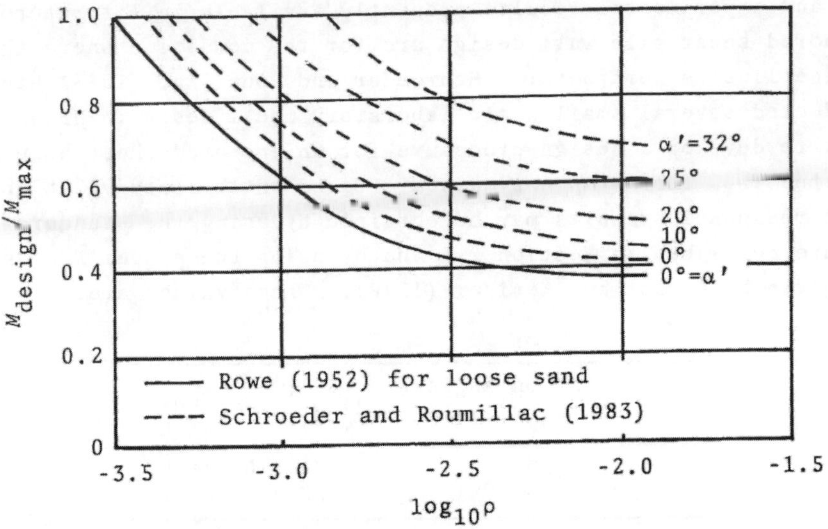

Figure 2.26. Plot of M_{design}/M_{max} against $\log_{10}\rho$ for moment reduction with sloping dredge line.

Lake (1952), and Packshaw (1954). According to the *CPD* method, the net pressure diagrams above and below the dredge line (as shown in Fig. 2.18) are replaced by rectangular pressure diagrams. This is demonstrated in Fig. 2.27, in which W_a is the width of the active

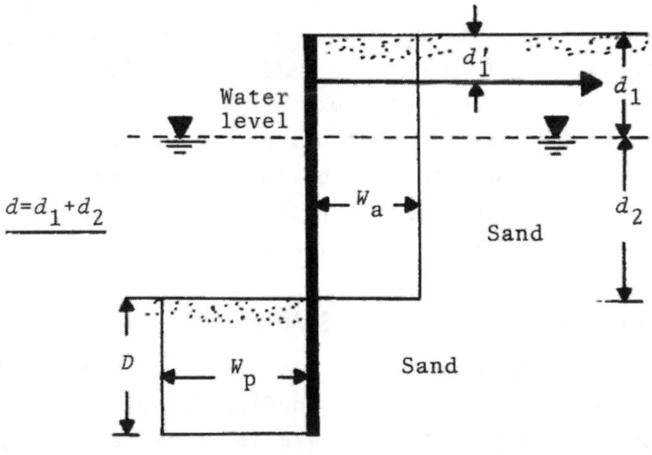

Figure 2.27. Computational pressure diagram method for design of anchored sheet pile wall--free earth support method.

pressure diagram above the dredge line and W_p is the width of the net passive pressure diagram below the dredge line. The relationships for W_a and W_p can be given as

$$W_a = CK_{aH}\gamma d \tag{2.79}$$

and

$$W_p \quad RCK_{aH}\gamma d \tag{2.80}$$

where K_{aH} = active pressure coefficient

γ = unit weight of soil

C, R = coefficients

The recommended values of C are as follows:

Type of soil	C	
	With surcharge	Without surcharge
Loose sand	1.00-1.25	0.8 -0.85
Medium sand	1.00-1.20	0.7 -0.75
Dense sand	0.75-0.85	0.55-0.65

In a similar manner, the recommended values of R are as follows:

Type of soil	R[a]
Loose sand	0.3 -0.5
Medium sand	0.55-0.65
Dense sand	0.60-0.75

[a] $R \quad \frac{W_p}{W_a} = \frac{d(d-2d_1')}{D(2d+D-2d_1')}$

Once the magnitudes of W_a and W_p are determined, the required depth of penetration D can be determined from the equation

$$D^2 + 2Dd(1-\beta') \quad (d^2/R)(1-2\beta') \quad 0 \tag{2.81}$$

where $\beta' = d_1'/d$

where d_1' = depth of the anchor rodmeasured from the ground surface

The anchor force per unit length of the wall can then be determined as

$$T = W_a(d - RD) \tag{2.82}$$

Based upon this type of design, the maximum design moment in the sheet pile is given by the expression

$$M_{max} \quad 0.5W_a d^2 \left[(1 - \frac{RD}{d})^2 - (\frac{2d_1'}{d})(1 - \frac{RD}{d})\right] \tag{2.83}$$

It may be noted that, for calculation of W_a and W_p, a single value of γ and a single value of K_{aH} have been used in Eqs. (2.79) and (2.80). However, if there is a variation of the effective unit weight with depth, a weighted average value of γ should be used. In a similar manner, if there is a variation of the soil friction angle ϕ and thus K_{aH}, a weighted average value of K_{aH} should be used.

When compared with the *conventional free earth support method* (*CFES*) of design, the *CPD* method of design yields the following:

1. The magnitude of D as obtained by the *CPD* method is about 1.25-1.5 times the depth of penetration obtained by the *CFES* method assuming F_s=1 for passive earth pressure (that is, $K_{pH}=K_{pF}$). This is consistent with the often-used method of using $K_{pH}=K_{pF}$ and determining D by the *CFES* method, and then increasing it arbitrarily by about 20-40%.
2. The maximum moment obtained by using Eq. (2.83) is about 0.6-0.75 M_{max} obtained by using the method outlined in Section 2.8. However, the magnitude of M_{max} obtained by using Eq. (2.83) is practically the same used for design via Rowe's moment reduction method (Section 2.10).
3. The magnitude of T obtained by using Eq. (2.82) is about 1.2-1.6 times the value of T obtained by using Eq. (2.65), which is consistent with the design values suggested by Casagrande (1973), Terzaghi (1954), and Tschebotarioff (1952).

Example 2.5. Refer to Fig. 2.18. Given q=0, d_1=3 m, d_2=5 m, d_1'= 1.5 m, d_2'=1.5 m, γ=16.5 kN/m^3, $\gamma_{sat(sand)}$=18.5 kN/m^3, ϕ=30°, and $\delta=2\phi/3$. Using the computational pressure diagram method, determine

a. the actual depth of penetration D
b. the maximum design moment M_{design}
c. the design value of the anchor force per unit length of the wall

Solution

Part a.

Effective unit weight of sand below the water table

$$\gamma' \quad \gamma_{sat(sand)} - \gamma_w = 18.5 - 9.81 = 8.69 \text{ kN/m}^3$$

Weighted average value of the unit weight of soil

$$\gamma_{av} = \frac{\gamma d_1 + \gamma' d_2}{d_1 + d_2} = \frac{(16.5)(3) + (8.69)(5)}{3+5} = \frac{49.5 + 43.45}{8}$$

$$= 11.62 \text{ kN/m}^3$$

For ϕ=30° and $\delta = 2\phi/3$, the magnitude of $K_{aH} = K_a \cos(2\phi/3)$. From Table 1.1, K_a=0.297. So

$$K_{aH} = (0.297)\cos(\tfrac{2}{3})(30) = 0.279$$

From Eq. (2.79)

$$W_a \quad CK_{aH}\gamma_{av}d$$

For loose sand without surcharge, assume $C \simeq 0.85$. So

$$W_a = (0.85)(0.279)(11.62)(8) = 22.05 \text{ kN/m}^2$$

Again, from Eq. (2.80)

$$W_p = RCK_{aH}\gamma_{av}d = RW_a$$

For loose sand, assume $R \simeq 0.4$ So

$$W_p = (0.4)(22.05) \quad 8.82 \text{ kN/m}^2$$

Now

$$\beta' = d_1'/d = 1.5/(3+5) = 0.1875$$

From Eq. (2.81)

$$D^2 + 2Dd(1-\beta') \quad (d^2/R)(1-2\beta') \quad 0$$

or

$$D^2 + [(2)(D)(8)(1-0.1875)] - [(8)^2/0.4][1-(2)(0.1875)] = 0$$

$$D^2 + 13D - 100 = 0$$

$$D \sim 5.5 \text{ m}$$

Check on the assumption of R:

$$R \quad \frac{W_p}{W_a} \quad \frac{d(d-2d_1')}{D(2d+D-2d_1')} \quad \frac{8[8-(2)(1.5)]}{5.5[(2)(8)+5.5-(2)(1.5)]}$$

$$\frac{(8)(5)}{5.5(16+5.5-3)} \sim 0.34$$

So, we can assume $R \simeq 0.35$ and recalculate D.

$$D^2 + 2Dd(1-\beta') - (d^2/R)(1-2\beta') \quad 0$$

$\beta' = 0.1875$; $d = 8$ m; $R \simeq 0.35$. So

$$D^2 + 13D - 114.3 = 0$$

$$D \simeq \underline{6 \text{ m}}$$

Part b. From Eq. (2.83)

$$M_{\text{design}} = 0.5W_a d^2 \left[(1 - \frac{RD}{d})^2 - \frac{2d'_1}{d}(1 - \frac{RD}{d})\right]$$

or

$$M_{\text{design}} \quad (0.5)(22.05)(8)^2 \left\{\left[1 \quad \frac{(0.35)(6)}{8}\right]^2 - \left[\frac{(2)(1.5)}{8}\right] \left[1 - \frac{(0.35)(6)}{8}\right]\right\}$$

$$= 705.6(0.544-0.277) = \underline{188.4 \text{ kN-m/m}}$$

Part c. From Eq. (2.82), anchor force

$$T = W_a(d-RD) \quad 22.05[8-(0.35)(6)] = \underline{130.1 \text{ kN/m}}$$

2.13 ANCHORED SHEET PILE WALL PENETRATING INTO CLAY--FREE EARTH SUPPORT METHOD

Figure 2.28 shows an anchored sheet pile wall penetrating into a clay layer having an undrained shear strength of c_u. The backfill above the dredge line is sand. The distribution of the horizontal pressure on the sheet pile wall from $z=0$ to $d(\simeq d_1+d_2)$ will be similar to that shown in Fig. 2.16, or

$$p_1 = \gamma d_1 K_{aH}$$

$$p_2 = (\gamma d_1 + \gamma' d_2)K_{aH}$$

For $z=d$ to $d+D$, the net passive pressure p_6 will from from left to right. Note that p_6 is defined by Eq. (2.49), or

$$p_6 = 4c_u - q'$$

where $q' \quad \gamma d_1 + \gamma' d_2$

For equilibrium

$$\Sigma \text{ Horizontal forces} = 0$$

or

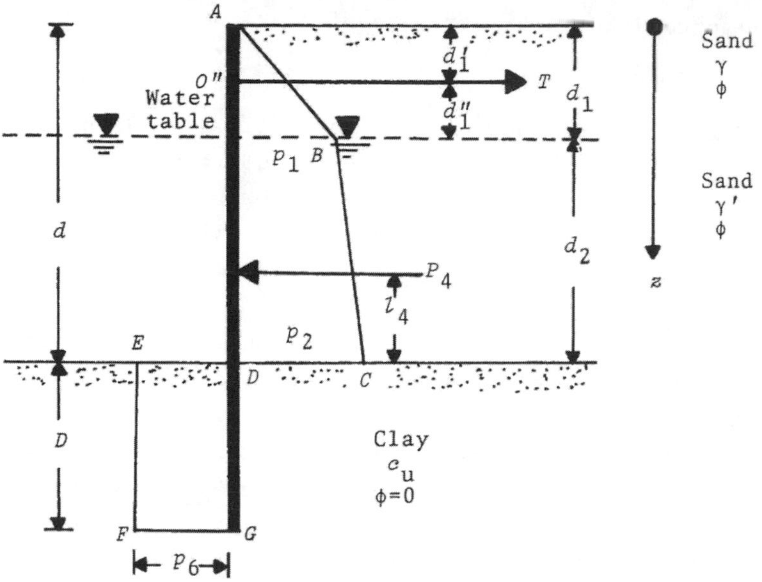

Figure 2.28. Free earth support method of design for sheet pile wall penetrating into clay.

$$P_4 - p_6 D = T \tag{2.84}$$

where P_4 area of the pressure diagram $ABCD$
T anchor force per unit length of the wall

Also, for equilibrium

$$\Sigma \text{ Moment of forces about } O'' = 0$$

So

$$P_4(d - d_1' - l_4) - (p_6 D)[d + (D/2) - d_1'] \quad 0$$

or

$$p_6 D^2 + 2p_6 D(d - d_1') - 2P_4(d - d_1' - l_4) = 0 \tag{2.85}$$

Equation (2.85) can be solved to determine the required depth of penetration. If the actual value of c_u is used to calculate p_6, then the theoretical depth of penetration should be increased by 40-50% in order to determine the design depth of penetration, or

$$D_{actual} = 1.4 \text{ to } 1.5 D_{theory} \tag{2.86}$$

As an alternative, initially a factor of safety should be assigned over the undrained shear strength of clay, or

$$c_{uF} = c_u/F_s$$

This value of c_{uF} may now be used to obtain p_6. Using this value of p_6, Eq. (2.86) may now be solved to determine the design depth of penetration.

The maximum moment of the sheet pile will occur between $z=d_1$ to $z=d+D$. This can be done by using the same procedure outlined in Section 2.8.

Rowe's moment reduction method for choosing sheet pile sections for the case where sheet piles penetrate into a sand layer has been described in Section 2.10. Rowe (1957) has also provided some experimental results for moment reduction of pile penetrating into a clay layer. These plots of M_{design}/M_{max} against the stability number (S_n) for various values of $\log_{10}\rho$ are shown in Fig. 2.29. Note that the flexibility number (ρ) has the same definition as given by Eq. (2.78). The stability number is defined as

$$S_n \quad 1.25c_u/q' \tag{2.86}$$

The term η is defined as

$$\eta = \frac{d}{d+D_{design}} \tag{2.87}$$

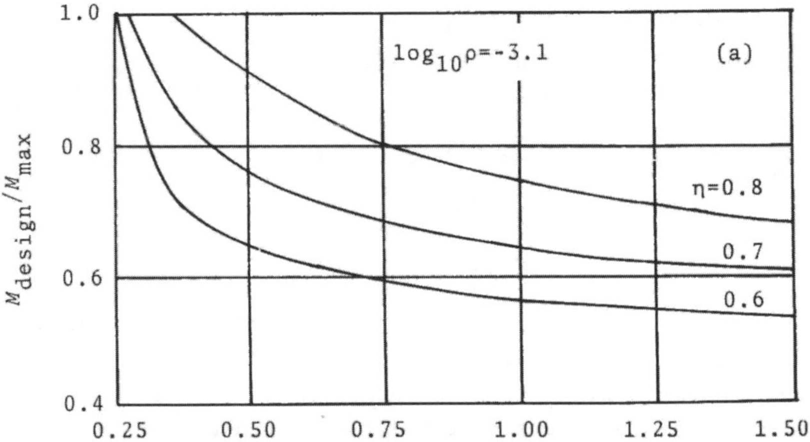

Figure 2.29. Plot of M_{design}/M_{max} vs. S_n for various values of $\log_{10}\rho$ (after Rowe, 1957).

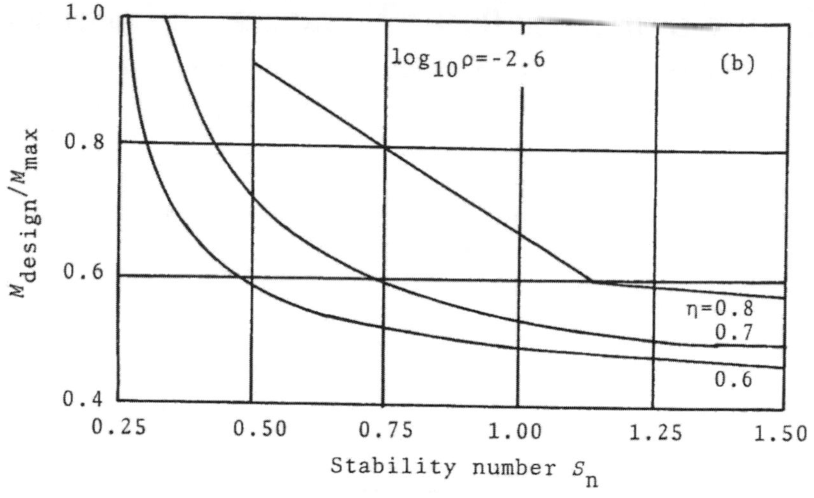

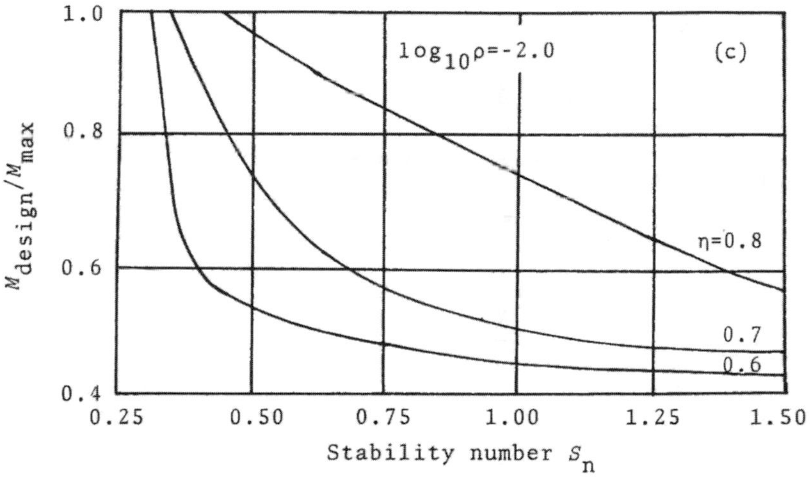

Figure 2.29. (Continued).

For a sheet pile wall, first the values of M_{max}, ρ, S_n, and η can be determined. For given values of S_n and η, the values of M_{design}/M_{max} for various values of $\log\rho$ can be obtained from Fig. 2.29 and plotted in a graph. Once this graph is obtained, the moment reduction procedure will be the same as described in Section 2.10.

2.14 FIXED EARTH SUPPORT METHOD FOR ANCHORED SHEET PILE WALLS

The differences between design procedures for anchored sheet pile walls by free and fixed earth support methods have been presented in Section 2.2 (Figs. 2.5 and 2.6). The details of the *fixed earth support analysis* will be discussed in this section. This type of analysis assumes that the bottom of the sheet pile is restrained from rotating as shown in Fig. 2.6 (that is, the toe of the sheet pile wall is fixed). Figure 2.30 shows a sheet pile wall

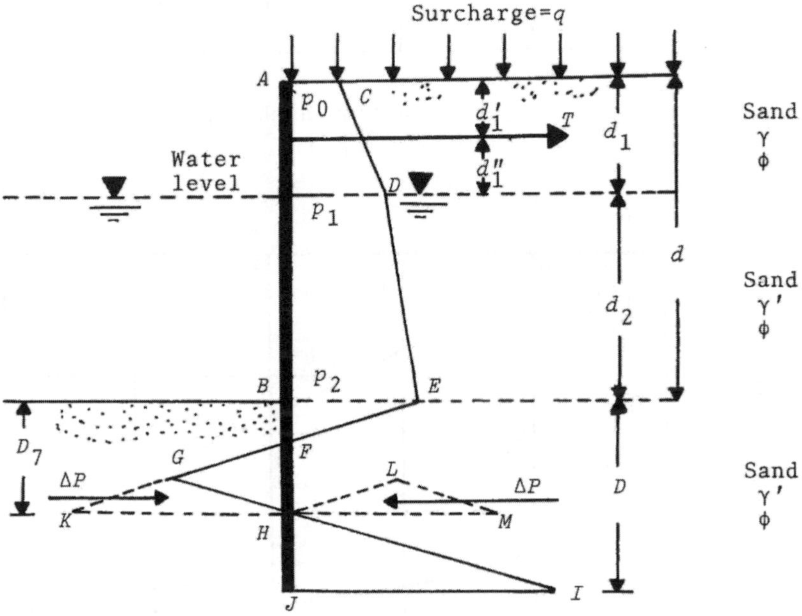

Figure 2.30. Sheet pile wall design by fixed earth support method.

having a granular soil backfill and penetrating into a granular soil. The approximate net horizontal earth pressure on the sheet pile wall is also shown in Fig. 2.30. For simplicity, we can add the areas *GKH* and *HLM* to the lateral pressure distribution diagram without changing the equilibrium. The area of each of these triangles (that is, *GHK* and *HLM*) is equal to ΔP. The problem can further be simplified by assuming that the net pressure distribution diagram on the sheet pile is as shown in Fig. 2.31. Note that the net pressure distribution diagram on the left-hand side of the sheet pile (represented by the triangle *FGH* in Fig. 2.30) has been replaced by the triangle *FKH* in Fig. 2.31. Similarly, the pressure

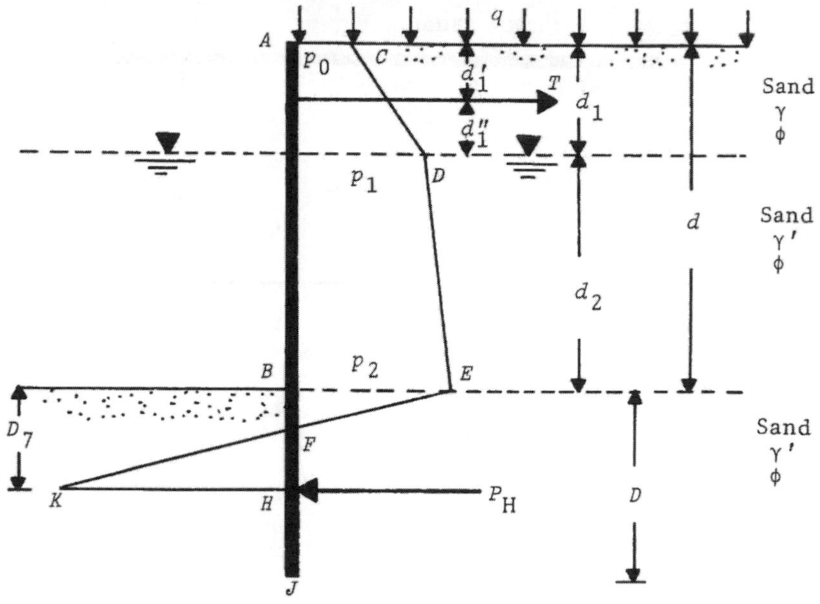

Figure 2.31. Simplification of net horizontal pressure distribution diagram as shown in Fig. 2.30.

distribution diagrams represented by *HLM* and *HIJ* on the right-hand side of the sheet pile as shown in Fig. 2.30 have been replaced by a single line load P_H (per unit length) at *H* in Fig. 2.30. The portion *HJ* of the sheet pile in Fig. 2.31 is assumed to be *fixed*. Terzaghi (1943) used a graphical solution to determine the depth D_7 (Fig. 2.31) which is referred to as the *deflection line method*. Once the depth D_7 is determined, we can approximate that the required depth of penetration $D \simeq 1.2D_7$. However, this method is laborious and is not generally used. Instead, a simplified method based on the above principle is now used. The development of this simplified method, called the *equivalent beam method*, is generally attributed to Blum (1930). The principles of this method will be discussed below.

The fixed end condition for the sheet piles (Fig. 2.31) described above can be compared to a loaded cantilever beam *RTSU* as shown in Fig. 2.32. Note that the support at *T* for the beam is equivalent to the *anchor load reaction* on the sheet pile (Fig. 2.31). It can be seen that the point *S* of the beam *RSTU* is the inflection point of the elastic line of the beam. If the beam is cut at *S* and a free support (reaction P_s) is provided at that point,

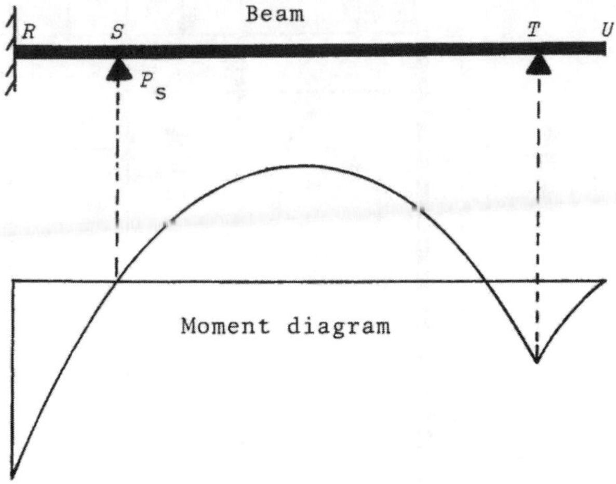

Figure 2.32. Equivalent cantilever beam concept.

the bending moment diagram for portion *STU* of the beam will remain unchanged. This beam *STU* will be equivalent for section *STU* of the beam *RSTU*.

The application of the above reasoning can now be applied to the sheet pile section shown in Fig. 2.33. Figure 2.33a shows the same sheet pile as shown in Fig. 2.31. Note that the net lateral pressure (net of active and passive) distribution diagram has been plotted in Fig. 2.31. However in Fig. 2.33a, the horizontal active and passive pressures are plotted on the right- and left-hand sides of the sheet pile, respectively, or active pressure

$$p_0 \quad qK_{aH}$$

$$p_1 \quad (q+\gamma d_1)K_{aH}$$

$$p_2 \quad (q+\gamma d_1+\gamma' d_2)K_{aH} \tag{2.88}$$

and

$$p_9 \quad (q+\gamma d_1+\gamma' d_2+\gamma' D_7)K_{aH}$$

and passive pressure

$$p_{10} = \gamma' D_7 K_{pH} \tag{2.89}$$

Let point *S* on the sheet pile be the point of *contraflexure* of the elastic line (compare with point *S* of the beam in Fig. 2.32) which is located at a depth *X* below the level of the dredge line. The

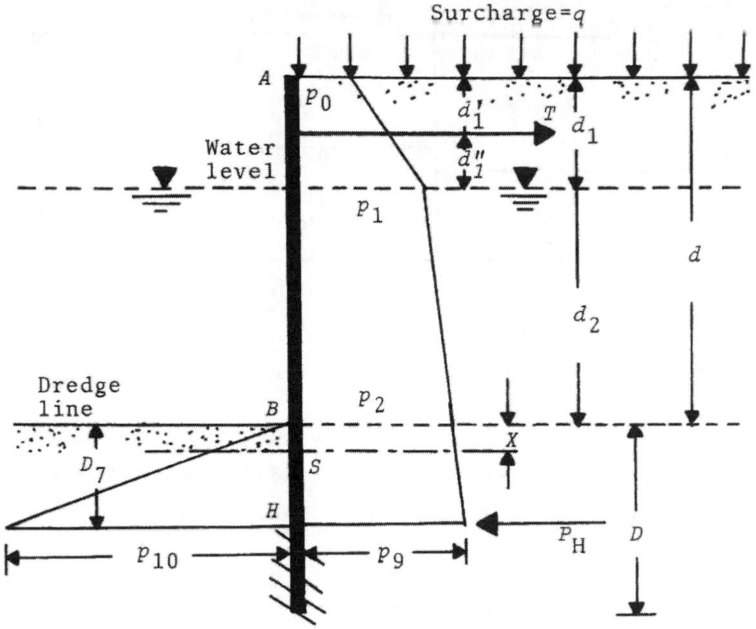

(a)

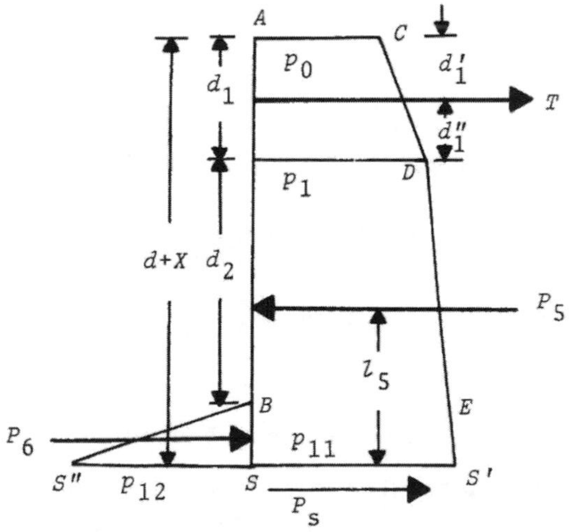

Note:

$\overline{SS}' = p_{11} = [q + \gamma d_1 + (d_2 + X)\gamma']K_{aH}$

$\overline{SS}'' = p_{12} = \gamma' X K_{pH}$

(b)

Figure 2.33. Equivalent beam method.

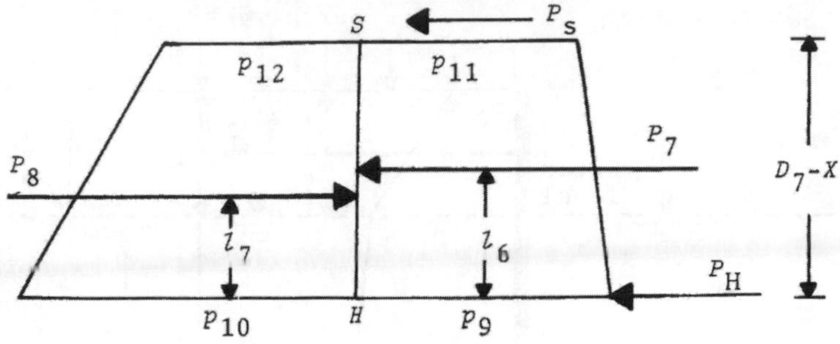

(c)

Figure 2.33. (Continued).

The elastic line method for sheet pile wall analysis has shown (Blum, 1930) that the variation of X/d with the soil friction angle will be as shown in Fig. 2.34. So, knowing d and ϕ, the magnitude

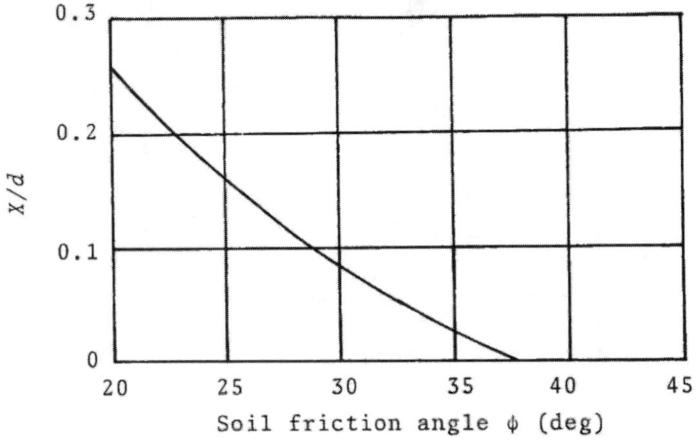

Figure 2.34. Plot of X/d vs. soil friction angle.

of X can be determined by using Fig. 2.34.

The following step-by-step procedure can now be used.

1. Referring to Fig. 2.33b, determine the force P_5 (per unit length of the wall) which is the area of the pressure diagram *ACDS'S* and also l_5 which defines the location of the *line of action of the resultant*.
2. Determine the force P_6 (per unit length of the wall) which is the area of the pressure diagram *BSS"* (Fig. 2.33b).

3. From Fig. 2.33b

$$\Sigma \text{ Moment of all forces about } S \quad 0$$

So

$$P_5 l_5 - T(d_2 + X + d_1'') - P_6 (X/3) = 0$$

Or; anchor force per unit length of wall

$$T = \frac{P_5 l_5 - P_6 (X/3)}{d_2 + X + d_1''} \tag{2.90}$$

4. Again, referring to Fig. 2.33b

$$\Sigma \text{ Horizontal forces} = 0$$

or

$$P_5 - P_6 - T - P_s = 0$$

So

$$P_s = P_5 - P_6 - T \tag{2.91}$$

5. Now, referring to Fig. 2.33c, determine the forces P_7 and P_8 (per unit length of the wall)

$$P_7 = \left(\frac{p_9 + p_{11}}{2}\right)(D_7 - X) \tag{2.92}$$

$$P_8 = \left(\frac{p_{10} + p_{12}}{2}\right)(D_7 - X) \tag{2.93}$$

6. Referring to Fig. 2.33c, determine the location of the line of action of the resultant forces P_7 and P_8, or

$$l_6 \quad \left(\frac{D_7 - X}{3}\right)\left(\frac{p_9 + 2p_{11}}{p_9 + p_{11}}\right) \tag{2.94}$$

and

$$l_7 \quad \left(\frac{D_7 - X}{3}\right)\left(\frac{p_{10} + 2p_{12}}{p_{10} + p_{12}}\right) \tag{2.95}$$

7. Taking the moment about H (Fig. 2.33c)

$$P_s (D_7 - X) + P_7 l_6 - P_8 l_7 \quad 0 \tag{2.96}$$

The preceding equation can now be solved to determine D_7.

8. The depth of penetration can now be given as

$$D \sim 1.2 D_7 \tag{2.97}$$

As in the case of an anchored sheet pile wall design by free earth support method, a factor of safety over K_{pH} can be applied initially and calculations made. In that case, the depth of penetration D, as given by Eq. (2.97), is the design depth of penetration. An alternative to this method is to use the actual K_{pH} in all calculations and then increase the theoretical depth of penetration [Eq. (2.97)] by about 20-30%.

It should be pointed out that the fixed earth support method is applicable only in the case where the sheet piles penetrate into a sand layer. This method of design should not be used when the surcharge is very large compared to the height of the wall measured above the dredge line. This also should not be used when the difference of water levels between the land and water sides is large. Cornfield (1969) has provided nomograms for design of sheet pile walls by the fixed earth support method of design for ϕ=30° and 35°.

Based upon several large scale model tests performed under the sponsorship of the Bureau of Yards and Docks of the United States Navy, Tschebotarioff (1973) proposed a *simplified equivalent beam method* of design for sheet pile walls. This method is also sometimes referred to as the *hinge-at-the-dredge line method*. According to this method, a hinge may be assumed to exist along the sheet pile wall at the level of the dredge line as shown in Fig. 2.35. For a factor of safety of 2 against *toe kick-out* of the sheet pile wall, the depth of penetration can be given as

$$D = 0.43[d+(q/\gamma)] \tag{2.98}$$

where q surcharge

γ = unit weight of soil

Note that the term q/γ in Eq. (2.96) is the equivalent height of soil in lieu of the surcharge q. The concept of equivalent beam is used to obtain the magnitude of T (anchor force per unit length of the wall).

2.15 OTHER OBSERVATIONS ON ANCHORED SHEET PILE WALL DESIGN

Tsinker (1983) has observed that the actual behavior of anchored sheet pile walls may be very different from that calculated theoretically. According to him, attempts to make precise mathematical evaluations of an anchored sheet pile walls can result in misleading and dangerous conclusions. Budin and Demina (1979) have made many large-scale field observations of sheet pile struc-

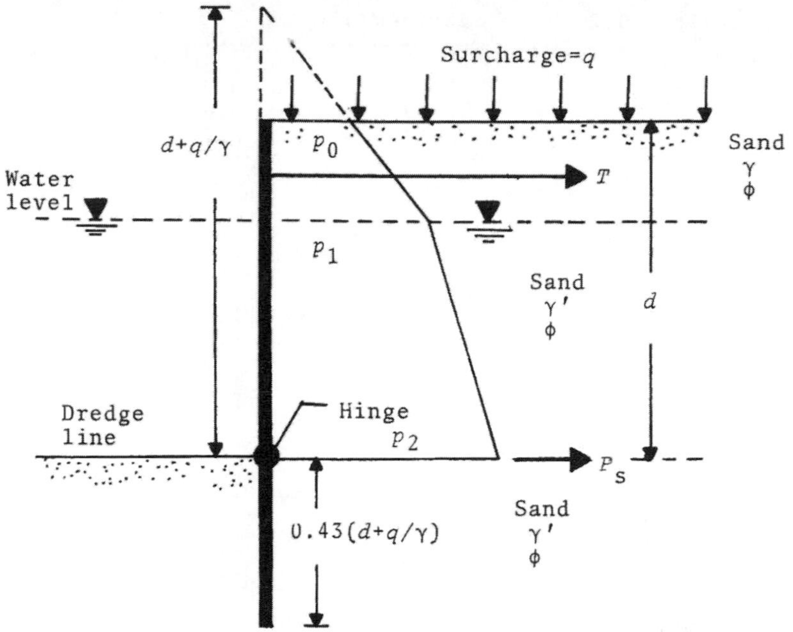

Figure 2.35. Hinge-at-the-dredge-line method of sheet pile wall design.

tures. Based on these observations, they have recommended that for design the following relationships for the design moment (M_{design}) on sheet piles and the design anchor force (T_{design}) may be used

$$M_{design} = M_{max}\lambda_m \tag{2.99}$$

$$T_{design} = T\lambda_a \tag{2.100}$$

where λ_m and λ_a are empirical factors

The values of λ_m and λ_a as suggested by Budin and Demina (1979) and later summarized by Tsinker are as follows. For notation in the following tabulations, refer to Figs. 2.8 and 2.9.

Values of λ_m:

For $d_1'/(d-d_1') \leq 0.25$:

$d_3/(d-d_1')$	λ_m
1	0.6
2/3	0.65
1/3	0.7
0	0.74

For $d_1'/(d-d_1') > 0.25$:

$d_3/(d-d_1')$	λ_m
1	0.56
2/3	0.6
1/3	0.65
0	0.7

Values of λ_a:

For $d_1'/(d-d_1') \leq 0.25$:

$d_3/(d-d_1')$	λ_a
1	1.65
2/3	1.6
1/3	1.5
0	1.4

For $d_1'(d-d_1') > 0.25$:

$d_3/(d-d_1')$	λ_a
1	1.7
2/3	1.65
1/3	1.55
0	1.45

Lasebnik (1961) has also suggested an empirical procedure for estimation of M_{design} and T_{design}. This will not be discussed here however the readers may refer to Tsinker (1983) for a summary of Lasebnik's conclusions.

References

Blum, H., 1955. Berechnung einfach oder nicht verankerter spundwande in Grundbau Taschenbuch. Berlin, Ernst, 1:328-333.
Blum, H., 1951. Beitrag zur berechnung von bohlwerken mit berucksichtgung der wandverformung. Berlin, Ernst.
Blum, H., 1930. Einspannungsverhaltnisse bei bohlwerken. Diss. Tech., Hochschule Braunschweig.
Brinch Hansen, J., 1953. Earth pressure calculation. Danish Tech. Press, Copenhagen, Denmark.
Budin, A.Y. and Demina, G.A. (1979). Quays handbook. Moscow (in Russian).
Casagrande, L., 1973. Comments on conventional design of retaining structures. J. Geotech. Eng. Div., ASCE, 99(SM2):181-198.
Caquot, A. and Kerisel, J., 1948. Tables for the calculation of passive pressure, active pressure, and bearing capacity of foundations. Gauthier-Villars, Paris, France.
Cornfield, G.M., 1975. Sheet pile structures. Foundation Engineering Handbook, H.F. Wintercorn and H.Y. Fang (Editors). Van Nostrand-Reinhold, New York, pp. 418-444.
Cornfield, G.M., 1969. Direct-reading nomograms for design of anchored sheet pile retaining walls. Civ. Eng. Public Works Review, August, London.
Jarquio, R., 1981. Total lateral surcharge pressure due to strip load. J. Geotech. Eng. Div., ASCE, 107(GT10):1424-1428.
Jumikis, A.R., 1971. Foundation Engineering. Intext Educational Publishers, Scranton, Pennsylvania.

Lasebnik, G.Y., 1969. Investigation of anchored sheet pile bulkheads. Ph.D. thesis, Kiev Politechnical Institute, Kiev, U.S.S.R.
Masih, R., 1984. Graphical solution for sheet pile embedment. J. Geotech. Eng., ASCE, 110(GT4):534-538.
Nataraj, M.S. and Hoadley, P.G., 1984. Design of anchored bulkheads in sand. J. Geotech. Eng., ASCE, 110(GT4):505-515.
Packshaw, S., 1954. Discussion on anchored bulkheads, by Terzaghi, K. Transactions, ASCE, 119:1301-1307.
Packshaw, S. and Lake, J.O., 1952. Correspondence on "Anchored sheet pile walls" by Rowe, P.W. Proc., Institute of Civil Engineers, London, England, 1(1):621-633.
Rowe, P.W., 1957. Sheet pile walls in clay. Proc., Institute of Civil Engineers, London, England, 7:629-654.
Rowe, P.W., 1952. Anchored sheet pile walls. Proc., Institute of Civil Engineers, London, England, 1(1):27-70.
Schroeder, W.J. and Roumillac, P., 1983. Anchored bulkheads with sloping dredge lines. J. Geotech. Eng., ASCE, 109(GT6):845-851.
Steenfelt, J.S. and Hansen, B., 1984. Sheet pile design of earth pressure for strip load. J. Geotech. Eng., ASCE, 110(GT7):976-986.
Terzaghi, K., 1954. Anchored bulkheads. Transactions, ASCE, 119:1243-1324.
Terzaghi, K., 1943. Theoretical soil mechanics. John Wiley and Sons, New York.
Tschebotarioff, G.P., 1973. Foundations, retaining structures, and earth structures. McGraw-Hill Book Company, New York.
Tschebotarioff, G.P., 1952. Soil mechanics, foundations and earth structures. McGraw-Hill Book Company, New York.
Tschebotarioff, G.P., 1949. Large scale model earth pressure tests on flexible bulkheads. Transactions, ASCE, 114:415-454.
Tsinker, G.P., 1983. Anchored sheet pile bulkheads: design practice. J. Geotech. Eng., ASCE, 109(GT8):1021-1038.

CHAPTER 3

Holding Capacity of Anchor Slabs and Helical Anchors

3.1 INTRODUCTION

In the construction of earth retaining structures such as sheet pile walls (Fig. 3.1a), it is necessary to use *vertical anchor slabs*. These anchors are usually made of reinforced concrete slabs and they are often arranged in a row as shown in Fig. 3.1b. In many instances, *horizontal anchor slabs* and also *helical anchors* (Fig. 3.1c and d) are used to ensure stability of foundations subjected to vertical uplifting forces. A typical example where such type of anchoring may be necessary is a transmission tower foundation. A relatively limited amount of research has been done so far to evaluate the holding capacity of anchors. The results of most of the related research work available at the present time will be summarized in this chapter.

3.2 ULTIMATE HOLDING CAPACITY OF SHALLOW VERTICAL ANCHOR SLABS IN SAND

Figure 3.2 shows an anchor slab with a height h and a width B measured at right angles to the cross section shown. The slab is embedded in a sand having a unit weight γ and angle of friction ϕ. P_u is the ultimate load at which the soil in front of the slab will fail. At failure, the anchor will be subjected to a *passive force* derived from the soil located in front of the slab and an *active force* due to the soil located behind the slab. The directions of the active and passive forces are also shown in Fig. 3.2. Thus, in general

$$P_u = \begin{pmatrix}\text{horizontal component}\\ \text{of the passive force}\end{pmatrix} - \begin{pmatrix}\text{horizontal component}\\ \text{of the active force}\end{pmatrix} \tag{3.1}$$

When the anchor is located at a *shallow depth* (that is, a smaller

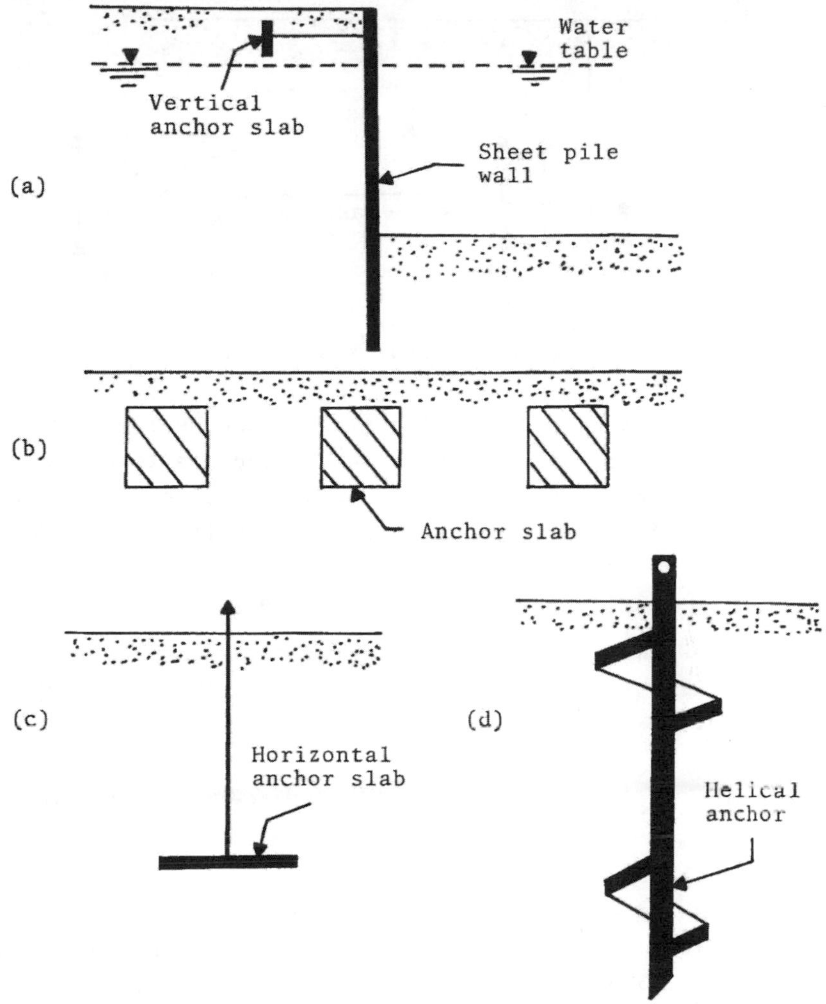

Figure 3.1. (a) Vertical anchor slab in a sheet pile wall; (b) Vertical anchor slabs in a row; (c) Horizontal anchor slab; (d) Helical anchor.

H/h), the failure surfaces in soil located in the front and back of the wall will extend to the ground surface. These types of anchors are referred to as *shallow anchors*. However, if the embedment ratio H/h is relatively large, *local shear failure* in soil around the anchor occurs at ultimate load, and such anchors may be referred to as *deep anchors*. For loose sands, anchors with $H/h \leq 4$ may be considered as shallow anchors. For dense sands, anchors with $H/h \leq$ about 6 behave as shallow anchors. Investigators such

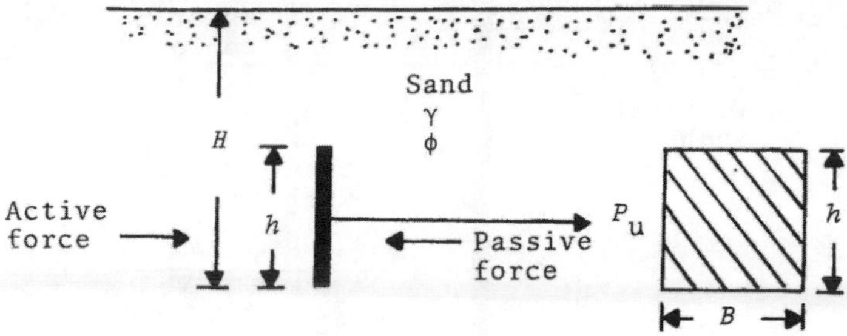

Figure 3.2. Shallow vertical anchor slab in sand.

as Teng (1962), Ovesen and Stromann (1972), and Neeley, Stuart, and Graham (1973) have proposed theoretical procedures to calculate the ultimate holding capacity of shallow anchor slabs embedded in sand. The principles of these procedures are given below.

3.2.1 Teng's Procedure for Shallow Anchors

Teng (1962) has suggested that for *continuous anchors* (plane strain case) with the embedment ratio H/h less than or equal to about 1.5 to 2

$$P_u = B(P_p - P_a) \tag{3.2}$$

where P_p Rankine passive force in front of the slab
P_a Rankine active force behind the slab

The distribution of the Rankine active and passive earth pressures is shown in Fig. 3.3. It has been shown in Chapter 1 that the

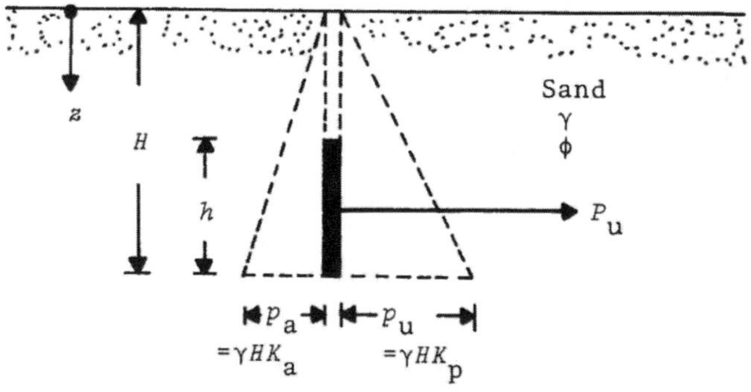

Figure 3.3. Shallow continuous anchor slab--Teng's theory for determination of ultimate load (1962).

Rankine active and passive pressures at a depth z below the ground surface can be given as

$$p_a \quad \gamma z \tan^2(45-\phi/2) = \gamma z K_a \tag{3.3}$$

and

$$p_p = \gamma z \tan^2(45+\phi/2) = \gamma z K_p \tag{3.4}$$

where p_a Rankine active pressure at a depth z

p_p Rankine passive pressure at a depth z

Hence

$$P_a = \int_0^H p_a dz = \tfrac{1}{2}\gamma H^2 \tan^2(45-\phi/2) \tag{3.5}$$

and

$$P_p \quad \int_0^H p_p dz = \tfrac{1}{2}\gamma H^2 \tan^2(45+\phi/2) \tag{3.6}$$

For an anchor slab at ultimate load with limited width-to-height ratio (B/h), a three-dimensional failure surface in soil will be observed. Hence, if the soil friction at the sides of the slab is taken into account, the ultimate load can be expressed by the following equation

$$P_u = B(P_p-P_a) + \tfrac{1}{3}K_o\gamma(\sqrt{K_p}+\sqrt{K_a})H^3\tan\phi \tag{3.7}$$

where K_o at-rest earth pressure coefficient

Teng (1961) has suggested an approximate value of $K_o \approx 0.4$ for use in Eq. (3.7).

Example 3.1. Refer to Fig. 3.3. Given H=2 m, h=1 m, B=1 m. For the soil, γ=17.2 kN/m^3, ϕ=30°. Estimate the holding capacity of the anchor.

Solution

$$K_a = \tan^2(45-\phi/2) = \tan^2(45-30/2) = \tfrac{1}{3}$$

$$K_p = \tan^2(45+\phi/2) = \tan^2(45+30/2) = 3$$

So

$$P_p = \tfrac{1}{2}\gamma H^2 K_p \quad \tfrac{1}{2}(17.2)(2)^2(3) = 103.2 \text{ kN/m}$$

$$P_a \quad \tfrac{1}{2}\gamma H^2 K_a \quad \tfrac{1}{2}(17.2)(2)^2(\tfrac{1}{3}) \simeq 11.47 \text{ kN/m}$$

$\sqrt{K_a} = \sqrt{\tfrac{1}{3}} = 0.577$; $\sqrt{K_p} \quad \sqrt{3} = 1.732$

H/h=2/1=2. Hence, Eq. (3.7) is applicable. Using $K_o \approx 0.4$

$$P_u = B(P_p - P_a) + \tfrac{1}{3}K_o\gamma(\sqrt{K_p}+\sqrt{K_a})H^3\tan\phi$$

$$= (1)(103.2-11.47) + \tfrac{1}{3}(0.4)(17.2)(1.732+0.577)(2)^3\tan 30$$

$$= 91.73+24.46 = \underline{116.19\ \text{kN}}$$

3.2.2 Ovesen and Stromann's Procedure for Shallow Anchors

The procedure for calculation of the ultimate holding capacity of anchors in sand as proposed by Ovesen and Stromann (1972) is primarily based on the model test results conducted by Ovesen (1964). In order to understand the procedure, let us consider a continuous anchor slab of height H, and let the top of the slab coincide with the ground surface as shown in Fig. 3.4. The ultimate load per unit length of the slab is equal to P'_u. The failure

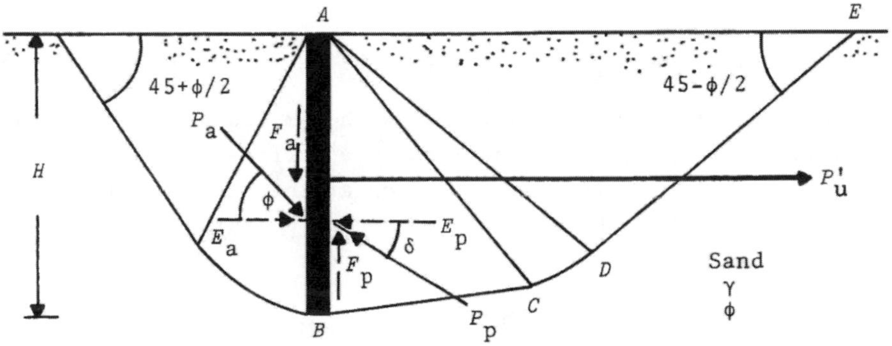

Figure 3.4. Failure surface in soil for basic case of a continuous anchor.

surface in soil at ultimate load is also shown in Fig. 3.4. For translation of the slab, the active force behind the slab can be calculated by using any ordinary active earth pressure theory applicable for a rough retaining wall. Thus

$$E_a \quad P_a\cos\phi \tag{3.8}$$

$$F_a \quad P_a\sin\phi \tag{3.9}$$

where P_a = active force per unit length of the wall

E_a = horizontal component of P_a

F_a = vertical component of P_a

ϕ = angle of soil friction

The passive failure surface in soil in front of the slab as assumed by Ovesen (1964) is shown in Fig. 4. It consists of a straight rupture line BC through the foot of the slab and a zone of rupture.

The zone of rupture consists of a Rankine passive zone *ADE* and a Prandtl radial shear zone *BCD*. The Rankine pressure zone consists of two systems of parallel rupture lines which intersect each other at an angle of $90 \pm \phi$. The Prandtl zone consists of straight lines passing through point *A* and a system of logarithmic spiral arcs with point *A* as their pole. This kind of failure in soil is re- rerred to as the *SfP* rupture. The horizontal and vertical components of the passive force derived in this manner can be given as

$$E_p \quad \tfrac{1}{2}\gamma H^2 K_{pH} \tag{3.10}$$

and

$$F_p \quad \tfrac{1}{2}\gamma H^2 K_{pH} \tan\delta \tag{3.11}$$

where E_p horizontal component of passive pressure per unit length of the slab
F_p = vertical component of passive pressure per unit length of the slab
K_{pH} = horizontal passive earth pressure coefficient
δ = slab-soil friction angle

The value of δ can be determined by means of the vertical equilibrium condition of the slab, or

$$F_p \quad W + F_a \tag{3.12}$$

where W = weight per unit length of the anchor slab
Substituting Eq. (3.11) into the preceding equation, we obtain

$$K_{pH}\tan\delta = \frac{W + F_a}{\tfrac{1}{2}\gamma H^2} \tag{3.13}$$

Figure 3.5 gives a plot of K_{pH} vs. $K_{pH}\tan\delta$ for various soil friction angles, ϕ. Once the value of K_{pH} is known, the magnitude of E_p can be easily estimated by using Eq. (3.10). Now, considering the horizontal equilibrium of the slab

$$P'_u = E_p - E_a$$

$$\quad \tfrac{1}{2}\gamma H^2 K_{pH} - E_a \tag{3.14}$$

This value of P'_u determined by Eq. (3.14) is referred to as the holding capacity for the basic case.

Now, consider a continuous anchor slab as shown in Fig. 3.6, for which $h \leq H$. This is referred to as the *strip case*. The ultimate holding capacity per unit length of the slab for the *strip*

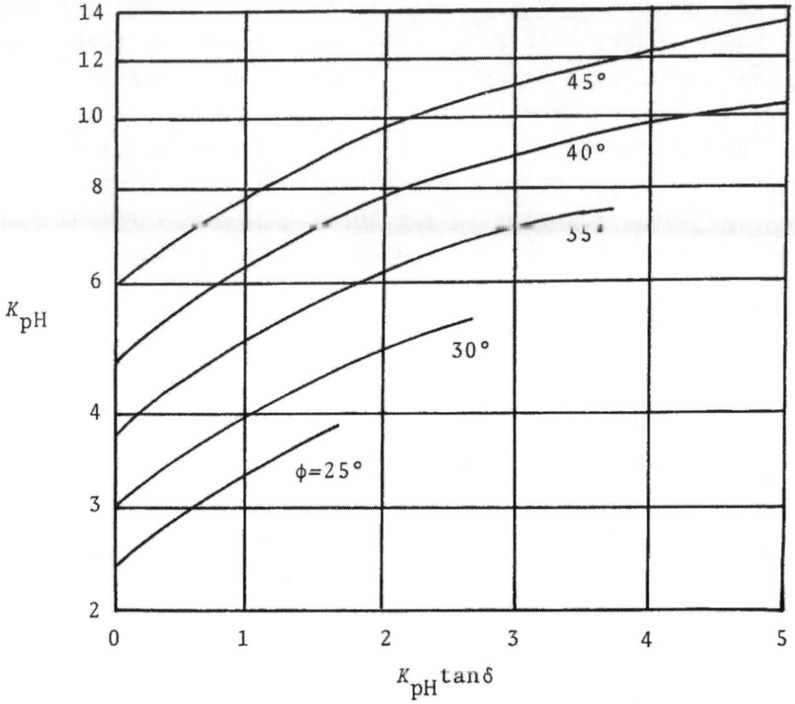

Figure 3.5. Variation of K_{pH} vs. $K_{pH}\tan\delta$ for various values of soil friction angle (after Ovesen and Stromann, 1972).

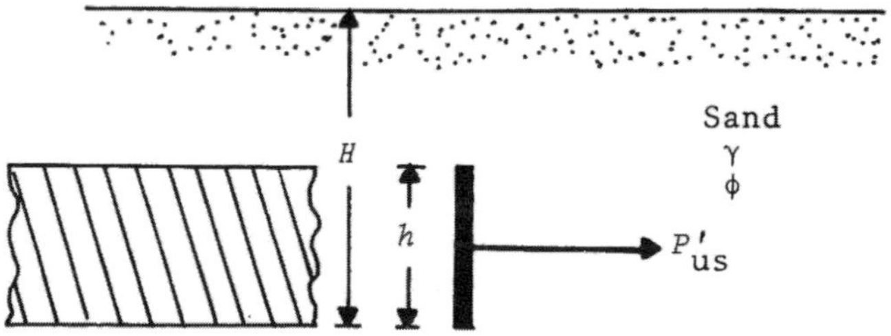

Figure 3.6. Strip case for a vertical anchor.

case is equal to P'_{us}. Based on experimental evidence, Ovesen and Stromann (1972) have given a nondimensional plot for the variation of P'_{us}/P'_{u} and h/H for an anchor slab embedded in loose and dense sand. This variation of P'_{us}/P'_{u} can be approximated as (Dickin and Leung, 1985)

$$R_{ov} = \frac{P'_{us}}{P'_u} = \frac{C_{ov}+1}{C_{ov}+(H/h)} \tag{3.15}$$

where C_{ov} = 19 for dense sand and 14 for loose sand

Hence with known values of P'_u and H/h, the magnitude of P'_{us} can be estimated. At this time, it needs to be pointed out that most anchor slabs used in actual construction cannot be considered to be equivalent to the strip case due to the fact that they have limited width-to-height ratios (B/h). Hence the actual holding capacity, P_u, of the anchor will be greater than $P'_{us}B$ due to a three-dimensional failure surface in soil. In order to take this into account, an *equivalent width* concept for the anchor may be introduced. This is shown in Fig. 3.7 in which B_e is the equivalent

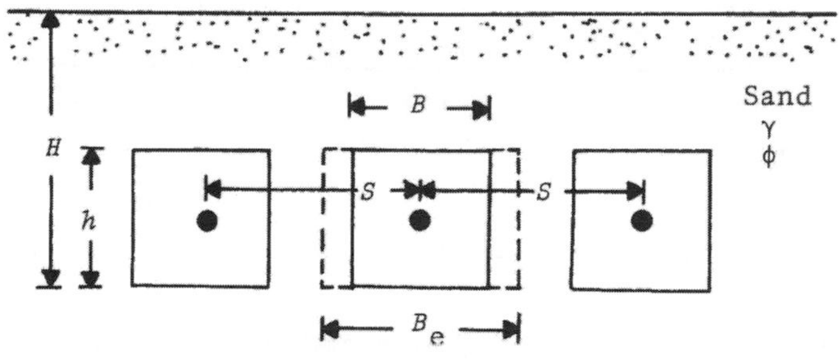

Figure 3.7. Equivalent width of a vertical anchor.

width. Note that B_e will also be a function of the center-to-center spacing of the slabs. Based on laboratory model and field tests, Ovesen and Stromann (1972) have given a nondimensional plot of $(B_e-B)/(H+h)$ vs. $(S-B)/(H+h)$ which is shown in Fig. 3.8. In this figure, note that S is the center-to-center spacing of the anchor slabs. Once the value of B_e is estimated from Fig. 3.8, the ultimate load for an anchor slab can be estimated as

$$P_u = P'_{us}B_e \tag{3.16}$$

Following is a step-by-step procedure to obtain P_u.

1. Determine H, h, B, and S for the anchor slabs and, also, the values of γ and ϕ for the soil.
2. Determine E_a and F_a [Eqs. (3.8) and (3.9)]. The value of P_a in Eqs. (3.8) and (3.9) can be determined from any standard graph or table given in Chapter 1.

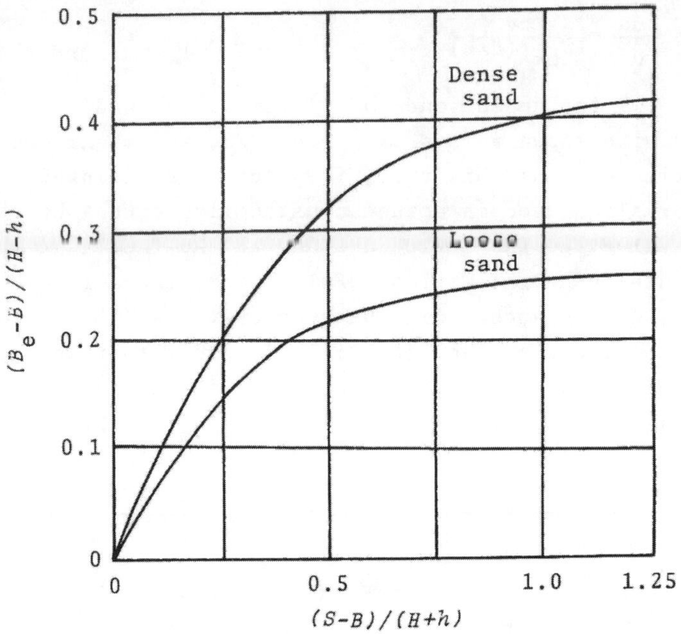

Figure 3.8. Variation of $(B_e-B)/(H-h)$ with $(S-B)/(H+h)$ (after Ovesen and Stromann, 1972).

3. Determine W for the anchor slab. Note that

$$W \quad Ht\gamma_c \tag{3.17}$$

where γ_c = unit weight of the slab material
t thickness of the slab

4. Calculate $K_{pH}\tan\delta$ from Eq. (3.13).
5. Determine the value of K_{pH} from Fig. 3.5.
6. Calculate P'_u from Eq. (3.14).
7. Using Eq. (3.15), estimate P'_{us} for the known value of H/h.
8. Using Fig. 3.8, obtain B_e.
9. Calculate the holding capacity of the anchor slab by using Eq. (3.16).

Example 3.2. Consider the vertical anchors given in Example 3.1. Assume that the anchors have a thickness of 0.15 m and are made of concrete. The unit weight of concrete γ_c=23.5 kN/m^3. Calculate the ultimate holding capacity of the anchor for S=1 m, 2 m, 3 m, and 4 m. Use the procedure suggested by Ovesen and Stromann (1972).

Solution

From Example 3.1, H=2 m, h=1 m, γ=17.2 kN/m^3, and ϕ=30°.

We will use Coulomb's earth pressure theory to calculate P_a for the basic case. From Eq. (1.32), using β=90°, δ=ϕ=30°, and α=0°, $K_a \simeq 0.296$. So

$$P_a = \frac{1}{2}\gamma H^2 K_a$$

Hence

$$P_a = \frac{1}{2}(17.2)(2)^2(0.296) = 10.182 \text{ kN/m}$$

From Eqs. (3.8) and (3.9)

$$E_a = P_a \cos\phi = (10.182)(\cos 30) \quad 8.817 \text{ kN/m}$$

$$F_a = P_a \sin\phi = (10.182)(\sin 30) = 5.091 \text{ kN/m}$$

From Eq. (3.17)

$$W = Ht\gamma_c$$

Given t = 0.15 m, so

$$W = (2)(0.15)(23.5) \quad 7.05 \text{ kN/m}$$

From Eq. (3.13)

$$K_{pH}\tan\delta = \frac{W+F_a}{\frac{1}{2}\gamma H^2} = \frac{7.05+5.091}{(\frac{1}{2})(17.2)(2)^2} = 0.353$$

For $K_{pH}\tan\delta = 0.353$ and ϕ=30°, the value of K_{pH} can be obtained from Fig. 3.5 to be about 3.35. So, using Eq. (3.14)

$$P'_u \quad \frac{1}{2}\gamma H^2 K_{pH} - E_a = (\frac{1}{2})(17.2)(2)^2(3.35) - 8.817 = 106.42 \text{ kN/m}$$

For the strip case, H/h=2/1=2.

In order to obtain P'_{us}, let us consider that the sand is loose. Now, from Eq. (3.15)

$$\frac{P'_{us}}{P'_u} = \frac{C_{ov}+1}{C_{ov}+(H/h)} \quad \frac{14+1}{14+2} \quad 0.938$$

So, P'_{us}=(0.938)(106.42)=99.82 kN/m.

Now, the following table can be prepared to calculate P_u for various center-to-center spacings of the anchors.

Note: There are several things to observe from this solution, and they are:

1. The value of P_u does not increase when $S \geq 3B$.
2. The value of P_u as calculated in Example 3.1 is for the

S (m)	B (m)	$S-B$ (m)	$H+h$ (m)	$\frac{S-B}{H+h}$	$\frac{B_e-B}{H+h}$ [a]	B_e (m)	$P_u=P'_{us}B_e$ [b] (kN)
1	1	0	3	0	0	1	99.82
2	1	1	3	0.333	0.09	1.27	126.77
3	1	2	3	0.67	0.24	1.72	171.7
4	1	3	3	1.0	0.24	1.72	171.7

[a]From Fig. 3.8 (for loose sand)
[b]P'_{us}=99.82 kN/m

condition $S \geq 3B$ which means that the interference due to the anchors located on both sides of this anchor does not exist. However, the value of P_u determined from Example 3.1 is smaller than that calculated in this problem.

3. Teng's method for determination of the ultimate holding capacity of anchors is conservative.

3.2.3 Analysis of Shallow Anchors By Biarez, Boucraut and Negre

Biarez, Boucraut, and Negre (1965) have presented the calculation methods for obtaining the ultimate load of strip vertical anchor slabs by means of limit analyses. The resistance of shallow anchors ($H/h<4$) is dependent on the anchor roughness and the self-weight of the anchor. Dickin and Leung (1985) have expressed the original expression of Biarez et al. (1965) in the following simplified form

$$M_{\gamma q} = (K_{pH}-K_{aH})\left(\frac{H}{h}-\frac{1}{2}\right) + \frac{K_{pH}\sin 2\phi}{2\tan(45+\phi/2)}\left(\frac{H}{h}-1\right)^2 \qquad (3.18)$$

In the preceding expression

$M_{\gamma q}$ = force coefficient = $P'_{us}/\gamma h^2$
P'_{us} = ultimate load per unit length of the anchor
K_{pH}, K_{aH} = horizontal components of passive and active earth pressure coefficients, respectively

An expression for determination of the ultimate load of shallow square anchors has also been given by Biarez et al. (1965). In a simplified form, it can be expressed as (Dickin and Leung, 1985)

$$M_{\gamma q(S)} = \frac{P_u}{\gamma B h^2} \quad M_{\gamma q} + \phi\left(\frac{h}{B}\right)\left(\sqrt{K_{pH}}-\sqrt{K_{aH}}\right)\left(\frac{H}{h}-\frac{2}{3}\right)$$

$$+ \frac{1}{2}(1+\phi)\left(\frac{h}{B}\right)(K_{pH})\sin 2\phi\left(\frac{H}{h} - 1\right) \quad (3.19)$$

where $M_{\gamma q(S)}$ force coefficient of square anchors

3.2.4 Meyerhof's Theory

Meyerhof (1973) analyzed the ultimate capacity of inclined anchors in sand--both for shallow and deep embedment conditions. According to his theory, the ultimate load per unit length (P_u') of a *shallow vertical strip anchor* slab can be given by the following relationship (with reference to Fig. 3.3)

$$P_u = \frac{1}{2}\gamma D_o^2 K_b B \quad (3.20)$$

where D_o = $H-h/2$

K_b = earth pressure coefficient

B = width of the anchor slab

The interpolated values of K_b from the work of Meyerhof (1973) are given in Fig. 3.9.

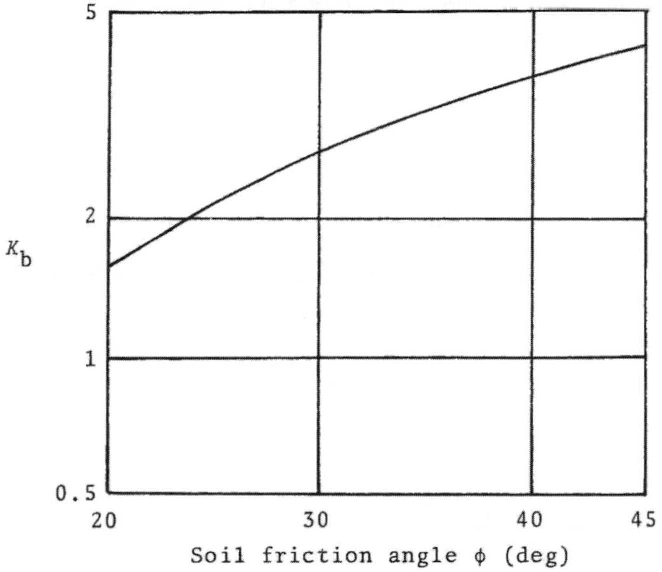

Figure 3.9. Meyerhof's values of K_b for use in Eq. (3.20).

3.2.5 Neeley, Stuart and Graham's Method

Neeley, Stuart and Graham (1973) have given a theoretical solution for estimation of the ultimate holding capacity of shallow vertical anchors in sand. The solution has been developed by considering only the passive pressure mobilized in front of the anchor slab. The nature of failure surface as obtained by this method is shown in Fig. 3.10. This method uses the equivalent free surface

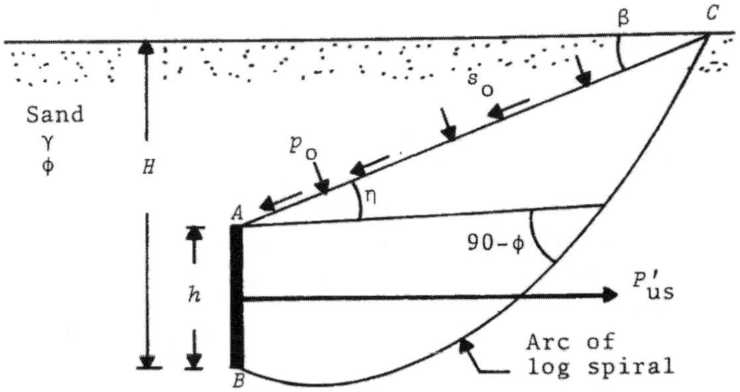

Figure 3.10. Holding capacity of a vertical strip anchor as analyzed by Neeley et al. (1973).

concept as proposed by Meyerhof (1951) for bearing capacity analysis of foundations. According to this theory, the shear stress along line *AC* (Fig. 3.10) may be given as

$$s_o = mp_o\tan\phi \tag{3.21}$$

where s_o and p_o = *mobilized* shear and normal stress on the *equivalent free surface AC*

m = degree of mobilization of shearing resistance on the equivalent free surface

The magnitude of m may vary between *zero* and *one*.

From Fig. 3.10, it can be seen that

$$\frac{H}{h} = \frac{1 + (\sin\beta\cos\phi)e^{\theta\tan\phi}}{\cos(\phi+\eta)} \tag{3.22}$$

With the above assumptions and using the *stress characteristic solution* procedure, Neeley et al. (1973) have expressed the ultimate holding capacity for plane strain case in the form of a nondimensional force coefficient, or

$$M_{\gamma q} = \frac{P'_{us}}{\gamma h^2} \tag{3.23}$$

where $M_{\gamma q}$ force coefficient

P'_{us} = ultimate load per unit width of the strip anchor slab

The force coefficient, $M_{\gamma q}$, is a function of H/h, ϕ, and m.

Das (1983) has suggested that the solutions of Neeley et al. (1973) can be expressed in a more convenient form as

$$N_q = \frac{P'_{us}}{h\gamma H} \tag{3.24}$$

where N_q = breakout factor

Using the solution of Neeley et al. (1973), the variation of N_q has been calculated and is shown in Fig. 3.11. In a similar manner,

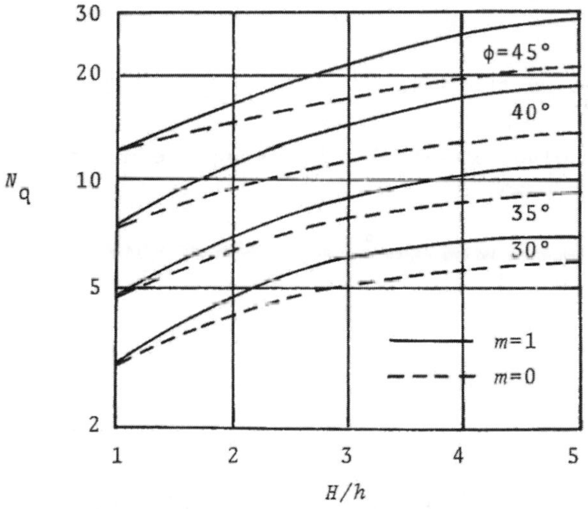

Figure 3.11. Variation of N_q [Eq. (3.24)] with H/h and ϕ--based on the analysis of Neeley et al. (1973).

for rectangular anchors, the breakout factor

$$N_{q(R)} = \frac{P_u}{Bh\gamma H} = N_q S_F \tag{3.25}$$

where S_F = shape factor

The variation of S_F with B/h and H/h is given in Fig. 3.12. It needs to be pointed out that the solution of Neeley et al. (1973) does not take into account the effect of interference of the

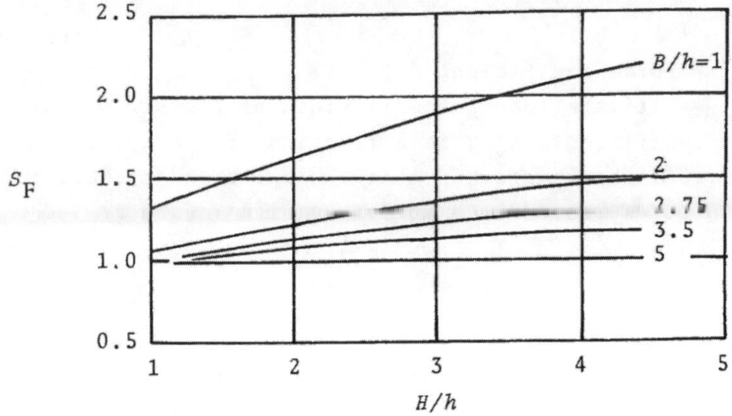

Figure 3.12. Variation of the shape factor, S_F (after Neeley et al., 1973).

anchors located on the sides.

Example 3.3. Solve Example 3.1 by using the method proposed by Neeley et al. (1973).

Solution

Given H=2 m, h=1 m, B=1 m, ϕ=30°. So, B/h=1; H/h=2. Referring to Fig. 3.11, for H/h=2

$$N_q \approx \begin{cases} 4.5 \text{ (for } m=1) \\ 4 \text{ (for } m=0) \end{cases}$$

From Fig. 3.12, for B/h=1 and H/h=2, $S_F \approx 1.65$. Hence, from Eq. (3.25)

$$P_u = N_q S_F B h \gamma H \begin{cases} (4.5)(1.65)(1)(1)(17.2)(2) = \underline{255.42 \text{ kN } (m=1)} \\ (4)(1.65)(1)(1)(17.2)(2) = \underline{227.04 \text{ kN } (m=0)} \end{cases}$$

Note: P_u=227.04 kN as obtained above with m=0 is the lower limit. The value of P_u obtained by this method with m=0 is somewhat higher than that obtained by the method of Ovesen and Stromann (1972) (Example 3.2) with S>about 3B (that is, with no interference due to the adjacent anchors).

3.2.6 Load-Displacement Relationship for Shallow Vertical Anchors in Sand

In many cases, it may be necessary to determine the *allowable holding capacity* of shallow vertical anchors for a limited horizontal displacement condition. Figure 3.13 shows the general nature of load vs. displacement plots for shallow anchors as observed by Neeley et al. (1973). From this figure, the following features may be observed.

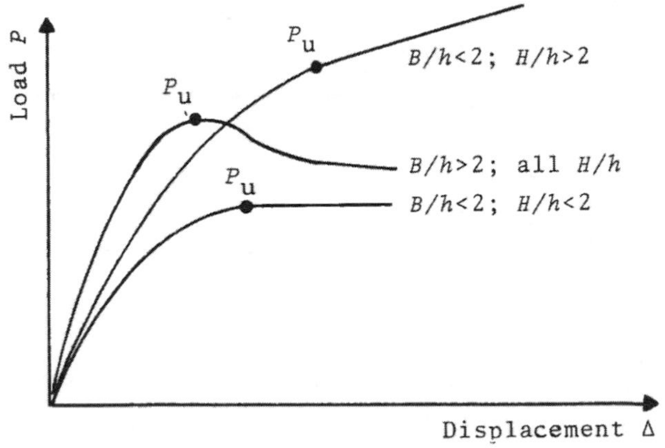

Figure 3.13. Nature of variation of load vs. displacement of shallow vertical anchor slab as observed by Neeley et al. (1973).

1. For anchor slabs with $B/h>2$, the load-displacement plot shows a peak value for ultimate load $P=P_u$ at a displacement of $\Delta=\Delta_u$.
2. For anchor slabs with $B/h<2$, two types of load-displacement plots are observed. When embedment ratio H/h is less than 2, the load P increases with displacement Δ and reaches a maximum value at $P=P_u$ with displacement of $\Delta=\Delta_u$. Beyond this point, the load on the anchor remains constant with further increase of displacement (that is, $P=P_u$ for $\Delta>\Delta_u$). However, when the embedment ratio of the anchors is greater than 2, a maximum load is never observed. The ultimate load may be defined as the point at which the load-displacement relationship becomes practically linear and a large displacement is observed with a small increase in load.

Figure 3.14 shows a nondimensional plot of the displacement with embedment ratio as observed by Neeley et al. (1973). These results were obtained from the laboratory experimental results conducted in medium dense sand. Similar laboratory model tests in loose and dense sands have also been conducted by the author. It

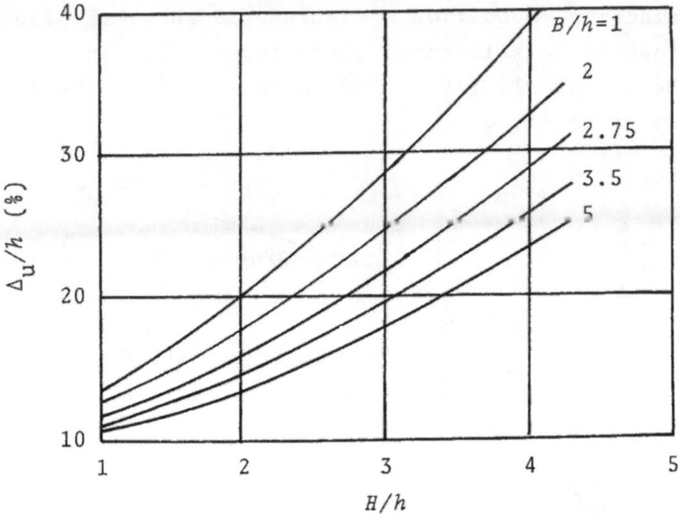

Figure 3.14. Nondimensional plot of Δ_u/h vs. H/h for various values of B/h (after Neeley et al., 1973).

appears that the general range of Δ_u as given in Fig. 3.14 is essentially in the correct range.

Based on their model tests, Das and Seeley (1975) proposed the following nondimensional relationship which holds good for $H/h \leq 5$ and $B/h \leq 5$

$$P/P_u \quad \frac{\Delta/\Delta_u}{[0.15+0.85(\Delta/\Delta_u)]} \tag{3.26}$$

where P load at a horizontal displacement of $\Delta \leq \Delta_u$

P_u = ultimate load at a horizontal displacement of Δ_u

If the ultimate load P_u can be estimated by using the theories presented in the preceding sections and the displacement Δ_u at the ultimate load can be obtained from Fig. 3.14, then an estimate of the anchor holding capacity P at any other displacement Δ can be calculated from Eq. (3.26). However, it needs to be pointed out that this estimated value of P is only approximate, since the value of Δ_u may vary somewhat from that shown in Fig. 3.24.

Example 3.4. Consider a vertical anchor in sand with H=5 ft, h=2 ft, and B=4 ft. For the soil, γ=105 lb/ft^3 and ϕ=35°. Determine:

a. the ultimate holding capacity, P_u
b. the displacement of anchor at ultimate load, Δ_u

c. the holding capacity of the anchor for a displacement of 1 in.

Solution

Part a. From Eq. (3.25)

$$N_q = \frac{P_u}{\gamma BhHS_F}$$

For this anchor, $H/h=5/2=2.5$; $B/h=4/2=2$. From Fig. 3.11, for $\phi=35°$ and $H/h=2.5$, $N_q \simeq 7.1$ (for $m=0$ to be conservative). Also from Fig. 3.12, for $B/h=2$ and $H/h=2.5$, $S_F \simeq 1.3$. Hence

$$7.1 = \frac{P_u}{\gamma AHS_F} = \frac{P_u}{(105)(2)(4)(5)(1.3)}$$

or

$$P_u \quad \underline{38,766 \text{ lb}}$$

Part b. For $B/h=2$ and $H/h=2.5$, the value of $\Delta_u/\Delta \simeq 0.21$ (Fig. 3.14). Hence

$$\Delta_u = (0.21)(h) = (0.21)(24 \text{ in.}) = \underline{5.04 \text{ in}}.$$

Part c. Given $\Delta=1$ in. So $\Delta/\Delta_u=1/5.04=0.198$. From Eq. (3.26)

$$P/P_u = \frac{\Delta/\Delta_u}{[0.15+0.85(\Delta/\Delta_u)]} \quad \frac{0.198}{[0.15+0.85(0.198)]} = 0.622$$

Hence

$$P = (0.622)(P_u) = (0.622)(38,766) = \underline{24,112 \text{ lb}}$$

3.2.7 Deep Vertical Anchors in Sand

As explained in Section 3.2, anchors can be defined as shallow when the failure surface of the soil at ultimate load extends up to the ground surface. For shallow anchors, the breakout factor N_q [or $N_{q(R)}$] increases with embedment ratio (H/h) up to a maximum value. The embedment ratio at which the maximum value of N_q [or $N_{q(R)}$] is attained can be defined as the critical embedment ratio, $(H/h)_{cr}$. For $H/h>(H/h)_{cr}$, an anchor behaves as a deep anchor, and the value of N_q [or $N_{q(R)}$] remains constant (Fig. 3.15). Very few laboratory or field tests available at this time have been extended to embedment ratios of $H/h>(H/h)_{cr}$. Meyerhof (1973) has indicated that for *square* anchors, $(H/h)_{cr}$ is about 4 in loose sand, and it

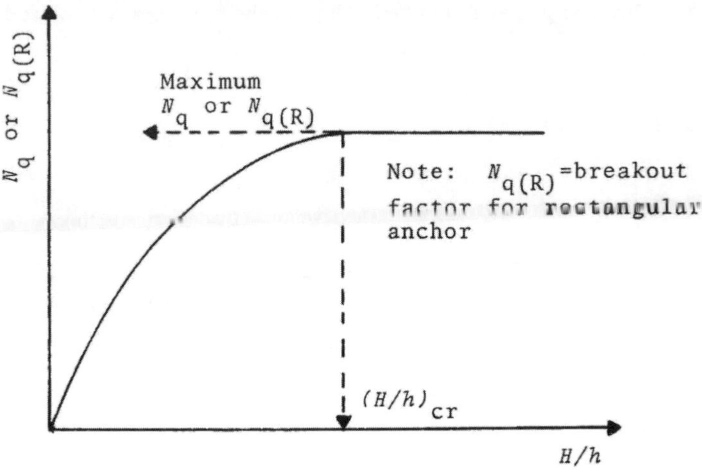

Figure 3.15. Nature of variation of N_q or $N_{q(R)}$ with H/h.

increases up to about 8 in dense sand. For *strip* anchors, the critical embedment ratio is about 50% more than the above-stated limits. Based on the above values of $(H/h)_{cr}$, Meyerhof (1973) has proposed the variation of N_q and $N_{q(R)}$ for deep anchors with the soil friction angle ϕ. This is shown in Fig. 3.16.

Ovesen (1964) has also given an approximate relationship for the holding capacity of *deep rectangular* anchors as

$$N_{q(R)} \quad \frac{P_u}{\gamma HBh} \quad (1.6+4.1\tan^4\phi)(1-\sin\phi)(e^{\pi\tan\phi})\tan^2(45+\phi/2)S_F \tag{3.27}$$

where S_F shape factor

The value of S_F is equal to 1.2 for square anchors and 1.0 for strip anchors. However, Ovesen has suggested that, due to the approximations involved in obtaining Eq. (3.27), the value of S_F may be taken as 1.0 for all anchors [that is, $N_q=N_{q(R)}$]. The variation of $N_q=N_{q(R)}$ as obtained from Eq. (3.27) (with $S_F=1$) is also shown in Fig. 3.16. Note that Ovesen's values of $N_q=N_{q(R)}$ are higher than those recommended by Meyerhof.

Based on some laboratory model test results, Das (1983) has suggested a conservative empirical correlation for the critical embedment ratio for square anchors which is of the form

$$(H/h)_{cr} \quad 5.5+0.166(\phi-30) \quad (\text{for } \phi\leq 30°\leq 45° \text{-- square anchor}) \tag{3.28}$$

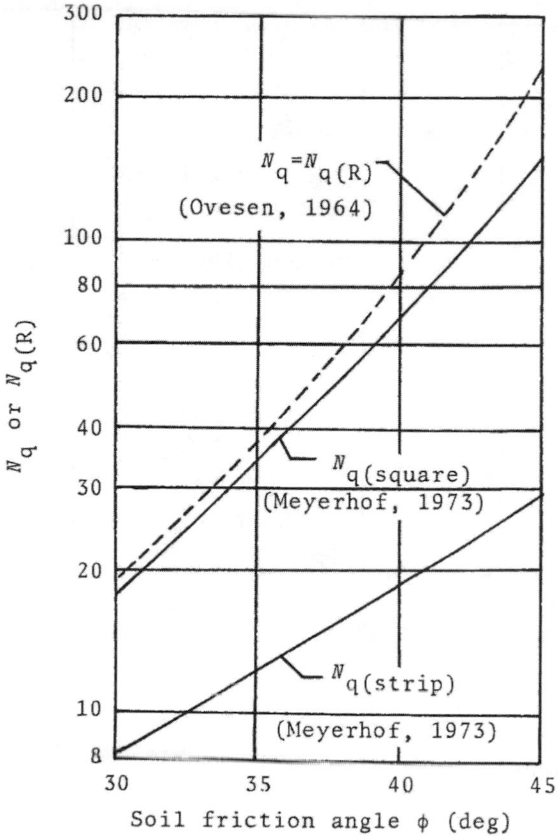

Figure 3.16. Breakout factor for deep vertical anchors.

Figure 3.17 shows a comparison of the laboratory model test results of Das (1983) and Akinmusuru (1978) for square anchors with the theories of Meyerhof (1973) and Ovesen (1964). The comparison shows that the model test results are somewhat lower than the predicted values.

3.2.8 Soil Friction Angle for Estimation of the Ultimate Anchor Capacity

It is entirely possible that in the field progressive failure in the soil mass around the anchor will take place at ultimate load. This means that the maximum resistance is not mobilized in the soil mass simultaneously at all points. In such a case, if the peak friction angle ϕ of the soil is used to estimate the ultimate

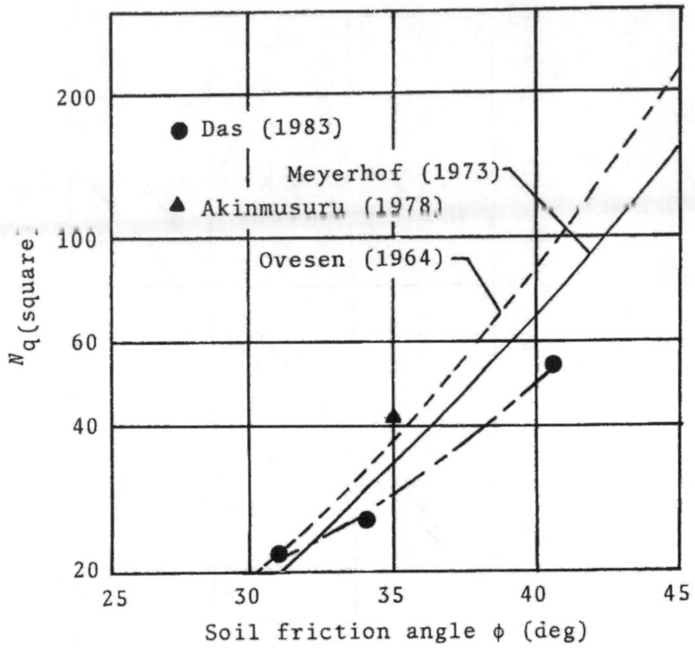

Figure 3.17. Comparison of model test results with theory for deep vertical anchors.

capacity of an anchor, it may result in serious overprediction. For that reason, caution needs to be used in selecting the friction angle of the sand. For continuous anchors, Dickin and Leung (1985) have suggested that the mobilized plane-strain friction angle ϕ_{mp} should be used in lieu of the peak plane-strain friction angle ϕ_{ps} for calculation of ultimate load, or

$$\phi_{mp} = \phi_{ps}(1-P_r) + P_r\phi_{cp} \tag{3.29}$$

where ϕ_{cp} = critical state friction angle
P_r = progressivity index (≈0.8)

3.3 VERTICAL ANCHOR SLABS IN CLAY

As in the case of anchor slabs in sand, the ultimate load of vertical anchors embedded in saturated clay (ϕ=0 concept) can be expressed in a nondimensional form as

$$N_c = \frac{P_u}{Bhc_u} \tag{3.30}$$

where N_c = breakout factor
P_u = ultimate load
B = length of anchor
h = anchor height
c_u = undrained cohesion of clay

Mackenzie (1955) has reported some laboratory model tests for strip vertical anchors in saturated clay, and these results are shown in Fig. 3.18 in a nondimensional form, that is, N_c vs. H/h.

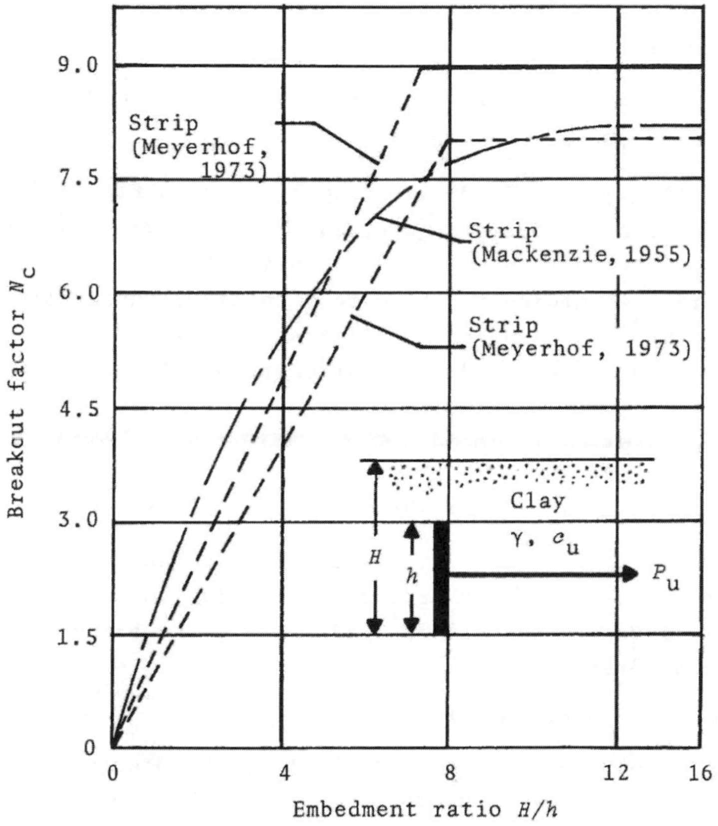

Figure 3.18. Variation of N_c vs. H/h based on the studies of Meyerhof (1973) and Mackenzie (1955).

It can be seen from this figure that N_c increases with the embedment ratio H/h up to a value of about 8.5 at $H/h \approx 11$. For $H/h \geq$ about 11, the value of N_c remains approximately constant. Based on this, the following general conclusions can be drawn:

1. Vertical anchors behave as *shallow* anchors up to a criti-

cal embedment ratio of $H/h=(H/h)_{cr}$. For shallow anchors, the breakout factor increases with embedment ratio.

2. For $H/h>(H/h)_{cr}$, the anchors behave as deep anchors and, for this case, the magnitude of $N_c=N_c^*$ remains constant.

Based on limited test results, Meyerhof (1973) has suggested the following:

For Strip Anchors:

$$N_c = 1.0(H/h) \leq 8 = N_c^* \tag{3.31}$$

For Square Anchors (that is, B=h)

$$N_c = 1.2(H/h) \leq 9 \quad N_c^* \tag{3.32}$$

For Circular Anchors (that is, h=anchor diameter):

$$N_c = 1.2(H/h) \leq 9 \quad N_c^* \tag{3.33}$$

Equations (3.31) and (3.32) are also plotted in Fig. 3.18,

More recently, Das, Tarquin, and Moreno (1985) have provided some laboratory test results for the ultimate capacity of vertital anchors. According to these test results, the critical embedment ratio of square and rectangular anchors will change with the magnitude of c_u (undrained shear strength of clay)

$$(H/h)_{cr(S)} = 4.7+2.9\times10^{-3}c_u \leq 7 \tag{3.34}$$

and

$$(H/h)_{cr(R)} = (H/h)_{cr(S)}[0.9+0.1(B/h)] \quad (\text{for } B/h \leq 4) \tag{3.35}$$

where $(H/h)_{cr(S)}$ and $(H/h)_{cr(R)}$ = critical embedment ratio of square and rectangular anchors, respectively. and c_u is in lb/ft^2.

For deep anchors, that is $H/h>(H/h)_{cr}$, the breakout factor $N_c=N_c^*$ can be given as

$$N_c^* = 9[0.825+0.175(h/B)] \tag{3.36}$$

For shallow anchors, that is $H/h \leq (H/h)_{cr}$, the breakout factor for square and rectangular anchors can be given by the empirical equation

$$N_c = \frac{N_c^*\left[\frac{H/h}{(H/h)_{cr}}\right]}{0.41+0.59\left[\frac{H/h}{(H/h)_{cr}}\right]} \tag{3.37}$$

Example 3.5. A square vertical anchor 0.5 m × 0.5 m in dimension is embedded in a saturated clay. Given H=1.5 m and c_u=for the clay is 47.9 kN/m^2. Determine the ultimate holding capacity of the anchor. Use the equations of Das, Tarquin, and Moreno (1985).

Solution

Given $B=h=0.5$ m.

Check for shallow anchor condition. From Eq. (3.34)

$$(H/h)_{cr(S)} = 4.7+2.9\times10^{-3}c_u \leq 7$$

$$c_u = 47.9 \text{ kN/m}^2 = 1000 \text{ lb/ft}^2$$

So

$$(H/h)_{cr(S)} \quad 4.7+(0.0029)(1000) = 4.7+2.9 = 7.6>7$$

So $(H/h)_{cr(S)} \quad 7$

For the case given in this problem, $H/h=1.5/0.5=3<7$. Hence, this is a shallow anchor.

From Eq. (3.37)

$$N_c \quad \frac{9(3/7)}{0.41+0.59(3/7)} = \frac{3.86}{0.663} = 5.82$$

So

$$P_u = N_c hBc_u = (5.82)(0.5)(0.5)(47.9) = \underline{69.7 \text{ kN}}$$

3.4 UPLIFT CAPACITY OF HORIZONTAL ANCHORS IN SAND

Figure 3.19 shows a horizontal anchor embedded in a sand at a depth H below the ground surface. The length and width of the

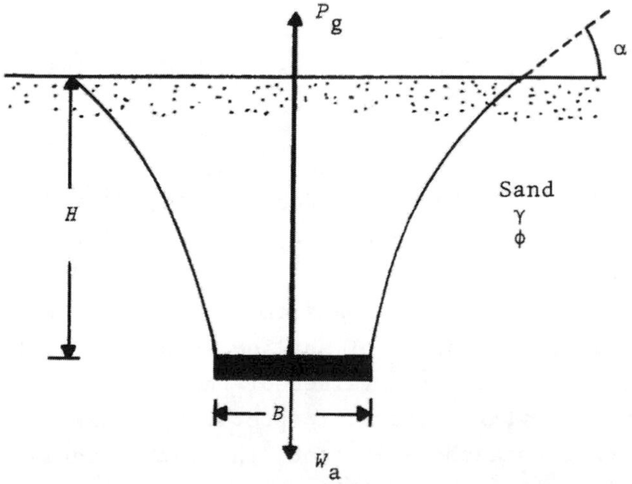

Figure 3.19. Parameters for a shallow horizontal anchor in sand

anchor are L and B, respectively. When the anchor has a relatively shallow embedment ratio (H/B) and is subjected to a gross ultimate load P_g, the failure surface in the soil will extend to the ground surface. The failure surface intersects the ground surface at an angle α. The value of α can vary from about 90° for very loose sand to about 45-ϕ/2 degrees for very dense sand with an average of about 90-ϕ/2 degrees. When the failure surface in soil extends up to the ground surface under the gross ultimate load, the anchor may be defined as a *shallow horizontal anchor*. However when the embedment ratio (H/B) is large, local shear failure in the soil around the anchor will occur, and the failure surface in soil will not extend to the ground surface. This type of anchor may be defined as a *deep horizontal anchor*.

The net ultimate uplift capacity (P_o) of an anchor can be given as

$$P_o \quad P_g - W_a \tag{3.38}$$

where W_a = effective self-weight of the anchor

Several investigators have proposed a number of theoretical solutions for obtaining P_o. An excellent review of these works has been given by Vesic (1971). He proposed that the net ultimate capacity P_o can be expressed by the following simple relation

$$P_o = (\gamma A H) N_q' \tag{3.39}$$

where γ = unit weight of soil

A = area of the anchor LB

N_q' = breakout factor for horizontal anchors

For shallow anchors, N_q' will increase with H/B. The value of N_q' is a function of several factors, or

$$N_q' = f(B/L, H/B, \phi) \tag{3.40}$$

However, beyond the critical embedment ratio, $H/B = (H/B)_{cr}$, the value of N_q' will remain constant (Fig. 3.20). This is called as *deep anchor condition*.

Based on the theory of expansion of cavities, Vesic (1971) has presented expressions for N_q' of shallow anchors. Tables 3.1 and 3.2 show Vesic's theoretical values of the breakout factor (N_q') for *strip* and *circular* anchors, respectively. Based on laboratory test experiments on anchors in sand, the author feels that the theory of uplift capacity of shallow anchors as presented by Meyerhof and Adams (1968) gives better agreement with the actual

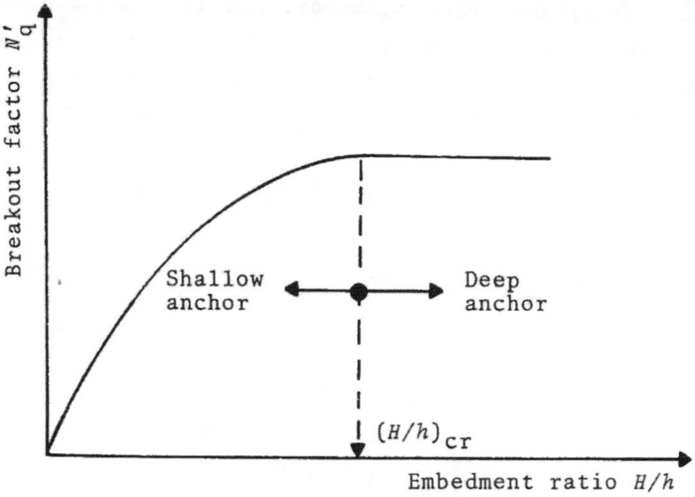

Figure 3.20. Definition of shallow and deep horizontal anchors in sand.

Table 3.1. Vesic's N'_q for Shallow Strip Anchors

Soil friction angle ϕ (deg)	H/B				
	0.5	1.0	1.5	2.5	5.0
0	1	1	1	1	1
10	1.09	1.16	1.25	1.42	1.83
20	1.17	1.33	1.49	1.83	2.65
30	1.24	1.47	1.71	2.19	3.38
40	1.3	1.58	1.87	2.46	3.91
50	1.32	2.04	1.96	2.6	4.20

Table 3.2. Vesic's N'_q for Shallow Circular Anchors

Soil friction angle ϕ (deg)	H/B				
	0.5	1.0	1.5	2.5	5.0
0	1	1	1	1	1
10	1.18	1.37	1.59	2.08	3.67
20	1.36	1.75	2.20	3.25	6.71
30	1.52	2.11	2.79	4.41	9.89
40	1.65	2.41	3.30	5.45	13.0
50	1.73	2.61	3.56	6.27	15.7

test results. For that reason, a more detailed discussion is included in this section. Meyerhof and Adams' theory (1968) can be easily explained by referring to the horizontal strip anchor shown in Fig. 3.21. At ultimate load, the stability of the soil wedge

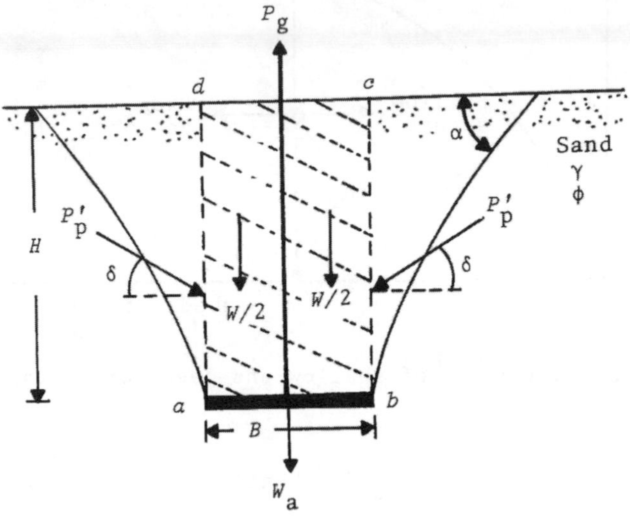

Figure 3.21. Derivation of Eq. (3.48).

abcd located immediately above the anchor can be expressed as

$$P'_o = 2P'_p \sin\delta + W \tag{3.41}$$

where P'_o = net ultimate uplift load *per unit length* of the anchor
P'_p = passive force per unit length along the faces *ab* and *cd*
W = weight of soil *per unit length* of the anchor
δ inclination of P'_p with the horizontal

Note that, in Eq. (3.41)

$$W \quad BH\gamma \tag{3.42}$$

The horizontal component of P'_p can be expressed as

$$P'_h \quad P'_p \cos\delta \quad \tfrac{1}{2} K_{pH} \gamma H^2 \tag{3.43}$$

where P'_h = horizontal component of P'_p
K_{pH} horizontal component of passive earth pressure coefficient

However

$$P'_p \sin\delta = P'_p (\cos\delta)(\tan\delta) = P'_h \tan\delta \quad \tfrac{1}{2} K_{pH} \gamma H^2 \tan\delta \tag{3.44}$$

Combining Eqs. (3.41) and (3.44)

$$P_o' = K_{pH}H^2\tan\delta+W \tag{3.45}$$

For an average value of $\alpha=90-\phi/2$, Meyerhof and Adams (1968) have shown that

$$\delta \approx \frac{2}{3}\phi \tag{3.46}$$

The corresponding passive earth pressure coefficient (K_{pH}) based on curved failure surface for $\delta\approx(2/3)\phi$ can be obtained from Caquot and Kerisel (1949). Furthermore, it is convenient to express $K_{pH}\tan\delta$ in the form

$$K_u'\tan\phi = K_{pH}\tan\delta \tag{3.47}$$

where K_u' = nominal uplift coefficient

The variation of the nominal uplift coefficient with the soil friction angle ϕ is shown in Fig. 3.22. Note that the value of K_u' does

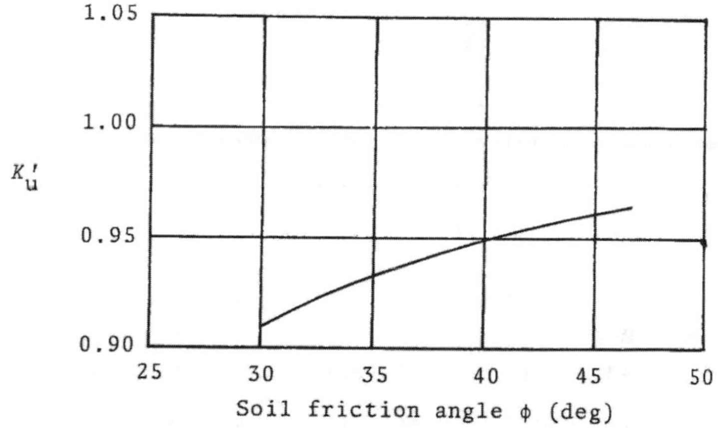

Figure 3.22. Variation of K_u' with soil friction angle.

not vary substantially with ϕ and, for all practical cases, may be taken as 0.95. Thus, from Eqs. (3.45) and (3.47)

$$P_o' = K_u'\gamma H^2\tan\phi+W \tag{3.48}$$

The preceding equation can be modified to determine the net uplift capacity of circular and rectangular anchors in the following manner:

Circular anchors:

$$P_o = (\pi/2)S_F'\gamma BH^2K_u'\tan\phi+W' \tag{3.49}$$

where B = diameter of the anchor

Rectangular anchors:

$$P_o = \gamma H^2(2S_F'B+L-B)K_u'\tan\phi+W' \tag{3.50}$$

where S_F' = shape factor

W' = weight of soil located above the anchor

The shape factor can be approximated as

$$S_F' = 1+m(H/B) \tag{3.51}$$

where m = a coefficient

The values of m as recommended by Meyerhof and Adams (1968) are as follows:

Friction angle ϕ (deg)	m
20	0.05
25	0.1
30	0.15
35	0.25
40	0.35
45	0.5
50	0.6

Linear interpolation may be used for obtaining m for intermediate friction angles.

Das and Seeley (1975) and Das and Jones (1982) have shown that Meyerhof and Adams' equation can be conveniently transformed to the form of Eq. (3.39). This can be done in the following manner.

For circular anchors: From Eqs. (3.49) and (3.51)

$$P_o = (\pi/2)[1+m(H/B)]\gamma BH^2K_u'\tan\phi + (\pi/4)B^2H\gamma$$

or

$$P_o = (\pi/4)B^2H\gamma\{2[1+m(H/B)](H/B)K_u'\tan\phi+1\} = A\gamma HN_q' \tag{3.52}$$

where A area of the anchor $= (\pi/4)B^2$

$$N_q' \quad \{2[1+m(H/B)](H/B)K_u'\tan\phi+1\} \tag{3.53}$$

For rectangular anchors:

$$P_o = \gamma H^2\{2[1+m(H/B)]B+L-B\}K_u'\tan\phi + BLH\gamma$$

or

$$F_o = DLH\gamma(\{[1+2m(H/B)](B/L)+1\}(H/B)K_u'\tan\phi+1)$$

$$= A\gamma HN_q' \qquad (3.54)$$

where A area of the anchor

N_q' breakout factor

$$= \{[1+2m(H/B)](B/L)+1\}(H/B)K_u'\tan\phi+1 \qquad (3.55)$$

The variation of N_q' for square and circular anchors is shown in Fig. 3.23.

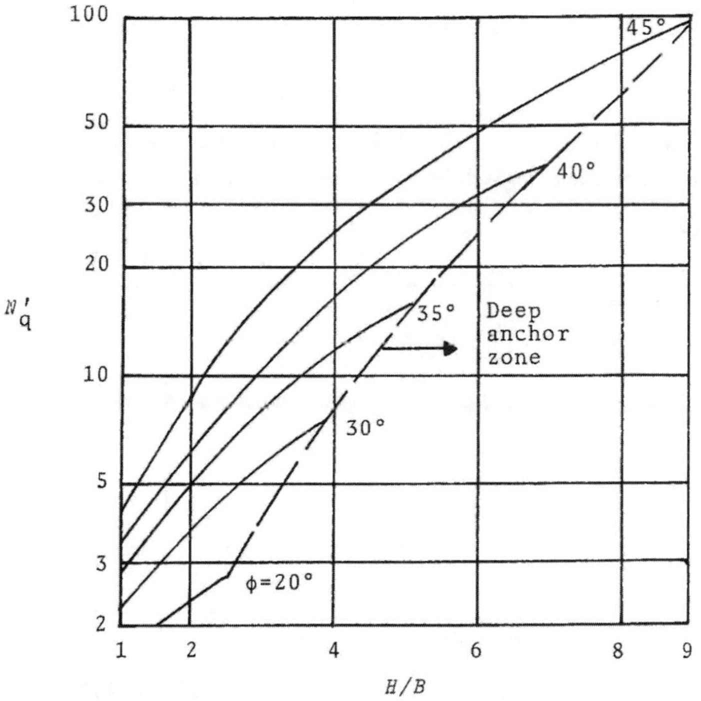

Figure 3.23. Plot of N_q' [Eq. (3.53)] for square and circular anchors.

Deep Horizontal Anchors

Equations (3.53) and (3.55) as presented above for estimation of the net ultimate uplift capacity are for shallow circular and rectangular anchors. It can be seen from these equations that, for a given value of ϕ, the magnitude of N_q' increases with embedment ratio. As has been previously mentioned in this section, the value of N_q' for deep anchors is practically constant beyond a critical embedment ratio $H/B = (H/B)_{cr}$. Hence, for deep anchors

$$P_o \quad A\gamma H N'_{q(deep)} \tag{3.56}$$

where $N'_{q(deep)}$ breakout factor for deep anchors

The relations for $N'_{q(deep)}$ can be obtained by substituting $(H/B)_{cr}$ in Eqs. (3.53) and (3.55) (Das and Jones, 1983), or

$$N'_{q(deep)} = (\{2[1+m(H/B)_{cr}]\}(H/B)_{cr}K'_u \tan\phi + 1)$$

(circular anchors) (3.57)

$$N'_{q(deep)} = (\{[1+2m(H/B)_{cr}](B/L)+1\}(H/B)_{cr}K'_u \tan\phi + 1)$$

(rectangular anchors) (3.58)

Meyerhof and Adams (1968) have recommended the following values of the critical embedment ratio for *square or circular* anchors $[(H/B)_{cr(S)}]$

Soil friction angle ϕ (deg)	$(H/B)_{cr(S)}$
20	2.5
25	3
30	4
35	5
40	7
45	9
48	11

Based on laboratory model test results, Das and Jones (1982) have suggested that, for rectangular anchors

$$(H/B)_{cr(R)} \quad (H/B)_{cr(S)}[0.133(L/B)+0.867] \le 1.4(H/B)_{cr(S)} \tag{3.59}$$

Step-by-Step Procedure to Estimate P_o

Based on the discussion presented above, following is a step-by-step procedure to estimate the net ultimate uplift capacity of horizontal anchors:

1. Determine H, L, and B for the anchor and ϕ for the soil.
2. Determine H/B for the anchor.
3. For a given value of ϕ, determine $(H/B)_{cr(S)}$.
4. Using Eq. (3.59), determine $(H/B)_{cr(R)}$.

5. If H/B (Step 2) is less than $(H/B)_{cr(R)}$, it is a shallow anchor. For this case, use Eq. (3.53) or (3.55) as the case may be to obtain P_o.
6. If $H/B>(H/B)_{cr(R)}$, use Eq. (3.57) or (3.58) as the case may be to obtain P_o.

Example 3.6. Refer to Fig. 3.19. For a horizontal anchor, given H=2 m, L=0.6 m, B=0.3 m, ϕ=35°, and γ=18 kN/m³. Estimate the net ultimate uplift capacity.

Solution

Given $H/B=2/0.3=6.67$

For $\phi=35°$, $(H/B)_{cr(S)}=5$.

From Eq. (3.59)

$$(H/B)_{cr(R)} = (H/B)_{cr(S)}[0.133(L/B)+0.867]$$

$$= 5[0.133(2)+0.867]\leq 1.4(H/B)_{cr(S)}=7$$

Since $H/B=6.67>(H/B)_{cr(R)}=5.67$, it is a deep anchor.
From Eq. (3.58)

$$N'_{q(deep)} = \{[1+2m(H/B)_{cr(R)}](B/L+1\}(H/B)_{cr}K'_u\tan\phi+1$$

$K'_u\simeq 0.95$; For ϕ=35°, m=0.25. So

$$N'_{q(deep)} \quad \{[1+2(0.25)(5.67)]0.5+1\}(5.67)(0.95)(\tan 35°)+1$$

$$12$$

$$P_o = N'_{q(deep)}A\gamma H = (12)(0.6\times 0.3)(18)(2) \quad \underline{77.76\text{ kN}}$$

Veesaert and Clemence's Theory for Horizontal Circular Anchors

Veesaert and Clemence (1977) assumed a failure surface in the shape of a truncated cone (Fig. 3.24) above a shallow circular anchor having a diameter B. With this type of failure surface, they determined the net ultimate uplift capacity of shallow circular anchors as

$$P_o \quad \gamma V_s+\pi\gamma K_o\tan\phi\cos^2(\phi/2)\left[\frac{BH^2}{2}+\frac{H^3\tan(\phi/2)}{3}\right] \qquad (3.60)$$

where V_s = volume of truncated cone above the anchor
K_o = coefficient of lateral earth pressure

The relationship for V_s is

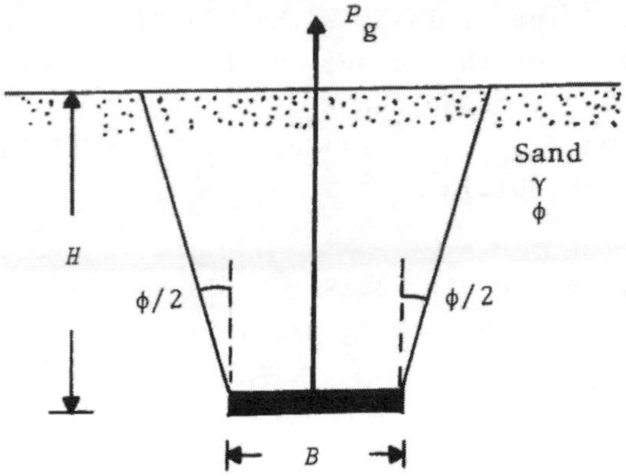

Figure 3.24. Assumption of the failure surface in sand for a circular horizontal anchor--Veesaert and Clemence's theory (1977).

$$V_s \quad (\pi/4)[B+H\tan(\phi/2)]^2 H \tag{3.61}$$

Substituting Eq. (3.61) into Eq. (3.60)

$$N'_q \quad \frac{P_o}{(\pi/4)B^2\gamma H} = \frac{P_o}{A\gamma H}$$

$$= [1+(H/B)\tan(\phi/2)]^2 + 4K_o\tan\phi\cos^2(\phi/2)[\tfrac{1}{2}(H/B) + (H/B)^2\frac{\tan(\phi/2)}{3}] \tag{3.62}$$

A comparison of laboratory and field test results shows that the value of K_o in the preceding equation may vary between 0.6 to 1.5 with an average of 1. Using an average value of K_o=1 in Eq. (3.62), the breakout factors have been calculated and are shown in Fig. 3.25. A comparison of the breakout factors shown in Figs. 3.23 and 3.25 shows the following:

1. For ϕ up to 35° with K_o=1, Eq. (3.62) yields higher values of N'_q (for similar H/B) than those obtained by using Eq. (3.53). For ϕ=40°, Eqs. (3.53) and (3.62) yield similar values.
2. For ϕ=40°, Eq. (3.62) yields lower values for a given H/B as compared to those obtained by using Eq. (3.53).

3.5 UPLIFT CAPACITY OF HORIZONTAL GROUP ANCHORS

In some instances, horizontal anchors are used in groups (Fig. 3.26). Unlike theoretical and experimental studies relating to the uplift capacity of single anchors, very few attempts have so

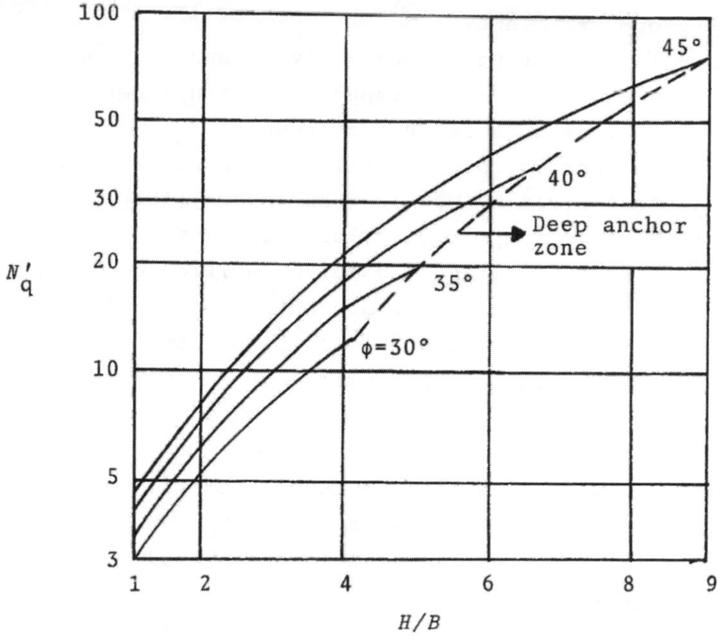

Figure 3.25. Variation of N_q' for shallow circular anchors [Eq. (3.62)].

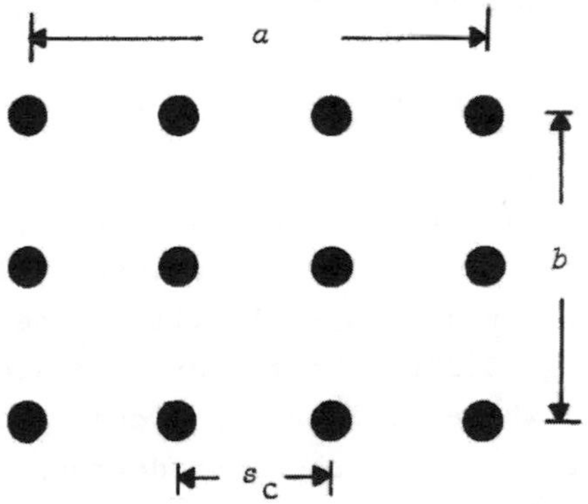

Figure 3.26. Plan of a group of circular horizontal anchors.

far been made to evaluate the ultimate uplift capacity of group anchors. In sand and clay, the group efficiency of horizontal anchors can be defined as

$$\eta = \frac{P_{o(g)}}{nP_o} \leq 1 \tag{3.63}$$

where η group efficiency

$P_{o(g)}$ net ultimate uplift capacity of anchor groups

P_o net ultimate uplift capacity of single anchors

n number of anchors in the group

The only theory presently available on group anchor uplift capacity is that given by Meyerhof and Adams (1968). According to this theory, for a group of circular anchors it is assumed that the passive earth pressure along the curved portion of the perimeter of the group is governed by the shape factor S_F' and the passive earth pressure along the straight portions is the same as for strip anchors. Thus, for a shallow anchor group

$$P_{o(g)} = \gamma H^2[a+b+S_F'(\pi/2)B]K_u'\tan\phi + W_g' \tag{3.64}$$

where a and b distances between centers of the corner anchors on length and width side

W_g' weight of the soil located immediately above the anchor group

Thus, by substituting Eqs. (3.50) and (3.64) into Eq. (3.63), we can obtain the theoretical group efficiency. Based on the experiences of the author, Meyerhof and Adams' theory only provides the general trend. Figure 3.27 shows some of the model test results of the author and the comparison with the preceding theory.

3.6 UPLIFT CAPACITY OF HORIZONTAL ANCHORS IN CLAY

Figure 3.28 shows a horizontal anchor in a saturated clay subjected to an uplifting force. The width of the anchor is B and the depth of embedment is equal to H. For anchors located at shallow depths, the failure surface in soil will extend to the ground surface as shown in Fig. 3.28. The failure surface in *soft clays* will be almost vertical; whereas, in *stiff clays*, it will intersect the ground surface at an angle of about 45 degrees. However, at larger embedment ratios (H/B), local shear failure occurs, and the failure surface in soil will not extend to the ground surface (*deep anchor condition*).

For undrained conditions in clay (ϕ=0), the gross ultimate holding capacity P_g can be expressed as (Vesic, 1971)

$$P_g = A(c_u N_c' + \gamma H) + W_a \tag{3.65}$$

where N_c' breakout factor

γ unit weight of soil

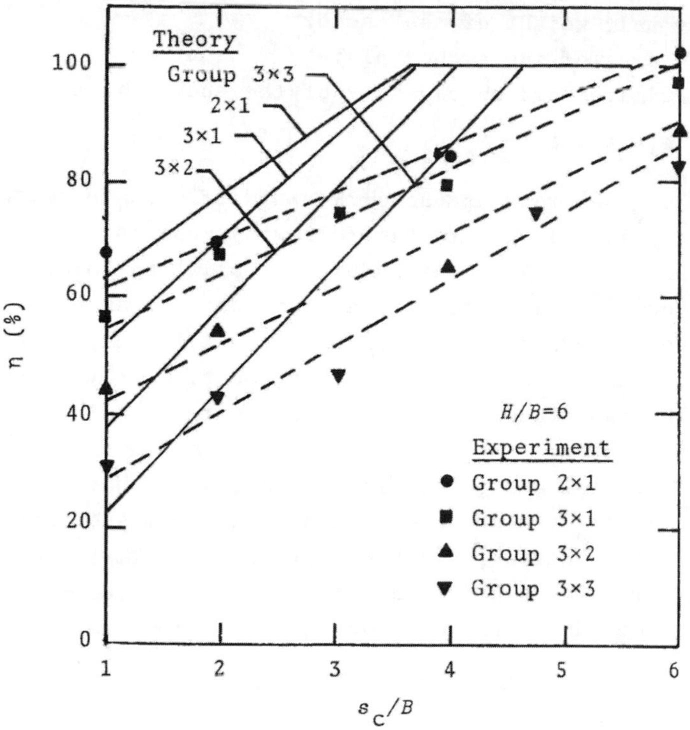

Figure 3.27. Comparison of laboratory group efficiency for horizontal anchors with theory.

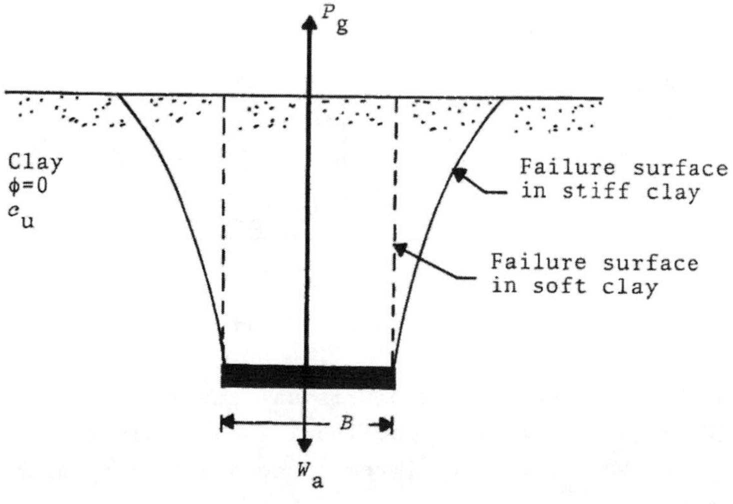

Figure 3.28. Horizontal anchor in clay.

W_a = self-weight of the anchor
A = area of the anchor plate

The net ultimate holding capacity of the anchor is now equal to

$$P_o = A(c_u N'_c + \gamma H) = P_g - W_a \tag{3.66}$$

The value of the breakout factor N'_c is difficult to determine theoretically, primarily due to the following reasons:

1. Due to the application of the uplifting load, premature tensile cracks appear in clay near the anchor.
2. Negative pore water pressure (suction) may develop below the anchor during the uplift, particularly in soft clays. The magnitude of negative pore water is difficult to determine.

For practical design considerations, it is safe to neglect the resistance of the anchor to uplift derived from suction. Without considering the suction, the breakout factor will increase with the embedment ratio for shallow anchors up to a maximum value (Fig. 3.29). The embedment ratio at which the maximum value of N'_c is ob-

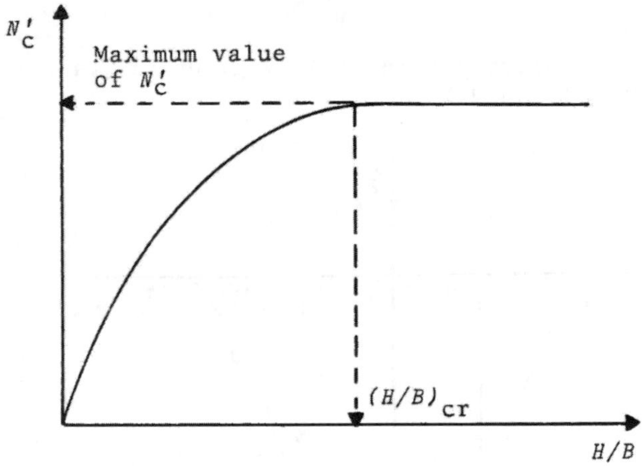

Figure 3.29. Nature of variation of N'_c with H/B.

tained may be referred to as the critical embedment ratio, $(H/B)_{cr}$. For $H/B > (H/B)_{cr}$, at ultimate load local shear failure will occur and, thus, the value of N'_c will remain constant (deep anchor condition).

Meyerhof (1973) has given the following relations for estimation of N'_c.

For shallow anchors:

$$N'_c = 0.6(H/B) \leq 8 \quad \text{(for strip anchors)} \tag{3.67a}$$

$$N'_c = 1.2(H/B) \leq 9 \quad \text{(for circular anchors)} \tag{3.67b}$$

For deep anchors:

$$N'_c = (N'_c)_{max} = 8 \quad \text{(for strip anchors)} \tag{3.68a}$$

$$N'_c \quad (N'_c)_{max} = 9 \quad \text{(for circular anchors)} \tag{3.68b}$$

From Eqs. (3.67a) and (3.67b), it can be seen that

$$(H/B)_{cr} = 8/0.6 = 13.33 \quad \text{(for strip anchors)} \tag{3.69a}$$

and

$$(H/B)_{cr} = 9/1.2 = 7.5 \quad \text{(for circular anchors)} \tag{3.69b}$$

Using the theory of expansion of cavities, Vesic (1971) has given the following values of N'_c for shallow anchor conditions.

	N'_c	
H/B	Strip anchors	Circular anchors
0	0	0
0.5	0.81	1.76
1	1.61	3.80
1.5	2.42	6.12
2.5	4.04	11.6
5.0	8.07	30.3

Das (1980) has compiled the existing experimental results for the ultimate holding capacity of horizontal anchors and has proposed the following step-by-step procedure to obtain P_o:

1. The critical embedment ratio $(H/B)_{cr}$ will vary with the value of c_u of the clay and also the length-to-width ratio L/B of the anchor (L=length and B=width). For that reason, first determine the value of c_u. Using the value of c_u obtain the critical embedment ratio of the anchor plate as

$$(H/B)_{cr(R)} = (H/B)_{cr(S)}[0.73+0.27(L/B)] \tag{3.70}$$

where $(H/B)_{cr(R)}$ critical embedment ratio for a rectangular anchor with dimensions of $L \times B$

$(H/B)_{cr(S)}$ = critical embedment ratio for a square

anchor with dimensions of $B \times B$

The experimentally determined empirical relation for $(H/B)_{cr(S)}$ can be expressed as

$$(H/B)_{cr(S)} = 0.107c_u + 2.5 \leq 7 \tag{3.71}$$

The unit of c_u in the above equation is in kN/m^2.

2. For a given anchor, obtain the embedment ratio H/B.
3. Determine

$$\alpha' = \frac{\overset{\text{Step 1}}{\overset{\downarrow}{(H/B)}}}{\underset{\text{Step 2}}{\underset{\uparrow}{(H/B)_{cr(R)}}}} \tag{3.72}$$

4. If $\alpha' \leq 1$, the anchor is a *shallow anchor*. For this case, refer to Fig. 3.30 and determine the value of the nondimensional parameter β' corresponding to α' determined in Step 3. Note that

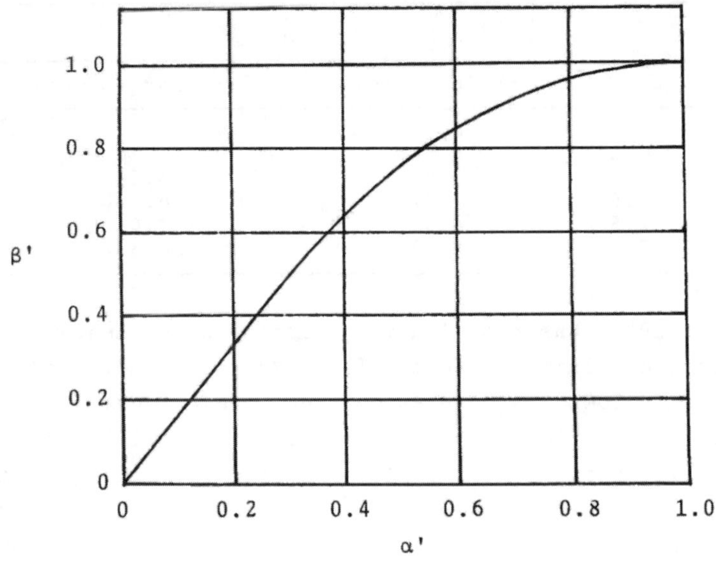

Figure 3.30. Plot of β' [Eq. (3.73)] vs. α' [Eq. (3.72)].

$$\beta' = \frac{N'_{c(R)}}{[N'_{c(R)}]_{max}} \tag{3.73}$$

where $N'_{c(R)}$ = breakout factor for rectangular anchor with dimensions of $L \times B$

$[N'_{c(R)}]_{max}$ = maximum value of the breakout factor for rectangular anchor with dimensions of $L \times B$ (deep anchor condition)

or

$$[N'_{c(R)}]_{max} = 7.56+1.44(B/L) \tag{3.74}$$

Using the values of β' from Fig. 3.30 and $[N'_{c(R)}]_{max}$ from Eq. (3.74), obtain $N'_{c(R)}$, With this value of $N'_{c(R)}$, obtain the net holding capacity of the anchor by using Eq. (3.66).

5. If the value of α' as calculated by using Eq. (3.72) (Step 3) is greater than one, it refers to a deep anchor condition. For this case

$$N'_{c(R)} = [N'_{c(R)}]_{max} = 7.56+1.44(B/L) \tag{3.75}$$

Hence, combining Eqs. (3.66) and (3.75)

$$P_o = BL\{[7.56+1.44(B/L)]c_u+\gamma H\} \tag{3.76}$$

Example 3.7. A horizontal anchor has dimensions ($L \times B$) of 2 ft×1 ft. The anchor is embedded in a clay having a c_u=10 lb/in^2 and γ= 120 lb/ft^3. Given H=5 ft. Determine the net ultimate uplift capacity of the anchor. Use the procedure developed by Das (1980).

Solution

For the anchor, B/L=1/2=0.5. Given c_u=10 lb/in^2=69 kN/m^2. From Eq. (3.71)

$$(H/B)_{cr(S)} \quad (0.107)(69)+2.5 = 9.8877$$

So

$$(H/B)_{cr(S)} = 7$$

Again, from Eq. (3.70)

$$(H/B)_{cr} \quad (H/B)_{cr(S)}[0.73+0.27(L/B)]$$

$$= 7[0.73+(0.27)(2)] - 8.89$$

Using Eq. (3.72)

$$\alpha' \quad \frac{(H/B)}{(H/B)_{cr(R)}} = \frac{(5/1)}{8.89} = 0.562$$

Since α' is less than one, it is a shallow anchor. Now, referring to Fig. 3.30, for α'=0.562, β'≈0.825, So

$$N'_{c(R)} = \beta'[N'_{c(R)}]_{max} = 0.825[N'_{c(R)}]_{max}$$

From Eq. (3.74)

$$[N'_{c(R)}]_{max} = 7.56+1.44(B/L) = 7.56+1.44(0.5) = 8.28$$

Hence

$$N'_{c(R)} = 0.825(8.28) \quad 6.831$$

$$P_o \quad BL[c_u N'_{c(R)}+\gamma H] = (1)(2)[(10)(6.831)+(120)(5)]$$

$$= 2(68.31+600) = \underline{1336.62 \text{ lb}}$$

3.7 UPLIFT CAPACITY OF HELICAL ANCHORS

Helical anchors are constructed of helical-shaped circular steel plates welded to a steel shaft at given spacings. These anchors (Fig. 1d) are driven into the ground by using truck-mounted augering equipment. Multihelix anchors can develop up to about 100,000 lbs or more resistance to uplift. Figure 3.31 shows the dimensions of a typical multihelix anchor.

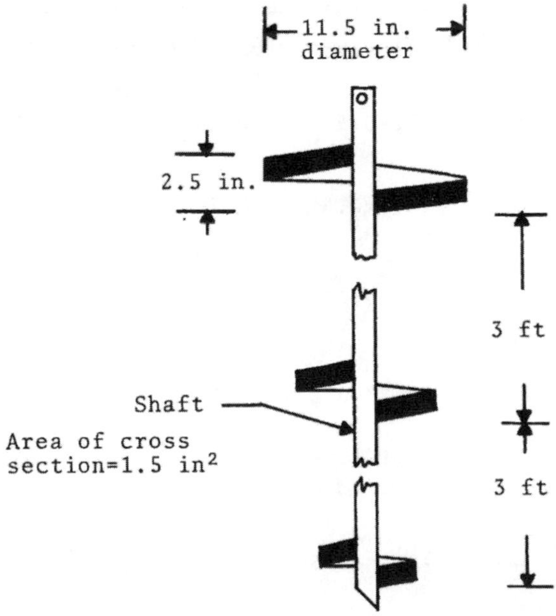

Figure 3.31. Dimensions of a typical multihelix anchor.

Helical Anchors in Sand

Figure 3.32 shows a helical anchor in sand. If the top helix is located at a shallow depth (that is, H_1/B_1 is relatively small),

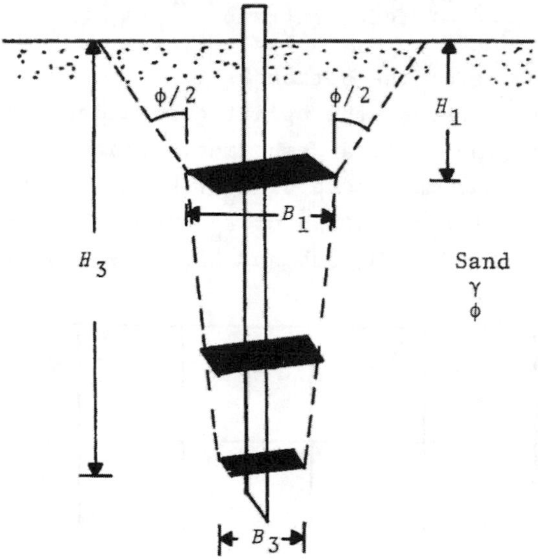

Figure 3.32. Helical anchor in sand.

the failure surface in soil can be idealized as shown in Fig. 3.32 (Mitsch and Clemence, 1985). In this case, the failure surface above the top helix is a truncated cone similar to that shown for circular plate anchors in Fig. 3.24. The net ultimate uplift capacity (P_o) for such a case can be given as (Mitsch and Clemence, 1985)

$$P_o = P_1 + P_2 \tag{3.77}$$

where P_1 = *bearing resistance* above the top helix

P_2 = *frictional resistance* along the perimeter of the soil cylinder between the top and bottom helix

The magnitude of P_1 can be expressed in the same form as Eq. (3.60), or

$$P_1 = (\pi/4)\gamma[B_1 + H_1\tan(\phi/2)]^2 H_1 + \pi\gamma K_u''\tan\phi\cos^2(\phi/2) \left[\frac{B_1H_1^2}{2} + \frac{H_1^3\tan(\phi/2)}{3}\right] \tag{3.78}$$

where K''_u = uplift coefficient
γ effective unit weight of soil
B_1 = diameter of the top helix

Similarly

$$P_2 = (\pi/2)\gamma\left(\frac{B_1+B_3}{2}\right)(H_3^2-H_1^2)K''_u\tan\phi \qquad (3.79)$$

where B_3 diameter of the bottom helix

For *shallow anchors*, the uplift coefficient increases linearly with H_1/B_1 up to the critical embedment ratio $(H_1/B_1)_{cr}$. For $H_1/B_1>(H_1/B_1)_{cr}$, the magnitude of K''_u remains practically constant. Figure 3.33 shows the variation of K''_u along with the limits of $(H_1/B_1)_{cr}$ as proposed by Mitsch and Clemence (1985).

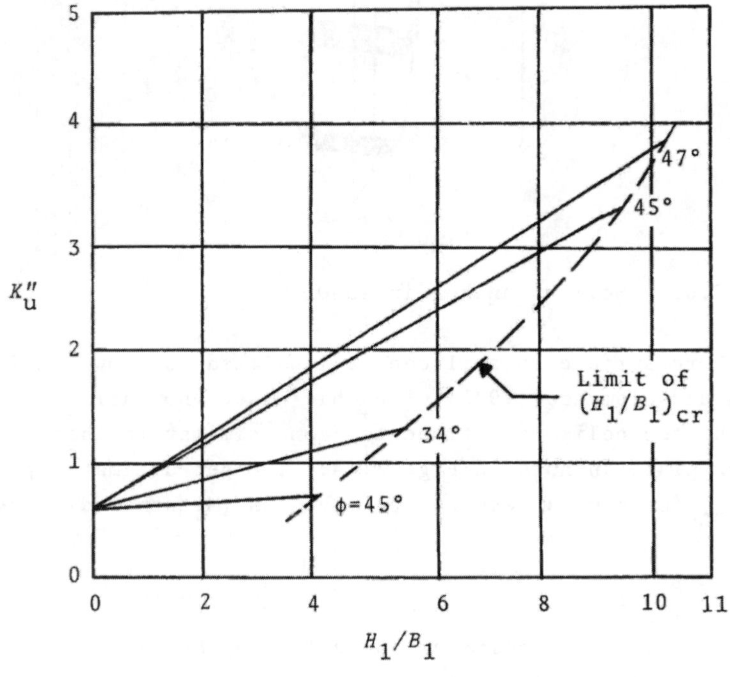

Figure 3.33. Variation of K''_u with H_1/B_1--based on the results of Mitsch and Clemence (1975).

For deep helical anchors, that is $H_1/B_1>(H_1/B_1)_{cr}$, the net ultimate uplift capacity can be given as (Mitsch and Clemence, 1985)

$$P_o = P_3+P_4+P_5 \qquad (3.80)$$

where P_3 = bearing resistance above the top helix

P_4 = frictional resistance along the perimeter of the soil cylinder between the top and bottom helix

P_5 = frictional resistance along the perimeter of the anchor shaft located above the top helix

The magnitude of P_3 can be obtained by substituting H_{cr} for H_1, and $K''_{u(max)}$ for K''_u in Eq. (3.78). The magnitudes of H_{cr}/B_1 and $K''_{u(max)}$ are given in Fig. 3.34, or

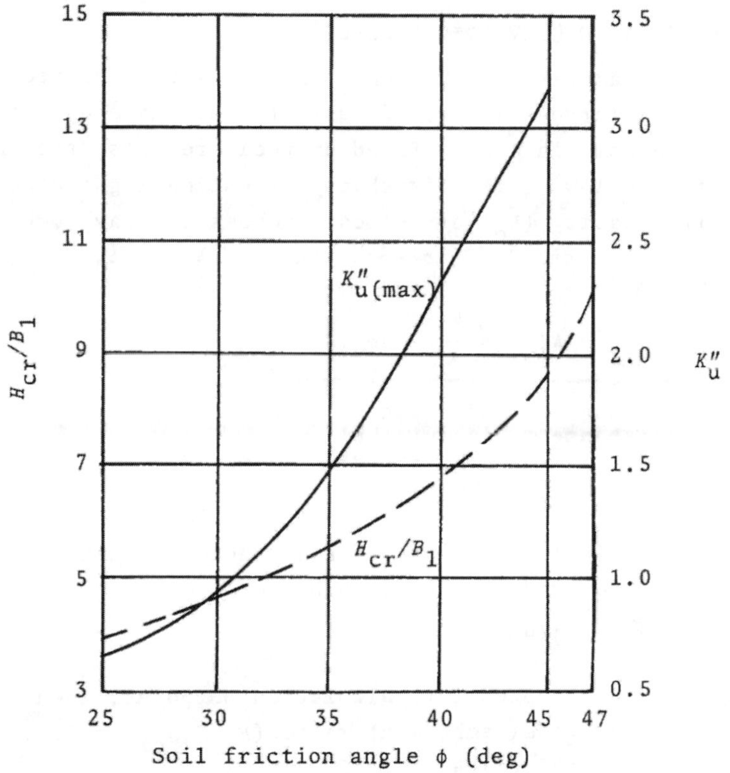

Figure 3.34. Variation of H_{cr}/B_1 and $K''_{u(max)}$ for use as derived from Fig. 3.33.

$$P_3 \quad (\pi/4)\gamma[B_1+H_{cr}\tan(\phi/2)]^2H_{cr} + \pi\gamma K''_{u(max)}\tan\phi\cos^2(\phi/2) \left[\frac{B_1H_{cr}^2}{2} + \frac{H_{cr}^3\tan(\phi/2)}{3}\right] \quad (3.81)$$

Similarly

$$P_4 \quad (\pi/2)\gamma\left(\frac{B_1+B_3}{2}\right)(H_3^2-H_1^2)K''_{u(max)}\tan\phi \tag{3.82}$$

and

$$P_5 \quad pH_1 \quad \frac{\gamma H_1}{2} K''_{u(max)}\tan\phi \tag{3.83}$$

where p perimeter of the cross section of the anchor shaft

In conducting large-scale field experiments, Mitsch and Clemence (1985) have shown that the preceding relationships agree well with the actual net uplifting load.

Helical Anchors in Clay (ϕ=0 concept)

Mooney, Adamczak, and Clemence (1985) have conducted several laboratory and large-scale field tests for ultimate uplift capacity of helical anchors in clay. Based on their results, they have recommended the following relationships for estimating the net ultimate uplift capacity (P_o) of helical anchors in clay (ϕ=0 concept) (Fig. 3.32). For *shallow anchors*, that is $H_1/B_1<(H_1/B_1)_{cr}$

$$P_o = \underbrace{(\pi/4)B_1^2 c_u N'_c}_{\text{Bearing resistance}} + \underbrace{\pi\left(\frac{B_1+B_3}{2}\right)c_u(H_3-H_1)}_{\text{Adhesion in soil in interhelical spaces}} \tag{3.84}$$

where N'_c breakout factor

The author feels that the breakout factor N'_c can be conservatively calculated by using Eq. (3.67b), or

$$N'_c = 1.2(H_1/B_1)\leq 9 \tag{3.85}$$

If Eq. (3.85) is accepted for calculation purposes, then it is obvious that the critical embedment ratio $(H_1/B_1)_{cr}$ is 7.5. Thus, for *deep anchors*, this is $H_1/B_1>7.5$

$$P_o = (\pi/4)B_1^2 c_u N'_{c(max)} + \pi\left(\frac{B_1+B_3}{2}\right)c_u(H_3-H_1) + pH_1 c_a \tag{3.86}$$

where p perimeter of the anchor shaft

c_a = adhesion

Mooney et al. (1985) have pointed out that the magnitude of c_a may vary from about $0.3c_u$ for very stiff clays to about $0.9c_u$ for soft clays.

References

Akinmusuru, J.O., 1978. Horizontally loaded vertical plate anchors in sand. J. Geotech. Engrg. Div., ASCE, 104(GT2):283-286.

Biarez, I., Boucraut, L.M., and Negre, R., 1965. Limiting equilibrium of vertical barriers subjected to translation and rotation forces. Proc., VI Intl. Conf. Soil Mech. Found. Engrg., Montreal, Canada, 2:368-372.

Caquot, A. and Kerisel, L., 1949. Traite de Mecanique des Sols. Gauthier-Villars, Paris, France.

Das, B.M., 1983. Holding capacity of vertical anchor slabs in granular soil. Proc., Coastal Structures '83, ASCE, pp. 379-392.

Das, B.M., 1980. A procedure for estimation of ultimate uplift capacity of foundations in clay. Soils and Foundations, (20)1: 77-82.

Das, B.M. and Jones, A.D., 1982. Uplift capacity of rectangular foundations in sand. Trans. Res. Rec. 884, National Academy of Sciences, Washington, D.C., pp. 54-58.

Das, B.M. and Seeley, G.R., 1975. Load-displacement relationships for vertical anchor plates. J. Geotech. Engrg. Div., ASCE, 101(GT7):711-715.

Das, B.M. and Seeley, G.R., 1975. Breakout resistance of horizontal anchors. J. Geotech. Engrg. Div., ASCE, 101(GT9):999-1003.

Das, B.M., Tarquin, A.J., and Moreno, R., 1985. Model tests for pullout resistance of vertical anchors in clay. Civ. Engrg. for Pract. Design Engrs., Pergamon Press, New York, New York, 4(2): 191-209.

Dickin, E. and Leung, C.F., 1985. Evaluation of design methods for vertical anchor plates. J. Geotech. Engrg., ASCE, 111(4):500-520.

Mackenzie, T.R., 1955. Strength of deadman anchors in clay. M.S. thesis, Princeton University, U.S.A.

Meyerhof, G.G., 1973. Uplift resistance of inclined anchors and piles. Proc., VIII Intl. Conf. Soil Mech. Found. Engrg., Moscow, U.S.S.R, 2.1:167-172.

Meyerhof, G.G., 1951. The ultimate bearing capacity of foundations. Geotechnique, 2(4):301-332.

Meyerhof, G.G. and Adams, J.I., 1968. The ultimate uplift capacity of foundations. Canadian Geotech. J., 5(4):225-244.

Mitsch, M.P. and Clemence, S.P., 1985. The uplift capacity of helix anchors in sand. Uplift Behavior of Anchor Foundations in Soil, S.P. Clemence (Editor), ASCE, pp. 26-47.

Mooney, J.S., Adamczak, S., and Clemence, S.P., 1985. Uplift capacity of helical anchors in clay and silt. Uplift Behavior of Anchor Foundations in Soil, S.P. Clemence (Editor), ASCE, pp. 48-72.

Neeley, W.J., Stuart, J.G., and Graham, J., 1973. Failure loads of vertical anchor plates in sand. J. Soil Mech. Found. Div., ASCE, 99(SM9):669-685.

Ovesen, N.K., 1964. Anchor slabs, calculation methods and model tests. Bull. 16, Danish Geotech. Inst., Copenhagen, Denmark.

Ovesen, N.K. and Stromann, H., 1972. Design method for vertical anchor slabs in sand. Proc., Specialty Conf. on Performance of Earth and Earth-Supported Structures, ASCE, 2.1, pp. 1481-1500.

Teng, W.C., 1962. Foundation Design. Prentice-Hall, Englewood Cliffs, New Jersey, U.S.A.

Veesaert, C.J. and Clemence, S.P., 1977. Dynamic pullout resistance of anchors. Proc., Intl. Soil Symposium on Soil-Structure Interaction, Rourkee, India, 1:389-397.

Vesic, A.S., 1971. Breakout resistance of objects embedded in ocean bottom. J. Soil Mech. Found. Div., ASCE, 97(SM9):1183-1205.

CHAPTER 4
Ultimate Bearing Capacity of Shallow Foundations

4.1 INTRODUCTION

Spread footings and mat foundations are generally classified as shallow foundations. These foundations distribute the loads from the superstructures to the soil on which they are resting. Failure of a shallow foundation may occur in two ways: (1) by shear failure of the soil supporting the foundation, and (2) by excessive settlement of the soil supporting the foundation. The first type of failure is generally called bearing capacity failure. The second type of failure may be due to excessive elastic or elastic and consolidation settlement. The procedures for calculation of elastic and consolidation settlements of a foundation can be found in many books (see Das, 1983) and will not be discussed in detail primarily due to lack of space. Instead, this chapter will summarize a number of bearing capacity theories available in literature at the present time.

4.2 TYPES OF BEARING CAPACITY FAILURE

Figure 4.1a shows a shallow foundation of width B located at a depth D_f and supported by a dense sand (or stiff clayey soil) extending to a great depth (relative to B). If this foundation is subjected to a load Q and the load is gradually increased, the load per unit area, $q=Q/A$ (A=area of the foundation), will increase and the foundation will undergo increasing settlement. When q becomes equal to q_u at foundation settlement $S=S_u$, the soil supporting the foundation will undergo sudden shear failure. The failure surface in the soil is shown in Fig. 4.1a along with the q vs. S plot in Fig. 4.1b. This type of failure is referred to as *general shear failure*, and q_u is called the *ultimate bearing capacity*. Note that in this type of failure a peak value $q=q_u$ is clearly noted.

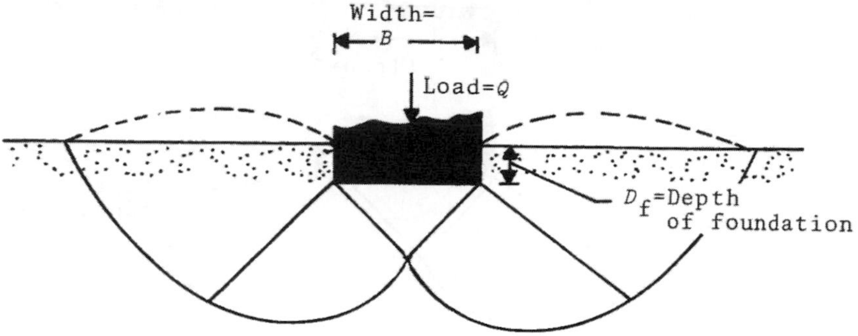

(a)

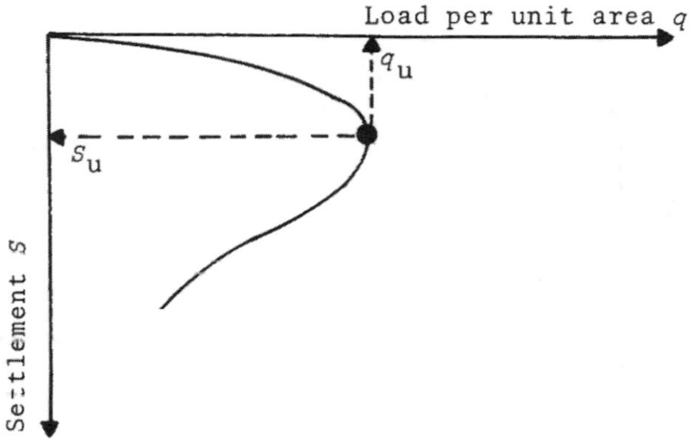

(b)

Figure 4.1. General shear failure in soil.

If the same foundation is now supported by a medium sand or clayey soil (Fig. 4.2a), the plot of q vs. S will be as shown in Fig. 4.2b. Note that the magnitude of q increases with settlement up to $q=q_{u(1)}$, which is usually referred to as the *first failure load* (Vesic, 1963). At this time, the developed failure surface in the soil will be as shown by the solid lines in Fig. 4.2a. However if the load on the foundation is further increased, the load-settlement curve proceeds in almost a linear manner (Fig. 4.2b) along with the gradual progress of the failure surface in the soil (shown by the broken lines in Fig. 4.2a) under the foundation. When q becomes equal to q_u (ultimate bearing capacity), the failure surface extends up to the ground surface. Beyond that, the plot of

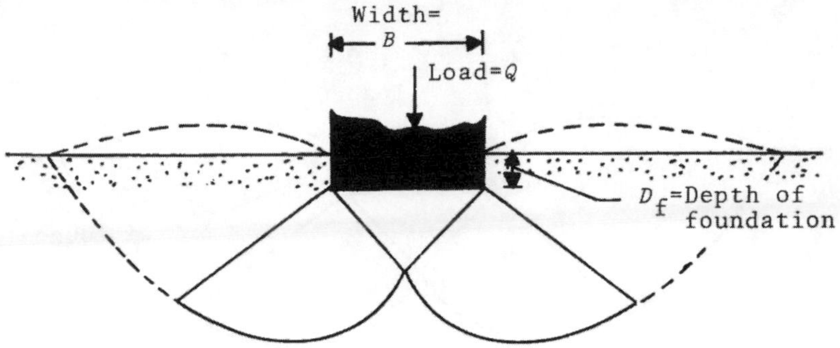

(a)

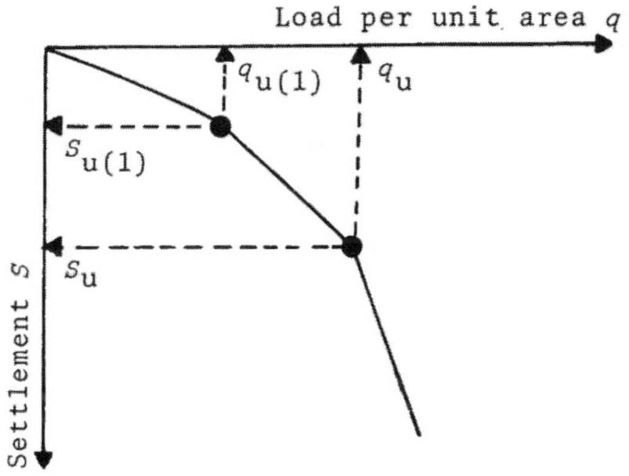

(b)

Figure 4.2. Local shear failure in soil.

q vs. S takes a linear shape and a peak load is never observed. This type of bearing capacity failure is called *local shear failure*.

Figure 4.3a shows the same foundation located on a loose sand or soft clayey soil. For this case, the load-settlement curve will be as shown in Fig. 4.3b. A peak value of load per unit area, q, is never observed. The ultimate bearing capacity q_u is defined as the point where $\Delta S/\Delta q$ becomes the largest and also practically constant. This type of failure in soil is referred to as *punching shear failure*. The failure surface never extends up to the ground surface.

It needs to be pointed out that the nature of bearing capacity failure in soil is not entirely dependent on the nature of *compaction*

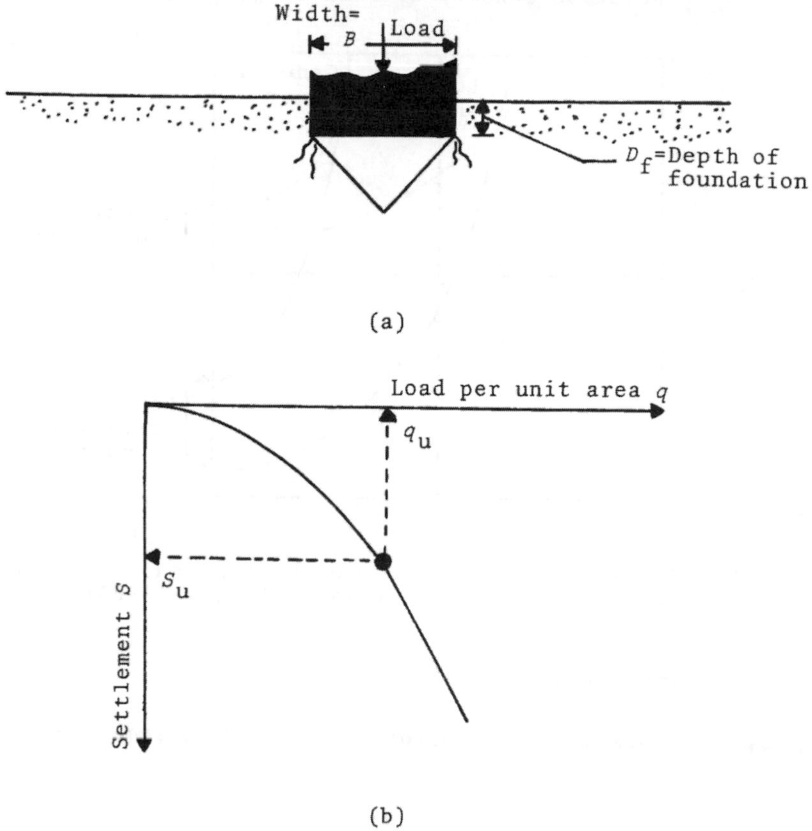

Figure 4.3. Punching shear failure in soil.

of the soil supporting the foundation. It also depends on the relative depth of the foundation, that is D_f/B. As an example, Fig. 4.4 shows the nature of bearing capacity failure of *circular* and *strip* foundations on sand as functions of the relative density and D_f/B. Figure 4.5 shows the plots of the variation of S_u (that is, settlement of the foundation at ultimate load) with the unit weight of soil (and thus relative density) for circular and long rectangular foundations located on sand with $D_f=0$ as obtained by Vesic (1963) from laboratory model tests. It needs to be pointed out that, in general, bearing capacity tests conducted with rectangular foundations with length-to-width ratios (L/B) greater than about 5 to 6 will approximately represent a plane-strain condition (that is, $L/B=\infty$). Also, based on laboratory and field tests, following

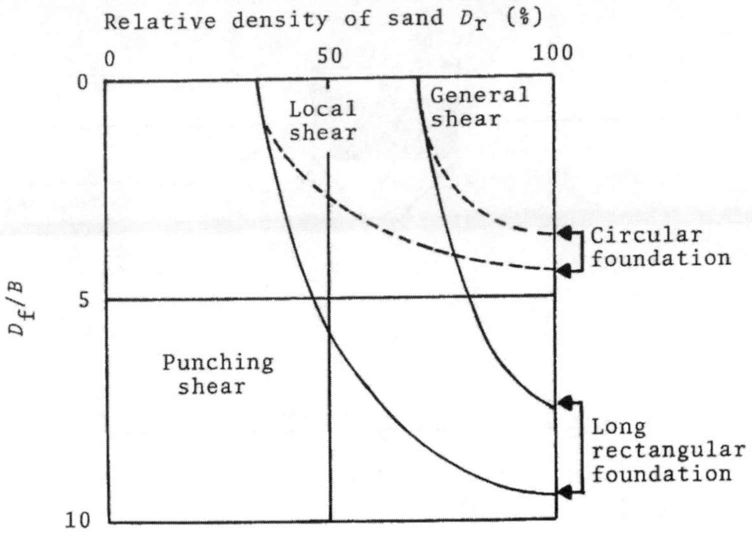

Figure 4.4. Variation of the nature of bearing capacity failure in sand with relative density and D_f/B (after Vesic, 1963).

is the approximate range of values of S_u.

Soil	D_f/B	S_u/B (%)
Sand	0	5 to 12
Sand	Large	25 to 28
Clay	0	4 to 8
Clay	Large	15 to 20

4.3 TERZAGHI'S BEARING CAPACITY THEORY

Terzaghi (1943) proposed a well-conceived theory for determination of the ultimate bearing capacity of shallow rough rigid *continuous* (strip) foundations supported by a homogeneous soil layer extending to a great depth. Terzaghi defined a shallow foundation as a foundation for which the width of the foundation, B, is equal to or less than its depth, D_f (Fig. 4.1a). The failure surface in soil at ultimate load, q_u, per unit area of the foundation, as assumed by Terzaghi is shown in Fig. 4.6. Referring to Fig. 4.6, the failure area in the soil under the foundation can be divided into three major zones. They are:

1. Zone *abc*. This is a triangular *elastic* zone located immediately below the bottom of the foundation. The inclination of

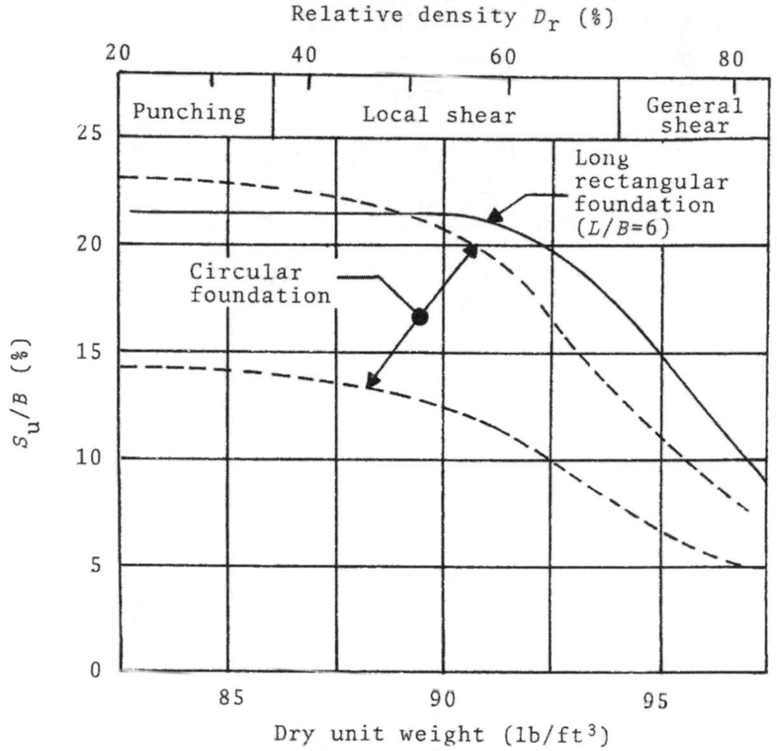

Figure 4.5. Settlement at ultimate load for surface foundation on sand (after Vesic, 1963).

of the sides *ac* and *bc* of the wedge with the horizontal is $\alpha=\phi$ (soil friction angle).

2. Zones *acd* and *bcf*. These zones are referred to as *Prandtl's radial shear zones*.

3. Zones *ade* and *bfg*. These zones are the Rankine passive pressure zones. The *slip lines* in these zones intersect the ground surface at $\pm(45-\phi/2)$ with the horizontal.

The lines *cd* and *cf* are arcs of a *log spiral* defined by the equation

$$r = r_0 e^{\theta \tan\phi} \tag{4.1}$$

The lines *ad*, *de*, *bf*, and *fg* are straight lines. Lines *de* and *fg* actually extend up to the ground surface. Terzaghi assumed that the soil located above the bottom of the foundation can be replaced by a surcharge $q=\gamma D_f$.

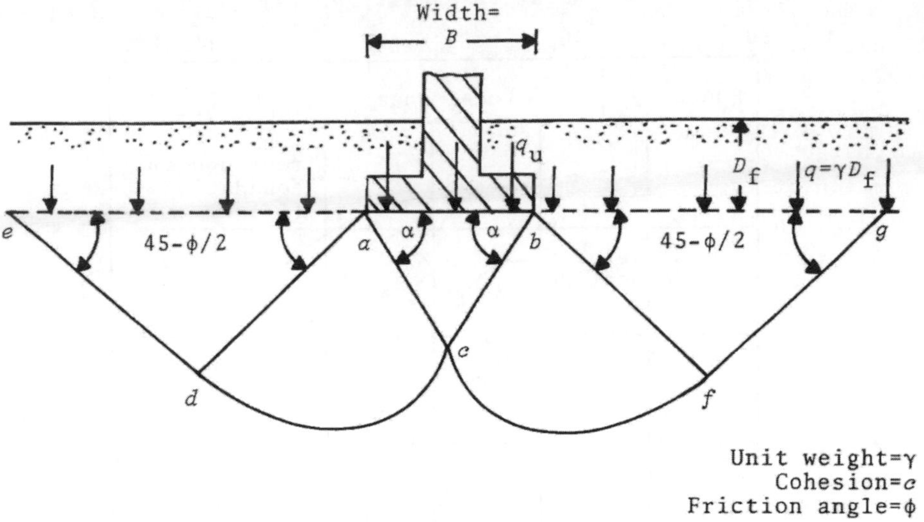

Figure 4.6. Failure surface in soil at ultimate load for a continuous rough rigid foundation as assumed by Terzaghi (1943).

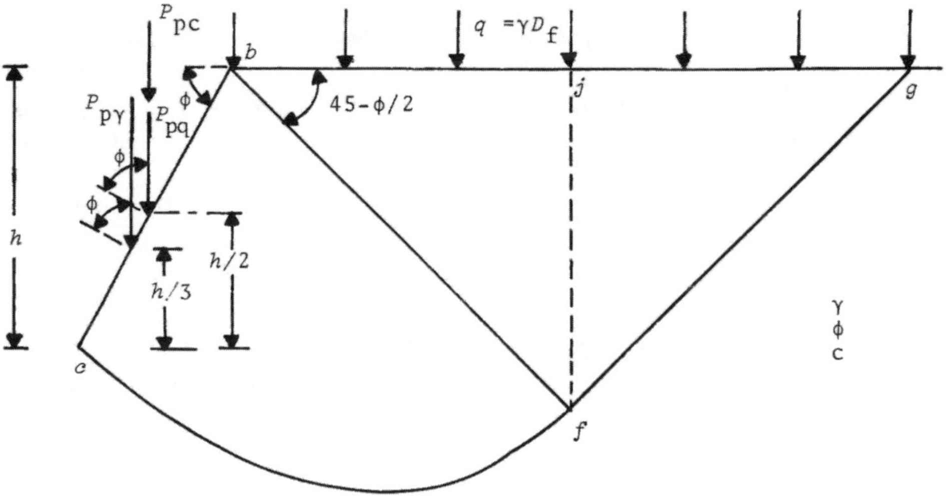

Figure 4.7. The passive force on the face *bc* of the wedge *abc* shown in Fig. 4.6.

The shear strength *s* of the soil can be given by the equation

$$s \quad \sigma' \tan\phi + c \tag{4.2}$$

where σ' = effective normal stress and c = cohesion

The ultimate bearing capacity q_u of the foundation can easily be determined if we consider the faces *ac* and *bc* of the triangular wedge *abc* and determine the passive force on each face required to cause failure. Note that the passive force P_p will be due to the surcharge $(q=\gamma D_f)$, cohesion (c), unit weight (γ), and angle of friction of the soil (ϕ). So referring to Fig. 4.7, the passive force P_p on the face *bc* per unit length of the foundation at right angles to the cross section is

$$P_p \quad P_{pq} + P_{pc} + P_{p\gamma} \tag{4.3}$$

where $P_{pq}, P_{pc}, P_{p\gamma}$ passive force contributions of q, c, and γ, respectively

It is important to note that the directions of P_{pq}, P_{pc}, and $P_{p\gamma}$ are vertical, since the face *bc* makes an angle ϕ with the horizontal, and P_{pq}, P_{pc}, and $P_{p\gamma}$ must make an angle ϕ to the normal drawn to *bc*. In order to obtain P_{pq}, P_{pc}, and $P_{p\gamma}$, we will use the method of superposition which will not be an exact solution.

<u>Determination of P_{pq} $(\phi\neq 0,\ \gamma=0,\ q\neq 0,\ c=0)$:</u>

As in Section 1.14, consider the free body diagram of the soil wedge *bcfj* as shown in Fig. 4.7 (also shown in Fig. 4.8a). For this case the center of the log spiral, of which *cf* is an arc, will be at point *b*. The forces *per unit length of the wedge bcfj* due to the surcharge q only are shown in Fig. 4.8a. They are:

1. P_{pq}
2. q--surcharge
3. $P_{p(1)}$--the Rankine passive force
4. F--the frictional resisting force along the arc *cf*

Note that $P_{p(1)}$ is the *Rankine passive force*, or

$$P_{p(1)} = qK_pH_d = qH_d\tan^2(45+\phi/2) \tag{4.4}$$

The line of action of the frictional force F will pass through point b which is the center of the log spiral (see Appendix B for the properties of log spirals). Now, taking the moment about point b

$$P_{pq}[(\tfrac{1}{2})(\tfrac{B}{2})] = q(\overline{bj})(\tfrac{\overline{bj}}{2}) + P_{p(1)}\,\frac{H_d}{2} \tag{4.5}$$

Let

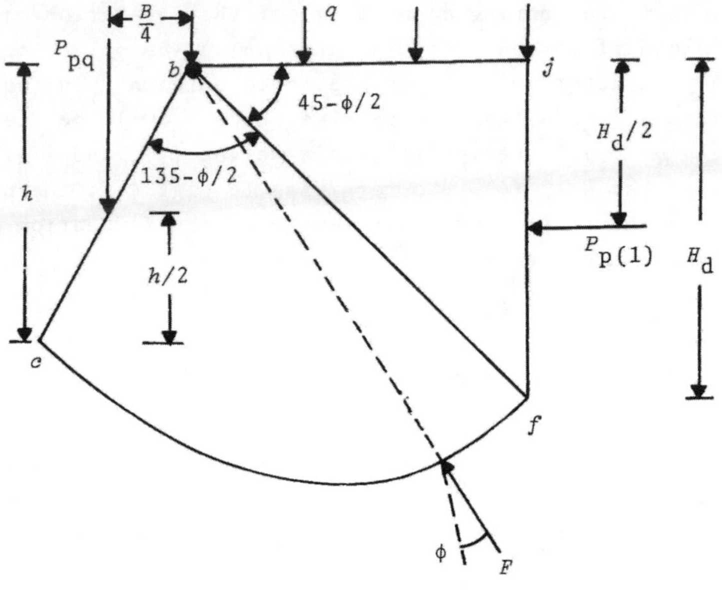

(a)

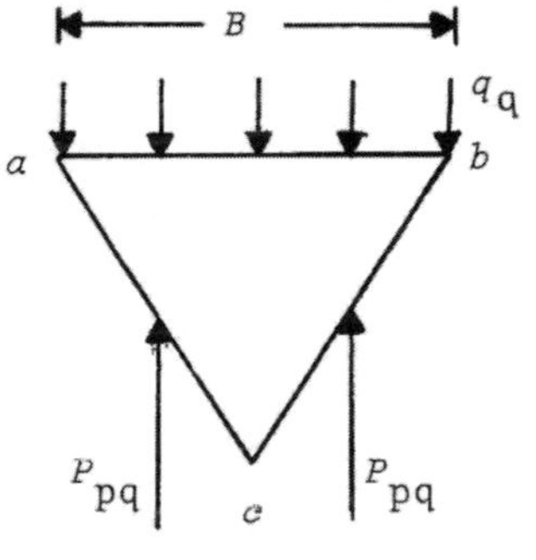

(b)

Figure 4.8. Determination of P_{pq} (γ=0, c=0, $\phi\neq 0$, $q\neq 0$).

$$\overline{bc} = r_0 = (\frac{B}{2})\sec\phi \tag{4.6}$$

From Eq. (4.1)

$$\overline{bf} = r_1 = r_0 e^{(3\pi/4-\phi/2)\tan\phi} \tag{4.7}$$

So

$$\overline{bj} \quad r_1 \cos(45-\phi/2) \tag{4.8}$$

and

$$H_{\mathrm{d}} \quad r_1 \sin(45-\phi/2) \tag{4.9}$$

Combining Eqs. (4.4), (4.5), (4.8), and (4.9)

$$\frac{P_{\mathrm{pq}}B}{2} = \frac{qr_1^2\cos^2(45-\phi/2)}{2} + \frac{qr_1^2\sin^2(45-\phi/2)\tan^2(45+\phi/2)}{2}$$

or

$$P_{\mathrm{pq}} = \frac{4}{B}[qr_1^2\cos^2(45-\phi/2)] \tag{4.10}$$

Now combining Eqs. (4.6), (4.7), and (4.10)

$$P_{\mathrm{pq}} \quad qB\sec^2\phi\left[e^{2(3\pi/4-\phi/2)}\right][\cos^2(45-\phi/2)]$$

$$= \frac{qBe^{2(3\pi/4-\phi/2)\tan\phi}}{4\cos^2(45+\phi/2)} \tag{4.11}$$

Now, referring to the stability of the elastic wedge *abc* under the foundation as shown in Fig. 4.8b

$$q_{\mathrm{q}}(B\times 1) \quad 2P_{\mathrm{pq}}$$

where q_{q} load per unit area on the foundation

or

$$q_{\mathrm{q}} \quad \frac{2P_{\mathrm{pq}}}{B} \quad (q)\underbrace{\left[\frac{e^{2(3\pi/4-\phi/2)\tan\phi}}{2\cos^2(45+\phi/2)}\right]}_{N_{\mathrm{q}}} = qN_{\mathrm{q}} \tag{4.12}$$

Determination of P_{pc} ($\phi\neq 0$, $\gamma=0$, $q=0$, $c\neq 0$):

Figure 4.9a shows the free body diagram for the wedge *bcfj* (also refer to Fig. 4.7). As in the case of P_{pq}, the center of the arc of the log spiral will be located at point *b*. The forces on the wedge which are due to cohesion *c* are also shown in Fig. 4.9a, and they are:

1. Passive force--P_{pc}.
2. $C = c(\overline{bc}\times 1)$.
3. Rankine passive force due to cohesion--$P_{\mathrm{p(2)}} = 2c\sqrt{K_{\mathrm{p}}}H_{\mathrm{d}}$ $2cH_{\mathrm{d}}\tan(45+\phi/2)$.
4. Cohesive force per unit area along arc *cf*--*c*.

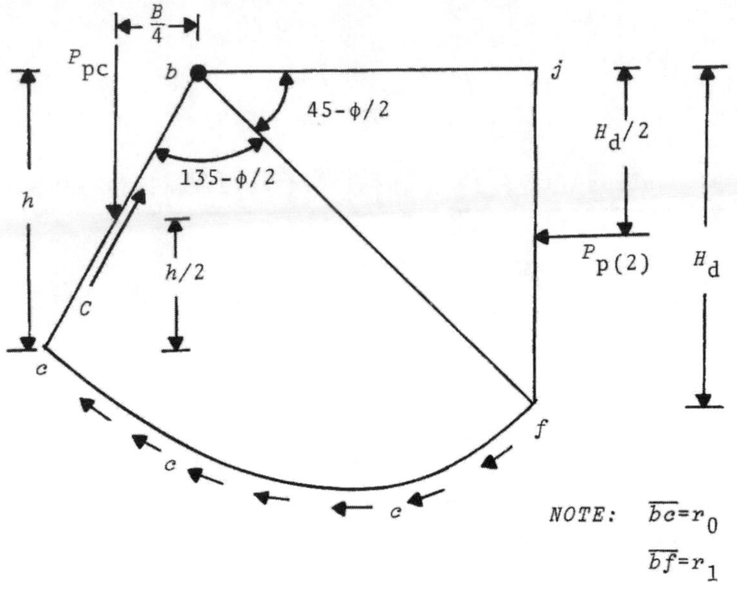

(a)

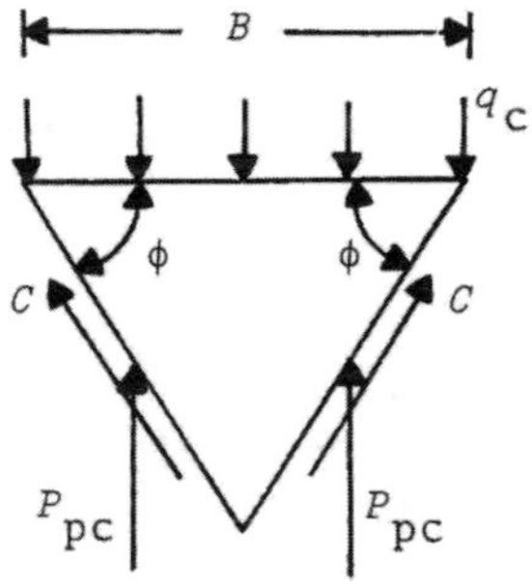

(b)

Figure 4.9. Determination of P_{pc} ($\gamma=0,\ c\neq 0,\ \phi\neq 0,\ q=0$).

Now taking the moment of all the forces about point b

$$P_{pc}\left(\frac{B}{4}\right) \quad P_{p(2)}\left[\frac{r_1 \sin(45-\phi/2)}{2}\right] + M_c \tag{4.13}$$

where M_c = moment due to cohesion c along arc cf
$r_1 = \overline{bf}$

From Eq. (1.77)

$$M_c = \frac{c}{2\tan\phi}(r_0^2 - r_1^2) \tag{4.14}$$

where $r_0 = \overline{bc}$

So

$$P_{pc}\left(\frac{B}{4}\right) = [2cH_d \tan(45+\phi/2)] \left[\frac{r_1 \sin(45-\phi/2)}{2}\right]$$

$$+ \left(\frac{c}{2\tan\phi}\right)(r_0^2 - r_1^2) \tag{4.15}$$

The relationships for H_d, r_0, and r_1 in terms of B and ϕ are given in Eqs. (4.9), (4.6), and (4.7), respectively. Combining Eqs. (4.6), (4.7), (4.9), and (4.15), and noting that $\sin^2(45-\phi/2)\times \tan(45+\phi/2) = \cos\phi/2$

$$P_{pc} = Bc(\sec^2\phi)\left[e^{2(3\pi/4-\phi/2)\tan\phi}\right]\left(\frac{\cos\phi}{2}\right)$$

$$+ \left(\frac{Bc}{2\tan\phi}\right)\sec^2\phi\left[e^{2(3\pi/4-\phi/2)\tan\phi} \quad 1\right] \tag{4.16}$$

Now, let us consider the equilibrium of the soil wedge *abc* (Fig. 4.9b)

$$q_c(B\times 1) = 2C\sin\phi + 2P_{pc}$$

or

$$q_c B = cB\sec\phi\sin\phi + 2P_{pc} \tag{4.17}$$

where q_c = load per unit area of the foundation

Combining Eqs. (4.16) and (4.17)

$$q_c = c\sec\phi e^{2(3\pi/4-\phi/2)} + \frac{c\sec^2\phi}{\tan\phi} e^{2(3\pi/4-\phi/2)\tan\phi}$$

$$- \frac{c\sec^2\phi}{\tan\phi} + c\tan\phi \tag{4.18}$$

or

$$q_c = c\ e^{2(3\pi/4-\phi/2)\tan\phi}\left[\sec\phi + \frac{\sec^2\phi}{\tan\phi}\right] - c\left[\frac{\sec^2\phi}{\tan\phi} - \tan\phi\right] \tag{4.19}$$

However

$$\sec\phi + \frac{\sec^2\phi}{\tan\phi} = \frac{1}{\cos\phi} + \frac{1}{\cos\phi\sin\phi} \qquad \cot\phi\left(\frac{1+\sin\phi}{\cos^2\phi}\right)$$

$$= \cot\phi\left[\frac{1}{2\cos^2(45+\phi/2)}\right] \tag{4.20}$$

Also

$$\frac{\sec^2\phi}{\tan\phi} \quad \tan\phi \quad \cot\phi(\sec^2\phi - \tan^2\phi) = \cot\phi\left(\frac{1}{\cos^2\phi} - \frac{\sin^2\phi}{\cos^2\phi}\right)$$

$$= \cot\phi\left(\frac{\cos^2\phi}{\cos^2\phi}\right) = \cot\phi \tag{4.21}$$

Substituting Eqs. (4.20) and (4.21) into Eq. (4.19)

$$q_c = \frac{c\, e^{2(3\pi/4-\phi/2)\tan\phi}\cot\phi}{2\cos^2(45+\phi/2)} - c\cot\phi$$

$$= c\cot\phi\underbrace{\left[\frac{e^{2(3\pi/4-\phi/2)\tan\phi}}{2\cos^2(45+\phi/2)} \quad 1\right]}_{N_c} = cN_c = c\cot\phi(N_q - 1) \tag{4.22}$$

Determination of $P_{p\gamma}$ ($\phi \neq 0$, $\gamma \neq 0$, $q = 0$, $c = 0$):

Figure 4.10a shows the free body diagram of the wedge *bcfj*. Unlike the free body diagrams shown in Figs. 4.8 and 4.9, the center of the log spiral of which *bf* is an arc is at a point *O* along the line *bf* and not at *b*. This is because the minimum value of $P_{p\gamma}$ has to be determined by means of several trials as done in Section 1.14. The point *O* is only one trial center. Per unit length of the wedge, the forces which need to be considered are:

1. $P_{p\gamma}$.
2. The weight of the wedge *bcfj*--*W*.
3. The resultant of the frictional resisting force acting along the arc *cf*--*F*.
4. The Rankine passive force--$P_{p(3)}$.

The Rankine passive force $P_{p(3)}$ can be given by the relation

$$P_{p(3)} = \tfrac{1}{2}\gamma H_d^2 \tan^2(45+\phi/2) \tag{4.23}$$

Also, note that the line of action of the force *F* will pass through *O*. Now, taking the moment about *O*

$$P_{p\gamma} l_p = W l_w + P_{p(3)} l_R$$

or

$$P_{p\gamma} \quad \frac{1}{l_p}[W l_w + P_{p(3)} l_R] \tag{4.24}$$

If a number of trials of this type are made by changing the location

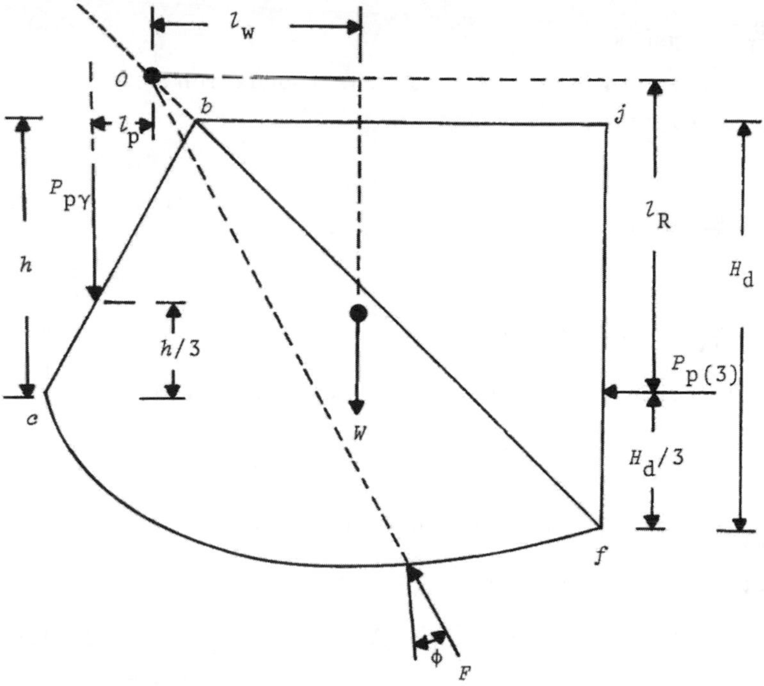

(a)

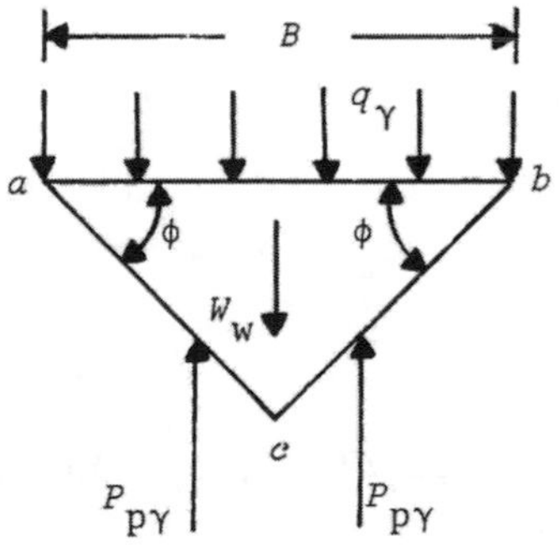

(b)

Figure 4.10. Determination of $P_{p\gamma}$ ($\phi \neq 0$, $\gamma \neq 0$, $q=0$, $c=0$).

of the center of the log spiral O along the line bf, then the minimum value of $P_{p\gamma}$ can be determined.

Now, consider the stability of the wedge abc as shown in Fig. 4.10b. We can write that

$$q_\gamma B = 2P_{p\gamma} - W_w \tag{4.25}$$

where q_γ = force per unit area of the foundation

W_w = weight of the wedge *abc*

However

$$W_w = \frac{B^2}{4}\tan\phi \tag{4.26}$$

So

$$q_\gamma = \frac{1}{B}\left(2P_{p\gamma} - \frac{B^2}{4}\tan\phi\right) \tag{4.27}$$

The passive force $P_{p\gamma}$ can be expressed in the form

$$P_{p\gamma} = \frac{1}{2}\gamma h^2 K_{p\gamma} = \frac{1}{2}\gamma\left(\frac{B\tan\phi}{2}\right)^2 K_{p\gamma} = \frac{1}{8}\gamma B^2\tan^2\phi K_{p\gamma} \tag{4.28}$$

where $K_{p\gamma}$ = passive earth pressure coefficient

Substituting Eq. (4.28) into Eq. (4.27)

$$q_\gamma = \frac{1}{B}\left(\frac{1}{4}\gamma B^2\tan^2\phi K_{p\gamma} - \frac{B^2}{4}\tan\phi\right) = \frac{1}{2}\gamma B\underbrace{\left(\frac{1}{2}K_{p\gamma}\tan^2\phi - \frac{\tan\phi}{2}\right)}_{N_\gamma}$$

$$= \frac{1}{2}\gamma B N_\gamma \tag{4.29}$$

Ultimate Bearing Capacity

The ultimate load per unit area of the foundation (that is, the ultimate bearing capacity q_u) for a soil with cohesion and friction can now be given as

$$q_u = q_q + q_c + q_\gamma \tag{4.30}$$

Substituting the relationships of q_q, q_c, and q_γ as given by Eqs. (4.12), (4.22), and (4.29) into Eq. (4.30) yields

$$q_u = cN_c + qN_q + \frac{1}{2}\gamma B N_\gamma \tag{4.31}$$

where N_c, N_q, N_γ = bearing capacity factors

and

$$N_q = \frac{e^{2(3\pi/4-\phi/2)\tan\phi}}{2\cos^2(45+\phi/2)} \tag{4.32}$$

$$N_c = \cot\phi(N_q - 1) \tag{4.33}$$

$$N_\gamma = \frac{1}{2}K_{p\gamma}\tan^2\phi - \frac{\tan\phi}{2} \tag{4.34}$$

Table 4.1 gives the variations of the bearing capacity factors with soil friction angle ϕ as given by Eqs. (4.32), (4.33), and (4.34). These variations are also plotted in Fig. 4.11. For foundations which are rectangular or circular in plan, a plane strain condition in soil at ultimate load does not exist. For that reason Terzaghi (1943) proposed the following relationships for square and circular foundations.

$$q_u \quad 1.3cN_c + qN_q + 0.4\gamma BN_\gamma$$

(for square foundations having a plan of $B \times B$) (4.35)

and

$$q_u = 1.3cN_c + qN_q + 0.3\gamma BN_\gamma$$

(for circular foundations having a diameter B) (4.36)

Table 4.1. Terzaghi's Bearing Capacity Factors [Eqs. (4.32), (4.33), and (4.34)]

Soil friction angle ϕ (deg)	N_c	N_q	N_γ
0	5.7	1.0	0
5	7.4	1.6	0.5
10	9.6	2.7	1.2
15	12.9	4.5	2.5
20	17.7	7.4	4.0
25	25.2	12.7	9.7
30	37.2	22.5	19.7
35	57.8	41.5	42.5
40	95.7	81.4	100.4
45	172.3	173.3	297.5

4.4 SOME OBSERVATIONS ON TERZAGHI'S BEARING CAPACITY THEORY

It is obvious from Section 4.3 that Terzaghi's bearing capacity theory has been obtained by assuming general shear failure in soil. However for local shear failure in soil, Terzaghi (1943) suggested the following relationships:

·Strip foundation ($B/L=0$; L=length of foundation)

$$q_u = c'N_c' + q'N_q' + \frac{1}{2}\gamma\ BN_\gamma' \tag{4.37}$$

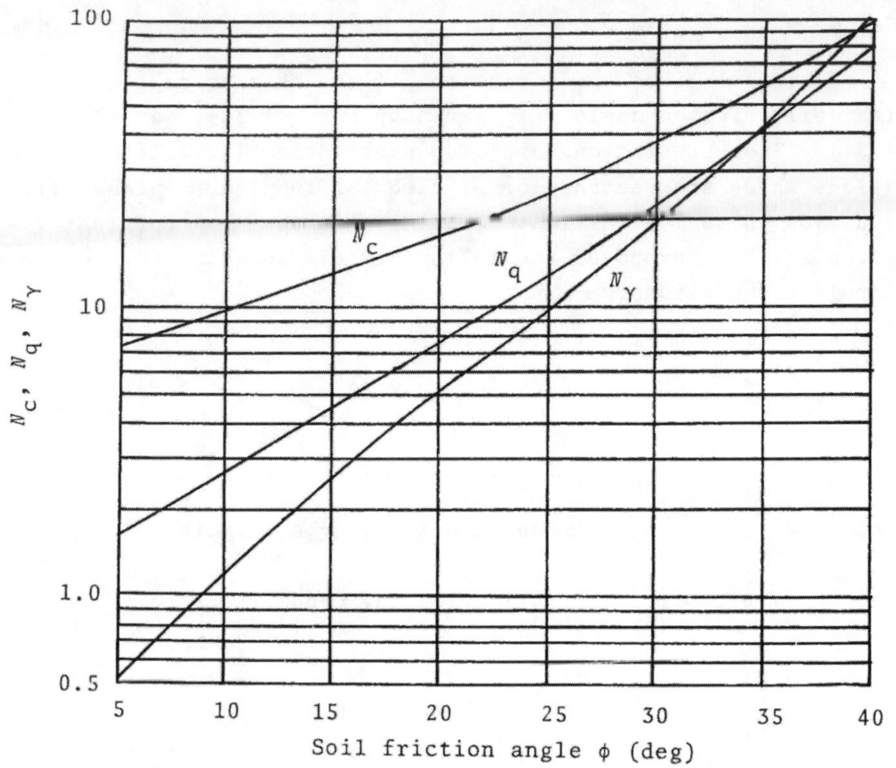

Figure 4.11. Terzaghi's bearing capacity factors--N_c, N_q, and N_γ.

Square foundation ($B=L$)

$$q_u = 1.3c'N'_c + qN_q + 0.4\gamma BN'_\gamma \tag{4.38}$$

Circular foundation (B=Diameter)

$$q_u \quad 1.3c'N'_c + qN_q + 0.3\gamma BN'_\gamma \tag{4.39}$$

where N'_c, N'_q, and N'_γ = modified bearing capacity factors
$c' = 2c/3$

The modified bearing capacity factors can be obtained by substituting $\phi'=\tan^{-1}(0.67\tan\phi)$ for ϕ in Eqs. (4.32), (4.33), and (4.34). The variations of N'_c, N'_q, and N'_γ with ϕ are shown in Table 4.2 and also in Fig. 4.12. Vesic (1973) has suggested that a better form of obtaining ϕ' for estimating N'_q and N'_γ for foundations on sand would be as follows

$$\phi' = \tan^{-1}(k\tan\phi) \tag{4.40}$$

Table 4.1. Terzaghi's Bearing Capacity Factors N'_c, N'_q, and N'_γ

Soil friction angle ϕ (deg)	N'_q	N'_q	N'_γ
0	5.7	1.0	0
5	6.7	1.4	0.2
10	8	1.9	0.5
15	9.7	2.7	0.9
20	11.8	3.9	1.7
25	14.8	5.6	3.2
30	19.0	8.3	5.7
35	25.2	12.6	10.1
40	34.9	20.5	18.8
45	51.2	35.1	37.7

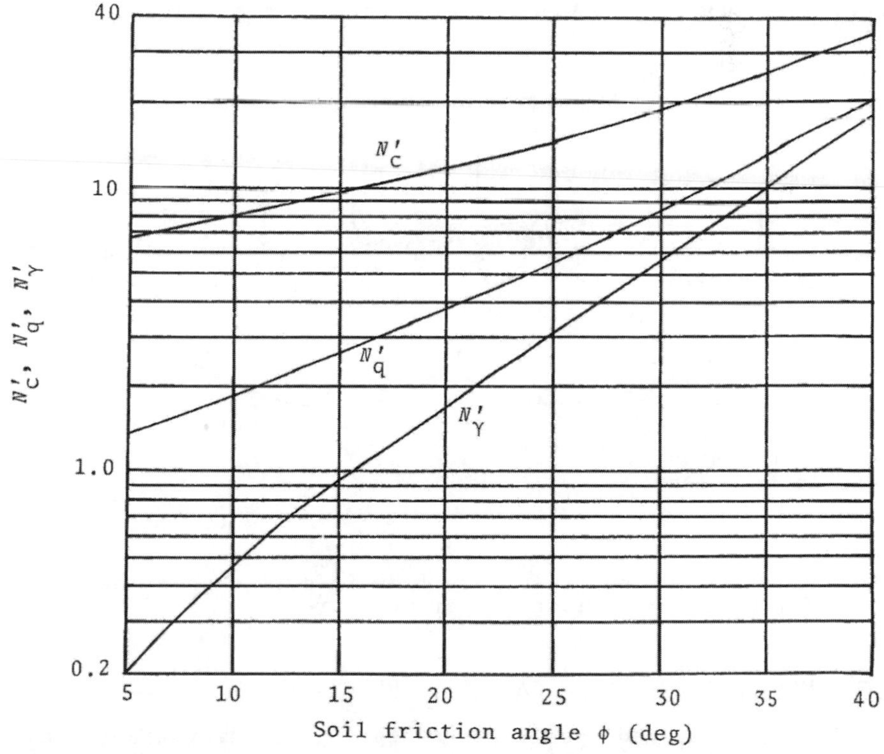

Figure 4.12. Terzaghi's bearing capacity factors for local shear failure--N'_c, N'_q, and N'_γ.

$$k = 0.67 + D_r - 0.75D_r^2 \quad (\text{for } 0 \le D_r \le 0.67) \qquad (4.41)$$

where D_r = relative density

Since Terzaghi's founding work, many experimental studies have been conducted for estimation of the ultimate bearing capacity of shallow foundations. Based on these studies, it appears that Terzaghi's assumption of the failure surface in soil at ultimate load is essentially correct. However, the angle α that the sides of the wedge *ac* and *bc* (Fig. 4.6) make with the horizontal is closer to $45+\phi/2$, and not ϕ as assumed by Terzaghi. In that case, the nature of the soil failure surface would be as shown in Fig. 4.13.

In Section 4.3, the method of superposition has been used to obtain the bearing capacity factors N_q, N_c, and N_γ. For derivation of N_c and N_q, the center of the arc of the log spiral *cf* is located at the edge of the foundation. However, for derivation of N_γ, it is not so. In effect, two different failure surfaces are used in deriving Eq. (4.31). However, it is on the safe side.

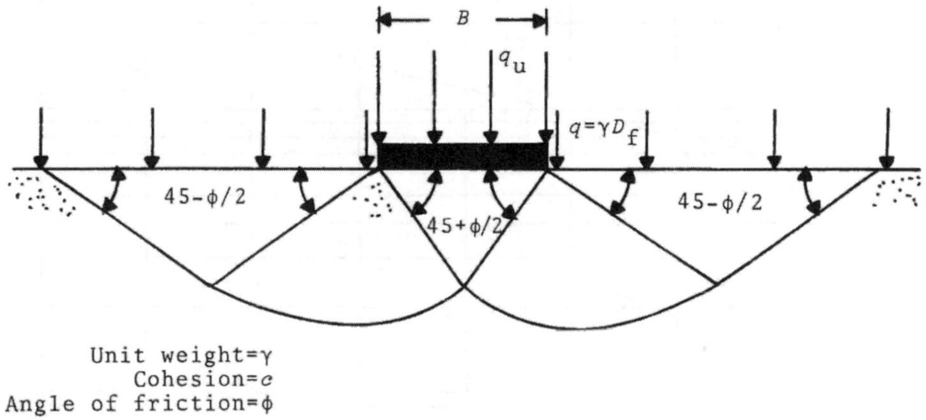

Figure 4.13. Modified failure surface in soil supporting a shallow foundation at ultimate load q_u per unit area.

4.5 MEYERHOF'S BEARING CAPACITY THEORY

In 1951, Meyerhof published a bearing capacity theory which can be applied to rough shallow and deep foundations. The failure surface at ultimate load under a continuous foundation as assumed by Meyerhof (1951) is shown in Fig. 4.14a and b. In this figure,

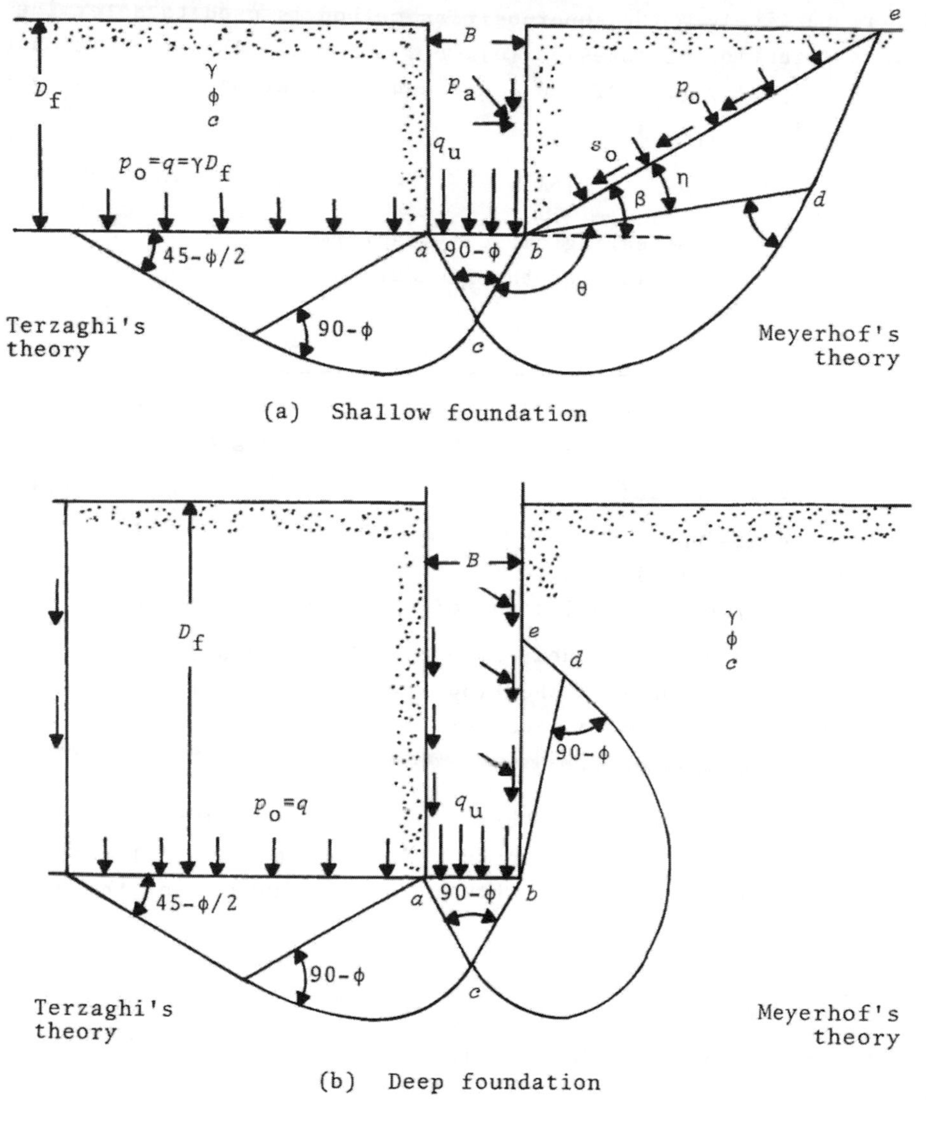

Figure 4.14. Slip line fields for rough strip foundations.

abc is the elastic triangular wedge shown in Fig. 4.13, *bcd* is the radial shear zone with *cd* being an arc of a log spiral, and *bdef* is a mixed shear zone in which the shear varies between the limits of radial and plane shear depending on the depth and roughness of the foundation. The plane *be* is referred to as an *equivalent free surface*. The normal and shear stresses on the plane *be* are p_o and

s_o, respectively. The superposition method is used to determine the contribution of cohesion (c), p_o, and γ and ϕ on the ultimate bearing capacity (q_u) of the *continuous* foundation and can be expressed as

$$q_u = cN_c + p_oN_q + \tfrac{1}{2}\gamma BN_\gamma \tag{4.42}$$

where N_c, N_q, N_γ = bearing capacity factors
B = width of the foundation

Derivations for N_c and N_q ($\phi \neq 0$, $\gamma = 0$, $p_o \neq 0$, $c \neq 0$):

For this case, the center of the log spiral [Eq. (4.1)] arc *cd* is taken at *b*. Also, it is assumed that along *be*

$$s_o \quad m(c + p_o\tan\phi) \tag{4.43}$$

where c = cohesion
ϕ soil friction angle
m degree of mobilization of shear strength ($0 \leq m \leq 1$)

Now, consider the linear zone *bde* (Fig. 4.15a). Plastic equilibrium requires that the shearing strength s_1 under the normal pressure p_1 is fully mobilized, or

$$s_1 \quad c + p_1\tan\phi \tag{4.44}$$

Figure 4.15b shows the Mohr's circle representing the stress conditions on the zone *bde*. Note that P is the *pole*. The traces of planes *bd* and *be* are also shown in this figure. For the Mohr's circle

$$R = \frac{s_1}{\cos\phi} \tag{4.45}$$

where R = radius of the Mohr's circle

Also

$$s_o = R\cos(2\eta+\phi) \quad \frac{s_1\cos(2\eta+\phi)}{\cos\phi} \tag{4.46}$$

Combining Eqs. (4.43), (4.44), and (4.46)

$$\cos(2\eta+\phi) = \frac{s_o\cos\phi}{c+p_1\tan\phi} = \frac{m(c+p_o\tan\phi)\cos\phi}{c+p_1\tan\phi} \tag{4.47}$$

Again, referring to the trace of the *de* plane (Fig. 4.15c)

$$s_1 = R\cos\phi$$

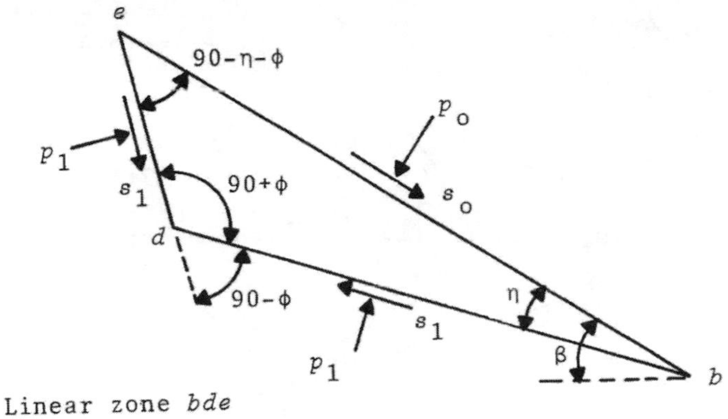

(a)

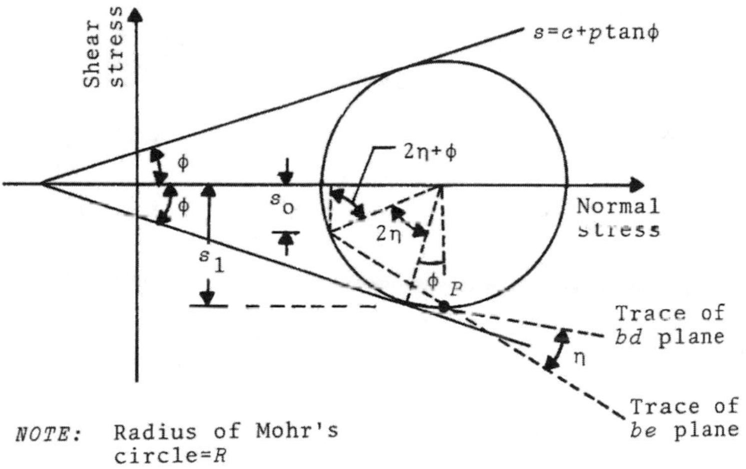

(b)

Figure 4.15. Determination of N_q and N_c.

$$R = \frac{c+p_1\tan\phi}{\cos\phi} \tag{4.48}$$

Note that

$$p_1 + R\sin\phi = p_o + R\sin(2\eta+\phi)$$

$$p_1 = R[\sin(2\eta+\phi)-\sin\phi] + p_o$$

$$\frac{c+p_1\tan\phi}{\cos\phi}[\sin(2\eta+\phi)-\sin\phi] + p_o \tag{4.49}$$

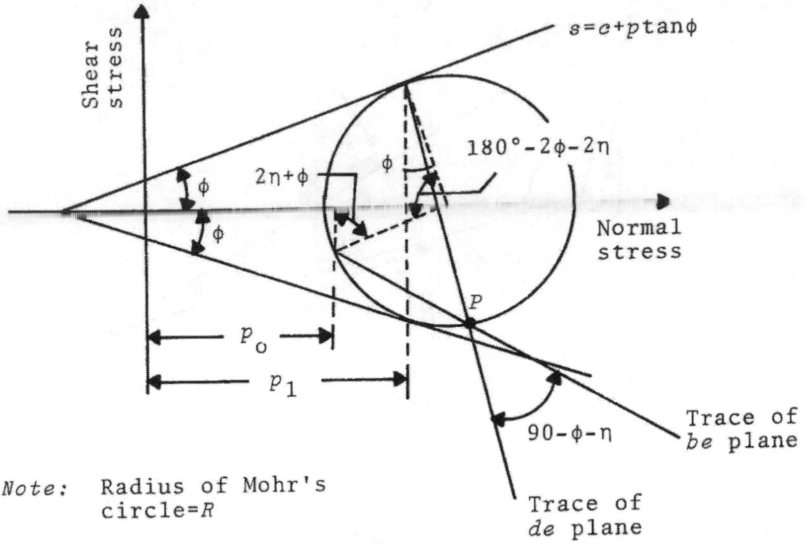

(c)

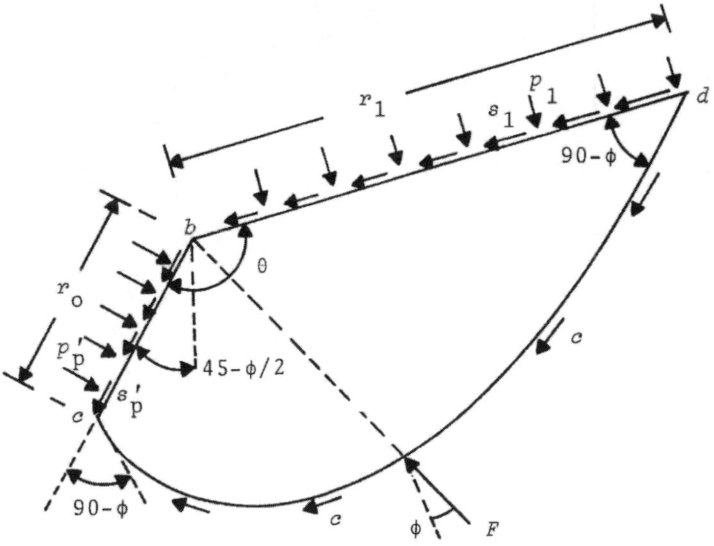

(d)

Figure 4.15. (Continued).

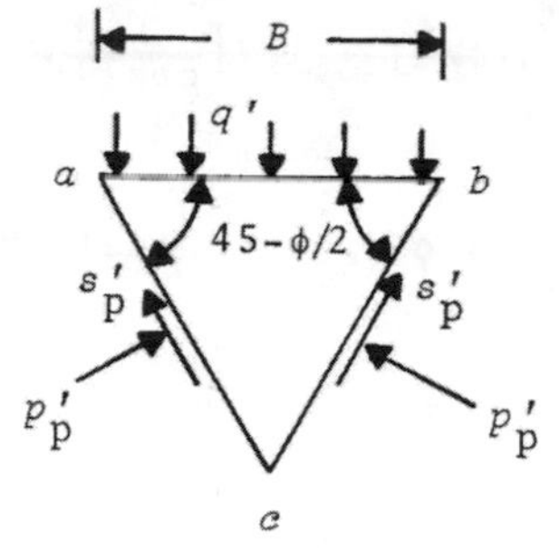

(e)

Figure 4.15. (Continued).

Figure 4.15d shows the free body diagram of the zone *bcd*. Note that the normal and shear stresses on the face *bc* are p'_p and s'_p, or

$$s'_p \quad c+p'_p\tan\phi$$

or

$$p'_p \quad (s'_p-c)\cot\phi \tag{4.50}$$

Taking the moment about *b*

$$p_1\left(\frac{r_1^2}{2}\right) - p'_p\left(\frac{r_0^2}{2}\right) + M_c = 0 \tag{4.51}$$

where $r_0 = \overline{bc}$

$$r_1 = \overline{bd} = r_0 e^{\theta\tan\phi} \tag{4.52}$$

As in Eq. (1.77)

$$M_c = \frac{c}{2\tan\phi}(r_1^2 - r_0^2) \tag{4.53}$$

Substitution of Eqs. (4.52) and (4.53) into Eq. (4.51) yields

$$p'_p = p_1 e^{2\theta\tan\phi} + c\cot\phi\left(e^{2\theta\tan\phi}-1\right) \tag{4.54}$$

Combining Eqs. (4.50) and (4.54)

$$s'_p \quad (c+p_1\tan\phi)e^{2\theta\tan\phi} \tag{4.55}$$

Figure 4.15e shows the free body diagram of the wedge *abc*. Resolving the forces in the vertical direction

$$2p'_p\left[\frac{B/2}{\cos(45+\phi/2)}\right]\cos(45+\phi/2) + 2s'_p\left[\frac{B/2}{\cos(45+\phi/2)}\right]\sin(45+\phi/2)$$

$$= q'B$$

where q' = load per unit area of the foundation

or

$$p'_p B + s'_p B\tan(45+\phi/2) = q'B$$

So

$$q' = p'_p + s'_p \cot(45-\phi/2) \tag{4.56}$$

Substitution of Eqs. (4.49), (4.50), and (4.55) into Eq. (4.56), and further simplification yields

$$q' = \underbrace{c\left\{\cot\phi\left[\frac{(1+\sin\phi)e^{2\theta\tan\phi}}{(1-\sin\phi)\sin(2\eta+\phi)} - 1\right]\right\}}_{N_c} + p_0\underbrace{\left[\frac{(1+\sin\phi)e^{2\theta\tan\phi}}{(1-\sin\phi)\sin(2\eta+\phi)}\right]}_{N_q}$$

$$= cN_c + p_oN_q \tag{4.57}$$

where N_c, N_q = bearing capacity factors

The bearing capacity factors will depend on the degree of mobilization of shear strength on the equivalent free surface m. This is because m controls η. From Eq. (4.47)

$$\cos(2\eta+\phi) = \frac{m(c+p_o\tan\phi)\cos\phi}{c+p_1\tan\phi}$$

For m=0, $\cos(2\eta+\phi)=0$, or

$$\eta = 45 - \phi/2 \tag{4.58}$$

For m=1, $\cos(2\eta+\phi)=\cos\phi$, or

$$\eta = 0 \tag{4.59}$$

Also, the factors N_c and N_q are influenced by the angle of inclination of the equivalent free surface β. From the geometry of Fig. 4.14a

$$\theta = 135° + \beta - \eta - \phi/2 \tag{4.60}$$

From Eq. (4.58), for m=0, the value of η is $(45-\phi/2)$. So

$$\theta = 90 + \beta \quad (\text{for } m=0) \tag{4.61}$$

Similarly, for m=1, since η=0 [Eq. (4.59)]

$$\theta = 135° + \beta - \phi/2 \quad (\text{for } m=1) \tag{4.62}$$

Figures 4.16 and 4.17 shows the variation of N_c and N_q with ϕ, β, and m. It is of interest to note that if we consider the surface foundation condition (as done in Figs. 4.8 and 4.9 for Terzaghi's bearing capacity equation derivation), then $\beta=0$ and $m=0$. So, from Eq. (4.61)

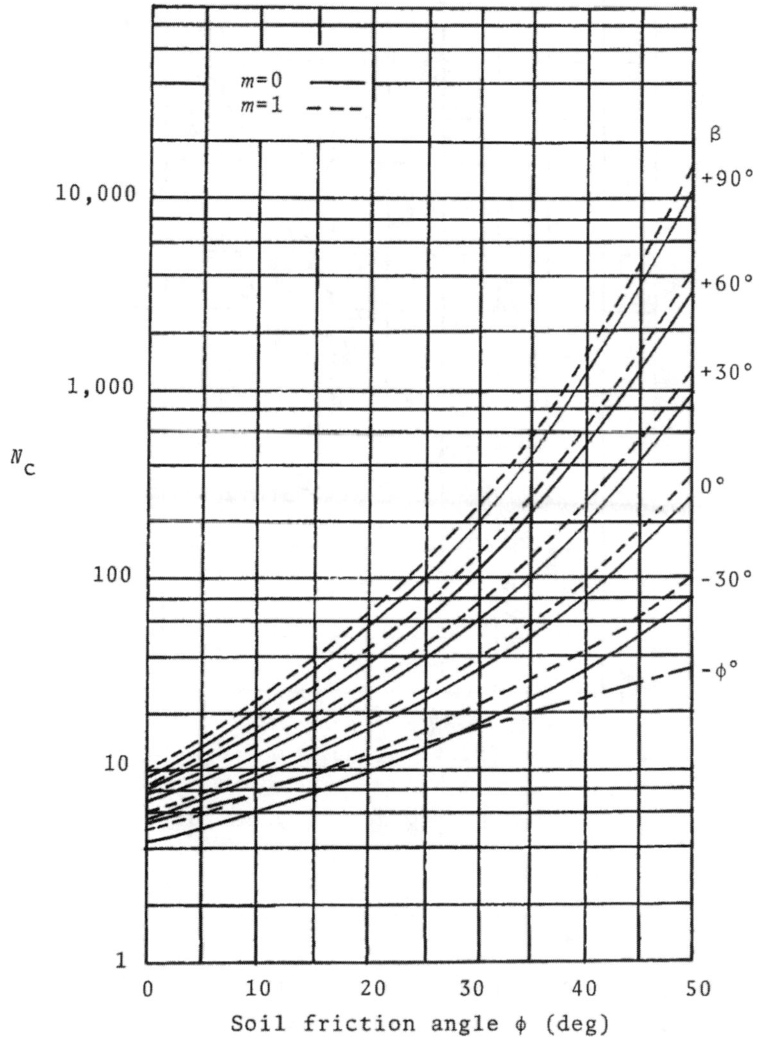

Figure 4.16. Meyerhof's bearing capacity factor--variation of N_c with β, ϕ, and m [Eq. (4.57)].

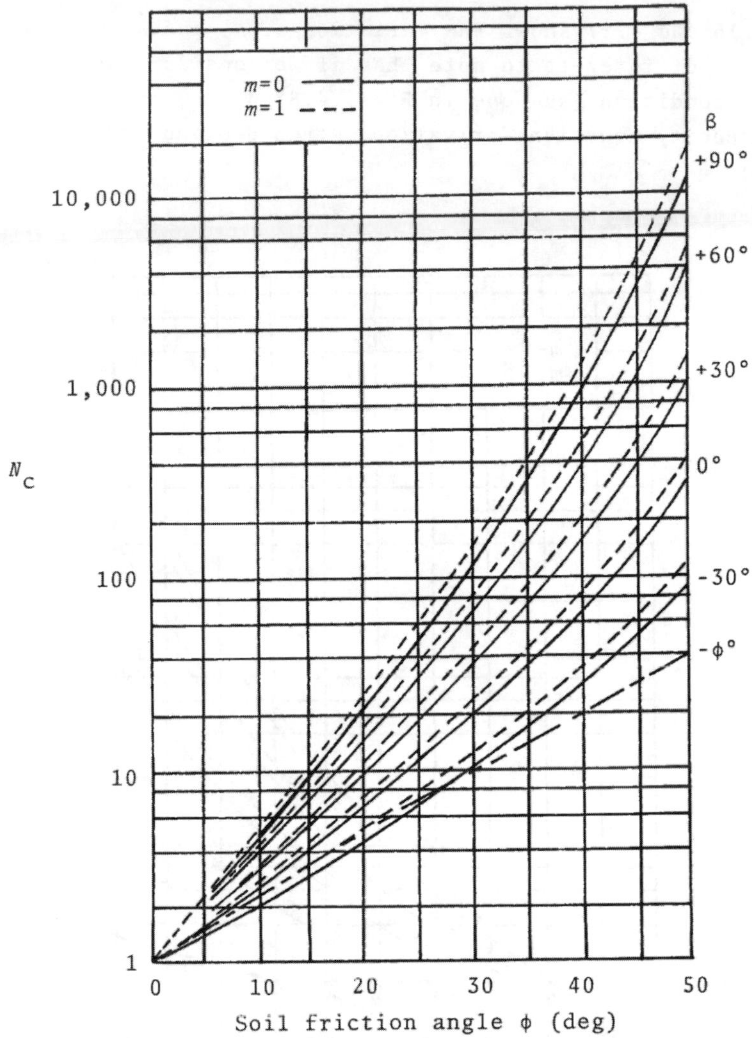

Figure 4.17. Meyerhof's bearing capacity factor--variation of N_q with β, ϕ, and *m* [Eq. (4.57)].

$$\theta = \frac{\pi}{2} \tag{4.63}$$

Hence for $m=0$, $\eta=45+\phi/2$, and $\theta=\phi/2$, the expressions for N_c and N_q are as follows (surface foundation condition)

$$N_q = e^{\pi \tan\phi}\left(\frac{1+\sin\phi}{1-\sin\phi}\right) \tag{4.64}$$

and

$$N_c = (N_q - 1)\cot\phi \tag{4.65}$$

Equations (4.64) and (4.65) are exactly the same as those derived by Reissner (1924) for N_q and Prandtl (1921) for N_c. For this condition, $p_o=\gamma D_f=q$. So, Eq. (4.57) becomes

$$q' \qquad \underset{\uparrow \atop \text{Eq. (4.65)}}{cN_c} + \underset{\uparrow \atop \text{Eq. (4.64)}}{qN_q} \tag{4.66}$$

Determination of N_γ ($\phi\neq0$, $\gamma\neq0$, $p_o=0$, $c=0$):

The determination of N_γ is done by trial and error as in the case of the derivation of Terzaghi's bearing capacity factor N_γ (Section 4.4). Following is a step-by-step approach with reference to Fig. 4.18a.

1. Choose a value for ϕ and a value for the angle β (such as +30°, +40°, -30°).

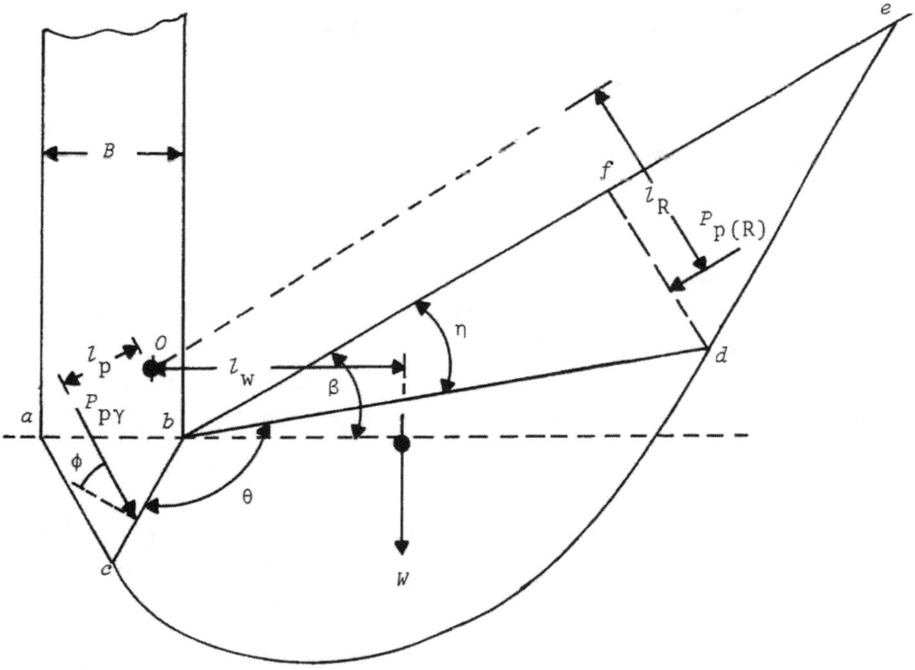

(a)

Figure 4.18. Determination of N_γ.

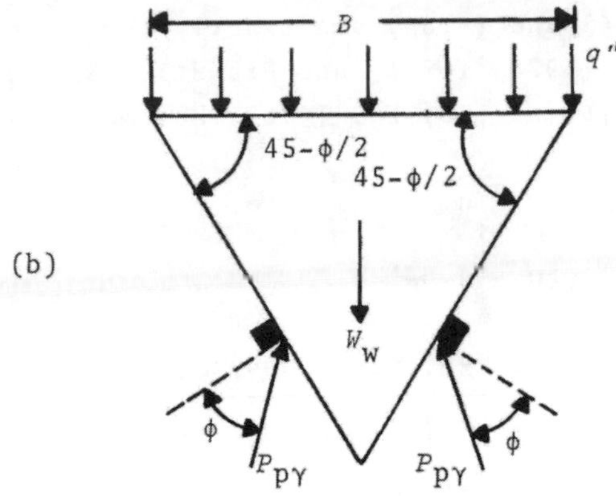

Figure 4.18. (Continued).

2. Choose a value for m (that is, m=0 or m=1).

3. Determine the value of θ from Eq. (4.61) or (4.62) for m=0 or m=1, as the case may be.

4. With known values of θ and β, draw lines *bd* and *be*.

5. Select a trial center such as *O* and draw an arc of a log spiral connecting points *c* and *d*. The log spiral follows the equation

$$r = r_0 e^{\theta \tan\phi}$$

6. Join points *d* and *e*. Note that lines *bd* and *be* make an angle of 90-φ due to the restrictions on slip lines in the linear zone *bde*. Hence the trial failure surface is not, in general, continuous at *d*.

7. Consider the trial wedge *bcdf*. Determine the following forces per unit length of the wedge at right angles to the cross section shown.

   a. Weight of the wedge *bcdf*--W.

   b. Rankine passive force on the face *df*--$P_{p(R)}$. This can be determined by using the procedure outlined in Section 1.12 for obtaining the passive pressure with inclined granular backfill.

8. Take the moment of the forces about the trial center of the log spiral *O*, or

$$P_{p\gamma} = \frac{W l_w + P_{p(R)} l_R}{l_p} \qquad (4.67)$$

where $P_{p\gamma}$ = passive force due to γ and ϕ only

Note that the line of action of $P_{p\gamma}$ acting on the face *bc* is located at two-thirds of the distance from *b*.

9. For given values of β, ϕ, and m, repeat Steps 5 through 8 to obtain the minimum value of $P_{p\gamma}$ by changing the location of point *O* (that is, the center of the log spiral).

10. Refer to Fig. 4.18b. Resolve the forces acting on the triangular wedge *abc* in the vertical direction, or

$$q'' = \frac{\gamma B}{2}\left[\frac{4P_{p\gamma}\sin(45+\phi/2)}{\gamma B^2} - \frac{1}{2}\tan(45+\phi/2)\right] \quad \frac{1}{2}\gamma B N_\gamma \qquad (4.68)$$

where N_γ = bearing capacity factor

q'' = force per unit area of the foundation

Note that W_w is the weight of the wedge *abc* in Fig. 4.18b.

The variation of N_γ (as determined in the above-mentioned manner) with β, ϕ, and m is given in Fig. 4.19.

Now, combining Eqs. (4.57) and (4.68), the ultimate bearing capacity of a continuous foundation (for the conditions $c\neq 0$, $\gamma\neq 0$, and $\phi\neq 0$) can be given as

$$q_u = q' + q'' \quad cN_c + p_o N_q + \frac{1}{2}\gamma B N_\gamma$$

The above equation is the same as given in Eq. (4.42). In a similar manner, for *surface foundation condition* (that is, $\beta=0$ and $m=0$) the ultimate bearing capacity of a continuous foundation can be given as

$$q_u \qquad \underset{\uparrow \atop \text{Eq. (4.66)}}{q'} \quad + \quad \underset{\uparrow \atop \text{Eq. (4.68)}}{q''}$$

$$\underset{\uparrow \atop \text{Eq. (4.65)}}{cN_c} \quad + \quad \underset{\uparrow \atop \text{Eq. (4.64)}}{qN_q} \quad + \frac{1}{2}\gamma B N_\gamma \qquad (4.69)$$

For shallow foundation design consideration, the ultimate bearing capacity relationship given by Eq. (4.69) is presently used. The variation of N_γ for surface foundation condition (that is, $\beta=0$ and $m=0$) as given in Fig. 4.19 can be approximated as (Meyerhof, 1963)

$$N_\gamma = \quad (\underset{\uparrow \atop \text{Eq. (4.64)}}{N_q} - 1)\tan(1.4\phi) \qquad (4.70)$$

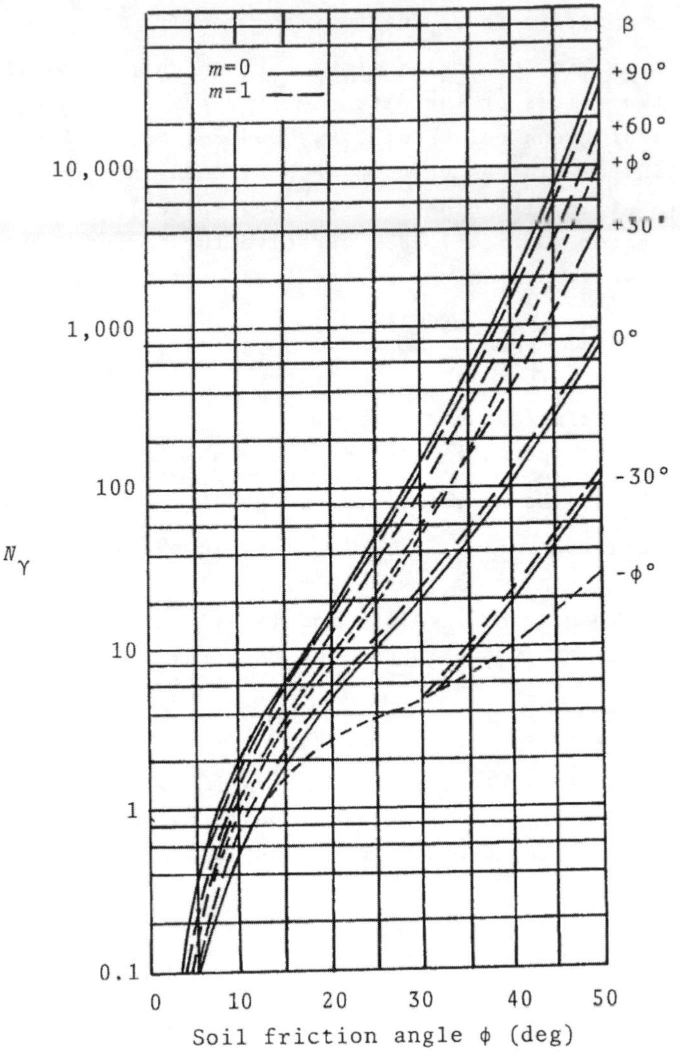

Figure 4.19. Meyerhof's bearing capacity factor--variation of N_γ with β, ϕ, and m [Eq. (4.68)].

Table 4.3 gives the variation of N_c and N_q as obtained from Eqs. (4.64) and (4.65). The variation of N_γ as obtained from Eq. (4.70) is plotted in Fig. 4.20.

Table 4.3. Variation of N_c and N_q [Eqs. (4.64) and (4.65)]

Soil friction angle ϕ (deg)	N_c	N_q
0	5.14	1.0
2	5.63	1.2
4	6.19	1.43
6	6.81	1.72
8	7.53	2.06
10	8.35	2.47
12	9.3	2.97
14	10.4	3.6
16	11.6	4.3
18	13.1	5.3
20	14.8	6.4
22	16.9	7.8
24	19.3	9.6
26	22.3	11.9
28	25.8	14.72
30	30.1	18.4
32	35.5	23.2
34	42.2	29.5
36	50.6	37.8
38	61.4	48.9
40	75.3	64.2
42	93.7	85.4
44	118.4	115.3
46	152.1	158.5
48	199.3	222.3
50	266.9	319.1

4.6 GENERAL BEARING CAPACITY EQUATION

The relationships for estimation of the ultimate bearing capacity presented in Sections 4.3 and 4.5 are for continuous (strip) foundations. They do not give the following:

1. The relationships for the ultimate bearing capacity of rectangular foundations (B=width; L=length)

2. The effect of load inclination ψ with respect to the vertical (Fig. 4.21)

3. The effect of the increase of q_u due to the increase of the foundation depth D_f

For these reasons, a general relationship can be written in the form

$$q_u = cN_c\lambda_{cs}\lambda_{ci}\lambda_{cd} + qN_q\lambda_{qs}\lambda_{qi}\lambda_{qd} + \tfrac{1}{2}\gamma BN_\gamma\lambda_{\gamma s}\lambda_{\gamma i}\lambda_{\gamma d} \tag{4.71}$$

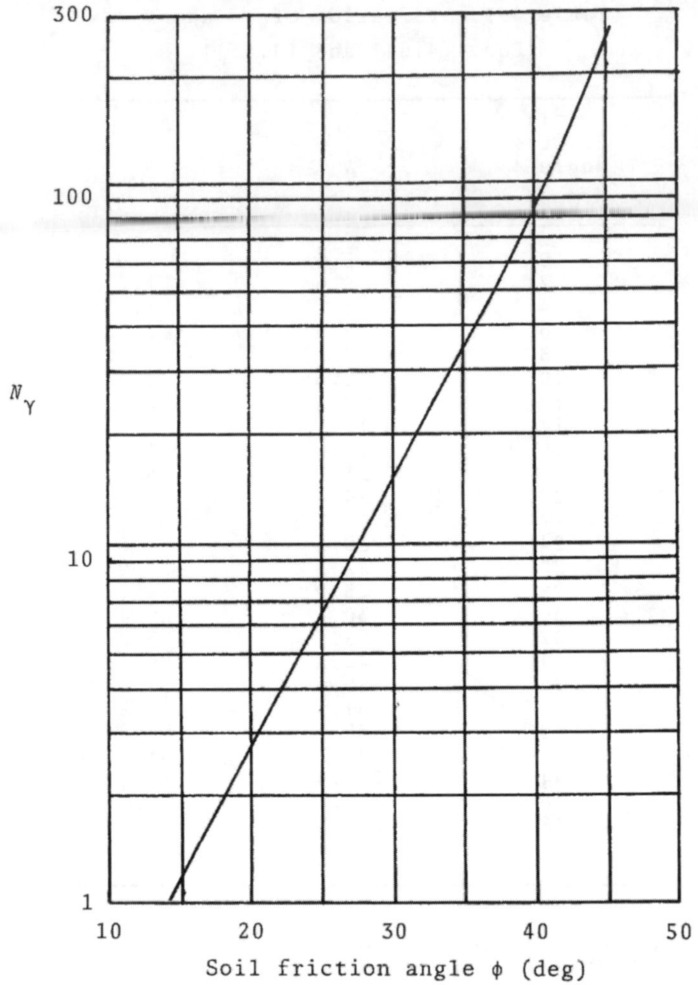

Figure 4.20. Meyerhof's bearing capacity factor N_γ [Eq. (4.70)].

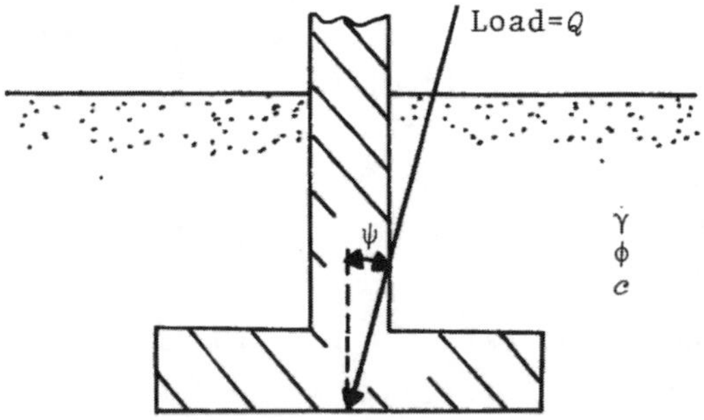

Figure 4.21. Inclined load on a foundation.

where $\lambda_{cs}, \lambda_{qs}, \lambda_{\gamma s}$ = shape factors
$\lambda_{ci}, \lambda_{qi}, \lambda_{\gamma i}$ = inclination factors
$\lambda_{cd}, \lambda_{qd}, \lambda_{\gamma d}$ = depth factors

Most of the shape, inclination, and depth factors available in the literature are now empirical and/or semiempirical. The factors are listed in Table 4.4.

Table 4.4. Summary of Shape, Inclination, and Depth Factors

Factor	Relationship	Reference
Shape	For $\phi=0°$: $\lambda_{cs}=1+0.2\left(\frac{B}{L}\right)$ $\lambda_{qs}=1$ $\lambda_{\gamma s}=1$ For $\phi\geq 10°$: $\lambda_{cs}=1+0.2\left(\frac{B}{L}\right)\tan^2(45+\phi/2)$ $\lambda_{qs}=\lambda_{\gamma s}=1+0.1\left(\frac{B}{L}\right)\tan^2(45+\phi/2)$	Meyerhof (1963)
	$\lambda_{cs}=1+\left(\frac{N_q}{N_c}\right)\left(\frac{B}{L}\right)$ [Note: Use Eq. (4.65) for N_c and Eq. (4.64) for N_q as given in Table 4.3] $\lambda_{qs}=1+\left(\frac{B}{L}\right)\tan\phi$ $\lambda_{\gamma s}=1-0.4\left(\frac{B}{L}\right)$	DeBeer (1970)
Inclination	$\lambda_{ci}=\lambda_{qi}=\left(1-\frac{\psi°}{90°}\right)^2$ $\lambda_{\gamma i}=\left(1-\frac{\psi}{\phi}\right)^2$	Meyerhof (1963)
	$\lambda_{qi}=\left(1-\frac{0.5Q_u\sin\phi}{Q_u\cos\psi+BLc\cot\phi}\right)^5$ $\lambda_{ci}=\lambda_{qi}-\left(\frac{1-\lambda_{qi}}{N_q-1}\right)$ ↑ Table 4.3 (for N_q)	Hansen (1970)

Table 4.4. (Continued).

Factor	Relationship	Reference
	$\lambda_{\gamma i}=1\left(\frac{0.7Q_u\sin\psi}{Q_u\cos\psi+BLc\cot\phi}\right)^5$ Q_u=ultimate load on foundation $=(B\times L)$	
Depth	For $\phi=0$: $\lambda_{cd}=1+0.2\left(\frac{D_f}{B}\right)$ $\lambda_{qd}=\lambda_{\gamma d}=1$ For $\phi>10°$: $\lambda_{cd}=1+0.2\left(\frac{D_f}{B}\right)\tan(45+\phi/2)$ $\lambda_{qd}=\lambda_{\gamma d}=1+0.1\left(\frac{D_f}{B}\right)\tan(45+\phi/2)$	Meyerhof (1963)
	For $D_f/B\leq 1$: $\lambda_{cd}=1+0.4\left(\frac{D_f}{B}\right)$ $\lambda_{qd}=1+0.2\tan\phi(1-\sin\phi)^2\left(\frac{D_f}{B}\right)$ $\lambda_{\gamma d}=1$ For $D_f/B>1$: $\lambda_{cd}=1+0.4\tan^{-1}\left(\frac{D_f}{B}\right)$ $\lambda_{qd}=1+0.2\tan\phi(1-\sin\phi)^2\tan^{-1}\left(\frac{D_f}{B}\right)$ $\lambda_{\gamma d}=1$ [Note: $\tan^{-1}\left(\frac{D_f}{B}\right)$ is in radians]	Hansen (1970)

Rao and Krisnamurthy (1972) have used Sokolovski's method of characteristics (1965) to determine theoretically the variation of $\lambda_{ci}N_c$, $\lambda_{qi}N_q$, and $\lambda_{\gamma i}N_\gamma$, and these are given in Fig. 4.22. These values, although somewhat lower, agree better with Meyerhof's values of $\lambda_{ci}N_c$, $\lambda_{qi}N_q$, and $\lambda_{\gamma i}N_\gamma$ than with those of Hansen.

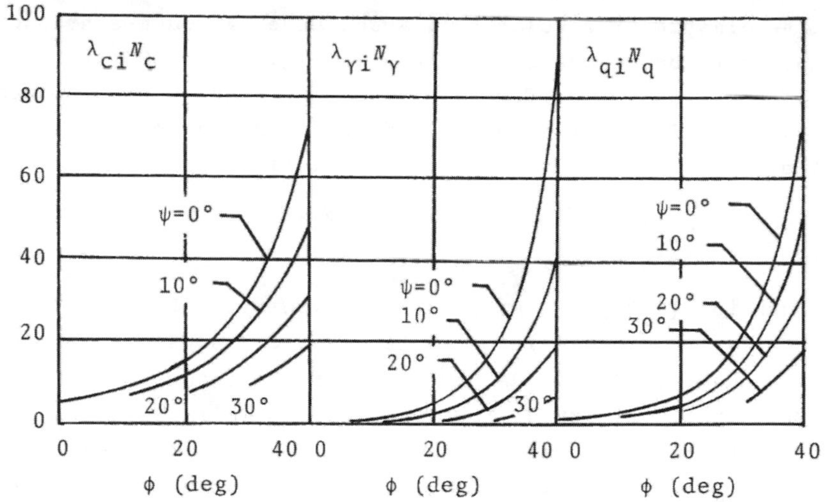

Figure 4.22. Variation of $\lambda_{ci}N_c$, $\lambda_{qi}N_q$, and $\lambda_{\gamma i}N_\gamma$ obtained from the method of characteristics (after Rao and Krisnamurthy, 1972).

4.7 OTHER SOLUTIONS FOR BEARING CAPACITY FACTORS

At this time, the general trend among geotechnical engineers is to accept the method of supposition as a proper means to estimate the ultimate bearing capacity of shallow rough foundations. For *rough continuous* foundations, the nature of failure surface in soil as shown in Fig. 4.13 has also found acceptance, and so have Prandtl's (1921) and Reissner's (1924) solutions for N_c and N_q, which are the same as Meyerhof's solution (1951) for surface foundations, or

$$N_q = e^{\pi \tan\phi}\left(\frac{1+\sin\phi}{1-\sin\phi}\right) \tag{4.64}$$

and

$$N_c = (N_q - 1)\cot\phi \tag{4.65}$$

There has, however, been considerable controversy over the theoretical values of N_γ. Hansen (1970) has proposed an approximate relationship for N_γ in the form

$$N_\gamma = 1.5 N_c \tan^2\phi \tag{4.72}$$

In the preceding equation, the relationship for N_c is that given by Prandtl's solution [Eq. (4.65)]. Caquot and Kerisel (1953) assumed that the elastic triangular soil wedge under a rough continuous foundation to be of the shape shown in Fig. 4.13. Using integration of Boussinesq's differential equation, they presented numerical values of N_γ for various soil friction angles ϕ. Vesic (1973) has approximated their solution in the form

$$N_\gamma = 2(N_q+1)\tan\phi \tag{4.73}$$

where N_q is given by Eq. (4.64) (Table 4.3).

Equation (4.73) has an error not exceeding 5% for $20°<\phi<40°$ as compared to the exact solution. Lundgren and Mortensen (1953) have developed numerical methods, by means of the theory of plasticity, for exact determination of rupture lines as well as the bearing capacity factor (N_γ) for particular cases. Figure 4.23 shows the nature of the rupture lines for this type of solution. Other solutions of N_γ generally referred to in literature are those of Balla (1962) and Chen (1975).

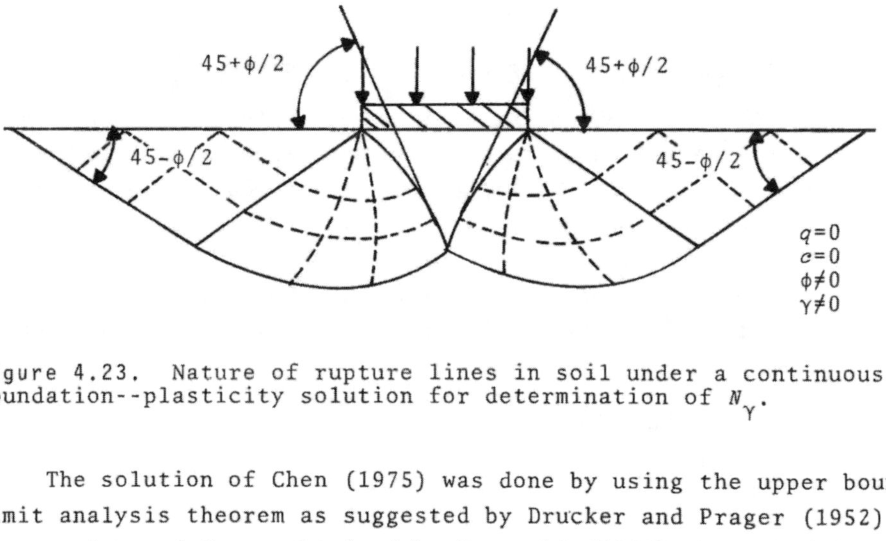

Figure 4.23. Nature of rupture lines in soil under a continuous foundation--plasticity solution for determination of N_γ.

The solution of Chen (1975) was done by using the upper bound limit analysis theorem as suggested by Drucker and Prager (1952). A comparison of N_γ as obtained by Terzaghi (1943), Meyerhof (1951, 1963), Caquot and Kerisel (1953) and approximated by Vesic (1973), Hansen (1970), and Lundgren and Mortensen (1953) are tabulated in Table 4.5. Figure 4.24 also shows a graphical comparison for these solutions. Meyerhof's and Hansen's theories are widely adopted in Canada and Europe. Vesic (1973) has recommended that Eq. (4.73) best fits the experimental results, and hence it should be used.

The primary reason for the development of several theories for

Table 4.5. A Comparison of N_γ Values

Soil friction angle ϕ (deg)	Terzaghi (1943)	Meyerhof Eq.(4.70)	Hansen Eq.(4.72)	Vesic Eq.(4.73)	Lundgren and Mortensen (1953)
0	0	0	0	0	0
5	0.5	0.1	0.1	0.45	0.17
10	1.2	0.4	0.4	1.22	0.46
15	2.5	1.1	1.2	2.65	1.4
25	9.7	6.8	6.9	10.88	6.92
30	19.7	15.7	15.1	22.4	15.32
35	42.4	37.1	33.9	48.03	35.19
40	100.4	93.7	79.5	109.41	86.46
45	297.5	262.7	200.8	271.76	215

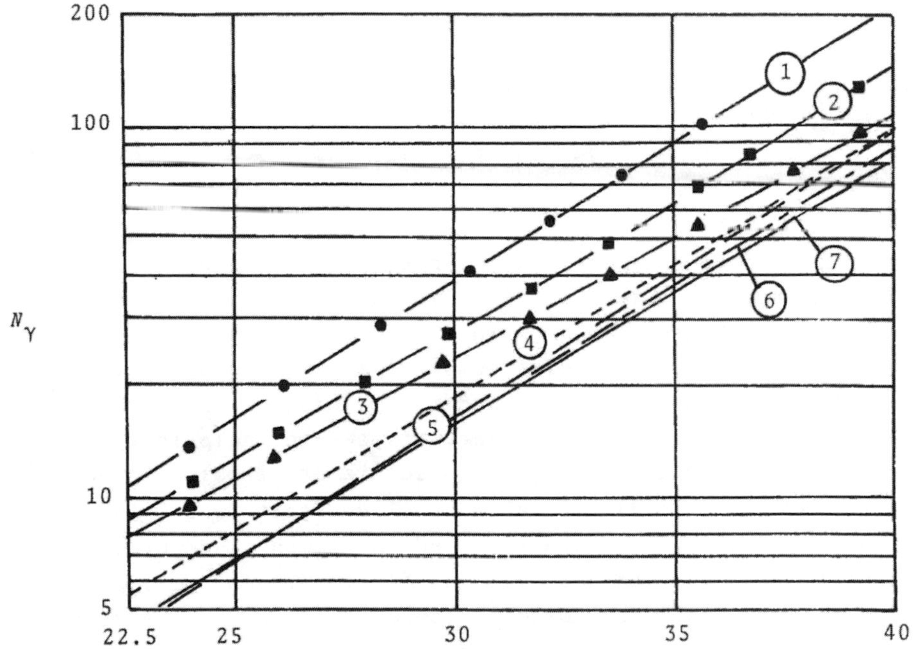

Figure 4.24. Comparison of bearing capacity factor N_γ. [*Note:* Curve 1--Balla (1962), Curve 2--Chen (1975), Curve 3--Vesic (1973), Curve 4--Terzaghi (1943), Curve 5--Meyerhof (1963), Curve 6--Lundgren and Mortensen (1953), Curve 7--Hansen (1970).]

N_γ and their lack of proper correlation with the experimental values lies in the difficulty in selecting a representative value of the soil friction angle ϕ for bearing capacity computation. The parameter ϕ depends on many factors such as intermediate principal stress condition, friction angle anisotropy (Section 4.9), and curvature of the Mohr-Coulomb failure envelope.

Ingra and Baecher (1983) have compared the theoretical solutions of N_γ with the experimental results obtained by several investigators for foundations with B/L=1 and 6 (B=width and L=length of the foundation), and their results are shown in Fig. 4.25a and b. Note that when *triaxial friction angles* are used to deduce experimental N_γ, the values of N_γ are substantially higher than those obtained theoretically. The regression lines shown in Fig. 4.25a and b represent the expected values of N_γ. The *expected values of variances* can be given as

$$E(N_\gamma)_{L/B=1} = \exp(-2.064+0.173\phi_t) \tag{4.74}$$

$$V(N_\gamma)_{L/B=1} \quad (0.0902)\exp(-4.128+0.346\phi_t) \tag{4.75}$$

$$E(N_\gamma)_{L/B=6} \quad \exp(-1.646+0.173\phi_t) \tag{4.76}$$

$$V(N_\gamma)_{L/B=6} \quad (0.0429)\exp(-3.292+0.345\phi_t) \tag{4.77}$$

where ϕ_t triaxial friction angle

It had been suggested in the past to use the plane strain soil friction angle (ϕ_p) instead of ϕ_t for bearing capacity estimation (for example, Hansen, 1970). To that effect, Vesic (1973) raised the issue that this type of assumption may help explain the differences between the theoretical and experimental results for long rectangular foundations. However, it does not help in the interpretation of results of tests with square or circular foundations. Ko and Davidson (1973) have also concluded that "when plane strain angles of internal friction are used in commonly accepted bearing capacity formulas, the bearing capacity for rough footings can be seriously overestimated for dense sands." To avoid the controversy Meyerhof (1963) has suggested the following

$$\phi \quad \left[1.1-0.1\left(\frac{B}{L}\right)\right]\phi_t \tag{4.78}$$

The problem of bearing capacity estimation becomes more complicated if the *scale effect* is taken into consideration. The scale effect, which has come into the limelight in more recent

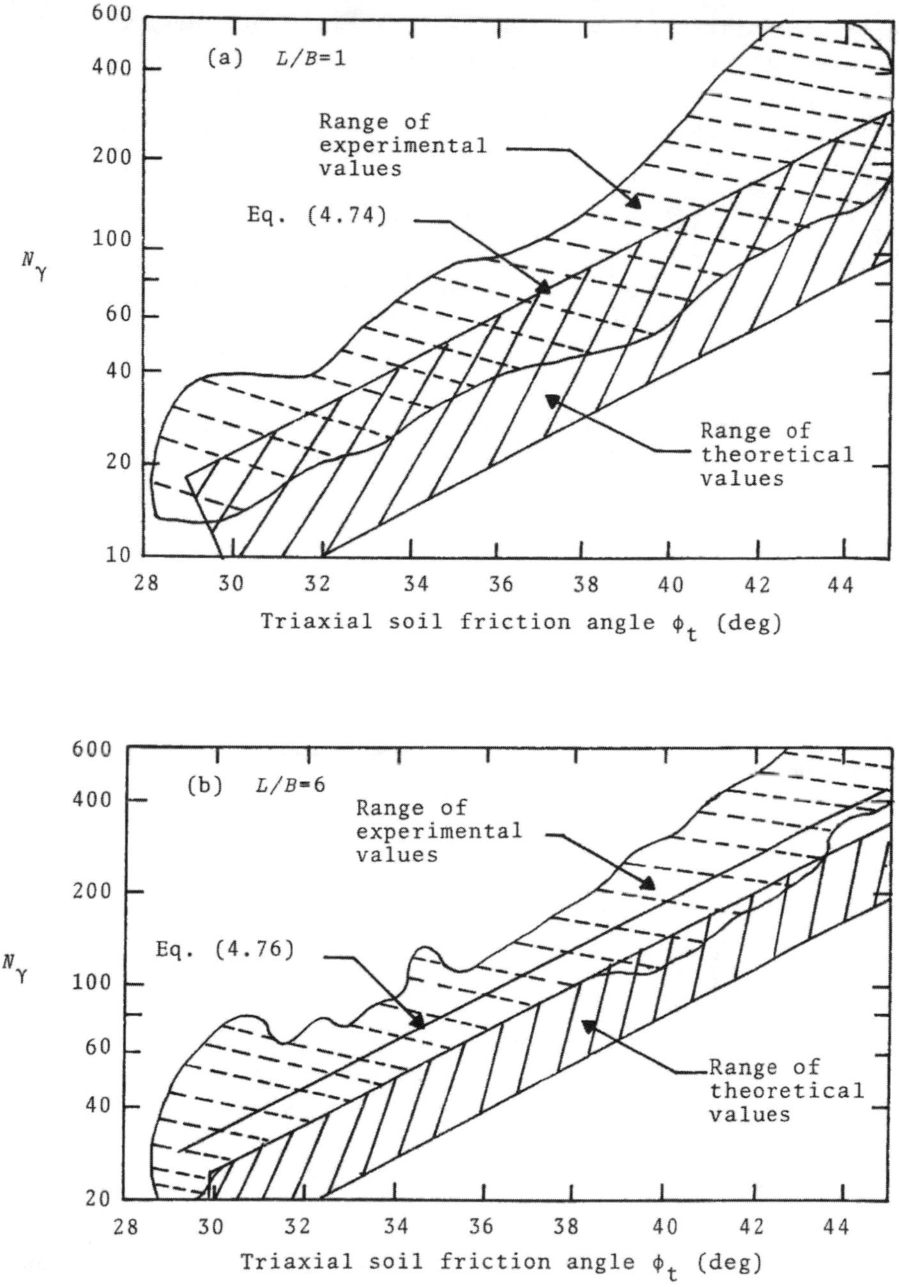

Figure 4.25. Variation of N_γ with triaxial friction angle--comparison of theoretical and experimental values for tests in sand (after Ingra and Baecher, 1983).

years, shows that the ultimate bearing capacity decreases with the increase of the size of the foundation (DeBeer, 1965). Figure 4.26 shows the nature of decrease of N_γ values with the width of the foundation B, which can be attributed primarily to the following reasons:

1. For larger size foundations, the rupture along the slip lines in soil is progressive, and the average shear strength mobilized (and so ϕ) along a slip line decreases with the increase of B.

2. The existence of zones of weakness in the soil under the foundation.

3. The curvature of the Mohr-Coulomb envelope.

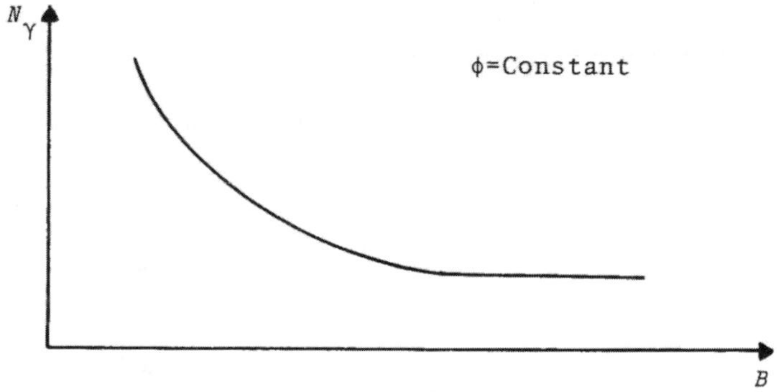

Figure 4.26. Scale effect--nature of variation of N_γ with B.

Example 4.1. A shallow foundation had a width of 0.6 m and a length of 1.2 m. Given D_f=0.6 m. The soil supporting the foundation has the following parameters: ϕ=25°, c=48 kN/m², and γ=18 kN/m³. Determine the allowable vertical load that the foundation can carry by using

a. Prandtl's value of N_c [Eq. (4.65)], Reissner's value of N_q [Eq. (4.64)], Vesic's value of N_γ [Eq. (4.73)] , and the shape and depth factors proposed by DeBeer and Hansen, respectively, (Table 4.4)

b. Meyerhof's values of N_c, N_q, and N_γ [Eqs. (4.65), (4.64) and (4.70)] and the shape and depth factors proposed by Meyerhof (1963) as given in Table 4.4.

Use a factor of safety of 3 over the ultimate bearing capacity.

Solution

From Eq. (4.71), for vertical loading condition

$$q_u = cN_c\lambda_{cs}\lambda_{cd} + qN_q\lambda_{qs}\lambda_{qd} + \tfrac{1}{2}\gamma BN_\gamma\lambda_{\gamma s}\lambda_{\gamma d}$$

(*Note:* $\lambda_{ci}=\lambda_{qi}=\lambda_{\gamma i}=1$.)

Part a. From Table 4.3, for $\phi=25°$, $N_c=20.72$ and $N_q=10.66$. Also, from Table 4.5, for $\phi=25°$, Vesic's value of $N_\gamma=10.88$.

DeBeer's shape factors are as follows:

$$\lambda_{cs} = 1+\left(\frac{N_q}{N_c}\right)\left(\frac{B}{L}\right) \quad 1+\left(\frac{10.66}{20.72}\right)\left(\frac{0.6}{1.2}\right) = 1.257$$

$$\lambda_{qs} = 1+\left(\frac{B}{L}\right)\tan\phi = 1+\left(\frac{0.6}{1.2}\right)\tan 25 \quad 1.233$$

$$\lambda_{\gamma s} = 1-0.4\left(\frac{B}{L}\right) = 1-(0.4)\left(\frac{0.6}{1.2}\right) = 0.8$$

Hansen's depth factors are as follows:

$$\lambda_{cd} = 1+(0.4)\left(\frac{D_f}{B}\right) = 1+(0.4)\left(\frac{0.6}{0.6}\right) \quad 1.4$$

$$\lambda_{qd} = 1+0.2\tan\phi(1-\sin\phi)^2\left(\frac{D_f}{B}\right)$$

$$= 1+(0.2)(\tan 25)(1-\sin 25)^2\left(\frac{0.6}{0.6}\right) = 1.031$$

$$\lambda_{\gamma d} \quad 1$$

So

$$q_u = (48)(20.72)(1.257)(1.4) + (0.6)(18)(10.66)(1.233)(1.031) + \left(\tfrac{1}{2}\right)(18)(0.6)(10.88)(0.8)(1)$$

$$= 1750.2 + 146.35 + 47 = 1943.55 \text{ kN/m}^2$$

Total allowable load =

$$\frac{(q_u)(BL)}{\text{Factor of safety, } F_s} = \frac{(1943.55)(0.6)(1.2)}{3} \approx \underline{466 \text{ kN}}$$

Part b. From Table 4.3, $N_c=20.72$ and $N_q=10.66$. From Table 4.5, Meyerhof's value of N_γ for $\phi=25°$ is 6.8. Now, referring to Table 4.4, Meyerhof's shape and depth factors are as follows:

$$\lambda_{cs} = 1+0.2\left(\frac{B}{L}\right)\tan^2(45+\phi/2)$$

$$1+(0.2)\left(\frac{0.6}{1.2}\right)\tan^2(45+25/2) = 1.246$$

$$\lambda_{qs} = \lambda_{\gamma s} = 1+0.1\left(\frac{B}{L}\right)\tan^2(45+\phi/2)$$

$$1+(0.1)\left(\frac{0.6}{1.2}\right)\tan^2(45+25/2) = 1.123$$

$$\lambda_{cd} \quad 1+0.2\left(\frac{D_f}{B}\right)\tan(45+\phi/2) = 1+(0.2)\left(\frac{0.6}{0.6}\right)\tan(45+25/2) = 1.314$$

$$\lambda_{qd} = \lambda_{\gamma d} = 1+0.1\left(\frac{D_f}{B}\right)\tan(45+\phi/2)$$

$$1+(0.1)\left(\frac{0.6}{0.6}\right)\tan(45+25/2) = 1.157$$

So

$$q_u = (48)(20.72)(1.246)(1.314) + (0.6)(18)(10.66)(1.123)$$
$$(1.157) + \left(\tfrac{1}{2}\right)(18)(0.6)(6.8)(1.123)(1.157)$$
$$= 1628.3 + 149.6 + 47.7 \quad 1825.6 \text{ kN/m}^2$$

Total allowable load =

$$\frac{(q_u)(BL)}{F_s} \quad \frac{(1825.6)(0.6)(1.2)}{3} = \underline{438 \text{ kN}}$$

4.8 EFFECT OF WATER TABLE

In the discussion presented in the preceding sections, it has been assumed that the water table is located below the failure surface in the soil supporting the foundation. However, if the water table is present close to the foundation, the terms q and γ in Eqs. (4.31), (4.35), (4.36), (4.37) to (4.39), Eq. (4.69) and Eq. (4.71) need to be modified. This can be explained with reference to Fig. 4.27, in which the water table is located at a depth d from the ground surface.

For $d=0$, the term $q=\gamma D_f$ should be changed to $q=\gamma' D_f$, and γ should be change to γ' (where γ'=effective unit weight of soil). For $0<d\leq D_f$, q should be equal to $\gamma d+\gamma'(D_f-d)$, and γ should be changed to γ'. Again, if $D_f\leq d\leq B$, $q=\gamma D_f$, and the term γ should be replaced by an average unit weight $\bar{\gamma}$, or

$$\bar{\gamma} \quad \gamma' + \left(\frac{d-D_f}{B}\right)(\gamma-\gamma') \tag{4.79}$$

For $d>D_f+B$, q is equal to γD_f and γ is equal to γ. This implies that the ground water table has no effect on the ultimate capacity.

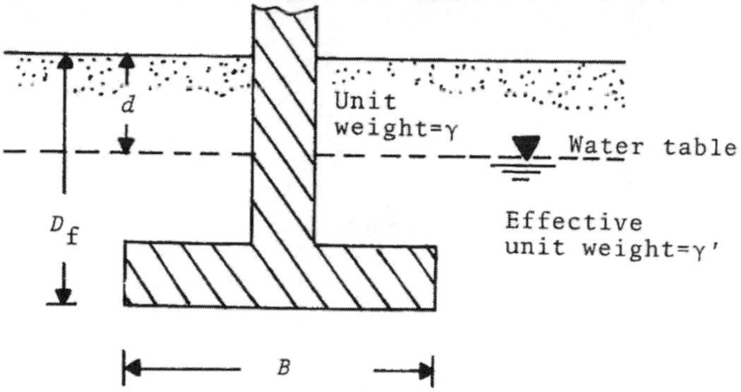

Figure 4.27. Effect of ground water table on the ultimate bearing capacity of shallow foundations.

4.9 BEARING CAPACITY OF FOUNDATIONS ON ANISOTROPIC SOIL EXTENDING TO A GREAT DEPTH

Foundations on Sand (c=0)

Most natural deposits of cohesionless soil have an inherent anisotropic structure due to their nature of deposition in horizontal layers. Initial deposition of the granular soil, and subsequent compaction in the vertical direction, causes the soil particles to take a preferred orientation. For a granular soil of this type, if the *direction of application of deviator stress* makes an angle i with the direction of deposition of soil (Fig. 4.28a), then the friction angle will be ϕ. The soil friction angle ϕ can be approximated in a form (Meyerhof, 1978)

$$\phi = \phi_1 - (\phi_1 - \phi_2)\left(\frac{i°}{90°}\right) \tag{4.80}$$

where ϕ_1 = soil friction angle with i=0°
ϕ_2 = soil friction angle wiht i=90°

Figure 4.28b shows a continuous (strip) rough foundation on an isotropic sand deposit. The slip lines in the soil at ultimate load are also shown in the figure. In the triangular zone (Zone 1) the soil friction angle will be $\phi=\phi_1$. However, the magnitude of ϕ will vary between the limits of ϕ_1 and ϕ_2 in Zone 2. In Zone 3, the effective friction angle of the soil will be equal to ϕ_2. Meyerhof (1978) has suggested that the ultimate bearing capacity of a continuous foundation on an anisotropic sand can be calculated by assuming an equivalent friction angle $\phi=\phi_{eq}$, or

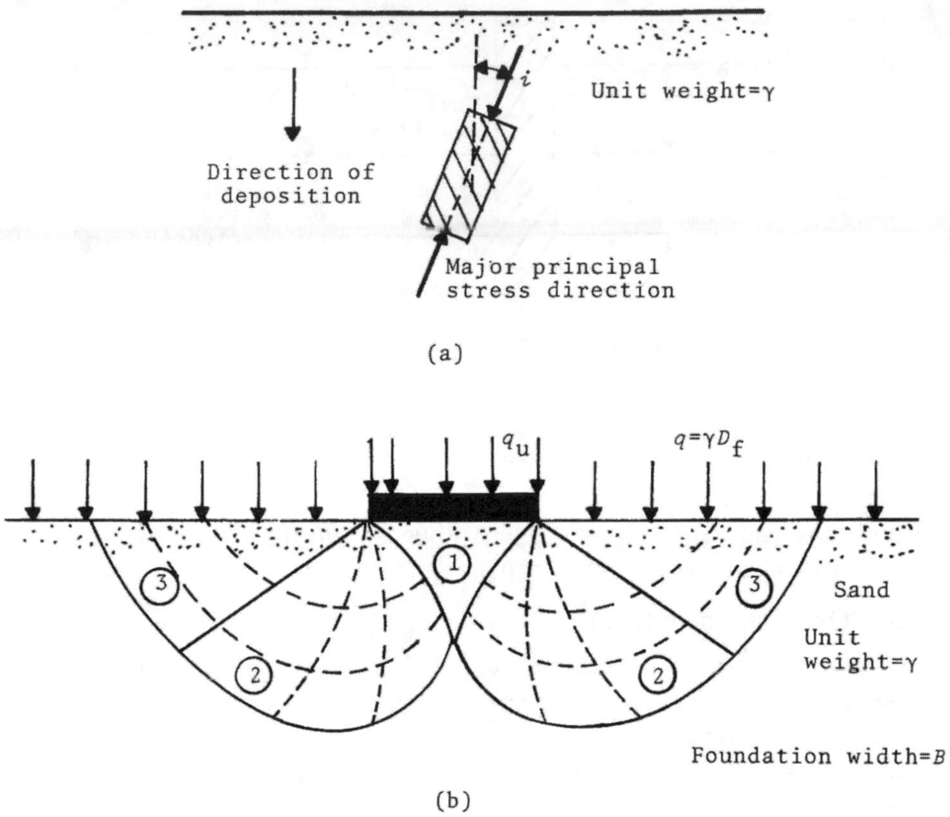

Figure 4.28. Continuous rough foundation on anisotropic sand deposit.

$$\phi_{eq} = \frac{(2\phi_1+\phi_2)}{3} = \frac{(2+m)\phi_1}{3} \tag{4.81}$$

$$\text{where } m = \text{friction ratio} = \frac{\phi_2}{\phi_1} \tag{4.82}$$

Once the equivalent friction angle is determined, the ultimate bearing capacity for vertical loading condition on the foundation can be expressed as (neglecting the depth factors)

$$q_u = qN_{q(eq)}\lambda_{qs} + \tfrac{1}{2}\gamma BN_{\gamma(eq)}\lambda_{\gamma s} \tag{4.83}$$

where $N_{q(eq)}$ and $N_{\gamma(eq)}$ = equivalent bearing capacity factors corresponding to the friction angle $\phi=\phi_{eq}$

In most cases, the value of ϕ_1 will be known. Figure 4.29 presents the plots of $N_{q(eq)}$ and $N_{\gamma(eq)}$ in terms of m and ϕ_1. Note that the soil friction angle ϕ-ϕ_{eq} has been used in Eqs. (4.64) and (4.73) for preparation of the graph. So, combining the relationships for shape factors (Table 4.4) given by DeBeer (1970)

$$q_u \quad qN_{q(eq)}\left[1+\left(\frac{B}{L}\right)\tan\phi_{eq}\right] + \frac{1}{2}\gamma BN_{\gamma(eq)}\left[1-0.4\left(\frac{B}{L}\right)\right] \tag{4.84}$$

Foundations on Saturated Clay (ϕ=0 concept)

As in the case of sand discussed in the preceding section, saturated clay deposits also exhibit anisotropic undrained shear strength properties. Figure 4.30a and b shows the nature of variation of the undrained shear strength of clays with respect to the direction of principal stress application (Davis and Christian, 1971). Note that the undrained shear strength plot shown in Fig. 4.30b is ellipitcal. However, the center of the ellipse does not match the origin. The geometry of the ellipse leads to the equation

$$\frac{b}{a} = \frac{c_{u(i=45°)}}{\sqrt{c_{u(i=0°)}c_{u(i=90°)}}} \tag{4.85}$$

A continuous foundation on a saturated clay layer (ϕ=0) whose directional strength variation follows Eq. (4.85) is shown in Fig. 4.30c. The failure surface in the soil at ultimate load is also shown in this figure. Note that in Zone I, the major principal stress direction is vertical. The direction of the major principal stress is horizontal in Zone III; however, it gradually changes from vertical to horizontal in Zone II. Using the stress characteristic solution, Davis and Christian (1971) have determined the bearing capacity factor $N_{c(i)}$ for the foundation. Or for a surface foundation

$$q_u = N_{c(i)}\left[\frac{c_{u(i=0°)}+c_{u(i=90°)}}{2}\right] \tag{4.86}$$

The variation of $N_{c(i)}$ with the ratio of a/b (Fig. 4.30b) is shown in Fig. 4.31. Note that when a=b, $N_{c(i)}$ becomes equal to N_c=5.14 [isotropic case--Eq. (4.65)].

In many practical conditions, the magnitude of $c_{u(i=0°)}$ and $c_{u(i=90°)}$ may be known, but not the magnitude of $c_{u(i=45°)}$. If such is the case, the magnitude of a/b [Eq. (4.85)] cannot be

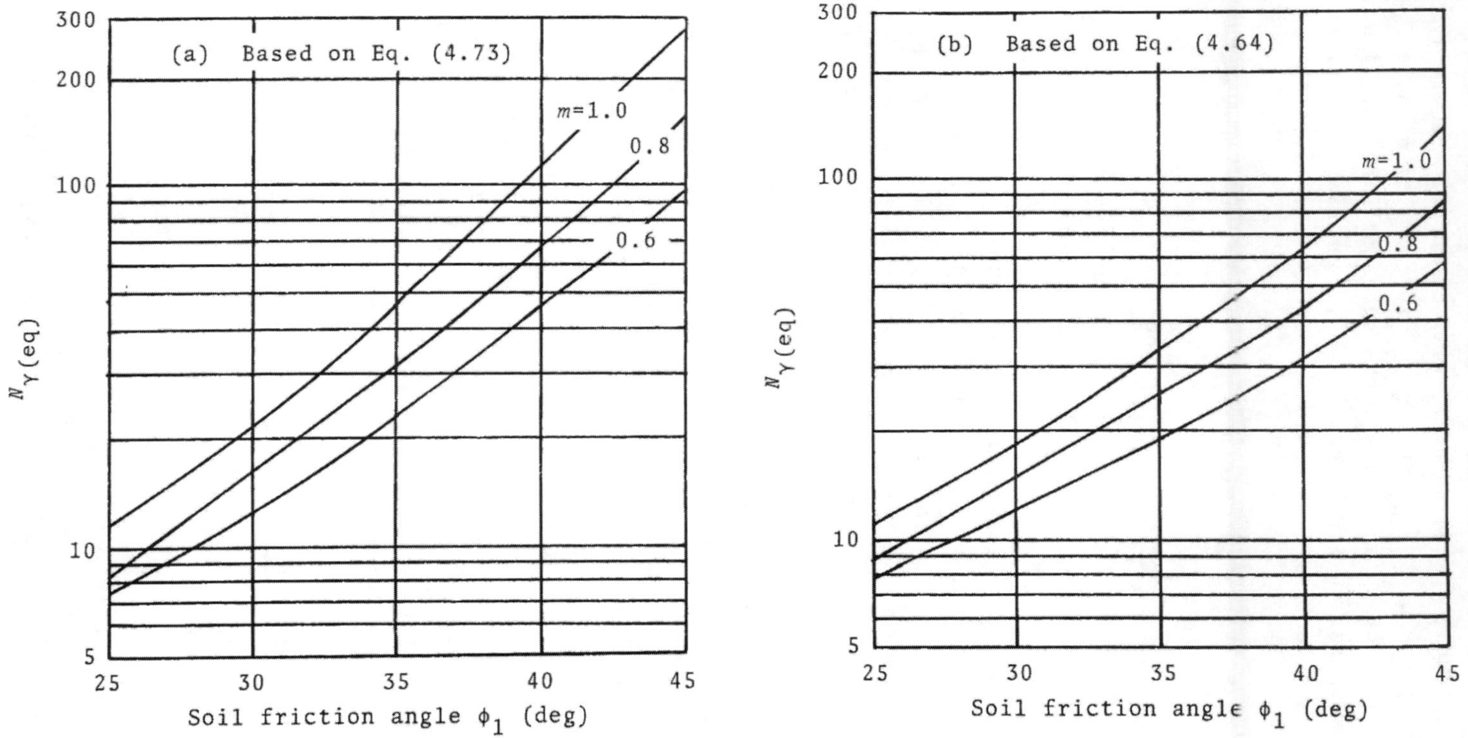

Figure 4.29. Variation of $N_{\gamma(eq)}$ and $N_{q(eq)}$ [Eq. (4.84)].

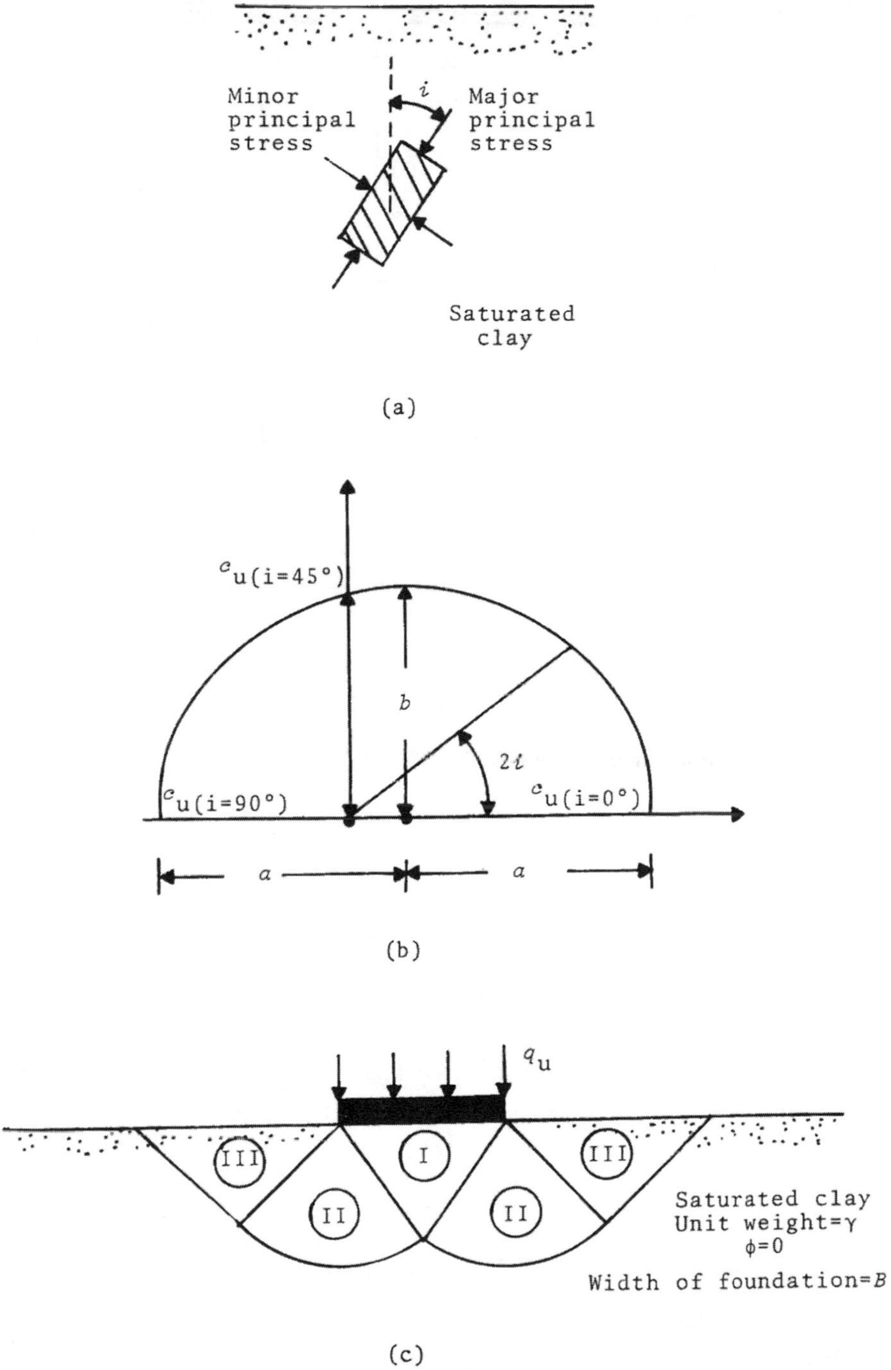

Figure 4.30. Bearing capacity of foundations on anisotropic saturated clay.

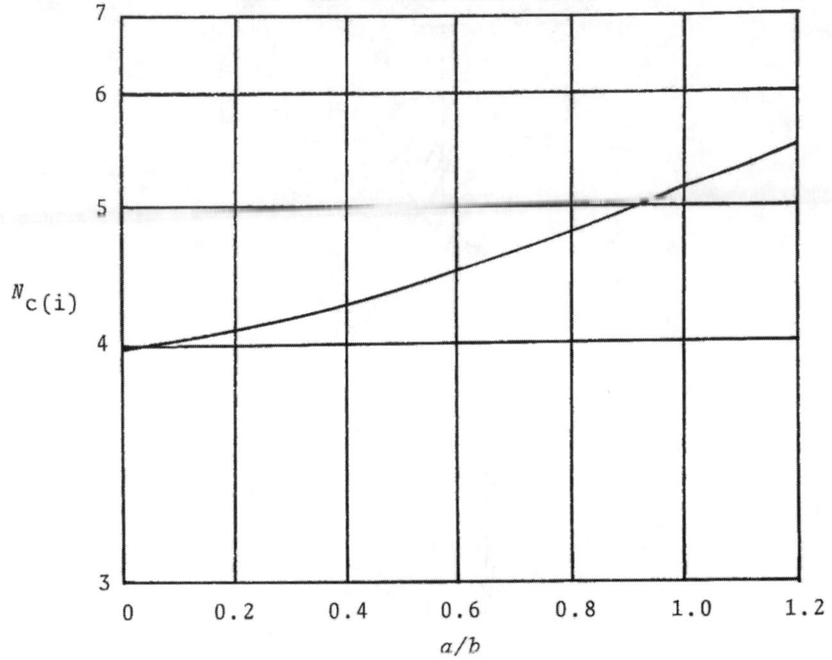

Figure 4.31. Variation of $N_{c(i)}$ with a/b--Eq. (4.86) (redrawn after Davis and Christian, 1971).

determined. For such conditions, the following approximate equation may be used.

$$q_u \simeq 0.9 \underset{\substack{\uparrow \\ =5.14 \\ [\text{Eq.}(4.65)]}}{N_c} \left[\frac{c_{u(i=0°)} + c_{u(i=90°)}}{2}\right] \tag{4.87}$$

The preceding equation, which has been suggested by Davis and Christian (1971), is based on the undrained shear strength results of several clays. So, in general, for a rectangular foundation with vertical loading condition

$$q_u = N_{c(i)}\left[\frac{c_{u(i=0°)} + c_{u(i=90°)}}{2}\right]\lambda_{cs}\lambda_{cd} + qN_q\lambda_{qs}\lambda_{qd} \tag{4.88}$$

For $\phi=0$ condition

$$N_q = 1$$
$$q = \gamma D_f$$

So

$$q_u = N_{c(i)}\left[\frac{c_{u(i=0°)}+c_{u(i=90°)}}{2}\right]\lambda_{cs}\lambda_{cd} + \gamma D_f\lambda_{qs}\lambda_{qd} \quad (4.89)$$

The desired relationships for the shape and depth factors can be taken from Table 4.4 and the magnitude of q_u can be estimated.

Foundations on c-ϕ Soil

The ultimate bearing capacity of shallow continuous foundations supported by anisotropic c-ϕ soil has been studied by Reddy and Srinivasan (1970) by using the method of characteristics. According to this analysis, it is assumed that the shear strength of a soil can be given as

$$s = \sigma'\tan\phi + c$$

However, it is assumed that the soil is anisotropic only with respect to cohesion. As mentioned previously in this section, the direction of the major principal stress (with respect to the vertical) along a slip surface in soil located below the foundation changes. In anisotropic soils, this will induce a change in the shearing resistance to the bearing capacity failure of the foundation. Reddy and Srinivasan (1970) assumed the directional variation of c at a given depth z below the foundation as (Fig. 4.32a)

$$c_{i(z)} = c_{H(z)} + [c_{V(z)}-c_{H(z)}]\cos^2 i \quad (4.90)$$

where $c_{i(z)}$ = cohesion at a depth z when the major principal stress is inclined at an angle i to the vertical (Fig. 4.32b)

$c_{H(z)}$ = cohesion at depth z for $i=90°$

$c_{V(z)}$ = cohesion at depth z for $i=0°$

The preceding equation is of the form as suggested by Casagrande and Carrillo (1944).

Figure 4.32b shows the nature of variation of $c_{i(z)}$ with i. The anisotropy coefficient K is defined as the ratio of $c_{V(z)}$ to $c_{H(z)}$.

$$K = \frac{c_{V(z)}}{c_{H(z)}} \quad (4.91)$$

In *overconsolidated* soils K is less than one, and for *normally consolidated* soils the magnitude of K is greater than one.

For many consolidated soils, the cohesion increases linearly with depth (Fig. 4.32c). Thus

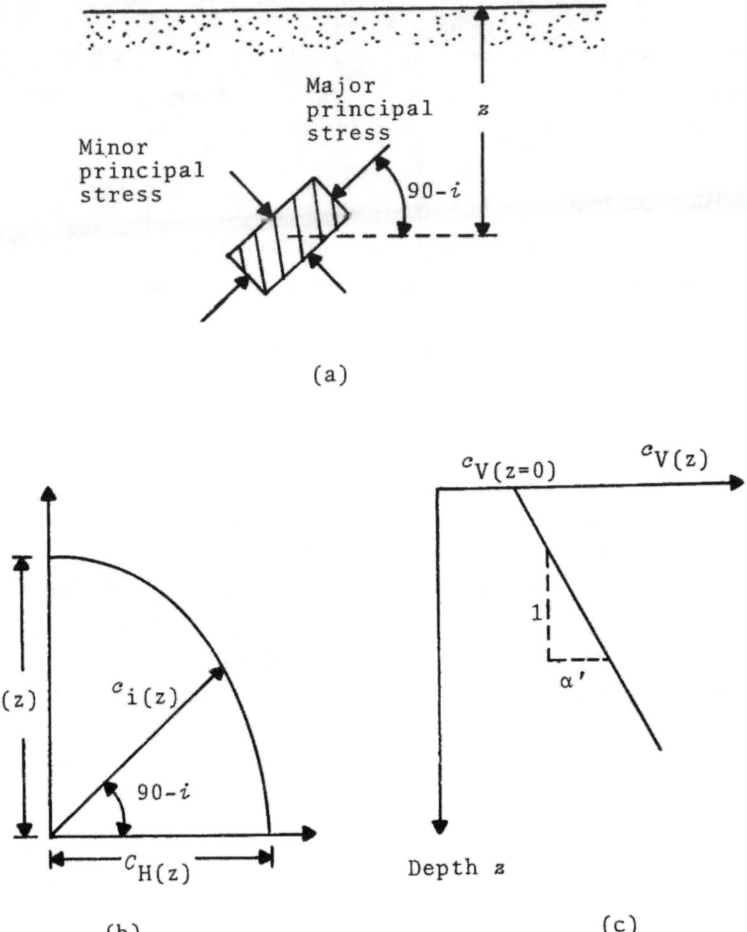

Figure 4.32. Anisotropic clay soil--assumptions for bearing capacity evaluation.

$$c_{V(z)} = c_{V(z=0)} + \alpha' z \tag{4.92}$$

where $c_{V(z)}, c_{V(z=0)}$ cohesion in the vertical direction at depths of z and z=0, respectively

α' = the rate of variation with depth z

According to this analysis, the ultimate bearing capacity of a *continuous* foundation may be given as

$$q_u = c_{V(z=0)} N_{c(i')} + q N_{q(i')} + \tfrac{1}{2}\gamma B N_{\gamma(i')} \tag{4.93}$$

where $N_{c(i')}, N_{q(i')}, N_{\gamma(i')}$ = bearing capacity factors

$$q = \gamma D_f$$

This equation is similar to Terzaghi's bearing capacity equation for continuous foundations [Eq. (4.31)].

The bearing capacity factors are functions of the parameters β_c and K, which can be defined as

$$\beta_c = \frac{\alpha' l}{c_{V(z=0)}} \tag{4.94}$$

where l = characteristic length $\dfrac{c_{V(z=0)}}{\gamma}$ (4.95)

Furthermore, $N_{c(i')}$ is also a function of the nondimensional width of the foundation B'

$$B' = \frac{B}{l} \tag{4.96}$$

The variations of the bearing capacity factors with β_c, B', ϕ, and K as determined by Reddy and Srinivasan (1970) are shown in Figs. 4.33, 4.34, and 4.35.

This study shows that the rupture surface in soil at ultimate load extends to a smaller distance below the bottom of the foundation for the case where the anisotropic constant K is greater than

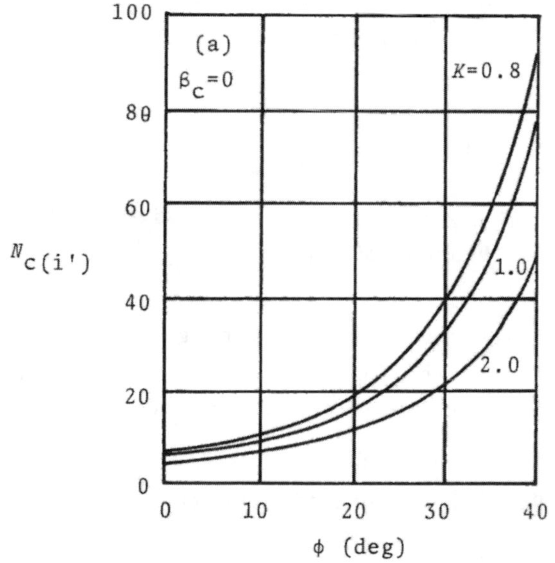

Figure 4.33. Influence of K on $N_{c(i')}$: (a) β_c=0; (b) β_c=0.2; and (c) β_c=0.4 (after Reddy and Srinivasan, 1970).

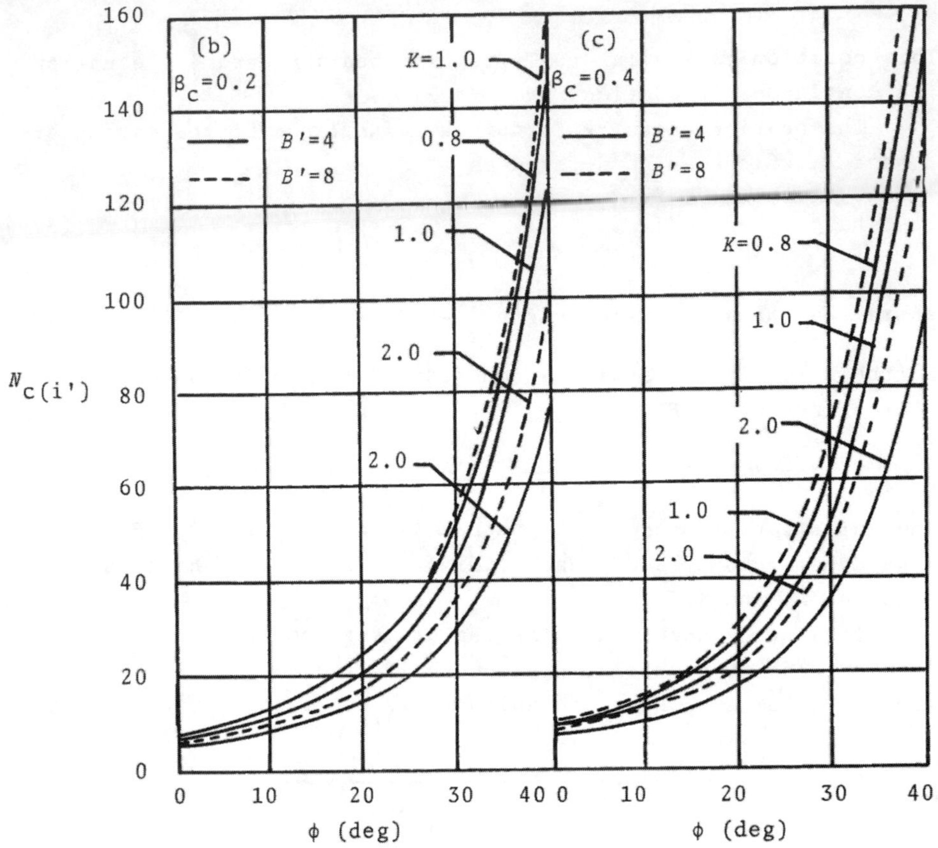

Figure 4.33. (Continued).

one. Also, when K changes from one to two with $\alpha'=0$, the magnitude of $N_{c(i')}$ is reduced by about 30-40%.

Example 4.2. Estimate the ultimate bearing capacity q_u of a continuous foundation with the following: B=9 ft, $c_{V(z=0)}$=250 lb/ft^2, α'=25 lb/ft^2/ft, D_f=3 ft, γ=110 lb/ft^3, and ϕ=20°. Assume K=2.

Solution

From Eq. (4.95), characteristic length =

$$l \quad \frac{c_{V(z=0)}}{\gamma} = \frac{250}{110} = 2.27$$

Nondimensional width =

$$B' \quad B/l = 9/2.27 = 3.96$$

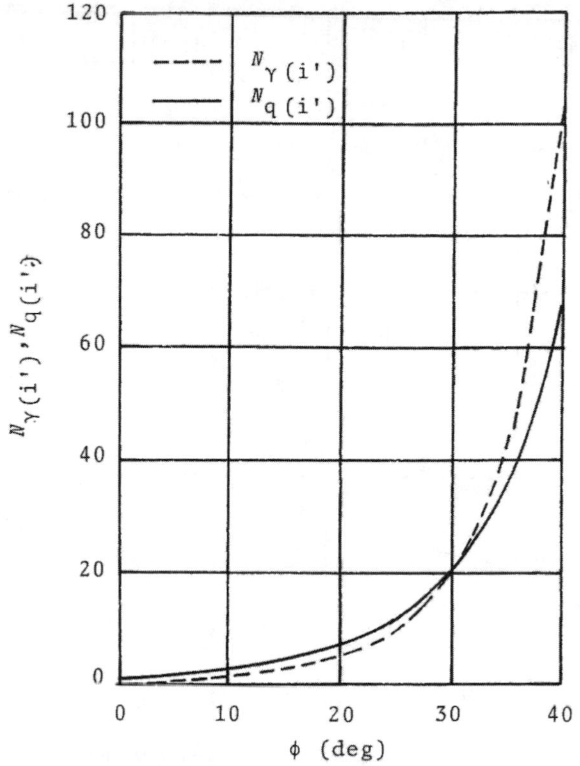

Figure 4.34. $N_{\gamma(i')}$ and $N_{q(i')}$ for K=1 and β_c=0 (after Reddy and Srinivasan, 1970).

Also

$$\beta_c = \frac{\alpha' l}{c_{V(z=0)}} \quad \frac{(25)(2.27)}{250} = 0.227$$

Now, referring to Figs. 4.33b and c and 4.35a and b, for ϕ=20°, β_c= 0.227, K=2, and B'=3.96 (by interpolation)

$$N_{c(i')} \simeq 14.5;\ N_{q(i')} \sim 6,\ \text{and}\ N_{\gamma(i')} \simeq 4$$

From Eq. (4.93)

$$q_u = c_{V(z=0)} N_{c(i')} + q N_{q(i')} + \tfrac{1}{2}\gamma B N_{\gamma(i')}$$

$$(250)(14.5) + (3)(110)(6) + (\tfrac{1}{2})\ (110)(10)(4)$$

$$\underline{7{,}805\ \text{lb/ft}^2}$$

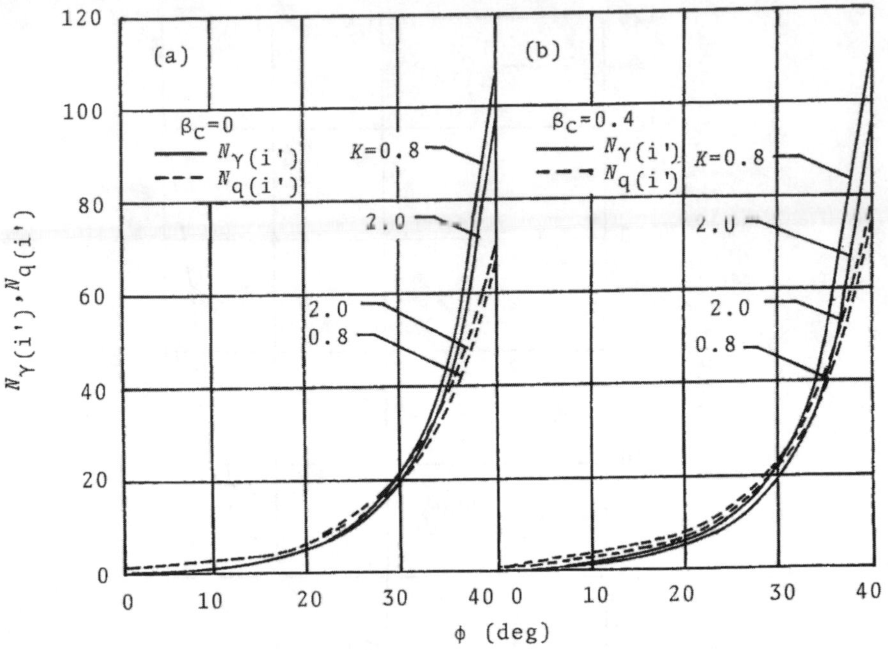

Figure 4.35. Influence of K on $N_{\gamma(i')}$ and $N_{q(i')}$ for (a) $\beta_c=0$ and β_c 0.4 (after Reddy and Srinivasan, 1970).

4.10 ULTIMATE BEARING CAPACITY DUE TO VERTICAL ECCENTRIC LOAD

Based on several laboratory model tests, Meyerhof (1953) has suggested a semiempirical procedure to determine the ultimate bearing capacity of shallow foundations due to eccentric loading condition. Eccentric loading of shallow foundations occurs when a vertical load Q is applied at a location other than the centroid of the foundation (Fig. 4.36a), or when a foundation is subjected to a centric vertical load of magnitude Q and a momentum M (Fig. 4.36b). In such a case, the load eccentricities may be given as

$$e_L = \frac{M_B}{Q} \tag{4.97a}$$

and

$$e_B \quad \frac{M_L}{Q} \tag{4.97b}$$

where e_L, e_B = load eccentricities, respectively, in the direction of *long* and *short* axes of the foundation

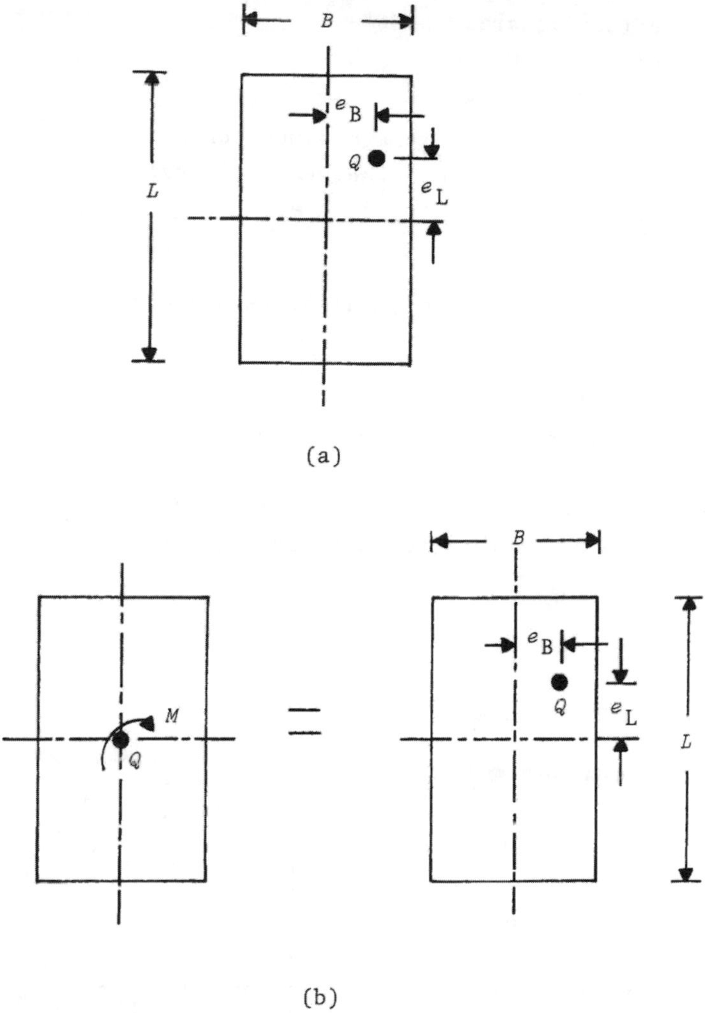

Figure 4.36. Eccentric load on shallow foundations.

M_B, M_L = moment about the short and long axes of the foundation, respectively

According to Meyerhof (1953), the ultimate bearing capacity q_u and ultimate load Q_u of an eccentrically load foundation (vertical load) can be given as

$$q_u = cN_c\lambda_{cs}\lambda_{cd} + qN_q\lambda_{qs}\lambda_{qd} + \tfrac{1}{2}\gamma B''N_\gamma\lambda_{\gamma s}\lambda_{\gamma d} \tag{4.98}$$

and

$$Q_u \quad (q_u)A'' \tag{4.99}$$

where A'' = effective area = $B''L''$

B'' = effective width

L'' = effective length

The *effective area* A'' is a minimum contact area of the foundation such that its *centroid coincides with that of the load.* For one-way eccentricity, that is, if $e_L=0$ (Fig. 4.37a), then

$$B'' = B-2e_B;\ L'' = L;\ A'' = B''L \tag{4.100a}$$

However, if $e_B=0$ (Fig. 4.37b), calculate $L''=L-2e_L$. The effective area is

$$A'' \quad B(L-2e_L) \tag{4.100b}$$

The effective width B'' is the smaller of the two values, that is B and $L-2e_L$.

For two-way eccentricities (that is, $e_L \neq 0$ and $e_B \neq 0$), four possible cases may arise (Highter and Anders, 1985). They are as follows:

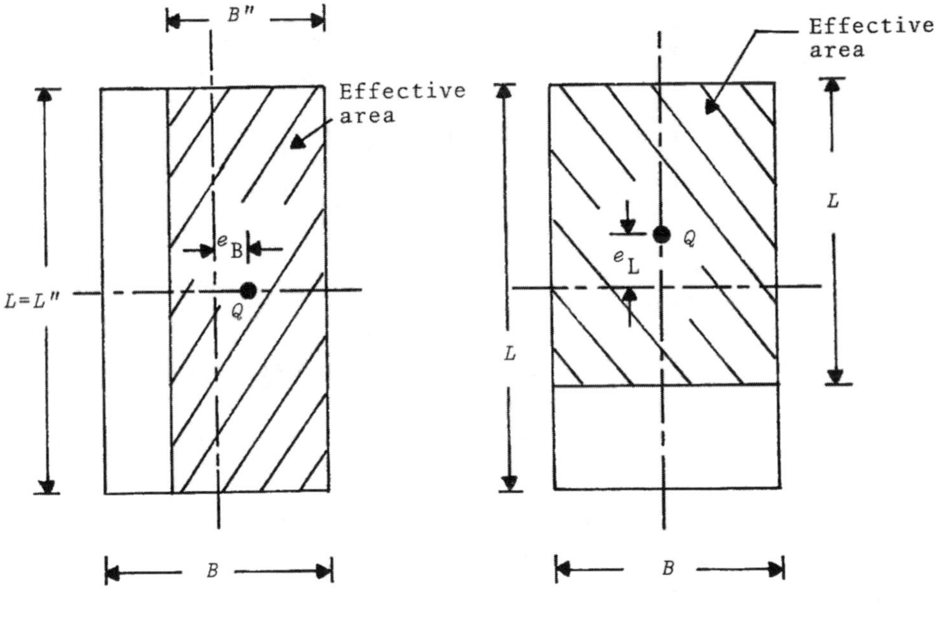

(a) (b)

Figure 4.37. Case of one-way eccentricity of the load on the foundation.

Case I ($e_L/L \geq 1/6$ and $e_B/B \geq 1/6$):

This case is shown in Fig. 4.38. For this, calculate

$$B_1 = B\left(1.5 - \frac{3e_B}{B}\right) \tag{4.101}$$

and

$$L_1 \quad L\left(1.5 - \frac{3e_L}{L}\right) \tag{4.102}$$

So, the effective area

$$A'' = \tfrac{1}{2}B_1 L_1 \tag{4.103}$$

The effective width B'' is equal to the smaller of B_1 and L_1 for use in Eq. (4.98)

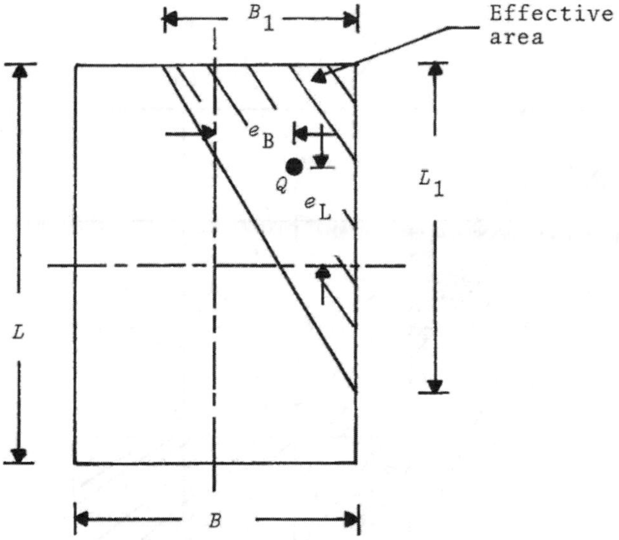

Figure 4.38. Effective area for the case of $e_L/L \geq 1/6$ and $e_B/B \geq 1/6$.

Case II ($e_L/L < 0.5$ and $0 < e_B/B < 1/6$):

This case is shown in Fig. 4.39. Knowing the magnitudes of e_L/L and e_B/B, the values of L_1/L and L_2/L (and thus L_1 and L_2) can be obtained from Fig. 4.39b. The effective area can now be given as

$$A'' = \tfrac{1}{2}(L_1 + L_2)B \tag{4.104}$$

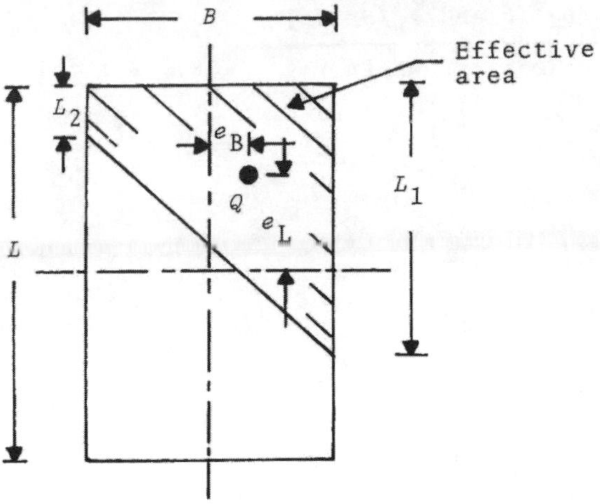

(a)

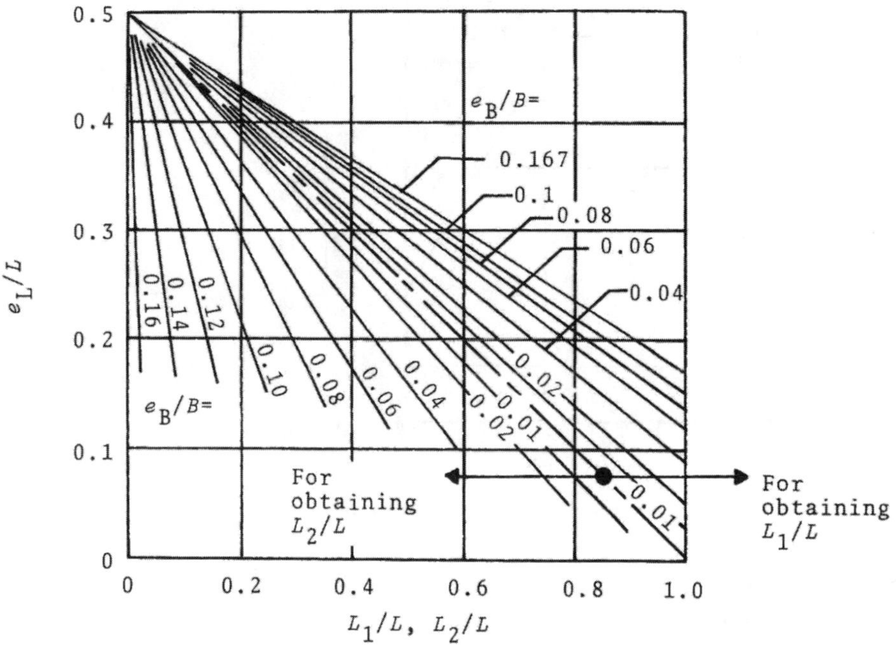

(b)

Figure 4.39. Effective area for the case of $e_L/L<0.5$ and $0<e_B/B<1/6$ (redrawn after Highter and Anders, 1985).

The effective length L'' is the larger of the two values L_1 and L_2. The effective width is equal to

$$B'' \quad \frac{A''}{L''} \tag{4.105}$$

Case III ($e_L/L<1/6$ and $0<e_B/B<0.5$):

Knowing the magnitudes of e_L/L and e_B/B, the magnitudes of B_1 and B_2 can be obtained from Fig. 4.40. So the effective area can be obtained as

$$A'' = \frac{1}{2}(B_1+B_2)L \tag{4.106}$$

In this case, the effective length is equal to

$$L'' = L \tag{4.107}$$

The effective width can be given as

$$B'' = \frac{A''}{L} \tag{4.108}$$

Case IV ($e_L/L<1/6$ and $e_B/B<1/6$):

For this case, the e_L/L curves sloping upward in Fig. 4.41 represent the values of B_2/B on the abscissa. Similarly, in the same figure the family of e_L/L curves which slope downward represent the values of L_2/L. Knowing B_2 and L_2, the effective area A'' can be calculated. For this case, $L''=L$ and $B''=A''/L''$.

In the case of circular foundations under eccentric loading (Fig. 4.42a), the eccentricity is always one way. The effective area A'' and the effective width B'' for a circular foundation are given in a nondimensional form in Fig. 4.42b.

Depending upon the nature of the load eccentricity and the shape of the foundation, once the magnitudes of the effective area and the effective width are determined, they can be used in Eqs. (4.98) and (4.99) to determine the ultimate load for the foundation. It needs to be pointed out that, in using Eq. (4.98)

1. The bearing capacity factors for a given friction angle are to be determined from those presented in Tables 4.3 and 4.5.

2. The shape factors are determined by using the relationships given in Table 4.4 by replacing B'' for B and L'' for L wherever they appear.

3. The depth factors are determined from the relationships

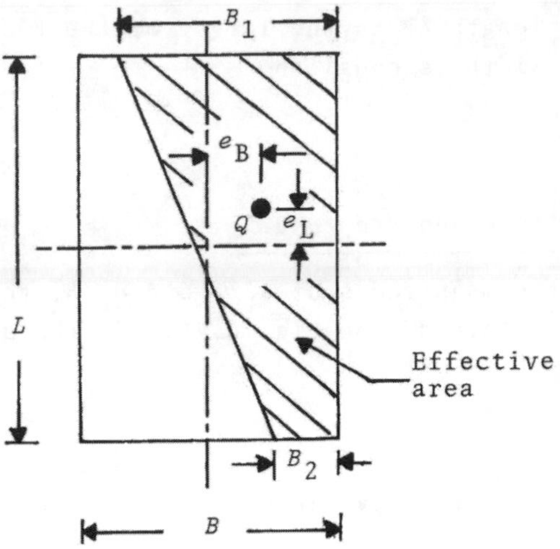

(a)

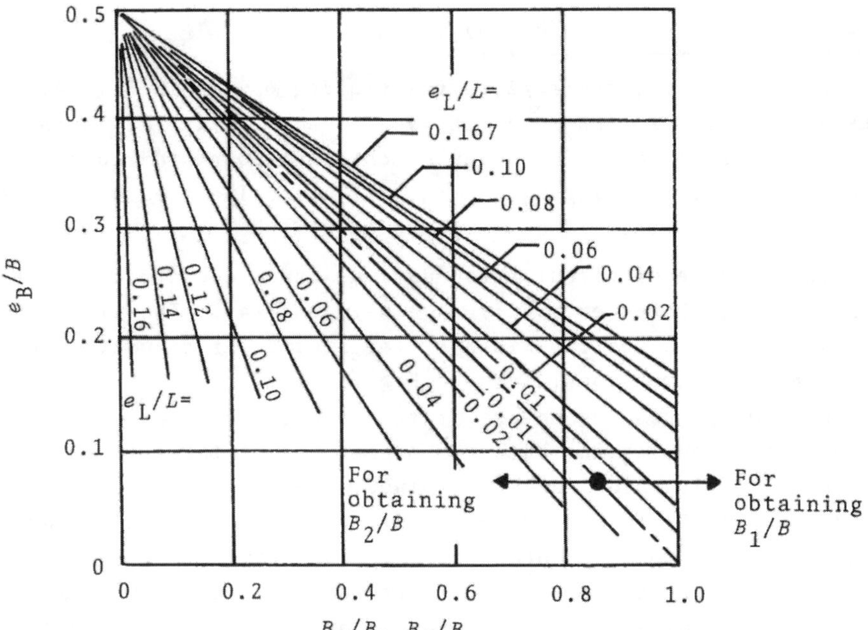

(b)

Figure 4.40. Effective area for the case of $e_L/L<1/6$ and $0<e_B/B<0.5$ (redrawn after Highter and Anders, 1985).

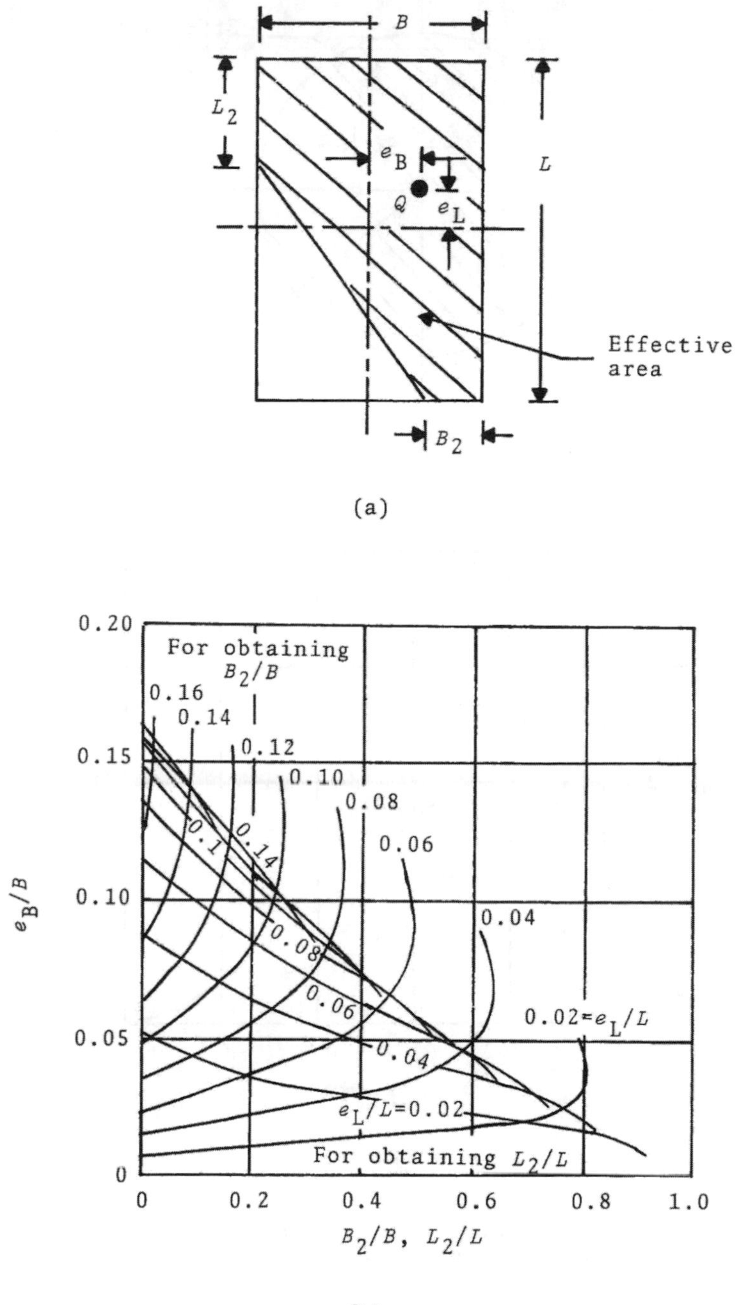

Figure 4.41. Effective area for the case of $e_L/L<1/6$ and $e_B/B<1/6$ (redrawn after Highter and Anders, 1985).

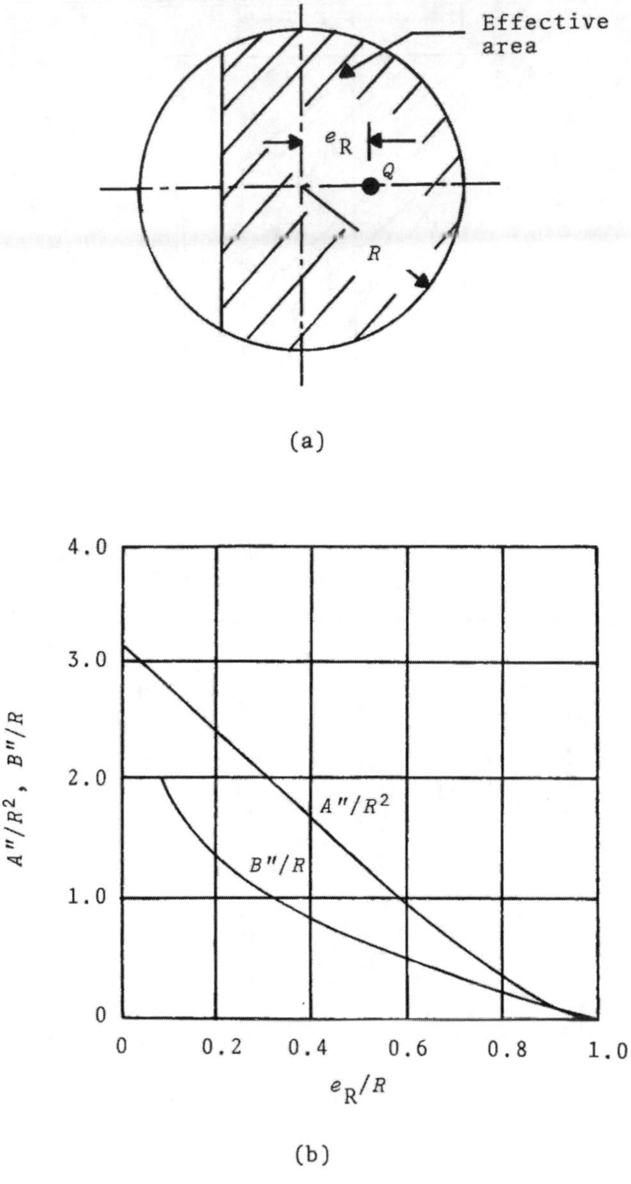

Figure 4.42. Normalized effective dimensions of circular foundations (after Highter and Anders, 1985).

given in Table 4.4. However, for calculation of depth factor, the term B is not replaced by B''.

Example 4.3. A shallow foundation measuring 4 ft × 6 ft is subjected to a centric load and a moment. If e_B=0.4 ft and e_L=1.2 ft, and the depth of the foundation is 3 ft, determine the allowable load the foundation can carry. Use a factor of safety of 4. For the soil, given:

Unit weight, γ	115 lb/ft^3
Friction angle, ϕ	35°
Cohesion, c	0

Solution

For this case

$$\frac{e_B}{B} = \frac{0.4}{4} = 0.1; \quad \frac{e_L}{L} = \frac{1.2}{6} = 0.2$$

For this type of condition, Case II as shown in Fig. 4.39 applies. Referring to this figure,

$$\frac{L_1}{L} \simeq 0.865, \text{ or } L_1 = (0.865)(6) = 5.19 \text{ ft}$$

$$\frac{L_2}{L} \simeq 0.22, \text{ or } L_2 \quad (0.22)(6) = 1.32 \text{ ft}$$

From Eq. (4.104)

$$A'' \quad \tfrac{1}{2}(L_1+L_2)B = \tfrac{1}{2}(5.19+1.32)(4) \quad 13.02 \text{ ft}^2$$

So

$$B'' = \frac{A''}{L''} = \frac{A''}{L_1} = \frac{13.02}{5.19} \sim 2.51 \text{ ft}$$

Since c=0

$$q_u = qN_q\lambda_{qs}\lambda_{qd} + \tfrac{1}{2}\gamma B''N_\gamma\lambda_{\gamma s}\lambda_{\gamma d}$$

From Table 4.3, for ϕ=25°, N_q=33.30. Also from Table 4.5, for ϕ=35°, Vesic's value for N_γ=48.03.

The shape factors as given by DeBeer (1970) are as follows:

$$\lambda_{qs} = 1+\left(\frac{B''}{L''}\right)\tan\phi = 1+\left(\frac{2.51}{5.19}\right)\tan 35 = 1.339$$

$$\lambda_{\gamma s} \quad 1-0.4\left(\frac{B''}{L''}\right) \quad 1-(0.4)\left(\frac{2.51}{5.19}\right) \quad 0.806$$

The shape factors as given by Hansen (1970) are as follows:

$$\lambda_{qd} = 1+0.2\tan\phi(1-\sin\phi)^2\left(\frac{D_f}{B}\right)$$

$$= 1+(0.2)(\tan 35)(1-\sin 35)^2\left(\tfrac{3}{4}\right) = 1.019$$

$$\lambda_{\gamma d} = 1$$

So

$$q_u = (115)(3)(33.3)(1.339)(1.019) + \left(\tfrac{1}{2}\right)(115)(2.51)(48.03)$$
$$(0.806)(1)$$

$$= 15{,}675 + 5{,}587 - 21{,}262 \text{ lb/ft}^2$$

So, allowable load on the foundation is

$$Q \quad \frac{q_u A''}{F_s} \quad \frac{(22{,}262)(13.02)}{4} = \underline{72{,}463 \text{ lb}}$$

Correlations of Purkayastha and Char

Purkayastha and Char (1977) have carried out stability analyses of eccentrically loaded *continuous foundations* ($e_B>0$, $e_L=0$) by using the method of slices as proposed by Janbu (1957). Based on their analyses, they have proposed that

$$R_k \quad 1 \quad \frac{q_{u(\text{eccentric})}}{q_{u(\text{centric})}} \tag{4.109}$$

where R_k = reduction factor
$q_{u(\text{eccentric})}$ = ultimate bearing capacity of eccentrically loaded continuous foundations
$q_{u(\text{centric})}$ = ultimate bearing capacity of centrally loaded continuous foundations

The magnitude of R_k can be expressed as

$$R_k = a\left(\frac{e_B}{B}\right)^k \tag{4.110}$$

where a and k are functions of the embedment ratio D_f/B (Table 4.6)

Table 4.6. Variations of a and k [Eq. (4.110)]

D_f/B	a	k
0	1.862	0.73
0.25	1.811	0.785
0.5	1.754	0.80
1.0	1.820	0.888

Hence, combining Eqs. (4.109) and (4.110)

$$q_{u(eccentric)} = q_{u(centric)}(1-R)$$

$$= q_{u(centric)}\left[1-a\left(\frac{e_B}{B}\right)^k\right] \tag{4.111}$$

Example 4.4. Consider a continuous foundation having a width of 2 m. If e_B=0.2 m and the depth of the foundation D_f=1 m, determine the ultimate load per meter length of the foundation. For the soil use ϕ=40°, γ=17.5 kN/m^3, and c=0.

Solution

Since c=0, L/B=0

$$q_{u(centric)} = qN_q\lambda_{qd} + \tfrac{1}{2}\gamma BN_\gamma\lambda_{\gamma d}$$

From Table 4,3, N_q=64.2. Also from Table 4.5, Meyerhof's N_γ=93.7. Again, from Table 4.4, Meyerhof's depth factors are as follows:

$$\lambda_{qd} = \lambda_{\gamma d} = 1+(0.1)\left(\frac{D_f}{B}\right)\tan(45+\phi/2)$$

$$1+(0.1)(\tfrac{1}{2})\tan(45+40/2) = 1.107$$

So

$$q_{u(centric)} = (1)(17.5)(64.2)(1.107) + (\tfrac{1}{2})(17.5)(2)(93.7)(1.1$$
$$(1.107)$$

$$= 1,243.7 + 1,815.2 = 3,058.9 \text{ kN/m}^2$$

According to Eq. (4.111)

$$q_{u(eccentric)} = q_{u(centric)}(1-R) = q_{u(centric)}\left[1-a\left(\frac{e_B}{B}\right)^k\right]$$

For D_f/B=1/2=0.5, from Table 4.6, a=1.754 and k=0.80. So

$$q_{u(eccentric)} \quad 3,058.9\left[1-1.754\left(\frac{0.2}{2}\right)^{0.8}\right] \approx 2,209 \text{ kN/m}^2$$

The ultimate load per unit length

$$Q \quad (2,209)(B)(1) = (2,209)(2)(1) = \underline{4,418 \text{ kN}}$$

Theory of Prakash and Saran

Prakash and Saran (1971) have provided a comprehensive mathematical formulation for estimation of the ultimate bearing capacity for rough *continuous foundations* under eccentric loading. Due to lack of space, the complete derivation of their theory will not be presented in this section; however, the general procedure will be outlined.

Figure 4.43a shows the assumed failure surface in a c-ϕ soil under a continuous foundation subjected to eccentric loading. Let Q_u be the ultimate load per unit length of the foundation of width

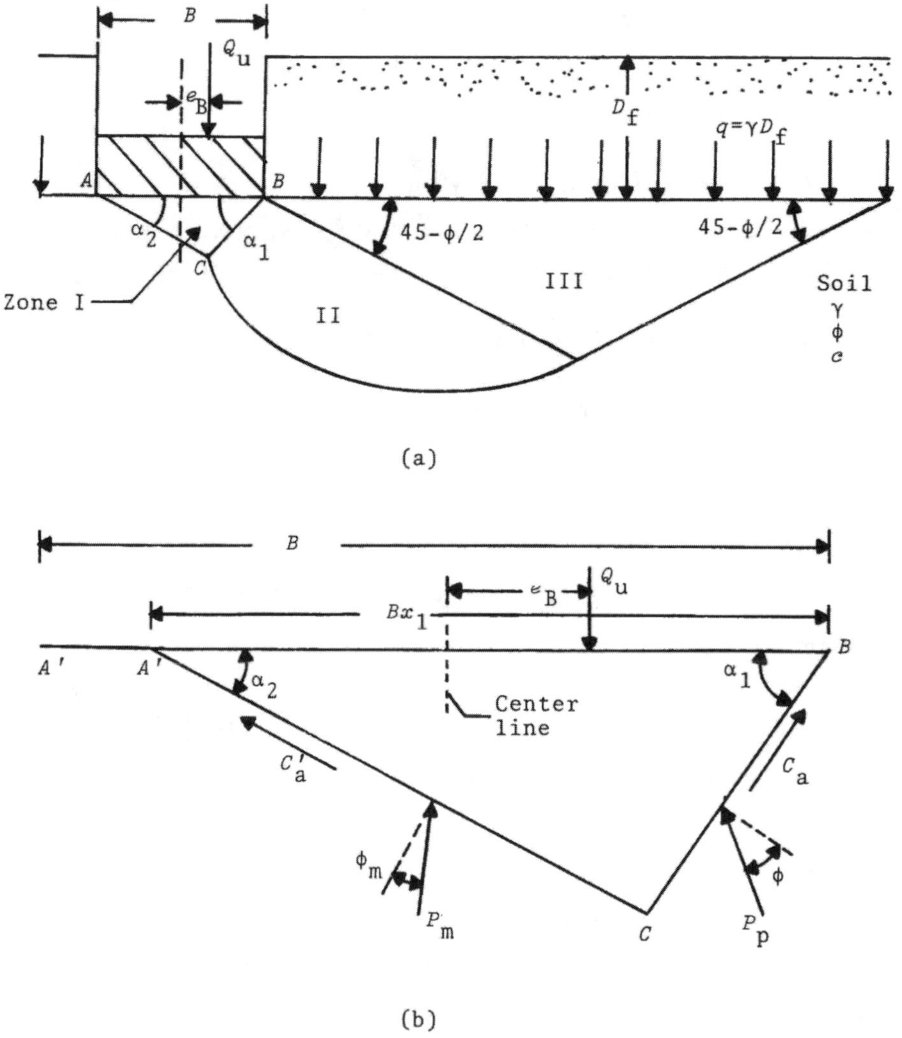

Figure 4.43. Bearing capacity theory of Prakash and Saran (1971) for eccentrically loaded rough continuous foundation.

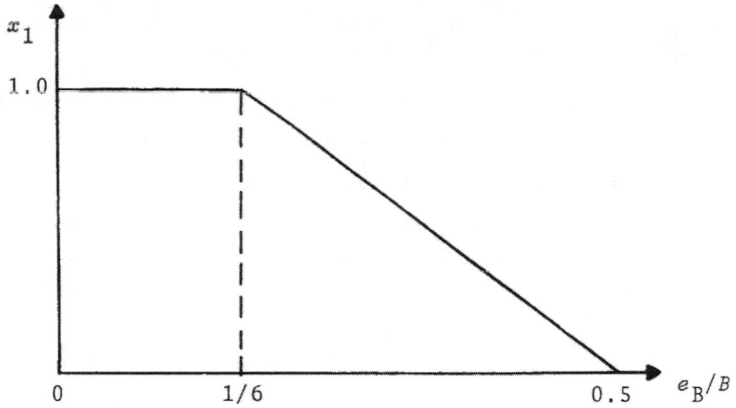

(c)

Figure 4.43. (Continued).

B with an eccentricity of e_B. In Fig. 4.43a, Zone I is an elastic zone with wedge angles of α_1 and α_2. Zones II and III are similar to those assumed by Terzaghi in Fig. 4.6 (that is, Zone II is a radial shear zone, and Zone III is a Rankine passive zone).

The bearing capacity expression can be developed by considering the equilibrium of the elastic wedge ABC located below the foundation (Fig. 4.43b). Note that in Fig. 4.43b the contact width of the foundation with the soil is equal to Bx_1. Neglecting the self weight of the wedge

$$Q_u = P_p \cos(\alpha_1 - \phi) + P_m \cos(\alpha_2 - \phi_m) + C_a \sin\alpha_1 + C'_a \sin\alpha_2 \qquad (4.112)$$

where P_p, P_m = passive forces per unit length of the wedge along the wedge faces BC and AC, respectively

ϕ = soil friction angle

ϕ_m = mobilized soil friction angle ($\leq\phi$)

C_a = adhesion along the wedge face BC = $\dfrac{cBx_1 \sin\alpha_2}{\sin(\alpha_1+\alpha_2)}$

C'_a = adhesion along the wedge face AC $\dfrac{mcBx_1 \sin\alpha_1}{\sin(\alpha_1+\alpha_2)}$

m = mobilization factor (≤ 1)

c cohesion

Equation (4.112) can be expressed in the form

$$q_u = \frac{Q_u}{(B \times 1)} = \frac{1}{2}\gamma B N_{\gamma(e)} + \gamma D_f N_{q(e)} + c N_{c(e)} \tag{4.113}$$

where $N_{\gamma(e)}, N_{q(e)}, N_{c(e)}$ = bearing capacity factors for an eccentrically loaded foundation

The above-stated bearing capacity factors will be functions of e_B/B, ϕ, and also the foundation contact factor x_1. In obtaining the bearing capacity factors, Prakash and Saran (1971) assumed the variation of x_1 as shown in Fig. 4.43c. Figure 4.44 shows the variation of $N_{\gamma(e)}$, $N_{q(e)}$, and $N_{c(3)}$ with ϕ and e_B/B. Note that for $e_B/B=0$, the bearing capacity factors coincide with those given by Terzaghi (1943) for a centrally loaded foundation (Table 4.1.)

For rectangular foundations with *one-way eccentricity* (that is $e_L/L=0$), the ultimate load may be expressed as

$$Q_u = q_u(BL) \quad (BL)\left[\frac{1}{2}\gamma B N_{\gamma(e)}\lambda_{\gamma s(e)} + \gamma D_f N_{q(e)}\lambda_{qs(e)} + c N_{c(e)}\lambda_{cs(e)}\right] \tag{4.114}$$

where $\lambda_{\gamma s(e)}, \lambda_{qs(e)}, \lambda_{cs(e)}$ = shape factors

The shape factors may be expressed by the following relationships

$$\lambda_{\gamma s(e)} \quad 1.0 + \left(\frac{2e_B}{B} - 0.68\right)\left(\frac{B}{L}\right) + \left(0.43 \quad \frac{3e_B}{2B}\right)\left(\frac{B}{L}\right)^2 \tag{4.115}$$

where L = length of the foundation

$$\lambda_{qs(e)} = 1 \tag{4.116}$$

and

$$\lambda_{cs(e)} = 1 + 0.2\left(\frac{B}{L}\right) \tag{4.117}$$

Prakash (1981) has also given the relationships for settlement of a given foundation under centric and eccentric loading conditions for an equal factor of safety F_s. They are as follows (Fig. 4.45)

$$\frac{S_e}{S_o} \quad 1.0 - 1.63\left(\frac{e_B}{B}\right) - 2.63\left(\frac{e_B}{B}\right)^2 + 5.83\left(\frac{e_B}{B}\right)^3 \tag{4.118}$$

and

$$\frac{S_m}{S_o} = 1.0 + 2.31\left(\frac{e_B}{B}\right) - 22.61\left(\frac{e_B}{B}\right)^2 + 31.54\left(\frac{e_B}{B}\right)^3 \tag{4.119}$$

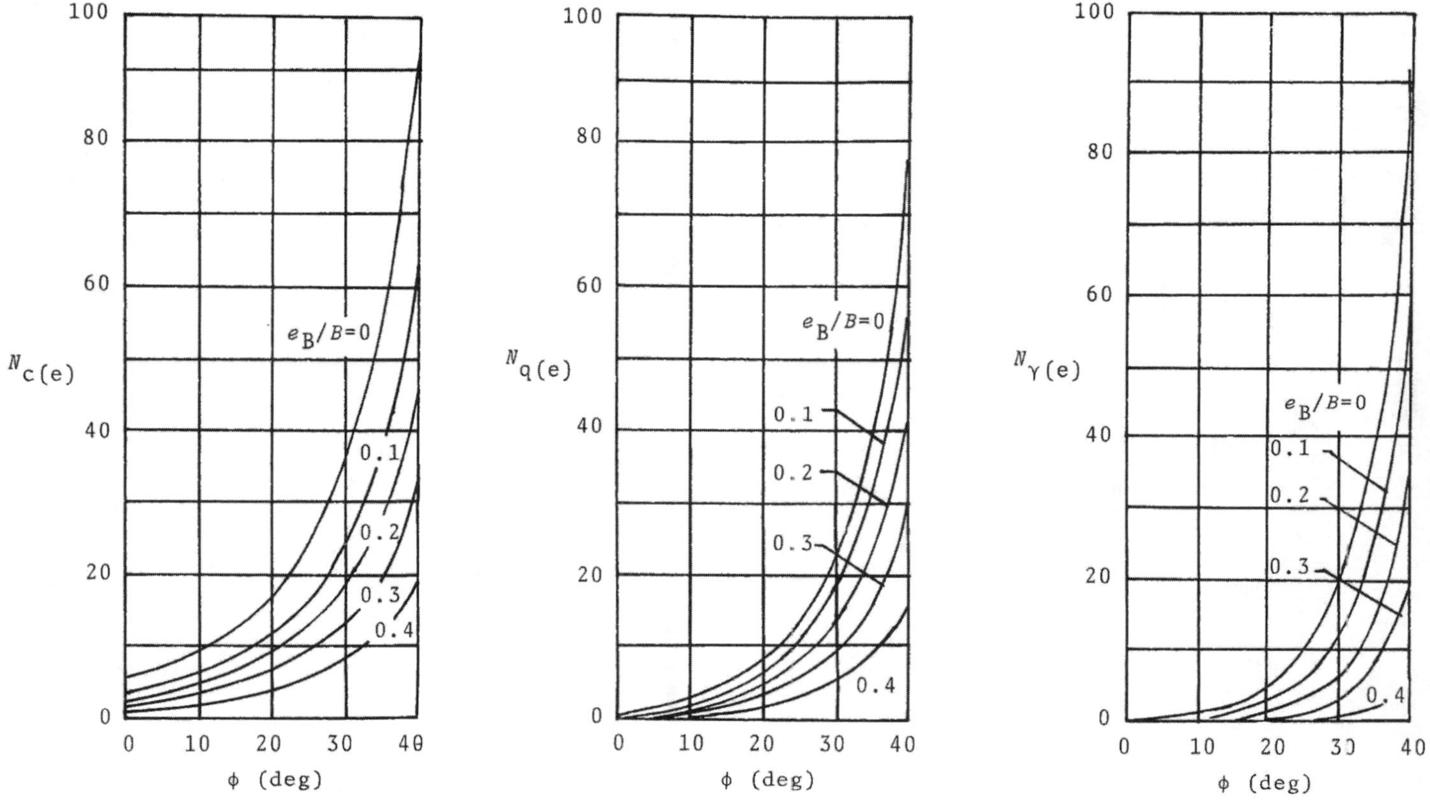

Figure 4.44. Variation of $N_{c(e)}$, $N_{q(e)}$, and $N_{\gamma(e)}$ with e_B/B and ϕ (after Prakash and Saran, 1971).

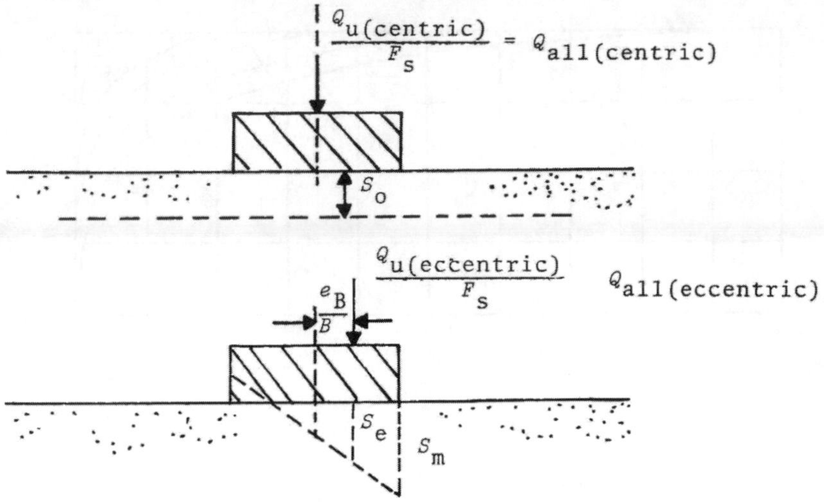

Figure 4.45. Notations for Eqs. (4.118) and (4.119).

where S_o settlement of a foundation under centric loading at

$$q_{all(centric)} = \frac{q_{u(centric)}}{F_s}$$

S_e, S_m settlements of the same foundation under eccentric loading at $q_{all(eccentric)}$ $\dfrac{q_{u(eccentric)}}{F_s}$

Example 4.5. Solve Example 4.4 by using the method of Prakash and Saran (1971).

Solution

From Eq. (4.113)

$$Q_u \quad (B\times1)\left[\tfrac{1}{2}\gamma B N_{\gamma(e)} + \gamma D_f N_{q(e)} + cN_{c(e)}\right]$$

Given $c=0$. For $\phi=40°$, $e_B/B = 0.2/2 = 0.1$. From Fig. 4.44, $N_{q(e)} = 56.09$ and $N_{\gamma(e)} = 71.8$. So

$$Q_u \quad (2\times1)\left[(\tfrac{1}{2})(17.5)(2)(71.8) + (17.5)(1)(56.09)\right]$$

$$= 2(1{,}256.5 + 981.6)$$

$$\sim \underline{4{,}476 \text{ kN}}$$

4.11 FOUNDATIONS SUPPORTED BY A SOIL WITH A RIGID ROUGH BASE AT A LIMITED DEPTH

The problem of bearing capacity of shallow rough foundations discussed in the preceding sections assume that the soil is homogeneous and extends to a great depth. However, it is possible that a rigid rough base is located at a shallow depth below the bottom of the foundation. In that case, the ultimate bearing capacity of the foundation will be altered to some extent. The problem can be explained more clearly by referring to Fig. 4.46. Figure 4.46a shows a shallow rough continuous foundation supported by a soil

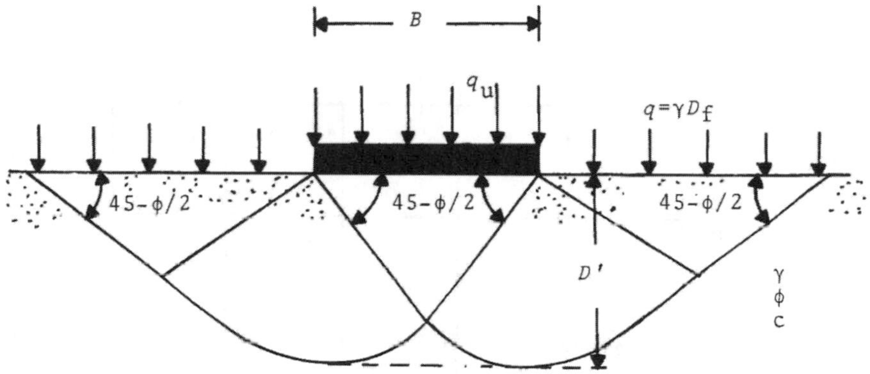

(a)

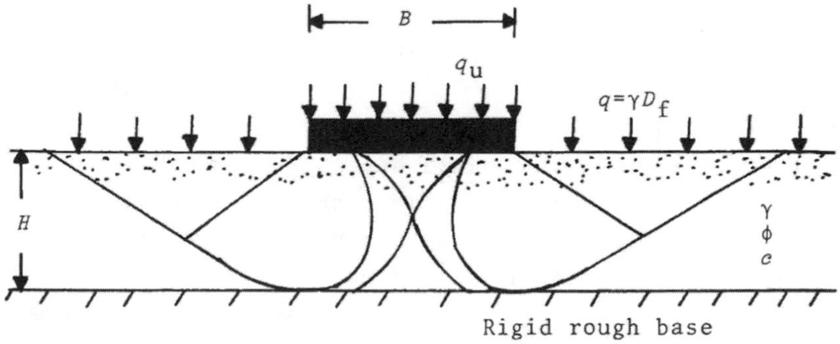

(b)

Figure 4.46. Failure surface under a rough continuous foundation: (a) Homogeneous soil extending to a great depth; (b) With a rough rigid base located at a shallow depth.

which extends to a great depth. The ultimate bearing capacity of this foundation can be expressed (neglecting the depth factors) as

$$q_u = cN_c + qN_q + \tfrac{1}{2}\gamma BN_\gamma \tag{4.120}$$

The procedure for determination of the bearing capacity factors N_c, N_q, and N_γ in homogeneous and isotropic soils has been outlined in Sections 4.3, 4.5, and 4.7. The extent of the failure zone in soil at ultimate load q_u is equal to D'. The magnitude of D' as obtained during the evaluation of the bearing capacity factors N_c (Prandtl, 1921) and N_q (Reissner, 1924) is given in a nondimensional form in Fig. 4.47. Similarly the magnitude of D' obtained during the evaluation of N_γ (Lundgren and Mortensen, 1953) is also given in Fig. 4.47.

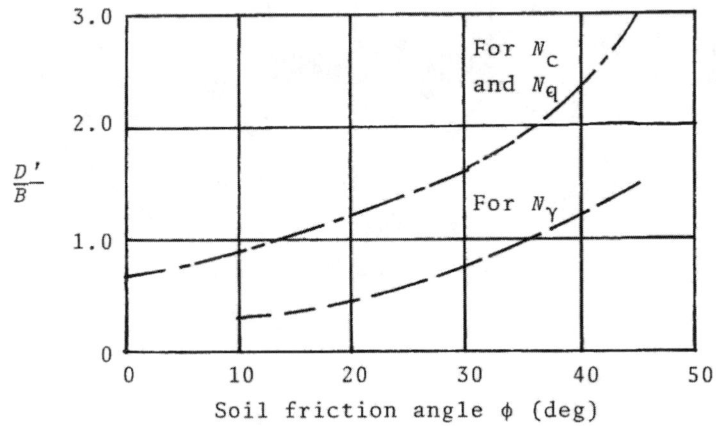

Figure 4.47. Variation of D'/B with soil friction angle.

Now, if a rigid rough base is located at a depth of $H<D'$ below the bottom of the foundation, full development of the failure surface in soil will be restricted. In such a case, at ultimate load the soil failure zone and the development of slip lines will be as shown in Fig. 4.46b. Mandel and Salencon (1972) have determined the bearing capacity factors for such a case by means of numerical integration using the theory of plasticity. According to Mandel and Salencon's theory (1972), the ultimate bearing capacity of a rough continuous foundation with a rigid rough base located at a shallow depth can be given by the relation

$$q_u = cN_c^* + qN_q^* + \tfrac{1}{2}\gamma BN_\gamma^* \tag{4.121}$$

where N_c^*, N_q^*, N_γ^* = modified bearing capacity factors

B = width of foundation

γ = unit weight of soil

Note that, for $H \geq D'$, $N_c^* = N_c$, $N_q^* = N_q$, and $N_\gamma^* = N_\gamma$ (Lundgren and Mortensen).

The variations of N_c, N_q, and N_γ with H/B and soil friction angle ϕ are given in Figs. 4.48, 4.49, and 4.50.

Neglecting the depth factors, the ultimate bearing capacity of rough circular and rectangular foundations on a sand layer (c=0) with a rough rigid base located at a shallow depth can be given as (Meyerhof, 1974)

$$q_u = qN_q^*\lambda_{qs}^* + \tfrac{1}{2}\gamma B N_\gamma^* \lambda_{\gamma s}^* \tag{4.122}$$

where $\lambda_{qs}^*, \lambda_{\gamma s}^*$ = modified shape factors

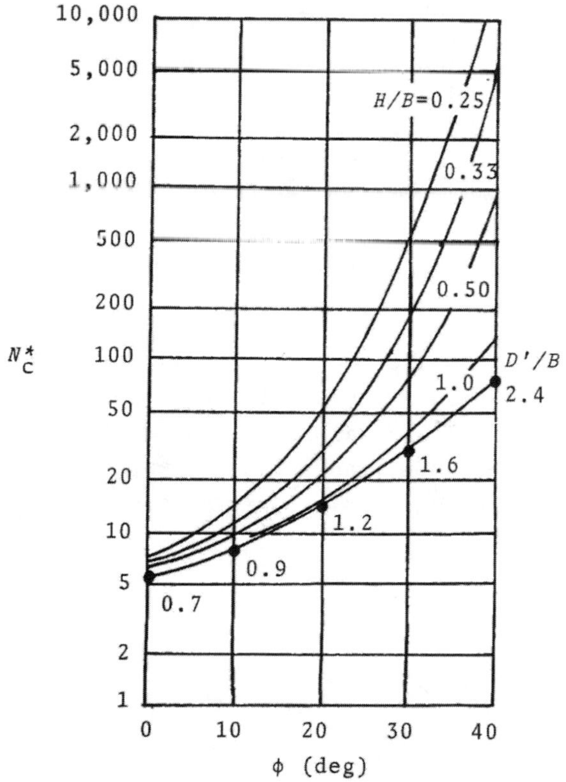

Figure 4.48. Mandel and Salencon's bearing capacity factor N_c^* [Eq. (4.121)].

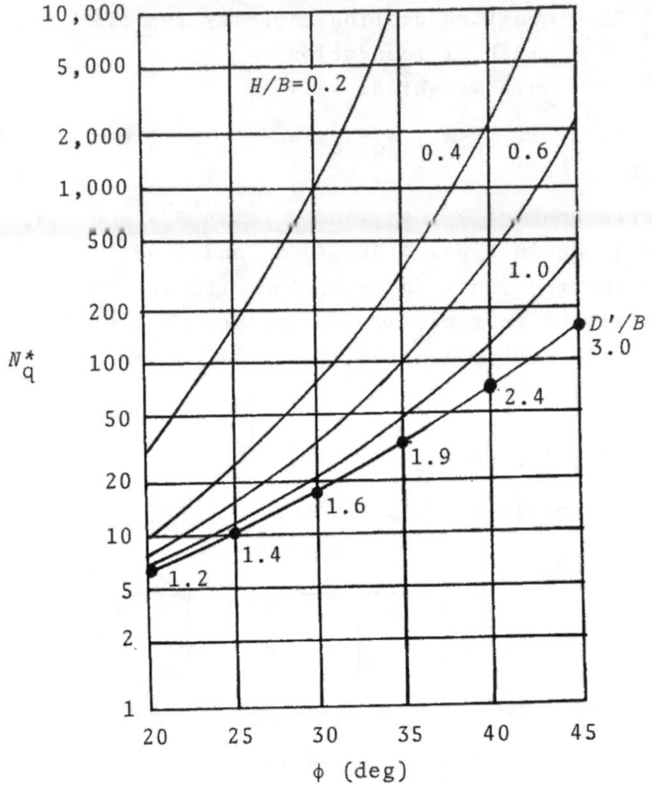

Figure 4.49. Mandel and Salencon's bearing capacity factor N_q^* [Eq. (4.121)].

The above-mentioned shape factors are functions of H/B and ϕ. Based on the works of Meyerhof and Chaplin (1953), and with simplifying the assumption that in radial planes the stresses and shear zones are identical to those in transverse planes, Meyerhof (1974) has evaluated the approximate values of λ_{qs}^* and $\lambda_{\gamma s}^*$ as

$$\lambda_{qs}^* = 1 \quad m_1(\frac{B}{L}) \tag{4.123}$$

and

$$\lambda_{\gamma s}^* = 1 - m_2(\frac{B}{L}) \tag{4.124}$$

where L length of the foundation

The variations of m_1 and m_2 with H/B and ϕ are given in Fig. 4.51a and b.

Milovic and Tournier (1971) and Pfeifle and Das (1978) have conducted laboratory tests to verify the theory of Mandel and

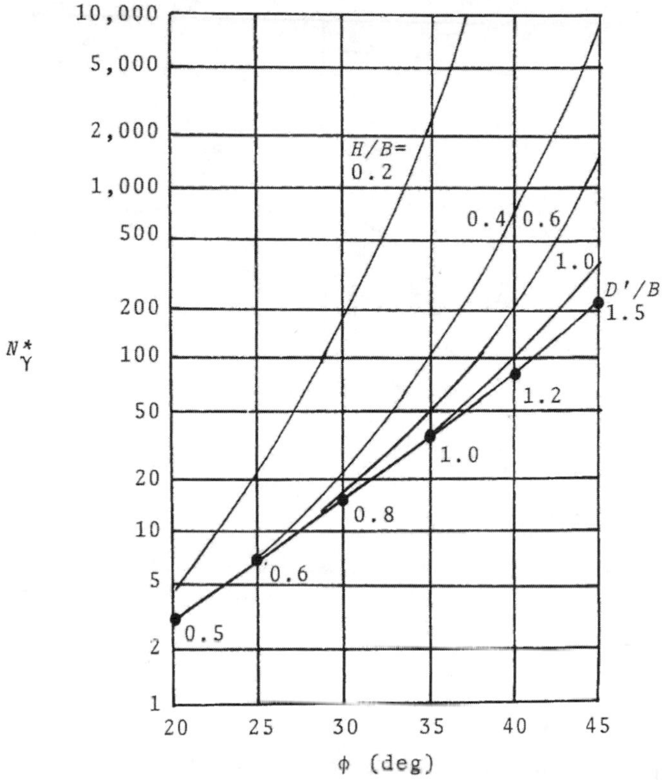

Figure 4.50. Mandel and Salencon's bearing capacity factor N_γ^* [Eq. (4.121)].

Salencon (1972). Figure 4.52 shows the comparison of the experimental variation of N_γ^* for a rough surface foundation ($D_f=0$) on a sand layer with theory (Pfeifle and Das, 1979). The angle of friction of the sand used for these tests was 43°. From Fig. 4.52, the following conclusions can be drawn:

1. The value of N_γ^* for a given foundation increases with the decrease of H/B.

2. The magnitude of $H/B=D'/B$ beyond which the presence of a rigid rough base has no influence on the N_γ^* value of a foundation is about 50-75% more than that predicted by the theory.

3. For H/B between 0.6 to about 1.9, the experimental values of N_γ^* are higher than those predicted theoretically.

4. For $H/B<$about 0.6, the experimental values of N_γ^* are substantially lower than those predicted by theory. This may be due

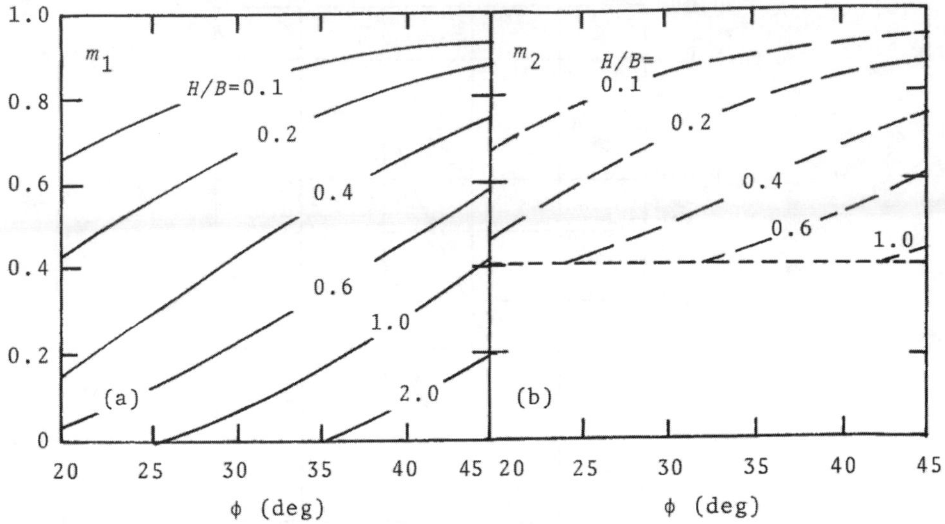

Figure 4.51. Variation of m_1 and m_2 (Meyerhof's values) for use in the modified shape factor equations [Eqs. (4.123) and (4.124)].

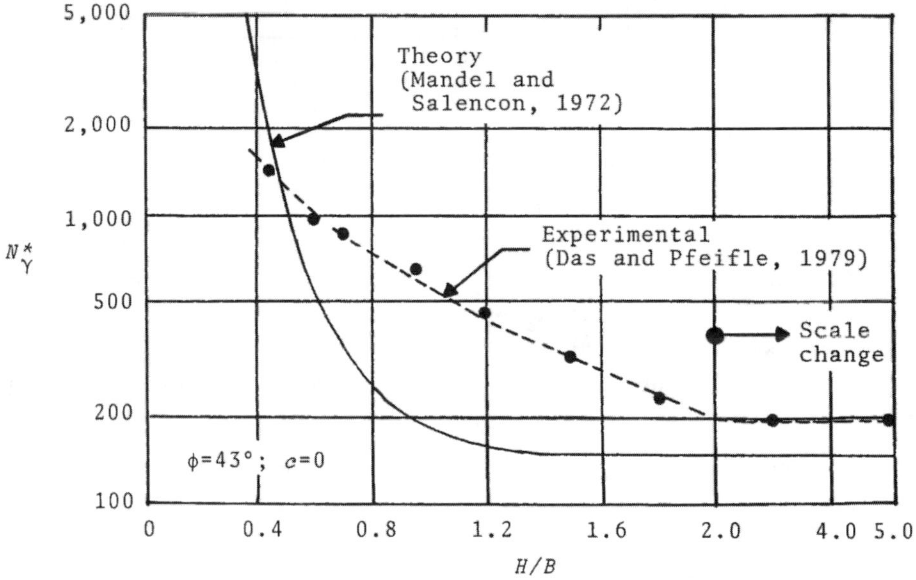

Figure 4.52. Comparison of theory with the experimental results for N^*_γ.

to two factors: (i) the crushing of sand grains at such high values of ultimate load, and (ii) the curvilinear nature of the actual failure envelope of soil at high normal stress levels.

4.12 BEARING CAPACITY OF FOUNDATIONS ON SLOPES

The bearing capacity of foundations on or near slopes is smaller than that obtained for foundations on level ground. This is primarily due to the fact that the soil zone involved in providing resistance at ultimate load is smaller on the downhill side of the slope (Fig. 4.53). Meyerhof (1957) has provided a solution for the determination of the ultimate bearing capacity of *rough continuous* foundations on or near a slope, which is an extension of his original bearing capacity theory (1951). According to the solution of Meyerhof

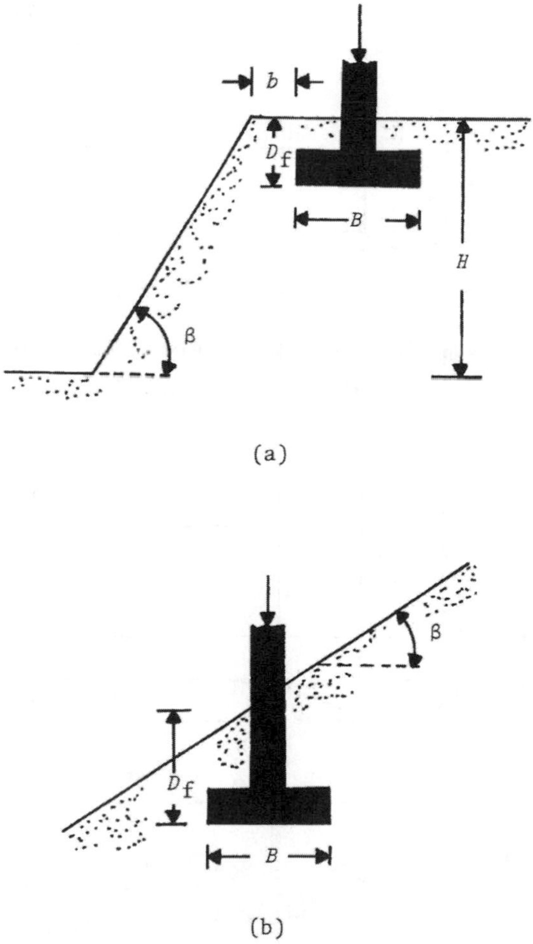

Figure 4.53. Continuous rough foundations: (a) Near a slope; and (b) On a slope.

$$q_u = cN_{cq} + \tfrac{1}{2}\gamma BN_{\gamma q} \tag{4.125}$$

where $N_{cq}, N_{\gamma q}$ = bearing capacity factors

For purely cohesion soils (that is, ϕ=0)

$$q_u \quad c_u N_{cq} \tag{4.126a}$$

For cohesionless soils (that is, c=0)

$$q_u = \tfrac{1}{2}\gamma BN_{\gamma q} \tag{4.126b}$$

The above equation is applicable to both cases--foundations on a slope (Fig. 4.53b) and foundations near a slope (Fig. 4.53a). The variations of N_{cq} and $N_{\gamma q}$ for use in Eqs. (4.126) and (4.127) are given in Figs. 4.54 and 4.55. In order to use Fig. 4.54, the following should be kept in mind.

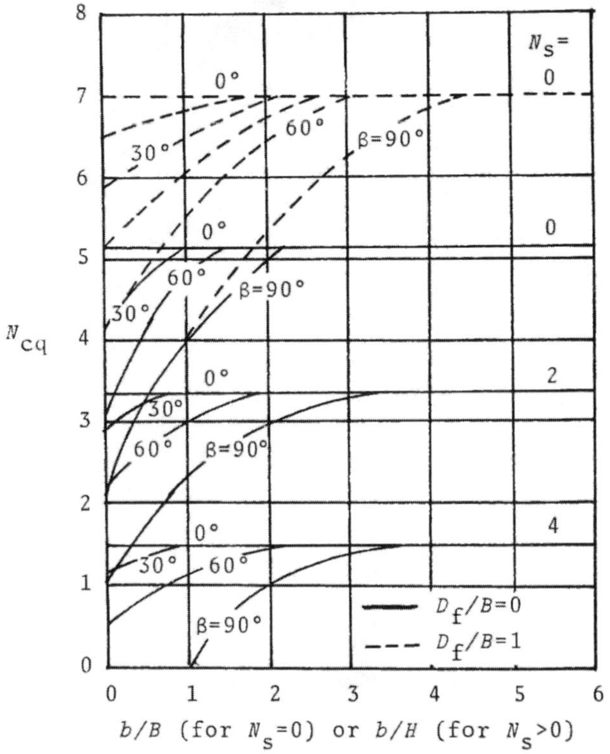

(a)

Figure 4.54. Meyerhof's bearing capacity factor N_{cq} for use in Eq. (4.126a): (a) For foundation near a slope (Fig. 4.53a); (b) For foundation on a slope (Fig. 4.53b).

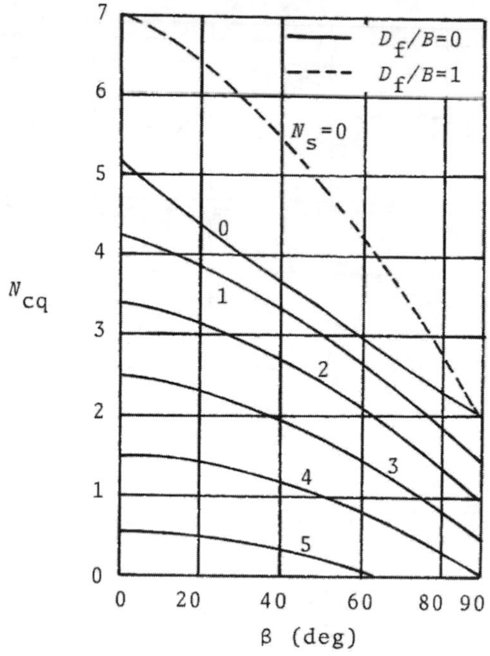

(b)

Figure 4.54. (Continued).

1. N_s stability factor $\frac{\gamma H}{c_u}$

where γ = unit weight of soil

H height of slope

c_u undrained cohesion

2. In order to obtain N_{cq} from Fig. 4.54a for foundations where $B<H$, use the curve for $N_s=0$. When necessary, interpolate for values of D_f/B between zero and one.

3. In order to obtain the value of N_{cq} from Fig. 4.54a for foundations where $B\geq H$, use the curve for calculated stability factor N_s.

4. Figure 4.55a and b are for cohesionless soils for use in Eq. (4.126b). When necessary, interpolate for values of D_f/B between zero and one.

Example 4.6. A continuous foundation having a width of 1 m is located on a slope. Referring to Fig. 4.53a, D_f=1 m, H=4 m, b=2 m,

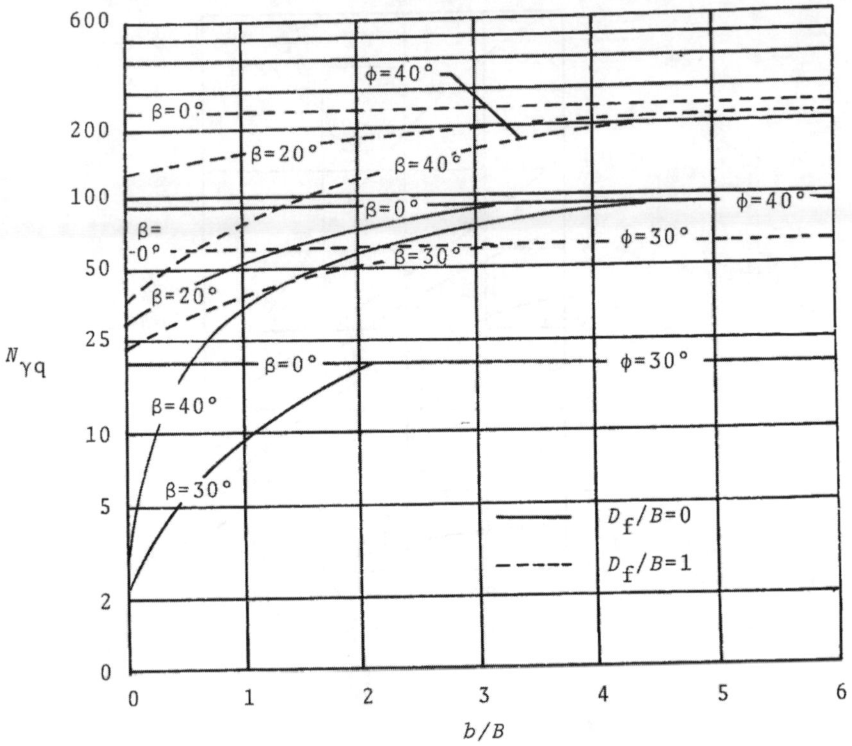

(a)

Figure 4.55. Meyerhof's bearing capacity factor $N_{\gamma q}$ for use in Eq. (4.126b): (a) For foundation near a slope (Fig. 4.53a); (b) For foundation on a slope (Fig. 4.53b).

$\gamma=16.8$ kN/m^3, $c=50$ kN/m^2, $\beta=60°$, and $\phi=0$. Determine the ultimate bearing capacity of the foundation.

Solution

Given $B=1$. For this case, $B>H$. So $N_s=0$.

$$\frac{b}{B} = \frac{2}{1} \quad 2$$

For $\beta=60°$, the value of N_{cq} from Fig. 4.54a is 6.35. So

$$q_u = c_u N_{cq} = (50)(6.35) = \underline{317.5 \text{ kN/m}^2}$$

4.13 BEARING CAPACITY OF FOUNDATIONS ON LAYERED SOIL

Most of the problems on bearing capacity treated so far in this chapter relate to foundations supported by soil deposits

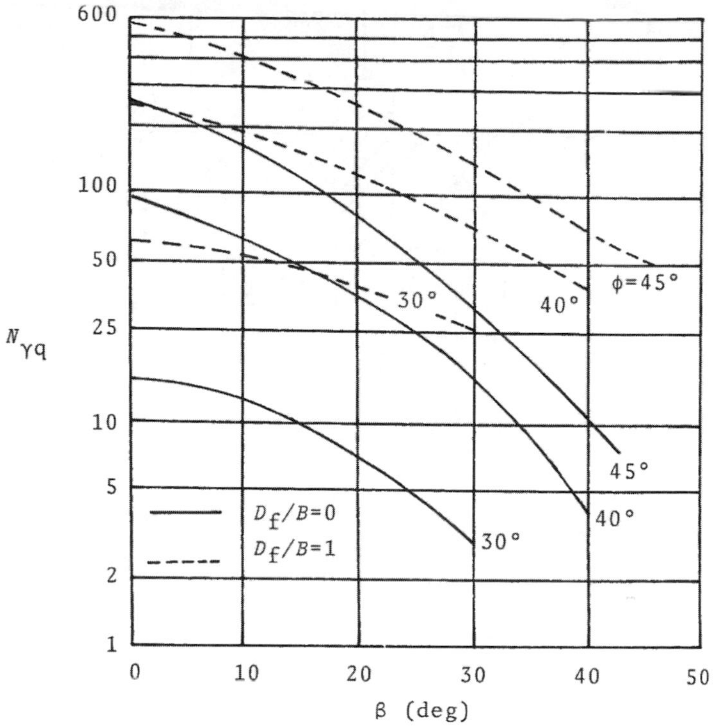

(b)

Figure 4.55. (Continued).

extending to great depths. In many circumstances it is possible that a foundation is located on a layered soil and the interface of the soil layering is at a shallow depth H as shown in Fig. 4.56. Depending on the ratio H/B, it is possible that the ultimate bearing capacity of the foundation will change. In this section, some of the possible cases for estimation of the ultimate bearing capacity of a shallow foundation on layered soil will be discussed in detail.

Foundations on Layered Clay (ϕ=0 concept)

Figure 4.57a shows a continuous surface foundation of width B located on a layered clay soil. The ultimate bearing capacity of this foundation can be given as

$$q_u = C_{u(V)-1} N_{c(L)} \tag{4.127}$$

where $C_{u(V)-1}$ = undrained shear strength of the top soil layer when

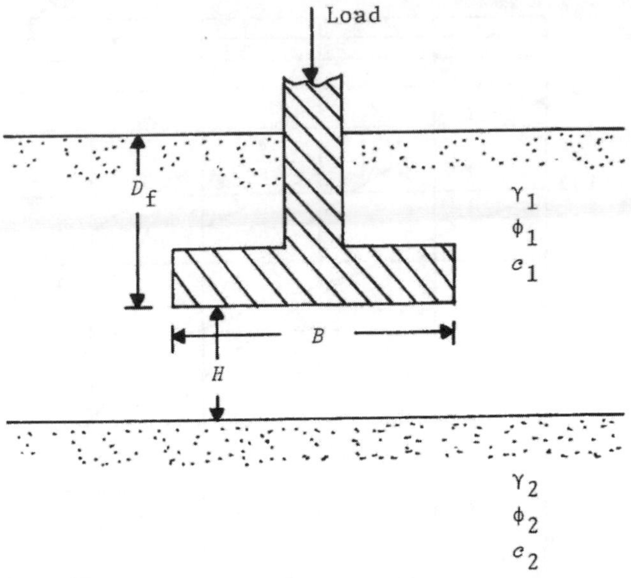

Figure 4.56. Shallow foundation on layered soil.

the major principal stress is vertical

$N_{c(L)}$ = bearing capacity factor

However, the bearing capacity factor $N_{c(L)}$ will be a function of H/B and $c_{u(V)-2}/c_{u(V)-1}$, or

$$N_{c(L)} \quad f\left[\frac{H}{B}, \frac{c_{u(V)-2}}{c_{u(V)-1}}\right] \tag{4.128}$$

where $c_{u(V)-2}$ undrained shear strength of the bottom clay layer when the major principal stress is vertical

Reddy and Srinivasan (1967) have developed a procedure to determine the variation of $N_{c(L)}$. In the development of their theory they assumed that (i) the failure surface is cylindrical as shown in Fig. 4.57a where the center of the trial failure surface is at O, and (ii) the undrained shear strength of the soil may be anisotropic and follows the Casagrande-Carrillo relationship (1954)

$$c_{u(i)} = c_{u(H)} + [c_{u(V)} - c_{u(H)}]\sin^2 i \tag{4.129}$$

where $c_{u(i)}$ undrained shear strength at a given depth when the major principal stress is inclined at an angle i with the horizontal [*Note:* Eqs. (4.90) and (4.129) are the same]

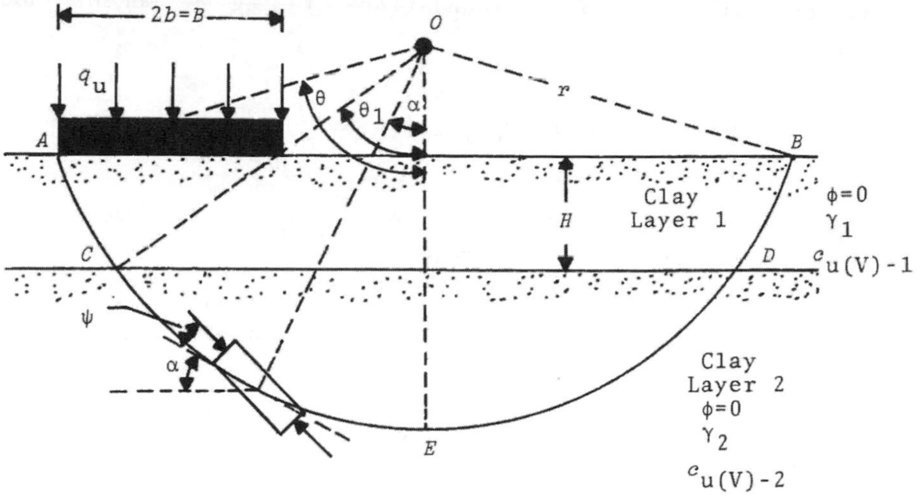

(a)

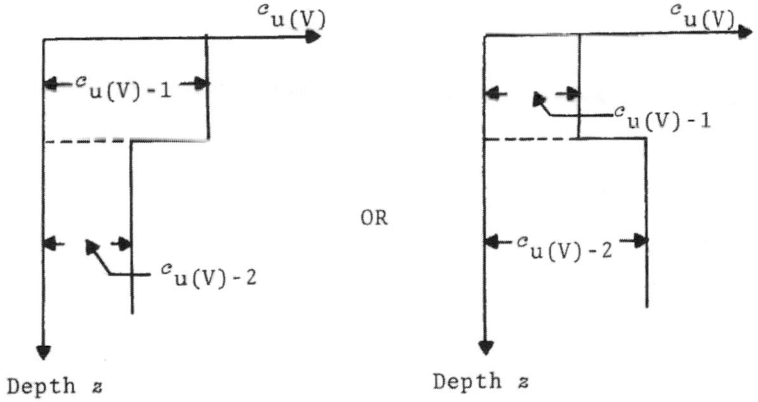

(b)

Figure 4.57. Assumptions in deriving the bearing capacity relationship of Reddy and Srinivasan (1967) for a continuous foundation on layered clay (anisotropic) soil ($\phi=0$).

$c_{u(V)}, c_{u(H)}$ = undrained shear strength for $i=90°$ and $0°$

Now, let us assume that the magnitude of $c_{u(V)}$ for the top clay layer [$c_{u(V)-1}$], and the bottom clay layer [$c_{u(V)-2}$] remains constant with depth z as shown in Fig. 4.57b.

In Fig. 4.57a, for equilibrium of the foundation (considering forces per unit length of the foundation), taking the moment about point O

$$2bq_u(r\sin\theta - b) \quad 2\int_{\theta_1}^{\theta} r^2[c_{u(i)-1}]d\alpha + 2\int_{0}^{\theta_1} r^2[c_{u(i)-2}]d\alpha \qquad (4.130)$$

where b = half-width of the foundation

r = radius of the trial circle

$c_{u(i)-1}, c_{u(i)-2}$ directional undrained shear strength for layers 1 and 2, respectively

As shown in Fig. 4.57a, let ψ be the angle between the failure plane and the direction of the major principal stress. Referring to Eq. (4.129)

Along arc AC

$$c_{u(i)-1} \quad c_{u(H)=1} + [c_{u(V)-1} - c_{u(H)-1}]\sin^2(\alpha+\psi) \qquad (4.131)$$

Along arc CE

$$c_{u(i)-2} = c_{u(H)-2} + [c_{u(V)-2} - c_{u(H)-2}]\sin^2(\alpha+\psi) \qquad (4.132)$$

Similarly, along arc DB

$$c_{u(i)-1} = c_{u(H)-1} + [c_{u(V)-1} - c_{u(H)-1}]\sin^2(\alpha-\psi) \qquad (4.133)$$

and, along arc ED

$$c_{u(i)-2} = c_{u(H)-2} + [c_{u(V)-2} - c_{u(H)-2}]\sin^2(\alpha-\psi) \qquad (4.134)$$

Note that for the portion of the arc AE, $i=\alpha+\psi$, and for the portion BE, $i=\alpha-\psi$. Now, let the anisotropy coefficient be defined as

$$K \quad \frac{c_{u(V)-1}}{c_{u(H)-1}} \quad \frac{c_{u(V)-2}}{c_{u(H)-2}} \qquad (4.135)$$

The magnitude of the anisotropy ratio K is less than one for over-consolidated clays, and $K>1$ for normally consolidated clays. Also, let

$$n = \left[\frac{c_{u(V)-2}}{c_{u(V)-1}}\right] - 1 \quad \left[\frac{c_{u(H)-2}}{c_{u(H)-1}}\right] - 1 \qquad (4.136)$$

where n a factor representing the relative strength of two clay layers

Combining Eqs. (4.130)(4.131), (4.132), (4.133), (4.134), and (4.136)

$$2bq_{\mathrm{u}}(r\sin\theta-b) = \int_{\theta_1}^{\theta} r^2\{c_{\mathrm{u(H)-1}}+[c_{\mathrm{u(V)-1}}-c_{\mathrm{u(H)-1}}]\sin^2(\alpha+\phi)\}d\alpha$$

$$+ \int_{\theta_1}^{\theta} r^2\{c_{\mathrm{u(H)-1}}+[c_{\mathrm{u(V)-1}}-c_{\mathrm{u(H)-1}}]\sin^2$$

$$(\alpha-\psi)\}d\alpha + \int_{0}^{\theta_1} r^2(n+1)\{c_{\mathrm{u(H)-1}}+[c_{\mathrm{u(V)-1}}$$

$$-c_{\mathrm{u(H)-1}}]\sin^2(\alpha+\psi)d\alpha + \int_{0}^{\theta_1} r^2(n+1)\{c_{\mathrm{u(H)-1}}$$

$$+[c_{\mathrm{u(V)-1}}-c_{\mathrm{u(H)-1}}]\sin^2(\alpha-\psi)\}d\alpha \qquad (4.137)$$

Or, combining Eqs. (4.135) and (4.137)

$$\frac{q_{\mathrm{u}}}{c_{\mathrm{u(V)-1}}} = \frac{r^2/b^2}{2K[(r/b)\sin\theta-1]}\Bigg\{2\theta+2n\theta_1+(K-1)\theta+n(K-1)\theta_1 \quad \frac{K-1}{2}$$

$$\left[\frac{\sin 2(\theta+\psi)}{2}+\frac{\sin 2(\theta-\psi)}{2}\right]-\frac{n(K-1)}{2}$$

$$\left[\frac{\sin 2(\theta_1-\psi)}{2} \quad \frac{\sin 2(\theta_1-\psi)}{2}\right]\Bigg\} \qquad (4.138)$$

where $\theta_1 \quad \cos^{-1}(\cos\theta+\frac{H}{r})$

From Eq. (4.127), note that

$$N_{\mathrm{c(L)}} \quad \frac{q_{\mathrm{u}}}{c_{\mathrm{u(V)-1}}} \qquad (4.139)$$

In order to obtain the minimum value of $N_{\mathrm{c(L)}}=(q_{\mathrm{u}})/[c_{\mathrm{u(V)-1}}]$, the theorem of maxima and minima need to be used, or

$$\frac{\partial N_{\mathrm{c(L)}}}{\partial\theta} = 0 \qquad (4.140)$$

and

$$\frac{\partial N_{\mathrm{c(L)}}}{\partial r} \quad 0 \qquad (4.141)$$

Equations (4.138), (4.140), and (4.141) will yield two relationships in terms of the variables θ and r/b. So, for given values of H/b, K, n, and ψ, the above relationships may be solved to obtain values of θ and r/b. These can be then be used in Eq. (4.138) to obtain the desired value of $N_{c(L)}$ (for given values of H/b, K, n, and ψ). Lo (1965) has shown that the angle ψ between the failure plane and the major principal stress for anisotropic soils can be taken as approximately equal to 35°.

The variations of the bearing capacity factor $N_{c(L)}$ obtained in this manner for K=0.8, 1 (isotropic case), 1.2, 1.4, 1.6, 1.8, and 2 are shown in Fig. 4.58.

It needs to be pointed out that, if a shallow foundation is located at a depth D_f (Fig. 4.59), the general ultimate bearing capacity equation [see Eq. (4.71)] will be of the form (ϕ=0 condition)

$$q_u = c_{u(V)-1} N_{c(L)} \lambda_{cs} \lambda_{cd} + q N_{q(L)} \lambda_{qs} \lambda_{qd}$$

where $q = \gamma_1 D_f$

$N_{q(L)} = 1$ for ϕ=0 condition

So

$$q_u = c_{u(V)-1} N_{c(L)} \lambda_{cs} \lambda_{cd} + q \lambda_{qs} \lambda_{qd} \qquad (4.142)$$

Proper shape and depth factors can be selected from Table 4.4.

Example 4.7. Refer to Fig. 4.59. For the foundation, given: D_f= 0.8 m, B=0.8 m, L =1.6 m, H=0.5 m, γ_1=17.8 kN/m³, γ_2=17.0 kN/m³, $c_{u(V)-1}$=45 kN/m², $c_{u(V)-2}$=30 kN/m², and anisotropy ratio K=1.4. Estimate the allowable load bearing capacity of the foundation with a factor of safety F_s=4.

Solution

From Eq. (4.142)

$$q_u = c_{u(V)-1} N_{c(L)} \lambda_{cs} \lambda_{cd} + (\gamma_1 D_f) \lambda_{qs} \lambda_{qd}$$

$H/b \quad H/(B/2) \quad 0.5/0.4 \quad 1.25;\ K = 1.4$

$$\frac{c_{u(V)-2}}{c_{u(V)-1}} = \frac{30}{45} \quad 0.667$$

So

$$n = 1 - 0.667 = 0.333$$

So, from Fig. 4.58d, the value of $N_{c(L)} \simeq 4.25$

We will use Meyerhof's shape and depth factors given in Table 4.4.

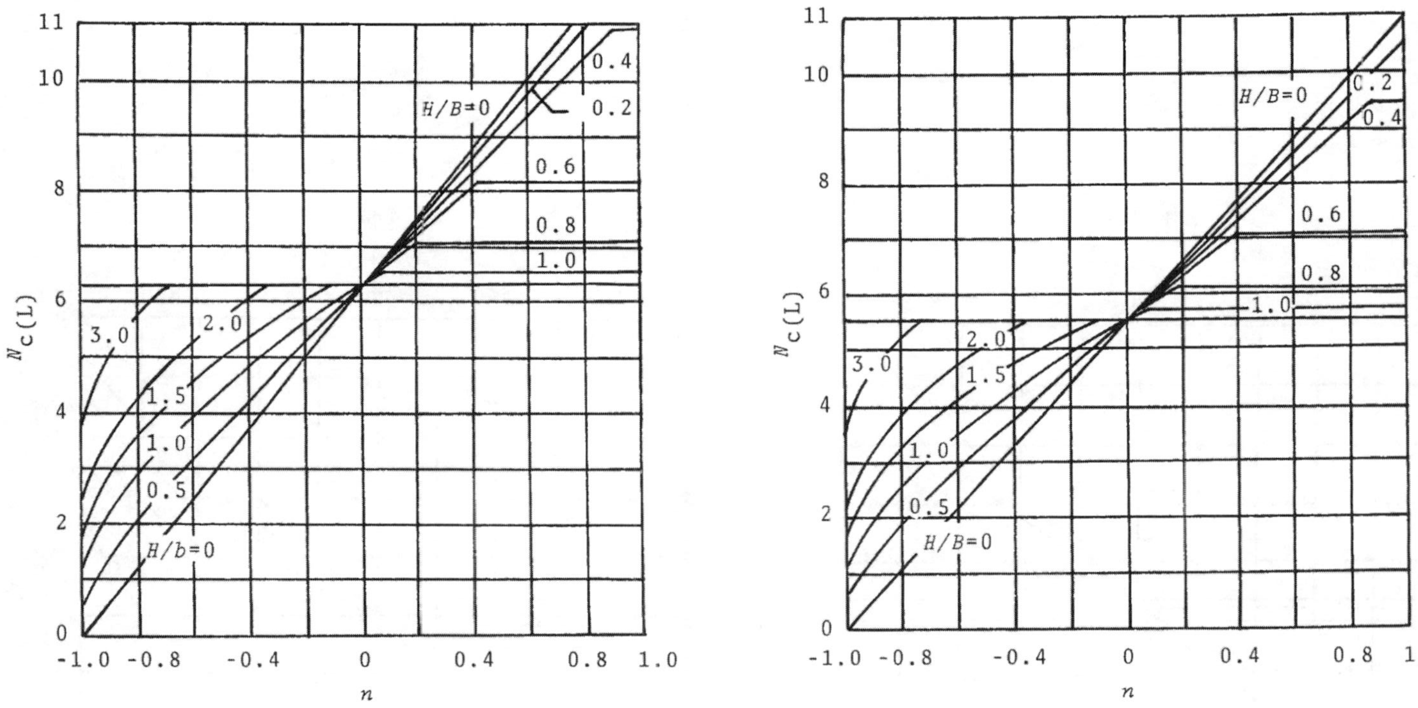

(a) K=0.8

(b) K=1.0

Figure 4.58. Variation of bearing capacity factor $N_{c(L)}$ (after Reddy and Srinivasan, 1967).

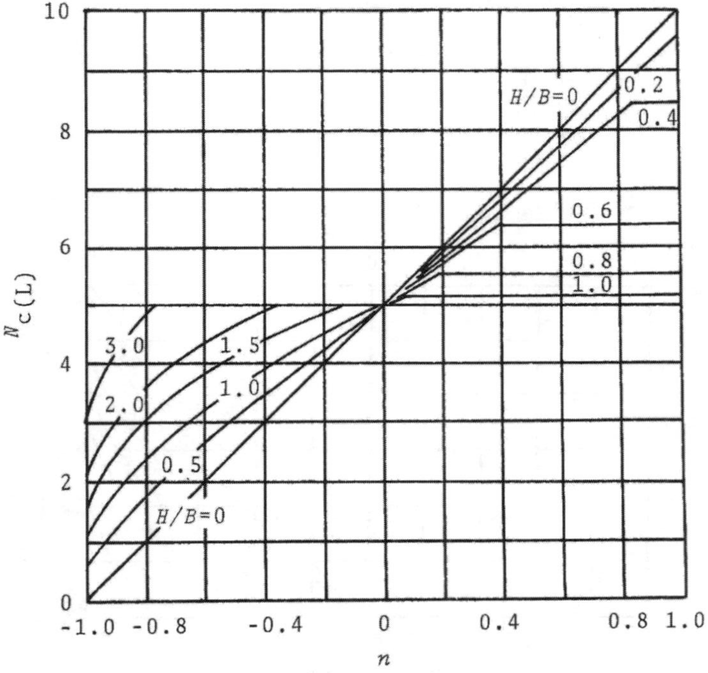

(c) K=1.2

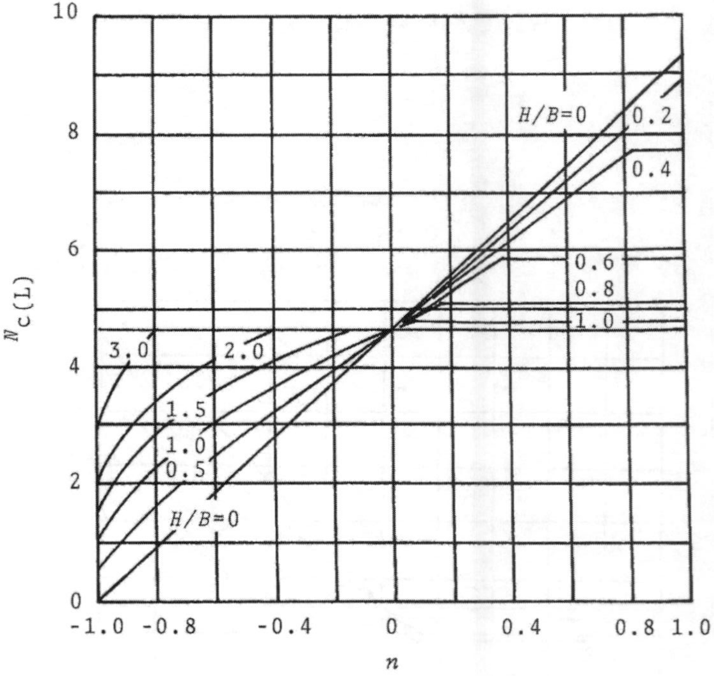

(d) K=1.4

Figure 4.58. (Continued).

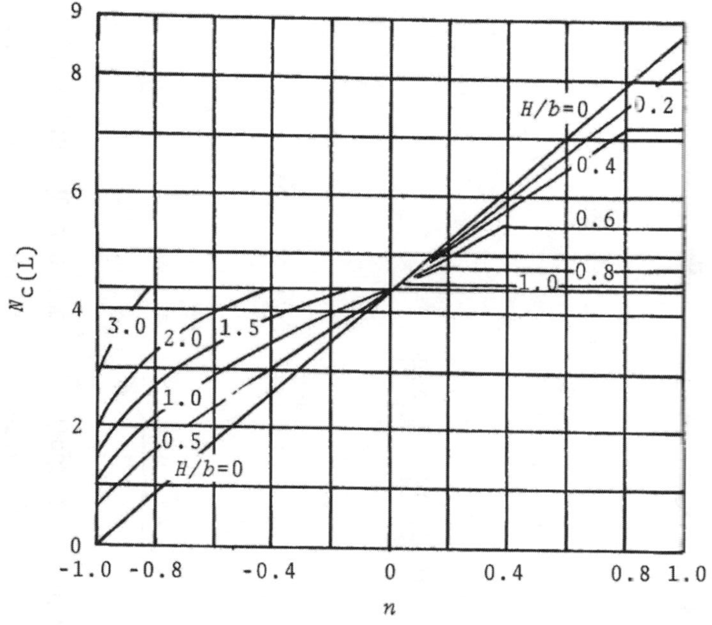

(e) $K=1.6$

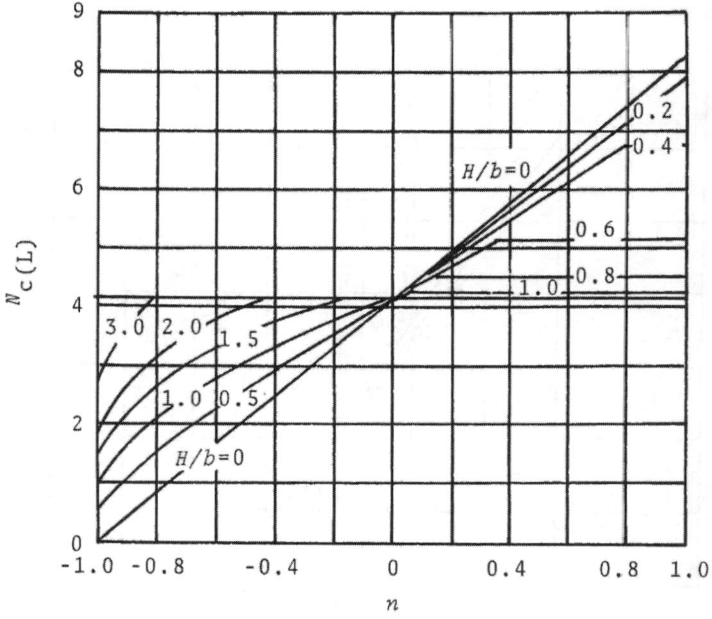

(f) $K=1.8$

Figure 4.58. (Continued).

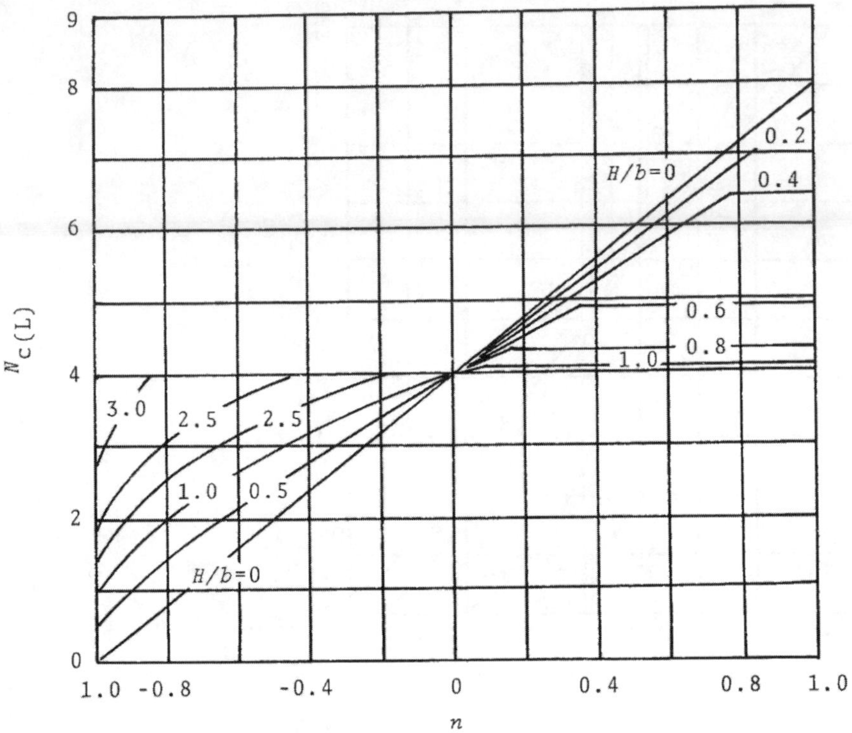

(g) K=2.0

Figure 4.58. (Continued).

$$\lambda_{cs} = 1+0.2\left(\frac{B}{L}\right) = 1+(0.2)\left(\frac{0.8}{1.6}\right) = 1.1$$

$$\lambda_{qs} = 1$$

$$\lambda_{cd} = 1+0.2\left(\frac{D_f}{B}\right) = 1+(0.2)\left(\frac{0.8}{0.8}\right) = 1.2$$

$$\lambda_{qd} = 1$$

So

$$q_u = (45)(4.25)(1.1)(1.2) + (17.8)(0.8)(1.0)(1.0)$$

$$= 252.45 + 14.25 = 266.69 \text{ kN/m}^2$$

$$q_{all} = \frac{q_u}{F_s} = \frac{266.69}{4} \quad \underline{66.67 \text{ kN/m}^2}$$

Meyerhof and Hanna's Theory--Strong Soil Underlain By Soft Soil

Meyerhof and Hanna (1978) developed a theory for estimating

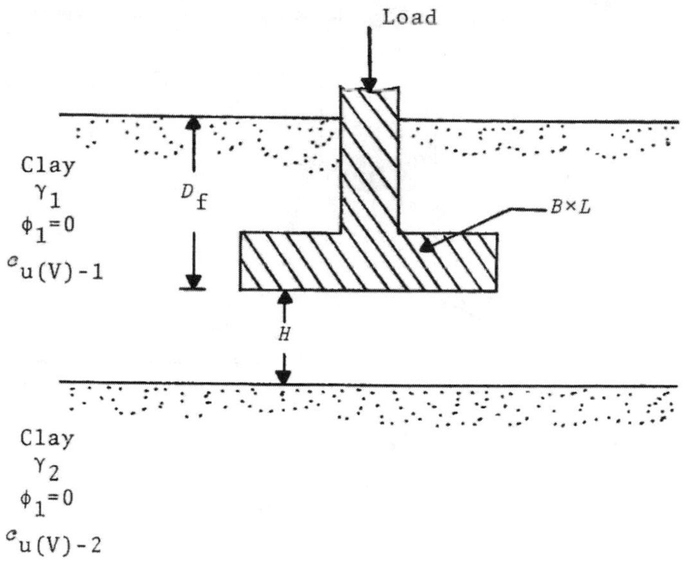

Figure 4.59. Rectangular foundation on anisotropic layered clay.

the ultimate bearing capacity of shallow rough continuous foundations supported by a strong soil layer underlain by a weaker soil layer as shown in Fig. 4.60. According to their theory, at ultimate load per unit area (q_u), the failure surface in soil will be as shown in Fig. 4.60. If the depth H is relatively small compared to the foundation width B, a punching shear failure will occur in the top soil (stronger) layer followed by a general shear failure in the bottom soil (weaker) layer. So, the ultimate bearing capacity of the continuous foundation can be given as

$$q_u = q_b + \frac{2(C_a + P_p \sin\delta)}{B} - \gamma_1 H \qquad (4.143)$$

where B = width of the foundation
γ_1 = unit weight of the stronger soil layer
C_a = adhesive force
P_p = passive force per unit length of the faces aa' and bb'
q_b bearing capacity of the bottom soil layer
δ inclination of the passive force P_p with the horizontal

Note that, in Eq. (4.143)

$$C_a \quad c_a H \qquad (4.144)$$

where c_a = adhesion

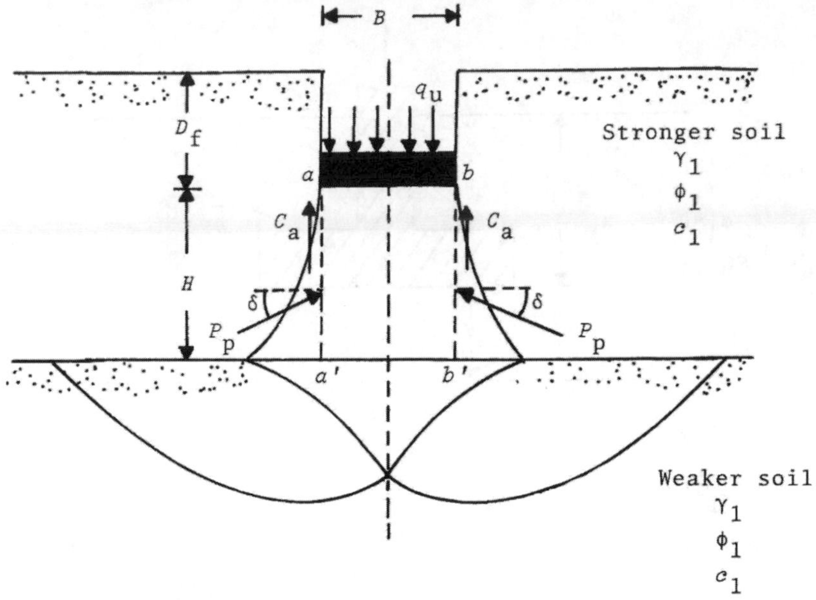

Figure 4.60. Shallow rough continuous foundation on layered soil--strong over weak.

$$P_p = \tfrac{1}{2}\gamma_1 H^2 \left(\frac{K_{pH}}{\cos\delta}\right) + (\gamma_1 D_f)(H)\left(\frac{K_{pH}}{\cos\delta}\right)$$

$$= \tfrac{1}{2}\gamma_1 H^2 \left(1 + \frac{2D_f}{H}\right)\left(\frac{K_{pH}}{\cos\delta}\right) \tag{4.145}$$

where K_{pH} = horizontal passive earth pressure coefficient

Also

$$q_b = c_2 N_{c(2)} + \gamma_1 (D_f + H) N_{q(2)} + \tfrac{1}{2}\gamma_2 B N_{\gamma(2)} \tag{4.146}$$

where c_2 cohesion of the bottom layer (weaker) or soil

γ_2 unit weight of the bottom layer of soil

$N_{c(2)}, N_{q(2)}, N_{\gamma(2)}$ = bearing capacity factors for the bottom soil layer (that is, with respect to the soil friction angle of the bottom soil layer--ϕ_2)

Combining Eqs. (4.143), (4.144), and (4.145)

$$q_u \quad q_b + \frac{2c_a H}{B} + (2)\left[\tfrac{1}{2}\gamma_1 H^2 \left(1 + \frac{2D_f}{H}\right)\right]\left(\frac{K_{pH}}{\cos\delta}\right)\left(\frac{\sin\delta}{B}\right) - \gamma_1 H$$

$$q_b + \frac{2c_a H}{B} + \gamma_1 H^2 \left(1 + \frac{2D_f}{H}\right) \frac{K_{pH}\tan\delta}{B} \quad \gamma_1 H \tag{4.147}$$

However, let

$$K_{pH}\tan\delta = K_s\tan\phi_1 \tag{4.148}$$

where K_s punching shear coefficient
So

$$q_u = q_b + \frac{2c_aH}{B} + \gamma_1H^2\left(1 + \frac{2D_f}{H}\right)\frac{K_s\tan\phi_1}{B} - \gamma_1H \tag{4.149}$$

The punching shear coefficient can be determined by using the passive earth pressure coefficient charts as proposed by Caquot and Kerisel (1949). Figure 4.61 gives the variation of K_s with q_2/q_1 and ϕ_1. Note that q_1 and q_2 are the ultimate bearing capacities of a continuous foundation of width B under vertical load on the surfaces of homogeneous thick beds of upper and lower soil. Or

$$q_1 = c_1N_{c(1)} + \tfrac{1}{2}\gamma_1BN_{\gamma(1)} \tag{4.150}$$

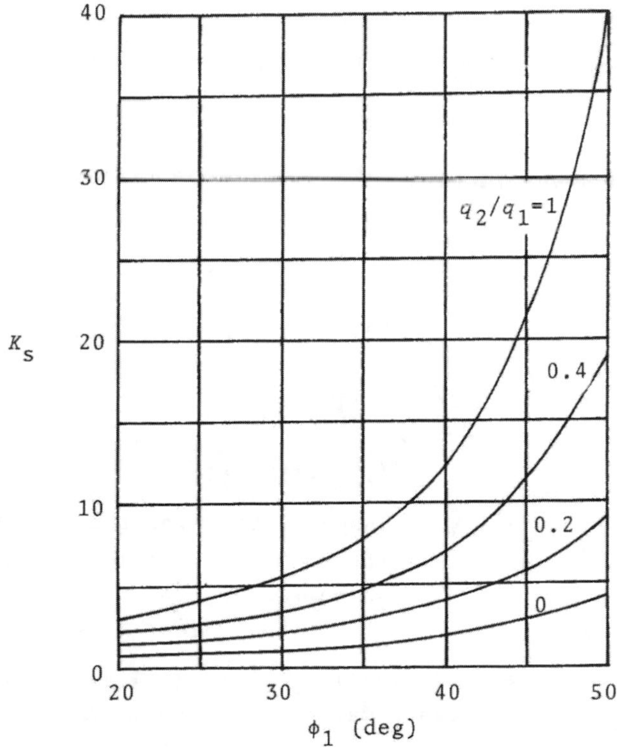

Figure 4.61. Variation of K_s with ϕ_1 and q_2/q_1 (after Meyerhof and Hanna, 1978).

where $N_{c(1)}, N_{\gamma(1)}$ bearing capacity factors for the soil friction angle ϕ_1

$$q_2 = c_2 N_{c(2)} + \tfrac{1}{2}\gamma_2 B N_{\gamma(2)} \tag{4.151}$$

However, if the height H (Fig. 4.60) is large compared to the width B, then the failure surface will be completely located in the upper stronger soil layer as shown in Fig. 4.62. In such case, the upper limit for q_u will be of the following form

$$q_u = q_t \quad c_1 N_{c(1)} + q N_{q(1)} + \tfrac{1}{2}\gamma_1 B N_{\gamma(1)} \tag{4.152}$$

Note that the preceding equation is similar to that given in Eq. (4.69). Hence, combining Eqs. (4.149) and (4.152)

$$q_u = q_b + \frac{2c_a H}{B} + \gamma_1 H^2 \left(1 + \frac{2D_f}{H}\right) \frac{K_s \tan\phi_1}{B} - \gamma_1 H \leq q_t \tag{4.153}$$

For rectangular foundations, the preceding equation can be modified as

$$q_u \quad q_b + \left(1 + \frac{B}{L}\right)\left(\frac{2c_a H}{B}\right) s_a + \left(1 + \frac{B}{L}\right)\gamma_1 H^2 \left(1 + \frac{2D_f}{H}\right)\left(\frac{K_s \tan\phi_1}{B}\right) s_s$$

$$\gamma_1 H \leq q_t \tag{4.154}$$

where s_a, s_s = shape factors

$$q_b \quad c_2 N_{c(2)} \lambda_{cs(2)} + \gamma_1 (D_f + H) N_{q(2)} \lambda_{qs(2)}$$

$$+ \tfrac{1}{2}\gamma_2 B N_{\gamma(2)} \lambda_{\gamma s(2)} \tag{4.155}$$

$$q_t - c_1 N_{c(1)} \lambda_{cs(1)} + \gamma_1 D_f N_{q(1)} \lambda_{qs(1)}$$

$$+ \tfrac{1}{2}\gamma_1 B N_{\gamma(1)} \lambda_{\gamma s(1)} \tag{4.156}$$

$\lambda_{cs(1)}, \lambda_{qs(1)}, \lambda_{\gamma s(1)}$ = shape factors for the top soil layer (friction angle=ϕ_1)--see Table 4.4

$\lambda_{cs(2)}, \lambda_{cs(2)}, \lambda_{\gamma s(2)}$ = shape factors for the bottom soil layer (friction angle=ϕ_2)--see Table 4.4

Based on the general equations [Eqs. (4.154), (4.155), and (4.156)], some special cases may be developed. They are as follows

Case I: Strong Sand Layer Over Weak Saturated Clay (ϕ_2=0)

For this case, c_1=0 and, hence, c_a=0. Also for ϕ_2=0, $N_{c(2)}$=

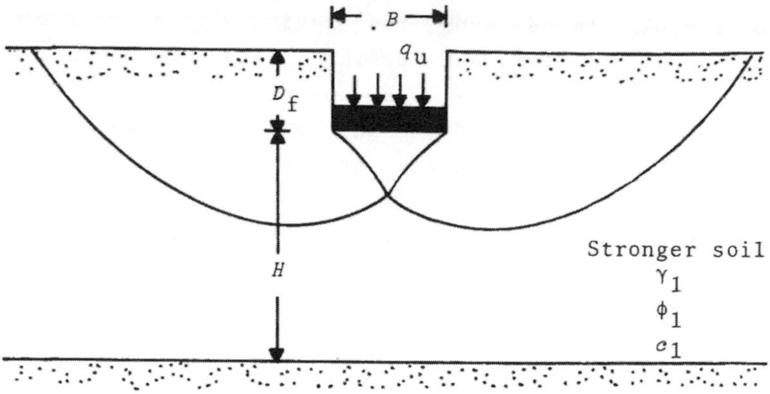

Figure 4.62. Continuous rough foundation on layered soil--H/B is large.

5.14, $N_{\gamma(2)}=0$, $N_{q(2)}=1$, $\lambda_{cs}=1+0.2(B/L)$, $\lambda_{qs}=1$ (shape factors are Meyerhof's values as given in Table 4.4). So

$$q_u = \left[1+0.2\left(\frac{B}{L}\right)\right](5.14)c_2 + \left(1+\frac{B}{L}\right)\gamma_1 H^2\left(1+\frac{2D_f}{H}\right)\frac{K_s \tan\phi_1}{B}\cdot s_s + \gamma_1 D_f \le q_t \tag{4.157}$$

$$\text{where } q_t = \gamma_1 D_f N_{q(1)}\left[1+0.1\left(\frac{B}{L}\right)\tan^2(45+\phi_1/2)\right] + \frac{1}{2}\gamma_1 B N_{\gamma(1)}\left[1+0.1\left(\frac{B}{L}\right)\tan^2(45+\phi_1/2)\right] \tag{4.158}$$

In Eq. (4.158), the relationships for the shape factors λ_{qs} and $\lambda_{\gamma s}$ are those given by Meyerhof (1963) as shown in Table 4.4. Note that K_s is a function of q_2/q_1 [Eqs. (4.150) and (4.151)]. For this case

$$\frac{q_2}{q_1} = \frac{c_2 N_{c(2)}}{\frac{1}{2}\gamma_1 B N_{\gamma(1)}} = \frac{5.14c_2}{0.5\gamma_1 B N_{\gamma(1)}} \tag{4.159}$$

So, once q_2/q_1 is known, the magnitude of K_s can be obtained from Fig. 4.61, which in turn can be used in Eq. (4.157) to determine the ultimate bearing capacity of the foundation q_u. The value of the shape factor s_s for strip foundations is equal to one. For square or circular foundations, the magnitude of s_s appears to vary between 1.1 to 1.27 (Hanna and Meyerhof, 1980). For conservative design, it may be taken to be equal to one.

Based on this concept, Hanna and Meyerhof (1980) have developed some alternate design charts for the determination of the punching shear coefficient K_s, and these are shown in Figs. 4.63 and 4.64. In order to use these charts, the following steps need to be followed.

1. Determine q_2/q_1.
2. With known values of ϕ_1 and q_2/q_1, determine the magnitude of δ/ϕ_1 from Fig. 4.63.
3. With known values of ϕ_1, δ/ϕ_1, and c_2, determine K_s from Fig. 4.64.

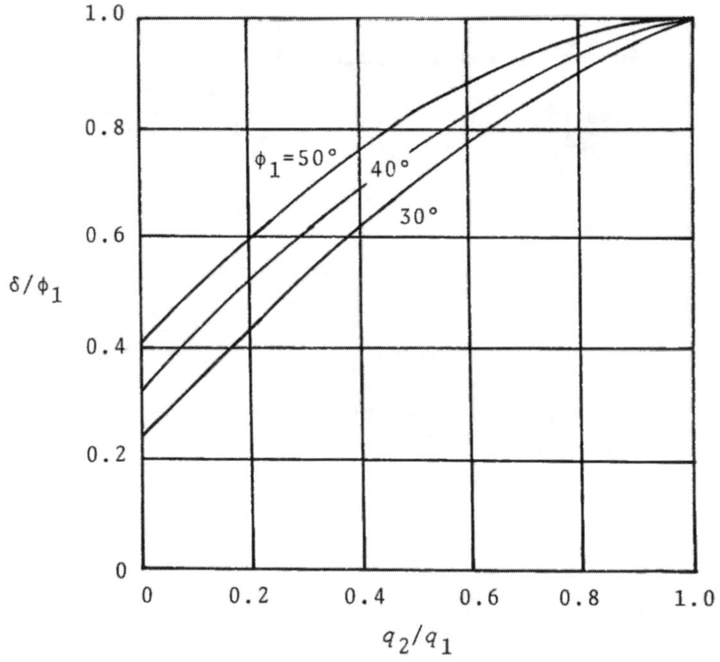

Figure 4.63. Variation of δ/ϕ_1 with q_2/q_1 and ϕ_1 for stronger sand layer over weaker clay layer (after Hanna and Meyerhof, 1980).

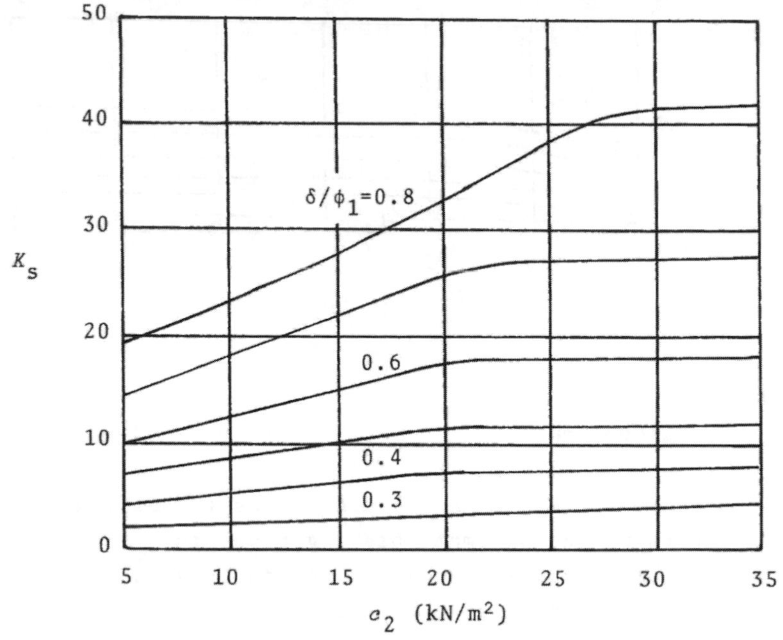

(a)

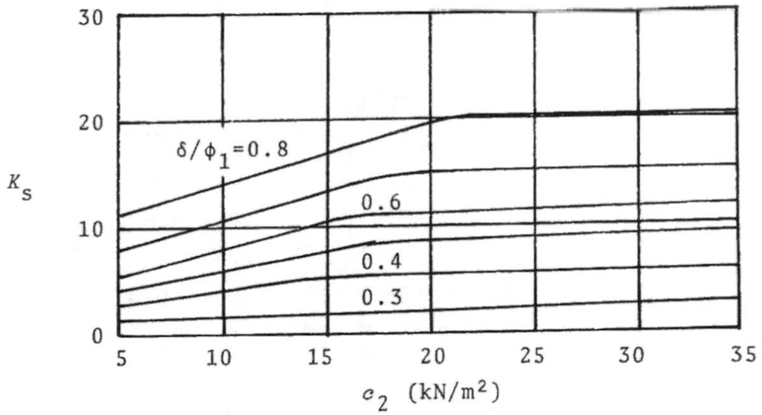

(b)

Figure 4.64. Coefficient of punching shear--strong sand over weaker clay: (a) ϕ_1=50°; (b) ϕ_1=45°; (c) ϕ_1=40° (after Hanna and Meyerhof, 1980).

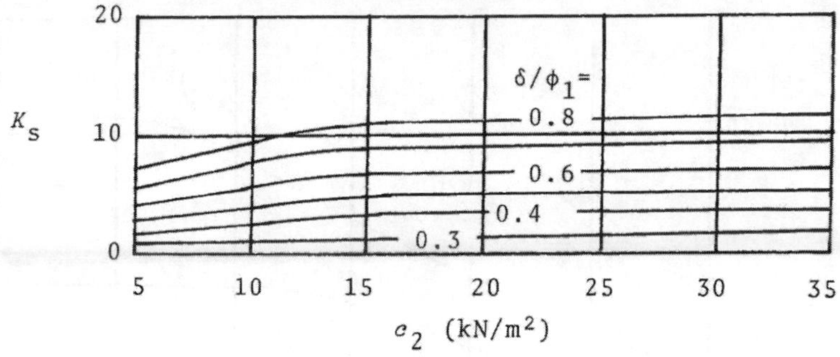

(c)

Figure 4.64. (Continued).

Example 4.8. Figure 4.65 shows a continuous foundation. (a) If H=1.5 m, determine the ultimate bearing capacity q_u. (b) At what minimum value of H/B will the clay layer not have any effect on the ultimate bearing capacity of the foundation?

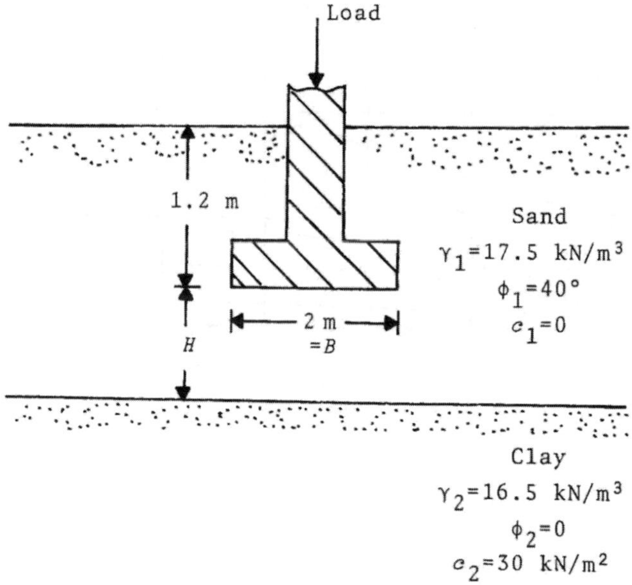

Figure 4.65.

Solution

Part a: For $B/L=0$ and with $s_s=1$, Eq. (4.157) takes the form

$$q_u = 5.14c_2 + \gamma_1 H^2\left(1 + \frac{2D_f}{H}\right)\frac{K_s \tan\phi_1}{B} + \gamma_1 D_f$$

$$= (5.14)(30) + (17.5)(H^2)\left[1 + \frac{(2)(1.2)}{H}\right]\frac{K_s \tan 40}{2} + (17.5)(1.2)$$

$$175.2 + 7.342H^2K_s\left(1 + \frac{2.4}{H}\right) \qquad \text{(a)}$$

For determination of K_s, we need to obtain q_2/q_1. From Eq. (4.159)

$$\frac{q_2}{q_1} = \frac{5.14c_2}{0.5\gamma_1 BN_{\gamma(1)}}$$

From Table 4.5, for $\phi_1=40°$, Meyerhof's value of $N_{\gamma(1)}$ is equal to 93.7. So

$$\frac{q_2}{q_1} = \frac{(5.14)(30)}{(0.5)(17.5)(2)(93.7)} = 0.094$$

Referring to Fig. 4.63, for $q_2/q_1=0.094$ and $\phi_1=40°$, the value of $\delta/\phi_1 \approx 0.42$. So with $\delta/\phi_1 \approx 0.42$, $c_2=30$ kN/m^2. Figure 4.64c gives the value of $K_s=3.89$. Substitution of this value of Eq. (a) gives

$$q_u = 175.2 + 28.56H^2\left(1 + \frac{2.4}{H}\right) \leq q_t \qquad \text{(b)}$$

$$q_t = \gamma_1 D_f N_{q(1)}\left[1+0.1\left(\frac{B}{L}\right)\tan^2(45+\phi_1/2)\right] + \frac{1}{2}\gamma_1 BN_{\gamma(1)}$$

$$\left[1+0.1\left(\frac{B}{L}\right)\tan^2(45+\phi_1/2)\right]$$

For continuous foundation, $B/L=0$. So

$$q_t = \gamma_1 D_f N_{q(1)} + \frac{1}{2}\gamma_1 BN_{\gamma(1)}$$

For $\phi_1=40°$, Meyerhof's value of $N_{\gamma(1)}=93.7$ (Table 4.5) and $N_q=64.2$ (Table 4.3). Hence

$$q_t \quad (17.5)(1.2)(64.2) + \left(\tfrac{1}{2}\right)(17.5)(2)(93.7)$$

$$= 1{,}348.2 + 1{,}639.75 \quad 2.987.95 \text{ kN/m}^2 \qquad \text{(c)}$$

If $H=1.5$ m is substituted into Eq. (b)

$$q_u = 175.2 + 28.56(1.5)^2\left(1+\frac{2.4}{1.5}\right) = 342.3 \text{ kN/m}^2$$

Since $q_u=342.3<q_t$, the ultimate bearing capacity is 342.3 kN/m^2

Part b. For the clay layer not to have any effect on the ultimate bearing capacity, the minimum value of H will be such that it will satisfy the relationship

$$q_u = 175.2 + 28.56H^2\left(1 + \frac{2.4}{H}\right) = q_t = 2{,}987.95$$

By trial and error, $H \simeq 6.2$ m. So

$$\frac{H}{B} \simeq \frac{6.2}{2} \quad \underline{3.1}$$

<u>Case II: Strong Sand Layer Over Weak Sand Layer</u>

For this case $c_1=0$ and $c_2=0$. Hence, referring to Eq. (4.154)

$$q_u = q_b + \left(1 + \frac{B}{L}\right)\gamma_1 H^2 \left(1 + \frac{2D_f}{H}\right)\left(\frac{K_s \tan\phi_1}{B}\right)s_s - \gamma_1 H \le q_t \qquad (4.160)$$

$$\text{where } q_b = \gamma_1(D_f+H)N_{q(2)}\lambda_{qs(2)} + \tfrac{1}{2}\gamma_2 BN_{\gamma(2)}\lambda_{\gamma s(2)} \qquad (4.161)$$

$$q_t = \gamma_1 D_f N_{q(1)}\lambda_{qs(1)} + \tfrac{1}{2}\gamma_1 BN_{\gamma(1)}\lambda_{\gamma s(1)} \qquad (4.162)$$

Using Meyerhof's shape factors given in Table 4.4

$$\lambda_{qs(1)} = \lambda_{\gamma s(1)} = 1 + 0.1\left(\frac{B}{L}\right)\tan^2(45+\phi_1/2) \qquad (4.163a)$$

and

$$\lambda_{qs(2)} \quad \lambda_{\gamma s(2)} \quad 1 + 0.1\left(\frac{B}{L}\right)\tan^2(45+\phi_2/2) \qquad (4.163b)$$

For conservative design, the magnitude of s_s can be taken to be equal to one for all B/L ratios. For this case

$$\frac{q_2}{q_1} \quad \frac{0.5\gamma_2 BN_{\gamma(2)}}{0.5\gamma_1 BN_{\gamma(1)}} \quad \frac{\gamma_2 N_{\gamma(2)}}{\gamma_1 N_{\gamma(1)}} \qquad (4.164)$$

Once the magnitude of q_2/q_1 is determined, the value of the punching shear coefficient K_s can be obtained from Fig. 4.61. Hanna (1981) has suggested that the friction angles as obtained from direct shear tests should be used.

More recently, Hanna (1981) has provided an improved design chart for estimation of the punching shear coefficient K_s in Eq. (4.160). In this development, he assumed that the variation of δ for the assumed vertical failure surface in the top strong sand layer will be of the nature as shown in Fig. 4.66, or

$$\delta_x = \eta\phi_2 + ax^2 \qquad (4.165)$$

where

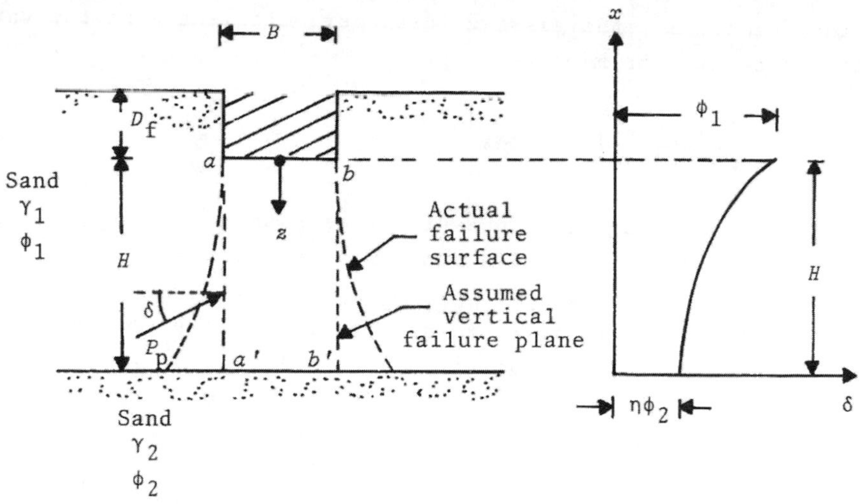

Figure 4.66. Variation of δ with depth for evaluation of K_s as assumed by Hanna (1981).

$$\eta = \frac{q_2}{q_1} \tag{4.166}$$

$$a = \frac{\phi_1 - (q_2/q_1)\phi_2}{H^2} \tag{4.167}$$

So

$$\delta_x = \left(\frac{q_2}{q_1}\right)\phi_2 + \left[\frac{\phi_1 - (q_2/q_1)\phi_2}{H^2}\right] x^2 \tag{4.168}$$

The preceding relationship means that at $x=0$ (that is, at the interface of the two soil layers)

$$\delta \quad \left(\frac{q_2}{q_1}\right) \phi_2 \tag{4.169}$$

and at the level of the foundation, that is, $x=H$

$$\delta \quad \phi_1 \tag{4.170}$$

Equation (4.165) can also be rewritten as

$$\delta_z = \left(\frac{q_2}{q_1}\right)\phi_2 + \left[\frac{\phi_1 - (q_2/q_1)\phi_2}{H^2}\right](H-z)^2 \tag{4.171}$$

where δ_z is the angle of inclination of the passive pressure with respect to the horizontal at a depth z measured from the bottom of the foundation. So, the passive force per unit length of the vertical surface aa' (or bb') is

$$P_p = \int_0^H \left[\frac{\gamma_1 K_{pH(z)}}{\cos\delta_z}\right](z+D_f)\,dz \tag{4.172}$$

where $K_{pH(z)}$ = horizontal component of the passive earth pressure coefficient at a depth z measured from the bottom of the foundation

The magnitude of P_p as expressed by Eq. (4.172), in combination with the expression δ_z given in Eq. (4.171), can be determined by means of a computer. In order to determine the average value of δ as given by Eq. (4.145), the following steps are taken:

1. Assume an average value of δ and obtain K_{pH} as given in the tables by Caquot and Kerisel (1949).

2. Calculate P_p from Eq. (4.145) using the average value of δ and K_{pH} obtained from Step 1.

3. Repeat Steps 1 and 2 until the magnitude of P_p obtained from Eq. (4.145) is the same as that calculated from Eq. (4.172).

4. The average value of δ, for which P_p calculated from Eqs. (4.145) and (4.172) are the same, is the value that needs to be used in Eq. (4.148) to calculate K_s.

Figure 4.67 gives the relationship for δ/ϕ_1 vs. ϕ_2 for various values of ϕ_1 obtained by the above procedure. Using Fig. 4.67, Hanna (1981) has given a design chart for K_s, and this is shown in Fig. 4.68.

Case III: Strong Clay Layer ($\phi_1=0$) Over Weak Clay Layer ($\phi_2=0$)

For this case, $N_{q(1)}$ and $N_{q(2)}$ are both equal to one, and $N_{\gamma(1)}=N_{\gamma(2)}=0$. Also, $N_{c(1)}=N_{c(2)}=5.14$. So, from Eq. (4.154)

$$q_u = \left[1+0.2\left(\frac{B}{L}\right)\right] c_2 N_{c(2)} + \left(1+\frac{B}{L}\right)\left(\frac{2c_a H}{B}\right)s_a + \gamma_1 D_f \le q_t \tag{4.173}$$

$$\text{where } q_t \quad \left[1+0.2\left(\frac{B}{L}\right)\right] c_1 N_{c(1)} + \gamma_1 D_f \tag{4.174}$$

The magnitude of the shape factor s_a may be taken as equal to one for conservative design. The magnitude of the adhesion c_a is a function of q_2/q_1. For this condition

$$\frac{q_2}{q_1} = \frac{c_2 N_{c(2)}}{c_1 N_{c(1)}} \quad \frac{5.14c_2}{5.14c_1} = \frac{c_2}{c_1} \tag{4.175}$$

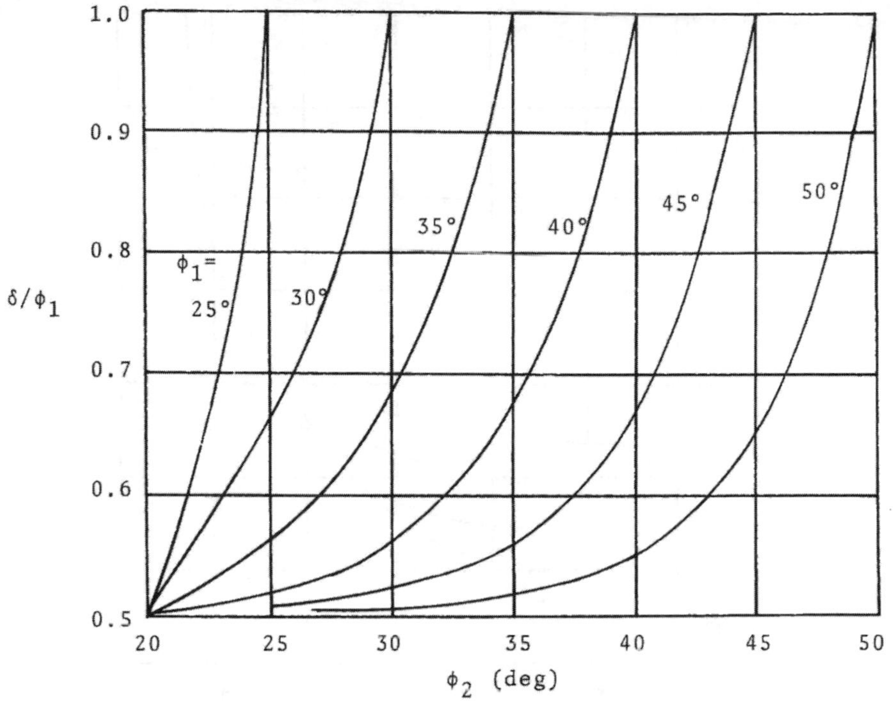

Figure 4.67. Variation of δ/ϕ_1 with ϕ_2 (after Hanna, 1981).

Figure 4.69 shows the theoretical variation of c_a with q_2/q_1 (Meyerhof and Hanna, 1978).

Example 4.9. Figure 4.70a shows a shallow foundation. Given: undrained shear strengh c_1 (for ϕ_1=0 condition)=80 kN/m^2, undrained shear strength c_2 (for ϕ_2=0 condition)=32 kN/m^2, γ_1=18 kN/m^3, D_f= 1 m, B=1.5 m, and L=3 m. Plot the variation of q_u with H/B.

Solution

From Eq. (4.175)

$$\frac{q_2}{q_1} = \frac{c_2}{c_1} = \frac{32}{80} \quad 0.4$$

From Fig. 4.69, for q_2/q_1=0.4, c_a/c_1=0.9. So

$$c_a = 0.9(80) = 72 \text{ kN/m}^2$$

Now, from Eq. (4.174)

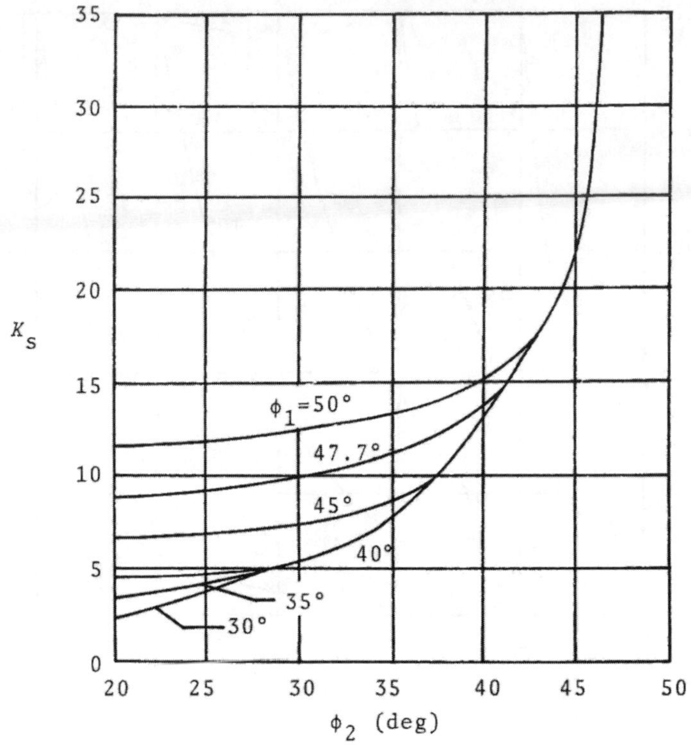

Figure 4.68. Design chart for determination of K_s--stronger sand overlying weaker sand (after Hanna, 1981).

$$q_t = \left[1+0.2\left(\frac{B}{L}\right)\right] c_1 N_{c(1)} + \gamma_1 D_f$$

$$= \left[1+(0.2)\left(\frac{1.5}{3}\right)\right](80)(5.14) + (18)(1) = 470.32 \text{ kN/m}^2$$

With s_s=1, Eq. (4.173) yields

$$q_u \quad \left[1+0.2\left(\frac{B}{L}\right)\right] c_2 N_{c(2)} + \left(1 + \frac{B}{L}\right)\left(\frac{2c_a H}{B}\right) + \gamma_1 D_f$$

$$(1+0.1)(32)(5.14) + (1.5)\left[(2)(32)\left(\frac{H}{B}\right)\right] + (18)(1)$$

$$= 198.93 + 96\left(\frac{H}{B}\right)$$

Note that q_u=198.93+96(H/B) gives a straight line variation. In order to determine the value of H/B at which q_u is equal to q_t, we set

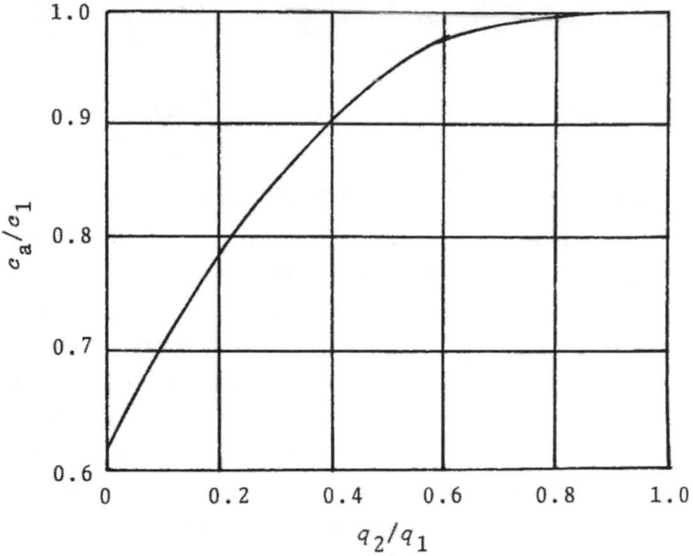

Figure 4.69. Variation of c_a/c_1 with q_2/q_1--Eq. (4.173) (after Meyerhof and Hanna, 1978).

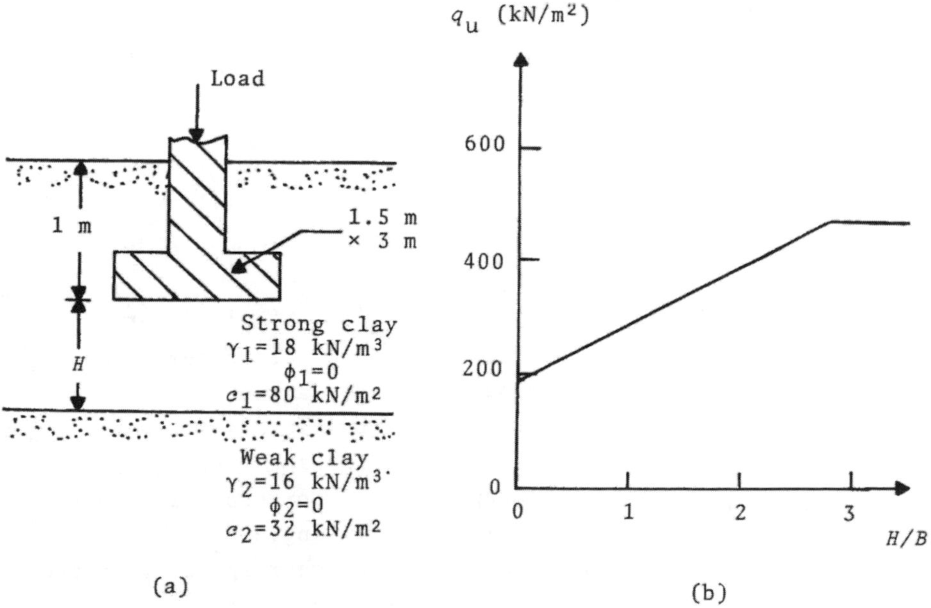

Figure 4.70.

$$q_t = 470.32 \text{ kN/m}^2 \quad 198.93+96\left(\frac{H}{B}\right)$$

So, H/B=2.83

Figure 4.70b shows the variation of q_u with H/B.

Weaker Soil Underlain by a Stronger Soil

In general, when a foundation is supported by a weaker soil layer which is underlain by a stronger soil at a shallow depth at shown in the left-hand side of Fig. 4.71a, the failure surface at ultimate load will pass through both soil layers. However, when the magnitude of H is relative large compared to the width of the foundation B, the failure surface at ultimate load will be fully located in the weaker soil layer (see the right-hand side of Fig. 4.71a). For estimation of the ultimate bearing capacity of such foundations, Meyerhof (1974) and Meyerhof and Hanna (1978) have proposed the following semiempirical relationship

$$q_u \quad q_t + (q_b - q_t)\left(1 - \frac{H}{H_f}\right)^2 \geq q_t \tag{4.176}$$

where H_f = depth of failure beneath the footing in the thick bed of the upper weak soil layer

q_t ultimate bearing capacity of the foundation in the thick bed of the upper soil layer

q_b ultimate bearing capacity of the foundation in the thick bed of the lower soil layer

So

$$q_t \quad c_1 N_{c(1)} \lambda_{cs(1)} + \gamma_1 D_f N_{q(1)} \lambda_{qs(1)} + \tfrac{1}{2}\gamma_1 N_{\gamma(1)} \lambda_{\gamma s(1)} \tag{4.177}$$

and

$$q_b = c_2 N_{c(2)} \lambda_{cs(2)} + \gamma_2 D_f N_{q(2)} \lambda_{qs(2)} + \tfrac{1}{2}\gamma_2 B N_{\gamma(2)} \lambda_{\gamma(2)} \tag{4.178}$$

where $N_{c(1)}, N_{q(1)}, N_{\gamma(1)}$ = bearing capacity factors corresponding to the soil friction angle ϕ_1

$N_{c(2)}, N_{q(2)}, N_{\gamma(2)}$ = bearing capacity factors corresponding to the soil friction angle ϕ_2

$\lambda_{cs(1)}, \lambda_{qs(1)}, \lambda_{\gamma s(1)}$ = shape factors corresponding to the soil friction angle ϕ_1

$\lambda_{cs(2)}, \lambda_{qs(2)}, \lambda_{\gamma s(2)}$ shape factors corresponding to the soil friction angle ϕ_2

Equations (4.176), (4.177), and (4.178) imply that the maximum and minimum limits of q_u are q_b and q_t, respectively (Fig. 4.71b).

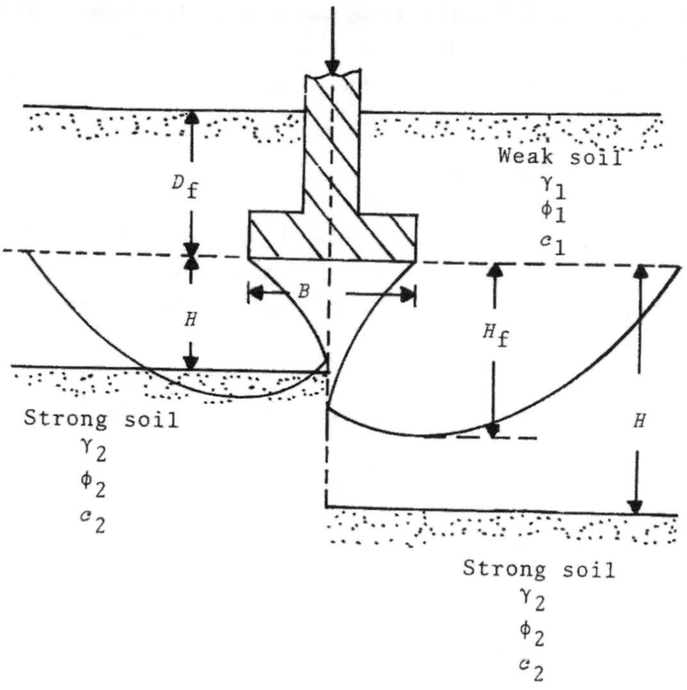

(a)

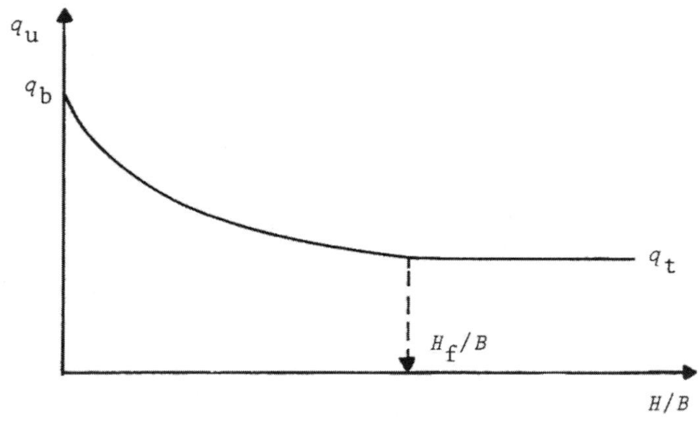

(b)

Figure 4.71. (a) Foundation on weaker soil layer underlain by stronger soil layer; (b) Nature of variation of q_u with H/B [Eq. 4.176)].

The magnitude of H_f/B varies from one for loose sand and clay to about 2 for dense sands (Meyerhof and Hanna, 1978). The author has conducted several laboratory model tests and found that a better approximation is obtained by modifying Eq. (4.176) as

$$q_u = q_t + (q_b - q_t)\left(1 - \frac{H}{2B}\right)^{1.8} \geq q_t \tag{4.179}$$

Based on several laboratory model tests, Hanna (1982) has proposed the following relationships for estimation of the ultimate bearing capacity q_u for a foundation resting on a *weak sand layer underlain by a strong sand layer*

$$q_u = \tfrac{1}{2}\gamma_1 B \lambda^*_{\gamma s} N_{\gamma(m)} + \gamma_1 D_f \lambda^*_{qs} N_{q(m)} \leq \tfrac{1}{2}\gamma_2 B \lambda_{\gamma s(2)} N_{\gamma(2)} + \gamma_1 D_f \lambda_{qs(2)} N_{q(2)} \tag{4.180}$$

where $N_{\gamma(2)}, N_{q(2)}$ = Meyerhof's bearing capacity factors with reference to the soil friction angle ϕ_2 (Tables 4.3 and 4.5)

$\lambda_{\gamma s(2)}, \lambda_{qs(2)}$ Meyerhof's shape factors (Table 4.4)

$1+0.1\left(\frac{B}{L}\right)\tan^2(45+\phi_2/2)$

$N_{\gamma(m)}, N_{q(m)}$ modified bearing capacity factors

$\lambda^*_{\gamma s}, \lambda^*_{qs}$ modified shape factors

The modified bearing capacity factors can be obtained as follows

$$N_{\gamma(m)} = N_{\gamma(2)} - \left[\frac{H}{H_{f(\gamma)}}\right][N_{\gamma(2)} - N_{\gamma(1)}] \tag{4.181}$$

$$N_{q(m)} = N_{q(2)} - \left[\frac{H}{H_{f(q)}}\right][N_{q(2)} - N_{q(1)}] \tag{4.182}$$

where $N_{\gamma(1)}, N_{q(1)}$ = Meyerhof's bearing capacity factors with reference to soil friction angle ϕ_1 (Tables 4.3 and 4.5)

The variation of $H_{f(\gamma)}$ and $H_{f(q)}$ with ϕ_1 is shown in Fig. 4.72. The relationships for the modified shape factors are the same as given in Eqs. (4.123) and (4.124). The term m_1 [Eq. (4.123)] can be determined from Fig. 4.51a by substituting $H_{f(q)}/B$ for D'/B and ϕ_1 for ϕ. Similarly, the term m_2 [Eq. (4.124)] can be obtained from Fig. 4.51b by substituting $H_{f(\gamma)}/B$ for D'/B and ϕ_1 for ϕ.

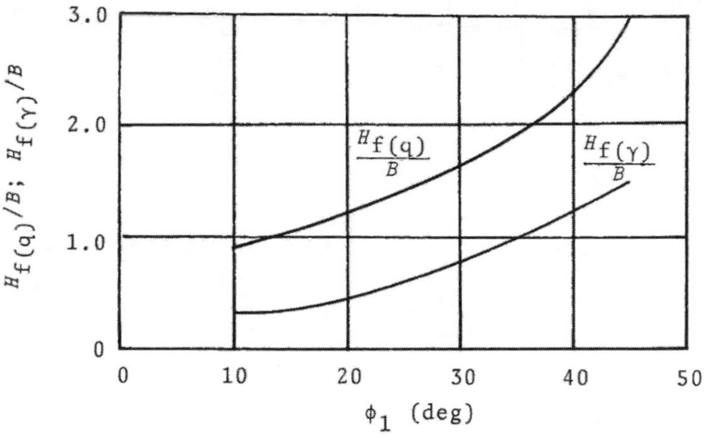

Figure 4.72. Variation of $H_{f(q)}/B$ and $H_{f(\gamma)}/B$ with ϕ_1 for use in Eqs. (4.181) and (4.182).

Example 4.10. A shallow foundation is shown in Fig. 4.73. Use Hanna's theory (1982) [Eq. (4.180)] and determine the ultimate bearing capacity q_u.

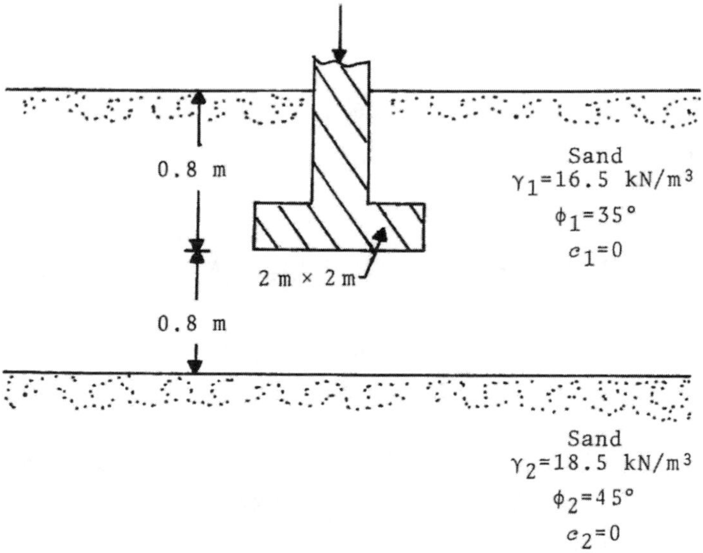

Figure 4.73.

Solution

H=0.5 m, ϕ_1=35°, ϕ_2=45°

From Fig. 4.72, for ϕ_1=35°

$$\frac{H_{f(\gamma)}}{B} = 1.0; \quad \frac{H_{f(q)}}{B} \quad 1.9$$

So, $H_{f(\gamma)}$=2.0 m, and $H_{f(q)}$=3.8 m. From Table 4.3 for ϕ_1=35° and ϕ_2=45°, $N_{q(1)}$=33.30 and $N_{q(2)}$=134.88. Similarly, from Table 4.5 for ϕ_1=35° and ϕ_2=45°, $N_{\gamma(1)}$=37.1 and $N_{\gamma(2)}$=262.7. Using Eqs. (4.181) and (4.182)

$$N_{\gamma(m)} = 262.7 \quad \left(\frac{0.5}{2}\right)(262.7-37.1) = 206.3$$

$$N_{q(m)} = 134.88 - \left(\frac{0.5}{3.8}\right)(133.88-33.3) = 121.5$$

From Eq. (4.180)

$$q_u = \tfrac{1}{2}\gamma_1 B\lambda^*_{\gamma s}N_{\gamma(m)} + \gamma_1 D_f\lambda^*_{qs}N_{q(m)}$$

From Eqs. (4.123) and (4.124) (*Note:* H/B=0.5/2=0.25 and ϕ_1=35°)

$$\lambda^*_{qs} \quad 1-m_1\left(\frac{B}{L}\right) \sim 1-0.73\left(\frac{2}{2}\right) \quad 0.27$$

and

$$\lambda^*_{\gamma s} \quad 1-m_2\left(\frac{B}{L}\right) = 1-0.72\left(\frac{2}{2}\right) = 0.28$$

So

$$q_u = (0.5)(16.5)(2)(0.28)(206.3) + (16.5)(0.8)(0.27)(121.5)$$

$$953.1 + 433 = 1{,}386 \text{ kN/m}^2$$

CHECK

$$q_u \quad q_b \quad \tfrac{1}{2}\gamma_2 B\lambda_{\gamma s(2)}N_{\gamma(2)} + \gamma_1 D_f\lambda_{qs(2)}N_{q(2)}$$

$$\lambda_{s\gamma(2)} = \lambda_{sq(2)} = 1+0.1\left(\frac{B}{L}\right)\tan^2(45+\phi_2/2)$$

So

$$q_u \quad (0.5)(18.5)(2)(1.583)(262.7) + (16.5)(0.8)(1.583)(134.88)$$

$$7{,}693.3 + 2{,}818.4 = 10{,}511.7 \text{ kN/m}^2$$

Hence, q_u <u>1,386 kN/m²</u>

4.14 CONTINUOUS FOUNDATION ON WEAK CLAY WITH A GRANULAR TRENCH

At the present time, there are several techniques used to improve the load bearing capacity and settlement of foundations on weak compressible soil layers. One of those techniques is the use of a granular trench under the foundation. Figure 4.74 shows a continuous rough foundation on a weak soil extending to a great depth with a granular trench. The width of the trench is equal to W and the width of the foundation is equal to B. The height of the trench is H. The width of the trench W can be smaller or larger than B. The parameters of the stronger trench material and the weak soil for bearing capacity calculation are as follows:

	Trench material	Weak soil
Angle of friction	ϕ_1	ϕ_2
Cohesion	c_1	c_2
Unit weight	γ_1	γ_2

Madhav and Vitkar (1978) have assumed a general failure mechanism in the soil under the foundation to analyze the ultimate bearing

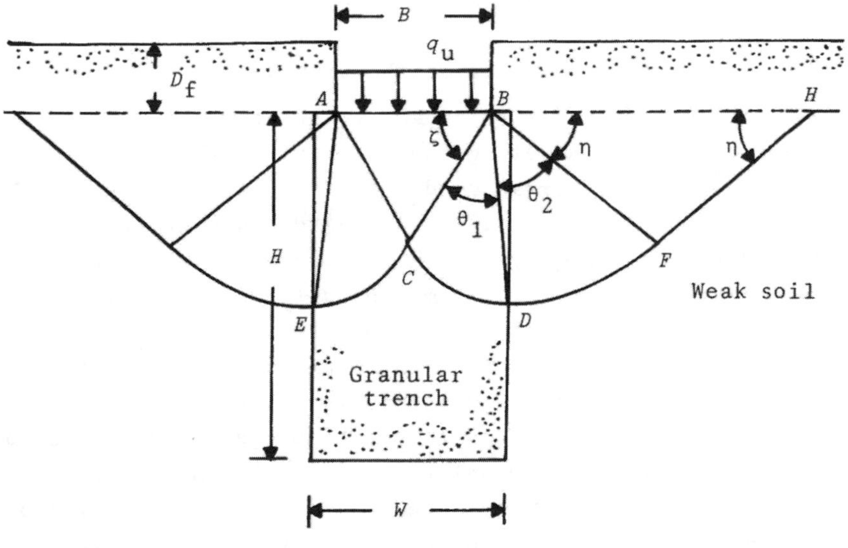

Figure 4.74. Continuous rough foundation on a weak soil with a granular trench.

capacity of the foundation by using the upper bound limit analysis as suggested by Drucker and Prager (1952). This is shown in Fig. 4.74. The failure zone in soil can be divided into subzones, and they are as follows:

1. An active Rankine zone *ABC* with a wedge angle of ζ
2. A mixed transition zone such as *BCD* bounded by angle θ_1. *CD* is the arc of a log spiral defined by the equation

$$r = r_0 e^{\theta \tan \phi 1}$$

where ϕ_1 = angle of friction of the trench material

3. A transition zone such as *BDF* with a central angle θ_2. *DF* is an arc of a log spiral defined by the equation

$$r \quad r_0 e^{\theta \tan \phi 2}$$

where ϕ_2 angle of friction of the weak soil

4. A Rankine passive zone like *BFH*.

Note that θ_1 and θ_2 are functions of ζ, η, W/B, and ϕ_1.

By using the upper bound limit analysis theorem, Madhav and Vitkar (1978) have expressed the ultimate bearing capacity of the foundation as

$$q_u = c_2 N_{c(T)} + D_f \gamma_2 N_{q(T)} + \left(\frac{\gamma_2 B}{2}\right) N_{\gamma(T)} \tag{4.183}$$

where $N_{c(T)}, N_{\gamma(T)}, N_{q(T)}$ bearing capacity factors with the presence of the trench

The variations of the bearing capacity factors [that is, $N_{c(T)}$, $N_{\gamma(T)}$, and $N_{q(T)}$] for purely granular trench soil (c_1=0) and soft saturated clay (total stress condition; ϕ_2=0 and c_2=c_u) as determined by Madhav and Vitkar (1978) are given in Fig. 4.75a, b, and c. The values of $N_{\gamma(T)}$ given in Fig. 4.75c are for γ_1/γ_2=1. In an actual case, the ratio γ_1/γ_2 may be different than one. However the error with this assumption is less than 10%.

At the present time, sufficient experimental results are not available in the literature to verify the above-mentioned theory of Madhav and Vitkar. Hamed, Das, and Echelberger (1986) have conducted several laboratory model tests to determine the variation of the ultimate bearing capacity of a strip foundation resting on a granular trench (sand; c_1=0) made in a saturated soft clay medium (c_2=c_u; ϕ_2=0). For these tests, the width of the foundation *B* was kept equal to the width of the trench *W*. For these tests, the

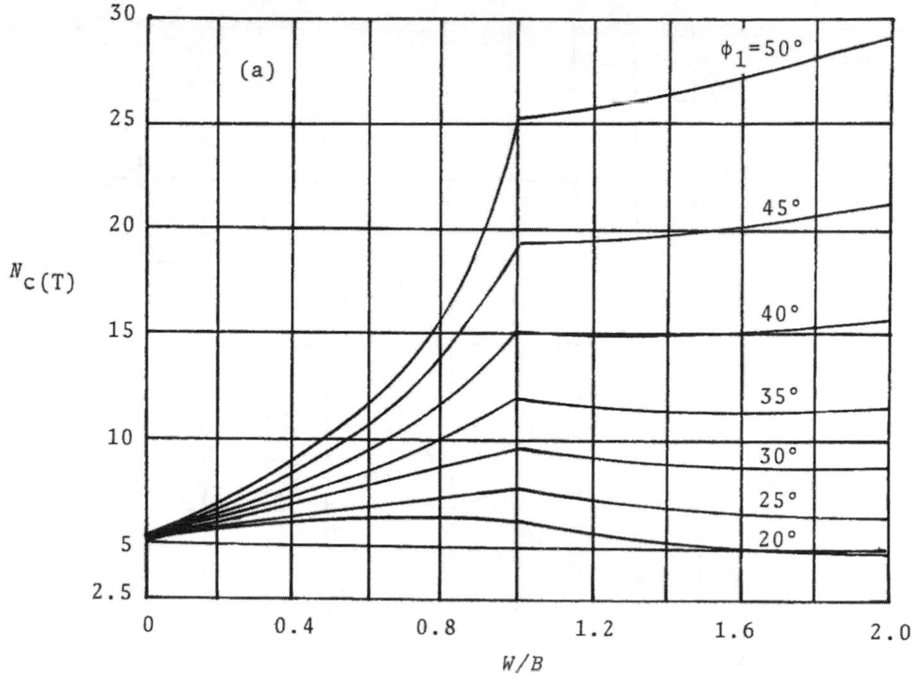

Figure 4.75. Madhav and Vitkar's bearing capacity factors--$N_{c(T)}$, $N_{q(T)}$, and $N_{\gamma(T)}$.

ratio of $H/B=H/W$ was varied. Figures 4.76 and 4.77 show the variation of the experimental q_u with H/B. In both cases, the experimental ultimate bearing capacity increases with H/B up to a maximum value and remains practically constant thereafter. The theoretical values of q_u as calculated by using Eq. (4.183) and Fig. 4.75a, b, and c are also shown in Figs. 4.76 and 4.77. From this comparison, the following major conclusions can be drawn:

1. Madhav and Vitkar's theory (1978) yields a higher and unsafe value of q_u compared to the experimental maximum value of q_u.

2. The minimum height H of the granular trench necessary to obtain the maximum value of q_u is about $3B$.

Since Eq. (4.183) yields unsafe values of the ultimate bearing capacity, Hamed, Das, and Echelberger (1986) have suggested a tentative procedure for estimation of the maximum q_u (for $c_1=0$, $c_2=c_u$ $\phi_2=0$, and $B=W$). This can be explained by referring to Fig. 4.78 in which A and B are two soil elements. For soil element A, the

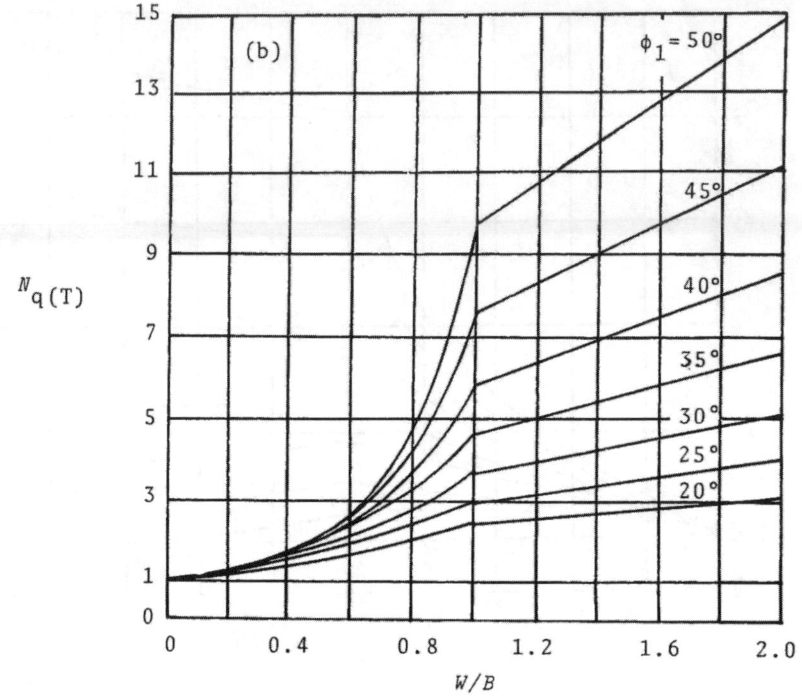

Figure 4.75. (Continued).

major principal stress is σ_I, and the minor principal stress is σ_{III}. So

$$\sigma_I \quad q_u = K_{p(1)}\sigma_{III} \tag{4.184}$$

where $K_{p(1)}$ = Rankine passive earth pressure coefficient
= $\tan^2(45+\phi_1/2)$

For soil element B, the major principal stress is σ_1 and the minor principal stress is σ_3. However

$$\sigma_3 \quad \gamma_2 D_f$$

and

$$\sigma_1 \quad \sigma_3 K_{p(2)} + 2c_u\sqrt{K_{p(2)}}$$

where $K_{p(2)}$ = Rankine passive earth pressure coefficient
= $\tan^2(45+\phi_2/2)$

For the present case, $\phi_2=0$, so $K_{p(2)}=1$. Hence

$$\sigma_1 \quad \sigma_3+2c_u \quad \gamma_2 D_f + 2c_u \tag{4.185}$$

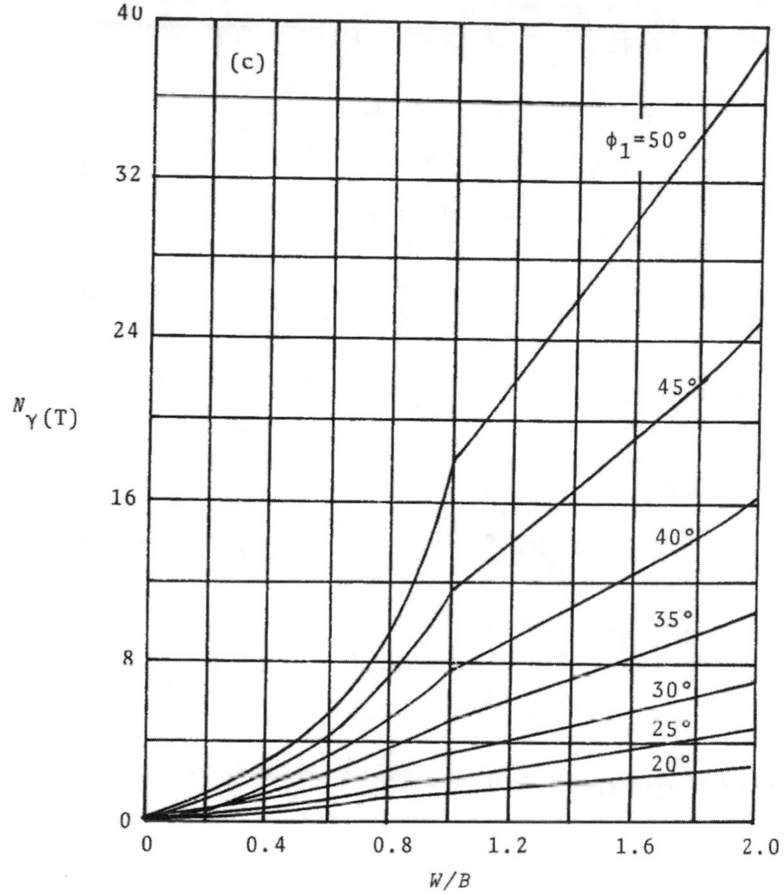

Figure 4.75. (Continued).

Again, referring to Fig. 4.78, it can be seen that at failure (which is of the bulging type)

$$\sigma_1 = \sigma_{III} \tag{4.186}$$

Now, combining Eqs. (4.184), (4.185), and (4.186)

$$q_u = K_{p(1)}(\gamma_2 D_f + 2c_u)$$

$$(\gamma_2 D_f + 2c_u)\tan^2(45+\phi_1/2) \tag{4.187}$$

With proper parameters, the ultimate bearing capacity calculated by using Eq. (4.187) has also been shown in Figs. 4.76 and 4.77. The comparision shows that the theoretical q_u [Eq. (4.187)]

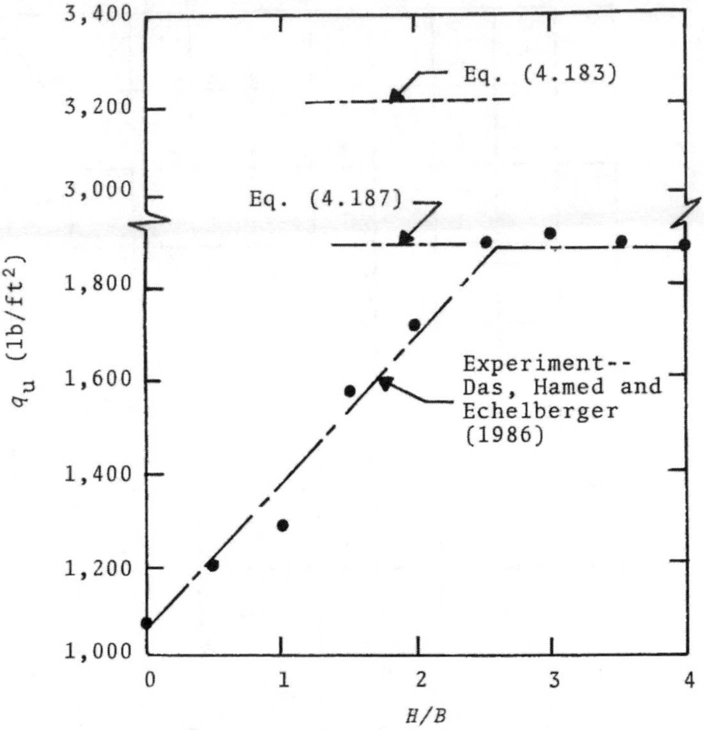

Figure 4.76. Ultimate bearing capacity of continuous foundation on soft clay with a granular trench. (*Note:* $B=W$, $\phi_1=40°$, $c_2=c_u=210$ lb/ft².)

is equal to or somewhat less than the experimental $q_{u(max)}$.

4.15 SHALLOW FOUNDATIONS ABOVE VOIDS

Under many circumstances, mining operations may leave underground voids at relatively shallow depths. Also in some instances, soluble bedrock dissolves at the interface of the soil and the bedrock leaving void spaces. Estimation of the ultimate bearing capacity of shallow foundations constructed over these voids, as well as stability, are gradually becoming important issues. Only a few studies have been reported in published literature so far. Baus and Wang (1983) have reported some experimental results for the ultimate bearing capacity of shallow rough continuous foundations located above voids as shown in Fig. 4.79. In this figure, it is assumed that the top of the rectangular void is located at a depth H below the bottom of the foundation. The void is continuous and

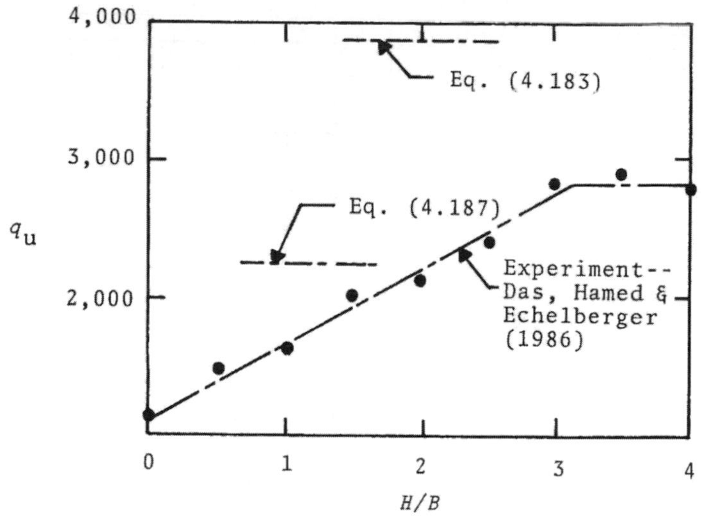

Figure 4.77. Ultimate bearing capacity of continuous foundation on soft clay with granular trench. (*Note:* $B=W$, $\phi=43°$, $c_2=c_u=210$ lb/ft^2.)

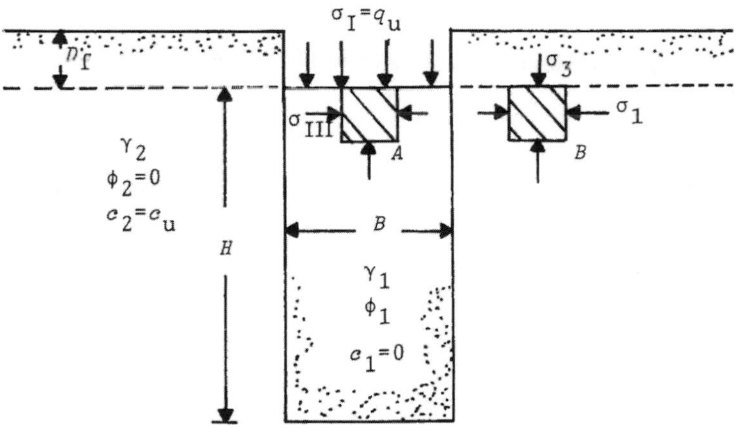

Figure 4.78. Derivation of Eq. (4.187).

has a cross-sectional dimension of $W'\times H'$. The laboratory model tests of Baus and Wang (1983) were conducted with a soil having the following properties.

Friction angle of soil, ϕ	13.5°
Cohesion	65.6 kN/m^2

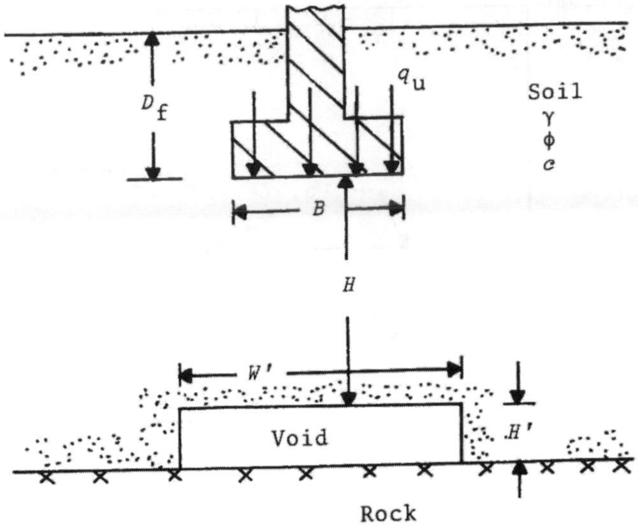

Figure 4.79. Shallow continuous rough foundation over a void.

Modulus in compression 4,670 kN/m^2
Modulus in tension 10,380 kN/m^2
Poisson's ratio = 0.28
Unit weight of compaction, γ = 18.42 kN/m^3

The results of their model tests are shown in a nondimensional form in Fig. 4.80. Note that the results of the tests which constitute Fig. 4.80 are for the case of D_f=0. From this figure, the following conclusions can be drawn:

1. For a given H/B, the ultimate bearing capacity decreases with the increase of the void width W'.

2. For any given W'/B, there is a critical H/B ratio beyond which the void has no effect. For W'/B=10, the value of the critical H/B is about 12.

Baus and Wang (1983) have conducted finite element analyses to compare the validity of their experimental findings. For the finite element analysis, the soil was treated as an *elastic-perfectly plastic material*. It was also assumed that, in the elastic range, Hooke's law is void, and in the perfectly plastic range the soil follows the von Mises yield criterion, or

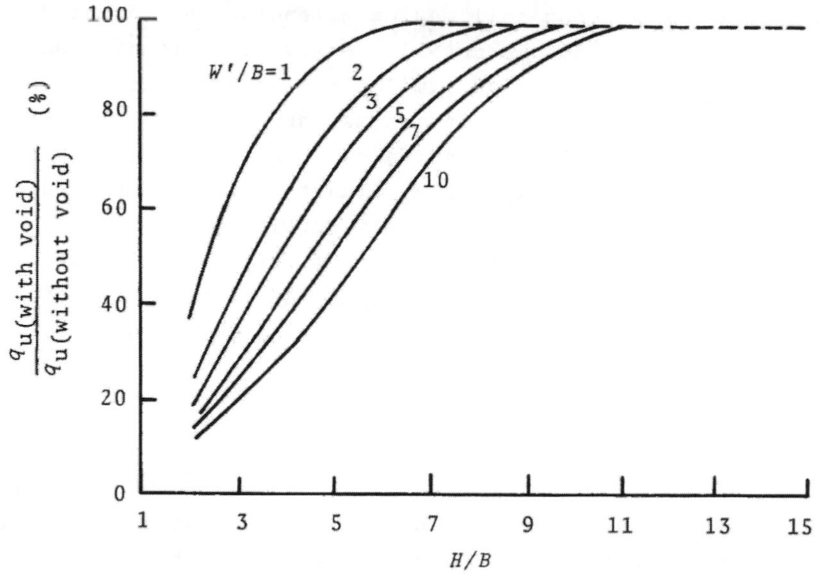

Figure 4.80. Experimental bearing capacity of a strip footing as a function of void size and location (after Baus and Wang, 1983).

$$f = \alpha I_1 + \sqrt{J_2} = k' \tag{4.188}$$

and

$$\dot{f} = 0 \tag{4.189}$$

where f = yield function

$$\alpha = \frac{\tan\phi}{(9+12\tan\phi)^{0.5}} \tag{4.190}$$

$$k' = \frac{3c}{(9+12\tan\phi)^{0.5}} \tag{4.191}$$

I_1 = first stress invariant
J_2 = second stress invariant

The relationships as shown in Eqs. (4.190) and (4.191) are based on the study of Drucker and Prager (1952). The results of the finite element analysis have shown good agreement with experiments.

4.16 INTERFERENCE OF SHALLOW CONTINUOUS FOUNDATIONS IN GRANULAR SOIL

In some of the earlier sections of this chapter, theories relating to the ultimate bearing capacity of single rough foundations resting in a homogeneous soil medium extending to a great depth have been discussed in some detail. However, if foundations are placed close to each other on similar soil conditions, the ultimate bearing capacity of each foundation may decrease due to the interference effect of the failure surface in the soil. This fact has been theoretically investigated by Stuart (1962). The results of this study will be summarized in this section.

Stuart (1962) assumed that the geometry of the rupture surface in the soil mass as assumed by Terzaghi and shown in Fig. 4.6 is correct. According to Stuart, the following conditions may arise (Fig. 4.81).

Case 1 (Fig. 4.81a):

If the center-to-center spacing of the two foundations is $x \geq x_1$ the rupture surface in the soil under each foundation will not overlap. So the ultimate bearing capacity of each continuous foundation can be given by Terzaghi's equation [Eq. (4.31)]. For $c=0$

$$q_u = qN_q + \tfrac{1}{2}\gamma BN_\gamma \tag{4.192}$$

where N_q, N_γ = Terzaghi's bearing capacity factors (Table 4.1)

Case 2 (Fig. 4.81b):

If the center-to-center spacing of the two foundations ($x=x_2<x_1$) are such that the Rankine passive zones just overlap, then the magnitude of q_u will still be given by Eq. (4.192). However, the foundation settlement at ultimate load will change (as compared to the case of an isolated foundation).

Case 3 (Fig. 4.81c):

This is the case where the center-to-center spacing of the two continuous foundations is $x=x_3<x_2$. Note that the triangular wedges in the soil under the foundation make angles of 180°-2ϕ at points d_1 and d_2. The arcs of the logarithmic spirals d_1g_1 and d_1e are tangent to each other at point d_1. Similarly, the arcs of the logarithmic spirals d_2g_2 and d_2e are tangent to each other at point d_2. For this case, the ultimate bearing capacity of each foundation can be given as ($c=0$)

$$q_u = qN_q\zeta_q + \tfrac{1}{2}\gamma BN_\gamma\zeta_\gamma \tag{4.193}$$

where ζ_q, ζ_γ efficiency ratios

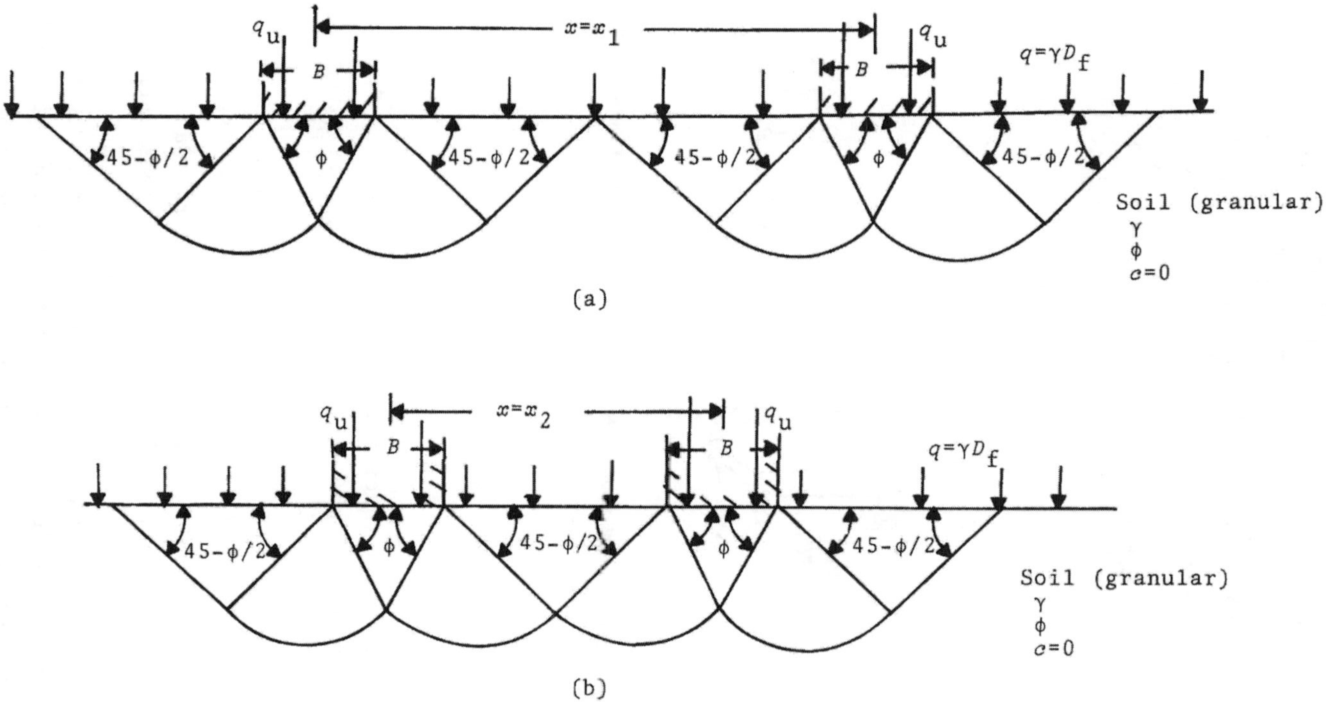

Figure 4.81. Stuart's assumptions for the failure surface in granular soil under two closely-spaced rough continous foundations.

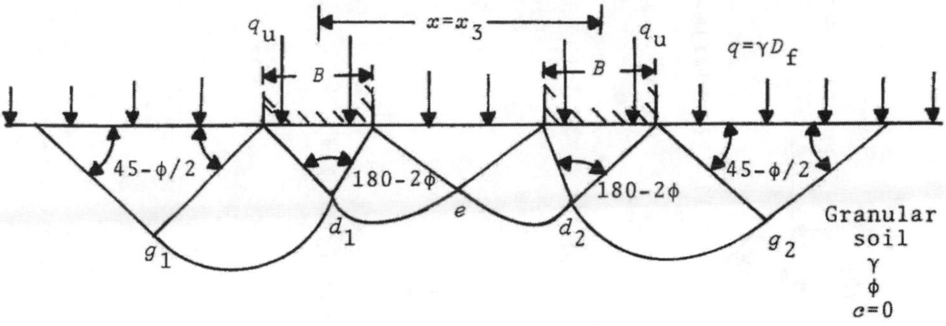

(c)

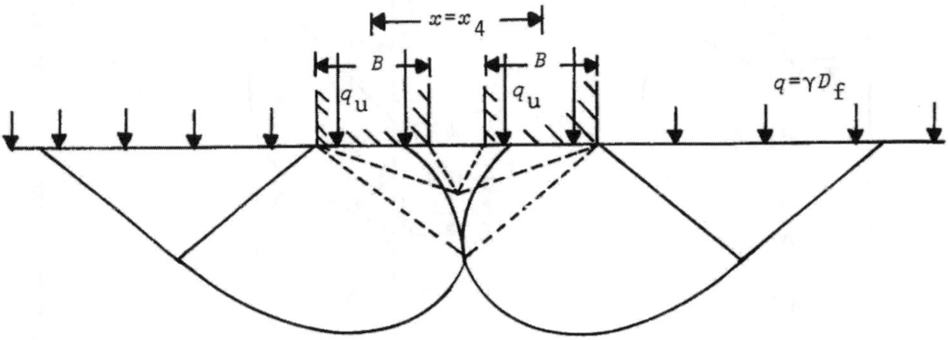

(d)

Figure 4.81. (Continued).

The efficiency ratios are functions of x/B and the soil friction angle ϕ. The theoretical variations of ζ_q and ζ_γ are given in Figs. 4.82 and 4.83.

<u>Case 4 (Fig. 4.81d)</u>:

If the spacing of the foundations is further reduced such that $x=x_4<x_3$, blocking will occur. The pair of foundations will act as a single foundation. The soil between the individual units forms an inverted arch which travels down with the foundations as the load is applied. When the two foundations touch, the zone of arching disappears and the system behaves as a single foundation with a width equal to $2B$. The ultimate bearing capacity for this case can be given by Eq. (4.192).

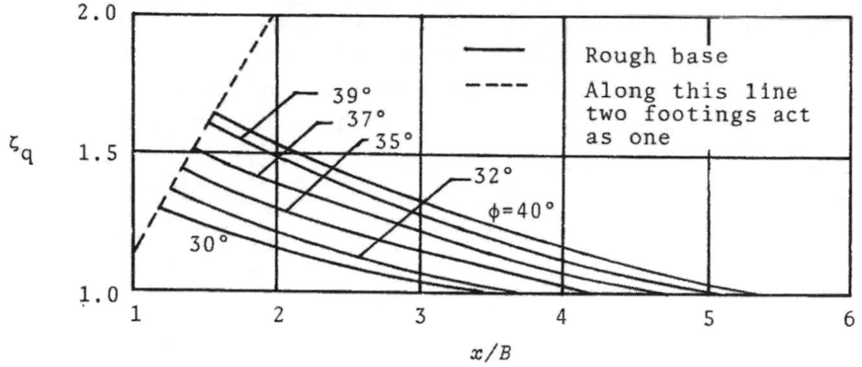

Figure 4.82. Stuart's interference factor ζ_q.

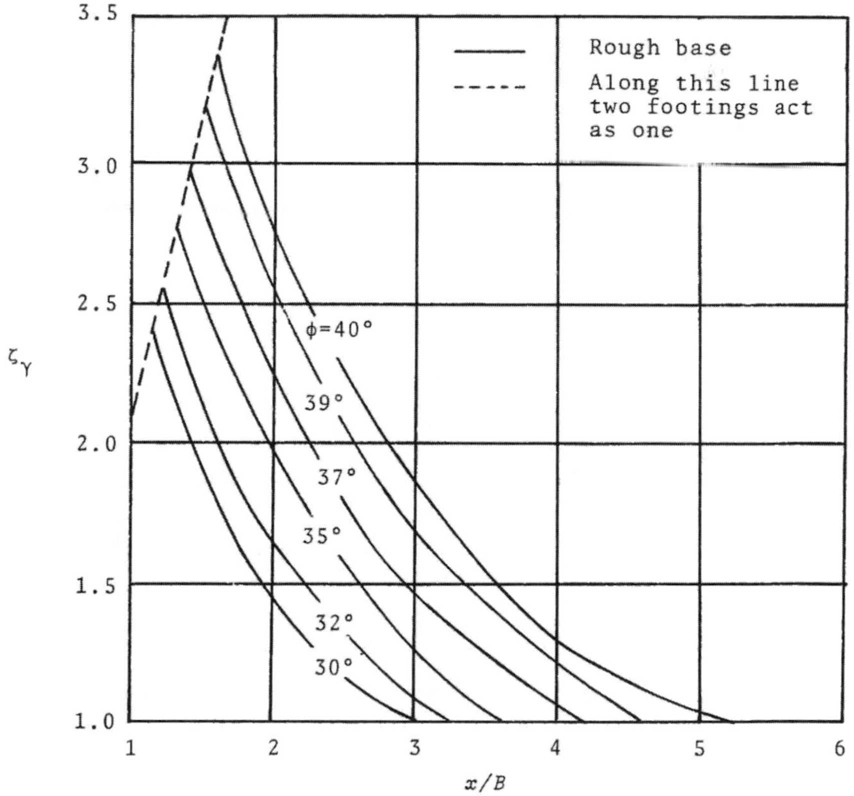

Figure 4.83. Stuart's interference factor ζ_γ.

Das and Larbi-Cherif (1983) conducted several laboratory model tests to determine the interference efficiency ratios (ζ_q and ζ_γ) of two rough continuous foundations resting on sand extending to a great depth. The sand used in the model tests was highly angular, and the tests were conducted at a relative density of about 60%. The angle of friction ϕ at this relative density of compaction was 39°. The load-displacement curves obtained from the model tests were of local shear type. The experimental variations of ζ_q and ζ_γ obtained from these tests are given in Figs. 4.84 and 4.85. From these figures, it may be seen that although the general trend of the experimental efficiency ratio variations are similar to those predicted by theory, there is a large variation in the magnitudes between the theory and the experimental results. Figure 4.86 shows the experimental variation of S_e/B with x/B (at ultimate load). The elastic settlement of the foundations decreases with the increase of the center-to-center spacing of the foundation and remains constant for $x >$ about $4B$.

4.17 FOUNDATION SETTLEMENT

It has been pointed out in Section 4.1 that, in the design of a foundation with required functional capability, one must ensure that the bearing capacity failure does not take place (that is, a proper factor of safety F_s is used over the ultimate bearing capacity to arrive at the allowable bearing capacity). In addition, it

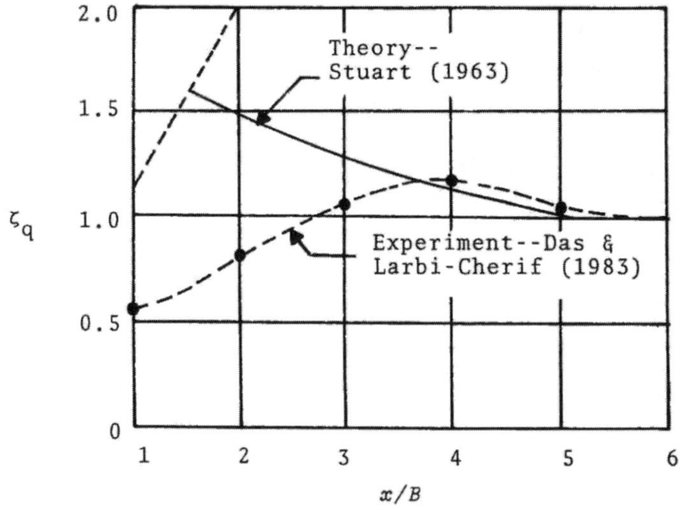

Figure 4.84. Comparison of experimental and theoretical ζ_q.

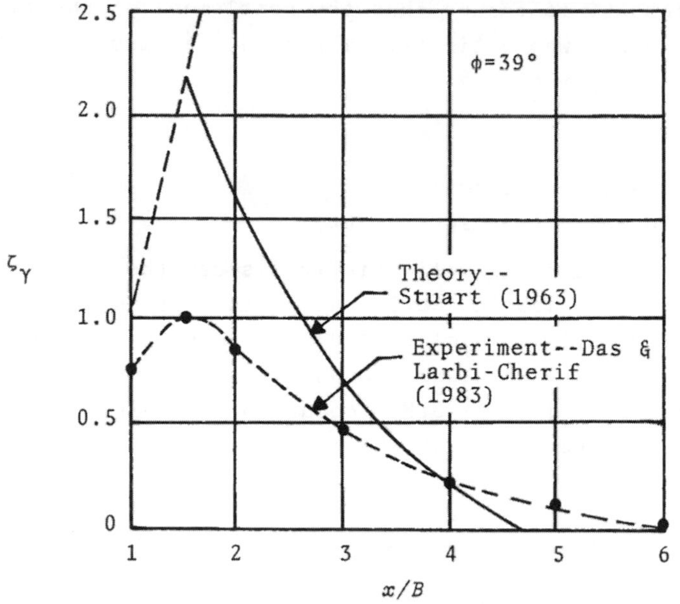

Figure 4.85. Comparison of experimental and theoretical ζ_γ.

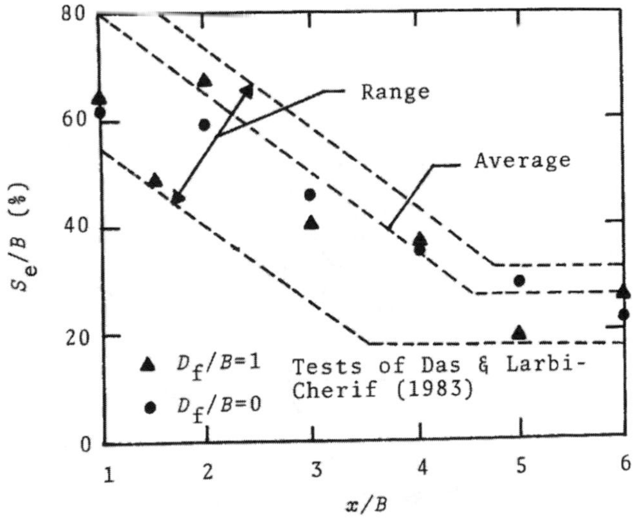

Figure 4.86. Variation of experimental elastic settlement (S_e/B) with center-to-center spacing of two continuous rough foundations.

will be necessary to ensure that the settlement of the foundation is within the allowable limits. The total settlement of the foundation S can be given as

$$S = S_e + S_c \tag{4.194}$$

where S_e = elastic settlement

S_c = consolidation settlement

In this section, we will briefly discuss the procedures for the evaluation of elastic settlement of foundations at allowable load. For foundations in clay (ϕ=0) extending to a great depth, let us consider the two square foundations with widths B_1 and B_2, respectively ($B_2 > B_1$). All other conditions remaining the same, the load will be of the nature shown in Fig. 4.87. This figures implies that

$$q_{u(B_1)} \simeq q_{u(B_2)} \tag{4.195}$$

where $q_{u(B_1)}, q_{u(B_2)}$ = ultimate bearing capacity of foundations with widths B_1 and B_2, respectively

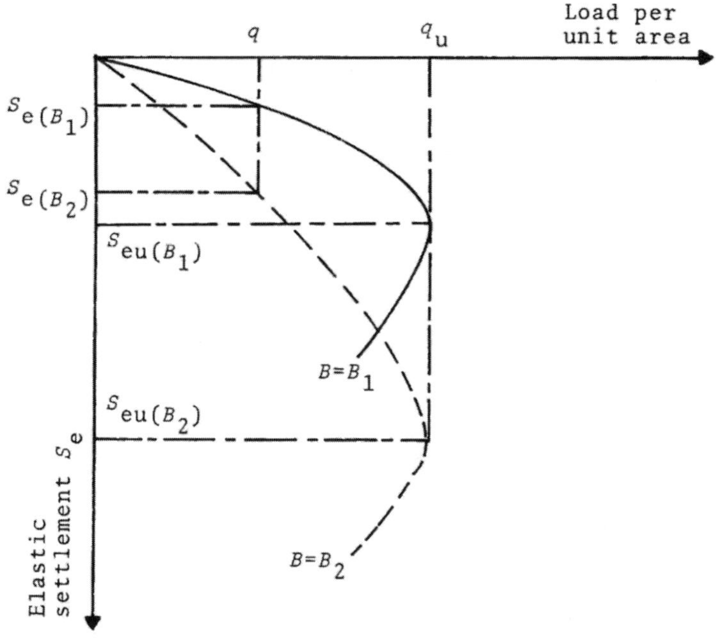

Figure 4.87. Nature of elastic settlement variation of two foundations in clay.

However

$$S_{eu(B_1)} \sim S_{eu(B_2)}\left(\frac{B_2}{B_1}\right) \qquad (4.196)$$

where $S_{eu(B_1)}, S_{eu(B_2)}$ = elastic settlements of foundations with widths B_1 and B_2, respectively, at ultimate load

In a similar manner, for a given load intensity ($q<q_u$) on the foundation

$$S_{e(B_2)} \simeq S_{e(B_1)}\left(\frac{B_2}{B_1}\right) \qquad (4.197)$$

Other conditions remaining the same, if the same two foundations are supported by a sand, then the load per unit area of the foundation q versus elastic settlement diagrams will be of the nature shown in Fig. 4.88. For this case

$$q_{u(B_2)} \sim q_{u(B_1)} \left(\frac{B_2}{B_1}\right) \qquad (4.198)$$

and

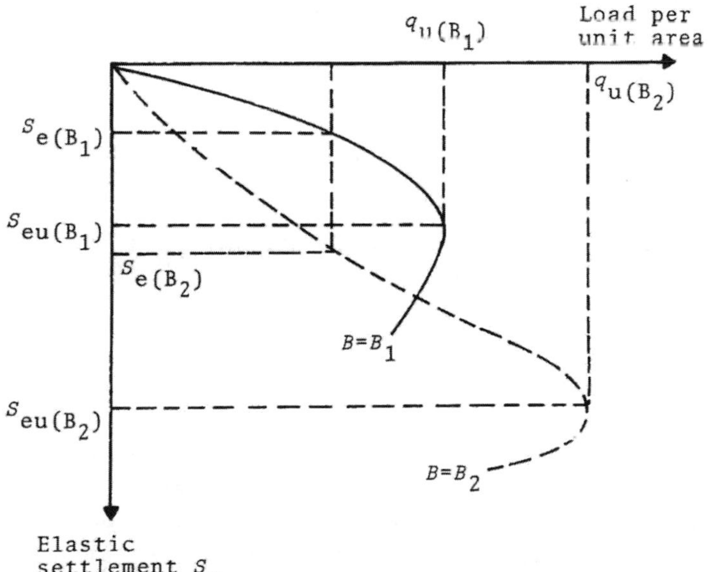

Figure 4.88. Nature of elastic settlement variation of two foundations in sand.

$$S_{eu(B_2)} \approx S_{eu(B_1)} \left(\frac{B_2}{B_1}\right)^2 \left(\frac{3.28B_2+1}{3.28B_1+1}\right)^2 \tag{4.199}$$

where B is in meters [Eq. (4.199)]

Also, for a given load intensity q ($<q_u$) on the foundation

$$S_{e(B_2)} = S_{e(B_1)} \left(\frac{B_2}{B_1}\right)^2 \left(\frac{3.28B_2+1}{3.28B_1+1}\right)^2 \tag{4.200}$$

where B is in meters [Eq. (4.200)]

The above relationships for load intensity q and elastic settlement S_e show that, in any soil, for a given allowable load intensity the level of settlement increases with the increase of the foundation width. So for smaller foundations (that is, smaller B), the ultimate bearing capacity may be the controlling factor. On the other hand, design of foundations having a larger B may be controlled by settlement criteria.

The elastic settlement calculations of foundations are, on many occasions, done by using the theory of elasticity. This is primarily based on Boussinesq's theory (1883). According to this theory, if a point load Q is located at the surface of a homogeneous semi-infinite medium (Fig. 4.89a), then the increase of vertical stress σ_z at a point A (x,y,z) can be given as

$$\sigma_z \quad \frac{3Q}{2\pi} \frac{z^3}{(x^2+y^2+z^2)^{5/2}} \tag{4.201}$$

The vertical settlement at that point due to the stress increases can be given as

$$S_e \quad \frac{Q(1+\mu)}{2E} \left[\frac{z^2}{(x^2+y^2+z^2)^{3/2}} + \frac{2(1-\mu)}{(x^2+y^2+z^2)^{1/2}}\right] \tag{4.202}$$

where μ Poisson's ratio
E modulus of elasticity

Now let us consider a flexible foundation having a width B and length L as shown in Fig. 4.89b. The foundation is subjected to an allowable load intensity q. The settlement at a point A below the center of the foundation due to the load on the foundation can be determined in the following manner. Consider an elementary area $dA=dxdy$ on the foundation. The total load on the elementary area can be given as

$$dQ \quad qdA = qdxdy \tag{4.203}$$

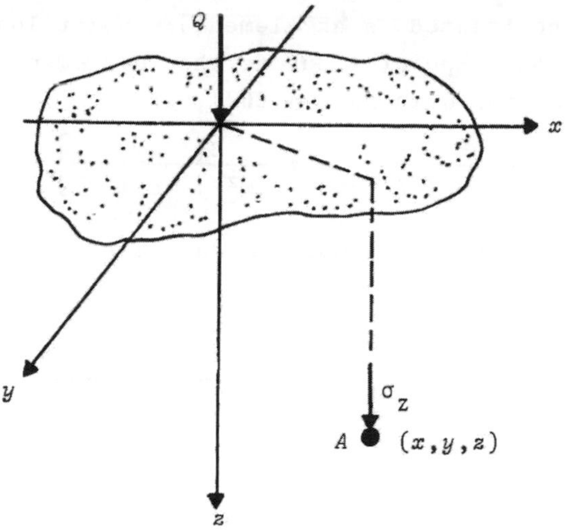

(a)

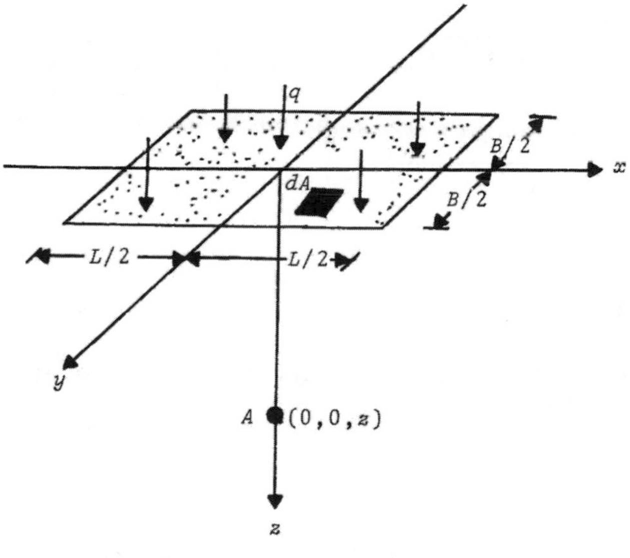

(b)

Figure 4.89. (a) Stress and settlement at a point due to a point load on the surface of an elastic, semi-infinite medium; (b) Calculation of elastic settlement below a rectangular foundation.

The load dQ can be treated as an elementary point load, and the vertical settlement at point A due to the load dQ can be determined by substituting dQ for Q in Eq. (4.202), or

$$dS_e = \frac{q\,dx\,dy}{2\pi E}\left[\frac{z^2}{(x^2+y^2+z^2)^{3/2}} + \frac{2(1-\mu)}{(x^2+y^2+z^2)^{1/2}}\right] \tag{4.204}$$

So, the elastic settlement at point A due to the entire foundation load can be expressed as

$$S_e = \int dS_e \quad \int_{y=-B/2}^{+B/2}\int_{x=-L/2}^{+L/2} \frac{q\,dx\,dy}{2\pi E}\left[\frac{z^2}{(x^2+y^2+z^2)^{3/2}} + \frac{2(1-\mu)}{(x^2+y^2+z^2)^{1/2}}\right] \tag{4.205}$$

The solution to the preceding equation can be given as (Harr, 1966)

$$S_e = \frac{Bq}{E}(1-\mu^2)\left[A_1 - \left(\frac{1-2\mu}{1-\mu}\right)A_2\right] \tag{4.206}$$

where
$$A_1 = \frac{1}{\pi}\left\{ ln\left(\frac{\sqrt{1+m^2+n^2}+m}{\sqrt{1+m^2+n^2}-m}\right) + m \cdot ln\left(\frac{\sqrt{1+m^2+n^2}+1}{\sqrt{1+m^2+n^2}-1}\right)\right\} \tag{4.207}$$

$$A_2 \quad \frac{n}{\pi}\tan^{-1}\left(\frac{m}{\sqrt{1+m^2+n^2}}\right) \tag{4.208}$$

$$m = \frac{L}{B} \tag{4.209}$$

$$n = \frac{2z}{B} \tag{4.210}$$

For settlement immediately below the center of the foundation, $z=0$. So $A_2=0$. Hence (Fig. 4.90)

$$S_{e(\mathrm{center})} = \frac{BA_1q}{E}(1-\mu^2) \tag{4.211}$$

The average elastic settlement $S_{e(\mathrm{av})}$ for a flexible foundation ($z=0$) can be given as

$$S_{e(\mathrm{av})} = \frac{BA_{1(\mathrm{av})}}{E}(1-\mu^2) \tag{4.212}$$

In a similar manner, the elastic settlement for a rigid foundation $S_{e(\mathrm{R})}$ (Fig. 4.91) can be expressed as ($z=0$)

$$S_{e(\mathrm{R})} = \frac{BA_{1(\mathrm{R})}}{E}(1-\mu^2) \tag{4.213}$$

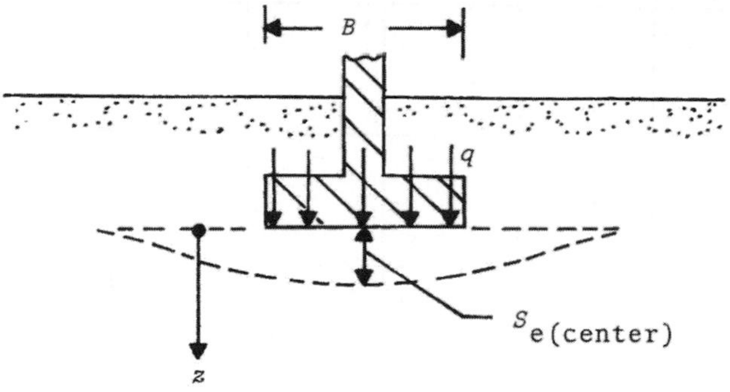

Figure 4.90. Elastic settlement below a flexible foundation.

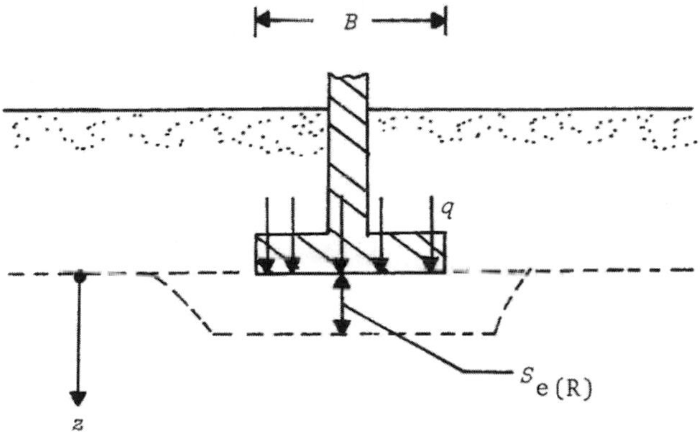

Figure 4.91. Elastic settlment below a rigid roundation.

The variation of A_1, $A_{1(av)}$, and $A_{1(R)}$ for calculation of foundation settlement (z=0) is given in Fig. 4.92.

Several empirical correlations for estimation of settlement of foundations on sand have been proposed by various investigators. Some of these are given below.

1. In 1956, Myerhof proposed the following empirical relationship for the *allowable bearing capacity* for a maximum foundation settlement of 1 in. (25.4 mm) (based on several field observations)

$$q_{all}\ (\text{kip/ft}^2) = \frac{N}{4} \quad (\text{for } B \leq 4 \text{ ft}) \tag{4.214}$$

and

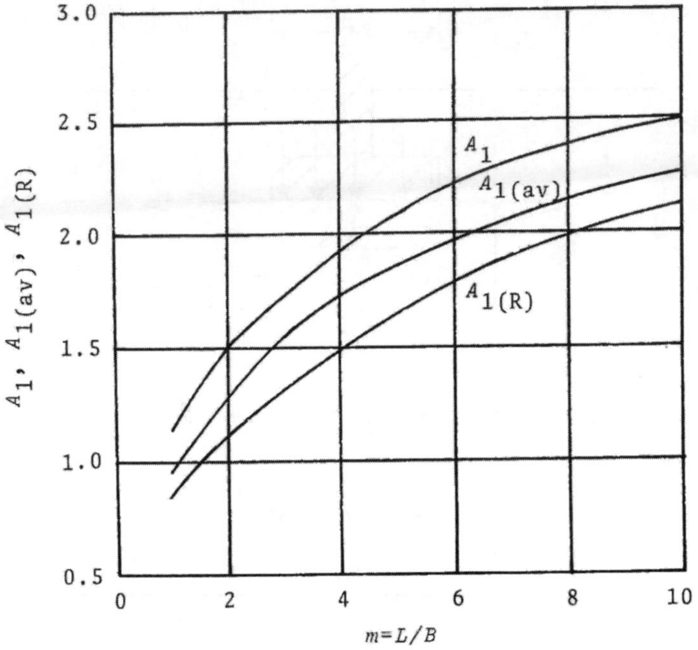

Figure 4.92. Variation of A_1, $A_{1(av)}$, and $A_{1(R)}$ with L/B.

$$q_{all}\ (\text{kip/ft}^2) = \frac{N}{6}\left(\frac{B+1}{B}\right)^2 \quad (\text{for } B>4 \text{ ft}) \tag{4.215}$$

where N = standard penetration resistance

2. Schultze and Sherif (1973) have analyzed some field test results. According to their analysis

$$S_e \propto q\sqrt{B} \tag{4.216}$$

where q load per unit area of the foundation

3. Based on the study of Schultze and Sherif (1973), Meyerhof (1974) has suggested the following relationships for estimation of elastic settlement of foundations in granular soil

$$S_e \quad \frac{q\sqrt{B}}{2N} \quad (\text{for sand and gravel}) \tag{4.217}$$

$$S_e \doteq \frac{q\sqrt{B}}{N} \quad (\text{for silty sand}) \tag{4.218}$$

where S_e is in inches, q is in ton/ft^2

N = standard penetration number

4. Hettler (1981) investigated the settlment of small-scale

footings on sand and presented a relation as

$$\frac{S_e}{B} = C\left(\frac{q}{\gamma B}\right)^{1.63} \tag{4.219}$$

where C = a constant which is a function of the relative density

γ = unit weight of sand

5. Zolkov (1974) has considered the effect of relative density on the elastic settlement of the foundation. According to him

$$\frac{S_e\ (\text{m})}{q\ (\text{kN/m}^2)} \quad (0.527\times10^{-6})(105-D_r)\sqrt{B\ (\text{m})} \qquad (\text{for } 45\% \le D_r \le 85\%)$$

where D_r = relative density

The readers are also referred to the semitheoretical procedure commonly referred to as the strain-influence factor method proposed by Schmertmann (1970) and Schmertmann and Hartman (1978) for calculation of elastic settlment of foundations on granular soil.

References

Balla, A., 1962. Bearing capacity of foundations. J. Soil Mech. Found. Div., ASCE, 89(SM5):13-34.

Baus, R.L. and Wang, M.C., 1983. Bearing capacity of strip footing above void. J. Geotech. Eng., ASCE, 109(GT1):1-14.

Boussinesq, J., 1883. Application des Potentials a L'Etude de L'Equilibre et due Mouvement des Solides Elastiques. Gauthier-Villars, Paris.

Caquot, A. and Kerisel, J., 1953. Sur le terme de surface dans le calcul des fondations en milieu pulverulent. Proc., III Intl. Conf. Soil Mech. Found. Eng., Zurich, Switzerland, 1:336-337.

Caquot, A. and Kerisel, J., 1949. Tables for the Calculation of Passive Pressure, Active Pressure, and Bearing Capacity of Foundations. Gauthier-Villars, Paris.

Casagrande, A. and Carrillo, N., 1944. Shear failure in anisotropic materials. In Contribution to Soil Mechanics 1941-53. Boston Soc. Civ. Engrs., pp. 122-135.

Chen, W.F., 1975. Limit Analysis and Soil Plasticity. Elsevier Publishing Co., New York.

Das, B.M., 1983. Advanced Soil Mechanics. McGraw-Hill, New York.

Das, B.M. and Larbi-Cherif, S., 1983. Bearing capacity of two closely-spaced shallow foundations on sand. Soils and Found., 23(1):1-7.

Davis, E. and Christian, J.T., 1971. Bearing capacity of anisotropic cohesive soil. J. Soil Mech. Found. Div., ASCE, 97(SM5): 753-769.

DeBeer, E.E., 1970. Experimental determination of the shape factors of sand. Geotechnique, 20(4):387-411.

DeBeer, E.E., 1965. Bearing capacity and settlement of shallow foundations on sand. Bearing Capacity and Settlement of Foundations, Proc., Symposium held at Duke University, Durham, N.C., pp. 15-34.

Drucker, D.C. and Prager, W., 1952. Soil mechanics and plastic analysis of limit design. Q. Appl. Math., 10:157-165.

Hamed, J.T., Das, B.M. and Echelberger, W.F., 1986. Bearing capacity of a strip foundation on granular trench in soft clay. Civ. Eng. Pract. Design Engrs., Pergamon Press, 5(5):359-376.
Hanna, A.M., 1982. Bearing capacity of foundations on a weak sand layer overlying a strong deposit. Canadian Geotech. J., 19(3): 392-396.
Hanna, A.M., 1981. Foundations on strong sand overlying weak sand. J. Geotech. Eng., ASCE, 107(GT7):915-927.
Hanna, A.M. and Meyerhof, G.G., 1980. Design charts for ultimate bearing capacity for sands overlying clays. Canadian Geotech J., 17(2):300-303.
Hansen, J.B., 1970. A revised and extended formula for bearing capacity. Bull. 28, Danish Geotech. Inst., Copenhagen.
Hansen, J.B., 1961. A general formular for bearing capacity. Bull. 11, Danish Geotech. Inst., Copenhagen.
Harr, M.E., 1966. Fundamentals of Theoretical Soil Mechanics. McGraw-Hill, New York.
Hettler, A., 1981. Verschiebungen starrer und elastischer grundungskorper in sand bei monotoner und zyklischer belastung. Dissertation, University of Karlsruhe.
Highter, W.H. and Anders, J.C., 1985. Dimensioning footings subjected to eccentric loads. J. Geotch. Eng., ASCE, 111(GT5): 659-665.
Ingra, T.S. and Baecher, G.B., 1983. Uncertainty in bearing capacity of sand. J. Geotech. Eng., ASCE, 109(GT7):899-914.
Janbu, N., 1957. Earth pressures and bearing capacity calculations by generalized procedure of slices. Proc., IV Intl. Conf. Soil Mech. Found. Eng., London, England, 2:207-212.
Ko, H.Y. and Davidson, L.W., 1973. Bearing capacity of footings in plane strain. J. Soil Mech. Found. Div., ASCE, 99(SM1):1-23.
Lo, K.Y., 1965. Stability of slopes in anisotropic soil. J. Soil Mech. Found. Div., ASCE, 91(SM4):85-106.
Lundgren, H. and Mortensen, K., 1953. Determination by the theory of plasticity of the bearing capacity of continuous footings on sand. Proc., III Intl. Conf. Soil Mech. Found Eng., Zurich, Switzerland, 1:409-412.
Madhav, M.R. and Vitkar, P.P., 1975. Strip footing on weak clay stabilized with a granular trench or pile. Canadian Geotech. J., 15(4):605-609.
Mandel, J. and Salencon, J., 1972. Force portante d'un sol sur une assise rigide (etude theorizue). Geotechnique, 22(1):79-93.
Meyerhof, G.G., 1978. Bearing capacity of anisotropic cohesionless soils. Canadian Geotech. J., 15(4):592-595.
Meyerhof, G.G., 1974. Ultimate bearing capacity of footings on sand layer overlying clay. Canadian Geotech. J., 11(2):224-229.
Meyerhof, G.G., 1974. General report: State-of-the-art of penetration testing in countries outside Europe. Proc., Eur. Symp. Pen. Test.
Meyerhof, G.G., 1963. Some recent research on the bearing capacity of foundations. Canadian Geotech. J., 1(1):16-26.
Meyerhof, G.G., 1957. The ultimate bearing capacity of foundations on slopes. Proc., IV Intl. Conf. Soil Mech. Found. Eng., London, England, 1:384-387.
Meyerhof, G.G., 1956. Penetration tests and bearing capacity of cohesionless soils. J. Soil Mech. Found. Div., ASCE, 82(SM1): 1-19.
Meyerhof, G.G., 1953. The bearing capacity of footings under eccentric and inclined loads. Proc., III Intl. Conf. Soil Mech. Found. Eng., Zurich, Switzerland, 1:440-443.
Meyerhof, G.G., 1951. The ultimate bearing capacity of foundations. Geotechnique, 2:301-331.
Meyerhof, G.G. and Chaplin, T.K., 1953. The compression and

bearing capacity of cohesive layers. Br. J. Appl. Phys., 4:20-26.
Meyerhof, G.G. and Hanna, A.M., 1978. Ultimate bearing capacity of foundations on layered soils under inclined load. Canadian Geotech. J., 15(4):565-572.
Milovic, D.M. and Tournier, J.P., 1971. Comportement de fondations reposant sur une coche compressible d'epaisseur limitee. Proc., Conf. Comportement des Sols Avant La Rupture, Paris, France.
Pfeifle, T.W. and Das, B.M., 1979. Bearing capacity of surface footings on sand layer resting on rigid rough base. Soils and Found., 19(1):1-11.
Prakash, S., 1981. Soil Dynamics. McGraw-Hill, New York.
Prakash, S. and Saran, S., 1971. Bearing capacity of eccentrically loaded footings. J. Soil Mech. Found. Div., ASCE, 97(SM1):95-117.
Prandtl, L., 1921. Uber die eindringungsfestigkeit plastisher baustoffe und die festigkeit von schneiden. Z. Ang. Math. Mech. 1(1):15-20.
Purkayastha, R.D. and Char, R.A.N., 1977. Stability analysis for eccentrically loaded footings. J. Geotech. Eng. Div., ASCE, 103(GT6):647-651.
Rao, N.S.V.K. and Krisnamurthy, S., 1972. Bearing capacity factors for inclined loads. J. Soil Mech. Found. Div., ASCE, 98(SM11): 1286-1290.
Reddy, A.S. and Srinivasan, R.J., 1970. Bearing capacity of footings on anisotropic soils. J. Soil Mech. Found. Div., ASCE, 96(SM6):1967-1986.
Reddy, A.S. and Srinivasan, R.J., 1967. Bearing capacity of footings on layered clays. J. Soil Mech. Found. Div., ASCE, 93(SM2): 83-99.
Reissner, H., 1924. Zum erddruckproblem. Proc., I Intl. Conf. Appl. Mech., Delft, The Netherlands, pp. 295-311.
Schmertmann, J.H., 1970. Static cone to compute settlement over sand. J. Soil Mech. Found. Div., ASCE, 96(SM3):1011-1043.
Schmertmann, J.H. and Hartman, J.P., 1978. Improved strain influence factor diagrams. J. Geotech. Eng. Div., ASCE, 104(GT8): 1131-1135.
Schultze, E. and Sherif, G., 1973. Prediction of settlements from evaluated settlement observations for sand. Proc., VIII Intl. Conf. Soil Mech. Found. Eng., Moscow, U.S.S.R., 1.3:225-230.
Solokovski, V.V., 1965. Statics of Granular Media. Pergamon Press New York.
Stuart, J.G., 1962. Interference between foundations with special reference to surface footing on sand. Geotechnique, 12(1):15-22.
Terzaghi, K., 1943. Theoretical Soil Mechanics. John Wiley, New York.
Vesic, A.S., 1973. Analysis of ultimate loads of shallow foundations. J. Soil Mech. Found. Div., ASCE, 99(SM1):45-73.
Vesic, A.S., 1963. Bearing capacity of deep foundations in sand. Highway Research Record 39, National Academy of Sciences, Washington, D.C., pp. 112-153.
Zolkov, E., 1974. The nature of a sand deposit and the settlements of shallow foundations. I Euro. Symp. Penetr. Test., Stockholm, Sweden, 2.2:421-431.

CHAPTER 5
Slope Stability

5.1 INTRODUCTION

Slope failure is defined as the failure of a soil mass forming a slope by downward and/or outward movement. Some existing slopes which have been stable for some time can experience failure due to various reasons, such as occurrence of earthquakes, increased loading on a slope at or close to its crest, removal of earth from the face of a slope to make it steeper, removal of earth close to the toe of a slope, gradual disintegration of the soil forming a slope, and increase of pore water pressure in the soil forming the slope.

The analysis of the *stability* of a slope or the evaluation of the *factor of safety* against slope failure needs a thorough knowledge of the *shear strength parameters* of the soil. In this chapter, it will be assumed that the readers are acquainted with them.

Due to the scope and limitations of this text, only the stability of *simple finite slopes* will be considered in this chapter.

Figure 5.1 shows a simple finite slope in which the ground

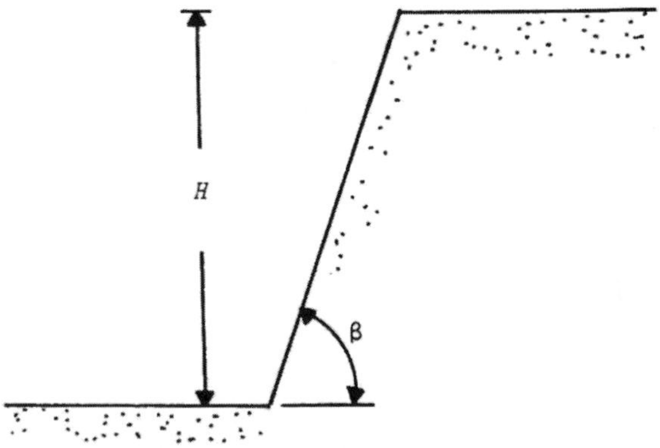

Figure 5.1. Simple finite slope.

surface is horizontal at its foot and at its top. In order to analyze the stability of this type of slope, the *critical failure surface* has been assumed by various investigators to be a *plane*, an *arc of a circle*, or an *arc of a logarithmic spiral*. Culmann's analysis (1866) for the stability of a simple finite slope is based on the assumption that the critical surface of possible sliding is a *plane*. This type of analysis has been discussed in Section 5.3. Although this method is simple, it gives good results only for the case where the slope is nearly vertical.

Around 1920, several investigations of slope failures were observed by a Swedish Geotechnical Commission. Based on their findings, this commission recommended that the failure surface of slopes can be approximated to be an *arc of a circle*. Figure 5.2 shows the various types of failure modes of a slope with the fail-

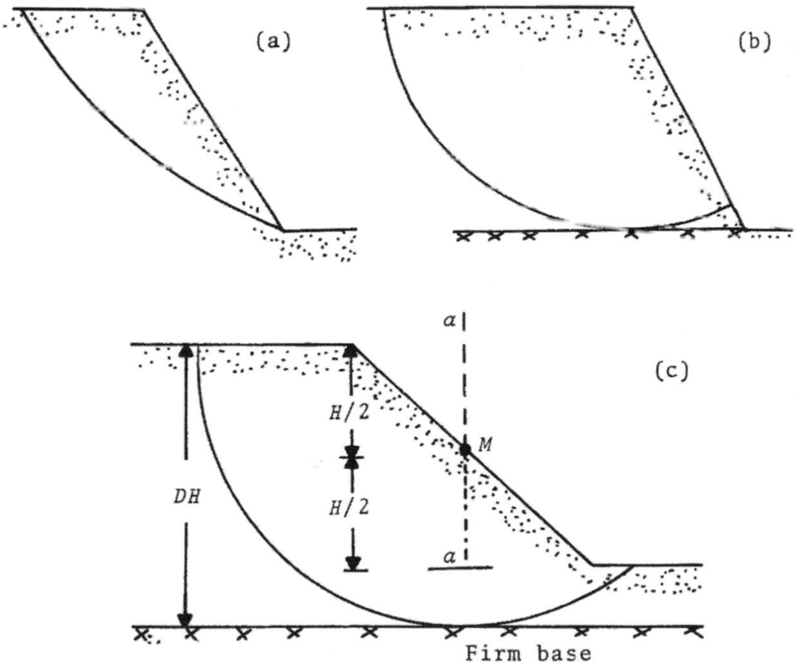

Figure 5.2. (a) Toe failure; (b) Slope failure; and (c) base failure.

ure surface being an arc of a circle. They are: (a) *toe failure*, (b) *slope failure*, and (c) *base failure*. In the case of *toe failure*, the failure surface passes through the toe of the slope. When the failure surface intersects the face of the slope, it is generally referred to as *slope failure*. However, if the soil beneath the toe cannot sustain the weight of the soil above it, the failure circle will pass under the toe of the slope and will intersect the ground surface at some point beyond the toe of the slope. This is referred to as *base failure*. Fellenius (1927) has shown that, for purely cohesive soils (ϕ=0), the center of the critical circle for base failure will be located along a vertical line passing through the midpoint of the slope (point *M* in Fig. 5.2c).

Several investigators have used the concept of the critical failure surface of a slope to be an *arc of a logarithmic spiral*. Spencer (1969) has shown that, for homogeneous slopes, the circular arc assumption is more critical than the assumption of the arc of a logarithmic spiral. However, Chen (1970) has suggested that both types of assumptions are equally correct.

5.2 FACTOR OF SAFETY--DEFINITION

The factor of safety with respect to strength along a surface of sliding can be defined as

$$F_s \quad s/\tau \tag{5.1}$$

where s shear strength of the soil
τ shear stress developed

The shear strength and shear stress can be given by the following relationships

$$s = c + \sigma'\tan\phi \tag{5.2}$$

and

$$\tau \quad c_d + \sigma'\tan\phi_d \tag{5.3}$$

where c, c_d cohesion and developed cohesion, respectively
σ' effective normal stress
ϕ, ϕ_d angle of friction and developed angle of friction, respectively

So

$$F_s = \frac{c + \sigma'\tan\phi}{c_d + \sigma'\tan\phi_d} \tag{5.4}$$

Other definitions for the factor of safety frequently used are

$$F_c \quad c/c_d \tag{5.5}$$

and

$$F_\phi = \frac{\tan\phi}{\tan\phi_d} \tag{5.6}$$

where F_c = factor of safety with respect to cohesion
F_ϕ = factor of safety with respect to friction

The concepts of F_s, F_c, and F_ϕ can be explained by means of Fig. 5.3. At a given point on the trial failure surface, let the normal

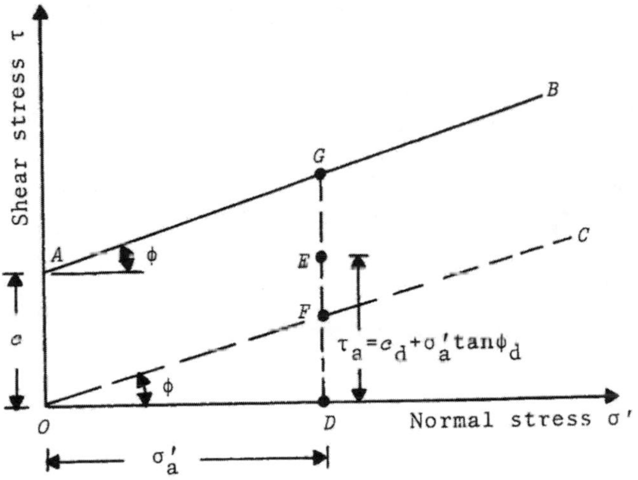

Figure 5.3. Definitions of F_s, F_c, and F_ϕ.

effective stress be equal to $\sigma_a' = OD$ and the shear stress be τ_a. Then, at that point

$$F_s = \frac{DG}{DE}$$

$$F_c = \frac{FG}{FE}$$

$$F_\phi = \frac{DF}{DE-FG}$$

It can be seen that when F_c is equal to F_ϕ it will also be equal to F_s, or

$$F_c = F_\phi = F_s$$

In all practical cases, the magnitude of F_s expressed is the average value of the factor of safety along the entire potential slip surface.

5.3 STABILITY OF FINITE SLOPES (c-ϕ SOIL)--PLANE FAILURE SURFACE

For determination of the factor of safety of finite slopes, Culmann (1866) assumed that the critical failure surface may be taken as a plane passing through the toe of a slope as shown in Fig. 5.4. In order to derive the relationship by Culmann's method,

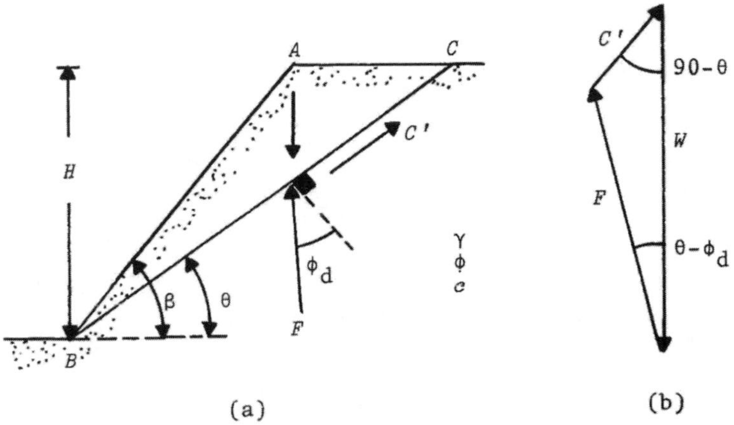

Figure 5.4. Stability of finite slope--plane failure surface.

let us consider a trial failure wedge *ABC* as shown in Fig. 5.4a. The forces acting on the sliding wedge (per unit length of the wedge at right angles to the cross section shown) are as follows

1. The weight of the wedge, *W*.--Based on the geometry of the slope

$$W = \tfrac{1}{2}\gamma X\left[\frac{H}{\sin\beta}\sin(\beta-\theta)\right] \tag{5.7}$$

where X = length of BC

γ = unit weight of soil

2. The force, C'.--This is the contribution of the cohesion. This force acts parallel to the plane BC. Thus

$$C' \quad c_d X \tag{5.8}$$

where c_d = developed cohesion

3. The force, *F*.--This is the resultant of normal and shear

forces (frictional components) acting on the plane *BC*. This force will be inclined at an angle ϕ_d to the normal drawn to *BC*.

Figure 5.4b shows the force triangle for the equilibrium of the wedge. From the force triangle, by using the law of sines, we can write

$$\frac{C'}{W} = \frac{\sin(\theta-\phi_d)}{\cos\phi_d} \tag{5.9}$$

Now, combining Eqs. (5.7), (5.8), and (5.9)

$$\frac{c_d X \sin\beta}{\frac{1}{2}\gamma X H[\sin(\beta-\theta)]} = \frac{\sin(\theta-\phi_d)}{\cos\phi_d}$$

or

$$\frac{c_d}{\gamma H} = \frac{\sin(\beta-\theta)\sin(\theta-\phi_d)}{2\cos\phi_d \sin\beta} \tag{5.10}$$

The most critical plane for failure will be the plane for which c_d will be maximum, or

$$\frac{\partial(c_d/\gamma H)}{\partial\theta} \quad \theta \tag{5.11}$$

The preceding equation yields

$$\theta = \theta_{cr} \quad \tfrac{1}{2}(\beta+\phi_d) \tag{5.12}$$

Substitution of Eq. (5.12) into Eq. (5.10) yields

$$\frac{\gamma H}{c_d} \quad N_s = \frac{4\sin\beta\cos\phi_d}{1-\cos(\beta-\phi_d)} \tag{5.13}$$

where N_s = stability number

Based on Eq. (5.13), the values of the stability number for various values of β and ϕ_d are given in Table 5.1.

The maximum height (H_{cr}) of the slope for critical equilibrium can be calculated by substituting ϕ for ϕ_d and c for c_d in Eq. (5.13), or

$$H_{cr} \quad \frac{4c}{\gamma}\left[\frac{\sin\beta\cos\phi}{1-\cos(\beta-\phi)}\right] \tag{5.14}$$

Slope With Water in Tensile Crack

In many cases, tensile cracks develop at the top of a slope. The cracks are filled with water as shown in Fig. 5.5. The

Table 5.1. Stability Numbers Based on Culmann's Analysis [Eq. (5.13)]

Slope angle β (deg)	ϕ_d (deg)	Stability number N_s
10	0	45.72
	5	181.84
15	0	30.38
	5	67.89
	10	267.93
20	0	22.69
	5	40.00
	10	88.68
	15	347.27
25	0	18.04
	5	27.92
	10	48.86
	15	107.48
	20	417.45
30	0	14.93
	5	21.27
	10	32.66
	15	56.70
	20	123.71
	25	476.34
40	0	10.99
	5	14.16
	10	18.90
	15	26.51
	20	40.06
	25	68.39
	30	146.57
50	0	8.58
	5	10.42
	10	12.89
50	15	16.37
	20	21.49
	25	29.64
	30	44.00
60	0	6.93
	5	8.09
	10	9.55
	15	11.42
	20	13.91
	25	17.36
	30	22.39
70	0	5.71
	5	6.49
	10	7.40
	15	8.51
	20	9.89
	25	11.63
	30	13.91
80	0	4.77
	5	5.29
	10	5.90
	15	6.59
	20	7.40
	25	8.37
	30	9.55
90	0	4.00
	5	4.37
	10	4.77
	15	5.21
	20	5.71
	25	6.28
	30	6.93

stability of the slope for such a case can be determined in the following manner.

Let the depth of the crack be equal to z_c and the depth of water in the crack be equal to z_1. Due to the water in the crack, let there be a linear distribution of pore water pressure varying from $z_c\gamma_w$ to zero along a trial failure surface *AB* as shown in Fig. 5.5 (γ_w=unit weight of water). So, the forces per unit length acting on the trial wedge *ABCD* are as follows

1. *W*--Weight (total) of the wedge

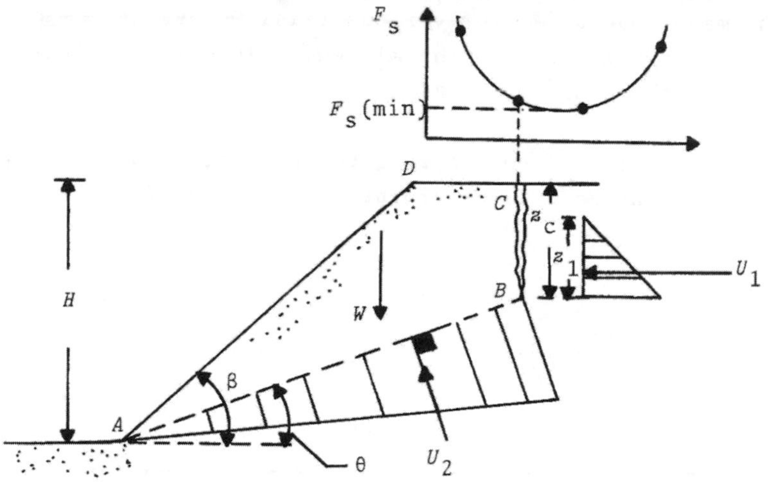

Figure 5.5. Slopes with water in tensile crack.

2. U_1--Force due to water in the crack
3. U_2--Force due to pore water pressure along AB

However

$$U_1 = \frac{1}{2}\gamma_w z_1^2$$

$$U_2 = \frac{1}{2}\gamma_w z_c X$$

where $X = AB \quad (H-z_c)\text{cosec}\theta$

So

$$U_2 \quad \frac{1}{2}\gamma_w z_c (H-z_c)\text{cosec}\theta$$

The sum of the components (F) of the above forces parallel to AB will tend to slide the wedge, or

$$F = W\sin\theta + U_1\cos\theta$$

The maximum resisting force which can be developed along AB is

$$R = cX + (W\cos\theta - U_1\sin\theta - U_2)\tan\phi$$

The factor of safety with respect to strength can then be given as

$$F_s \quad \frac{R}{F} = \frac{cX + (W\cos\theta - U_1\sin\theta - U_2)\tan\phi}{W\sin\theta + U_1\cos\theta} \tag{5.15}$$

The magnitude of F_s for various trial wedges can be calculated by varying the value of θ. The minimum value of F_s is the factor of safety of the slope with respect to strength.

Example 5.1. For a given slope, given H=21 m, c=25 kN/m², γ=16.5 kN/m³, ϕ=25°, and β=30°. Determine F_s. Use Eq. (5.13).

Solution

From Eq. (5.13)

$$H \quad \frac{4c_d}{\gamma}\left[\frac{\sin\beta\cos\phi_d}{1-\cos(\beta-\phi_d)}\right]$$

Now the following table can be prepared.

Assumed F_s	$c_d=\frac{c}{F_s}$	$\phi_d=\tan^{-1}(\frac{\tan\phi}{F_s})$	H Eq. (5.13)
2	12.5	13.12	34.23
2.5	10	10.57	21.28*

*Actual height is 21 m. So

$F_s \sim \underline{2.5}$

STABILITY ANALYSIS WITH CURVED FAILURE SURFACE

5.4 STABILITY OF CLAY SLOPES (ϕ=0 CONDITION)

The factor of safety of a slope along a *circular failure surface* may be determined as

$$F_s = \frac{\text{Resisting moment per unit length, } M_r}{\text{Disturbing moment per unit length, } M_d} \tag{5.16}$$

Referring to Fig. 5.6, AB is a slope making an angle β with the horizontal. The undrained shear strength of the soil constituting the slope can be given as

$$s \quad c_u \tag{5.17}$$

The point O_1 is the center of a *trial circle* with a radius R. CD is the circular trial failure arc. The force resisting the slope from sliding is due to the undrained cohesion along the *trial failure surface*. So, the magnitude of the resisting force will be equal to $R\theta c_u$. Hence the resisting moment M_r about O_1 can be given as

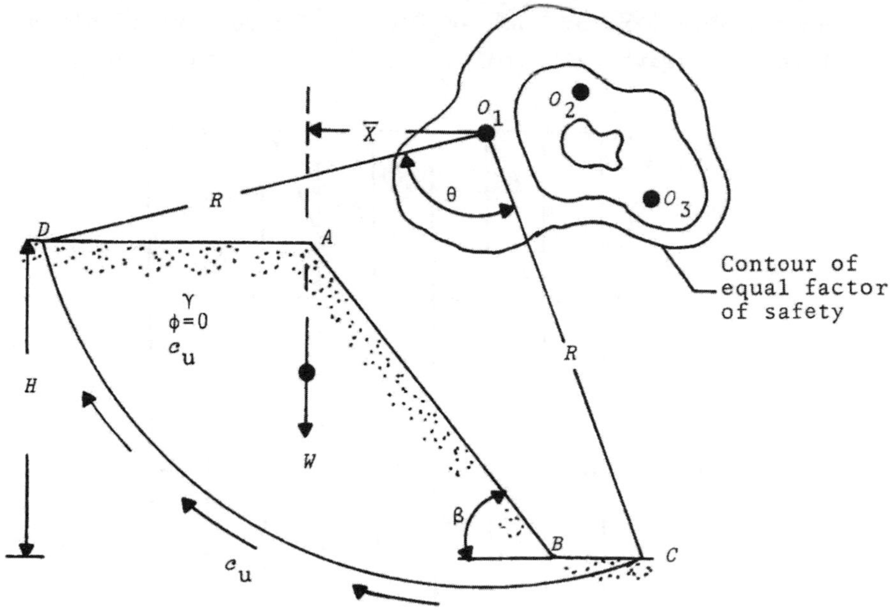

Figure 5.6. Stability of clay slopes (ϕ=0 condition).

$$M_r \quad R^2\theta c_u \tag{5.18}$$

The disturbing moment is due to the weight of the soil (W) located above the probable failure surface CD. So

$$M_d \quad W\overline{X} \tag{5.19}$$

where $\overline{X}$ = distance between O_1 and the centroid of the soil mass located above the trial failure surface CD

Hence

$$F_s = \frac{R^2\theta c_u}{W\overline{X}} \tag{5.20}$$

The factor of safety calculated in the preceding equation is for one trial failure surface. In order to determine the critical failure surface along which the value of F_s is minimum, one has to make several trials with centers such as O_2, O_3, . . . (Fig. 5.6). The values of F_s determined from such calculations can be used to obtain contours of equal factors of safety. The minimum value of F_s is the factor of safety of the slope against failure with respect to strength. The process of obtaining the factor of safety by the trial method just described is tedious. Taylor (1937) has presented the results of such analyses. The results of his study are shown in a graphical form in Fig. 5.7. In this figure

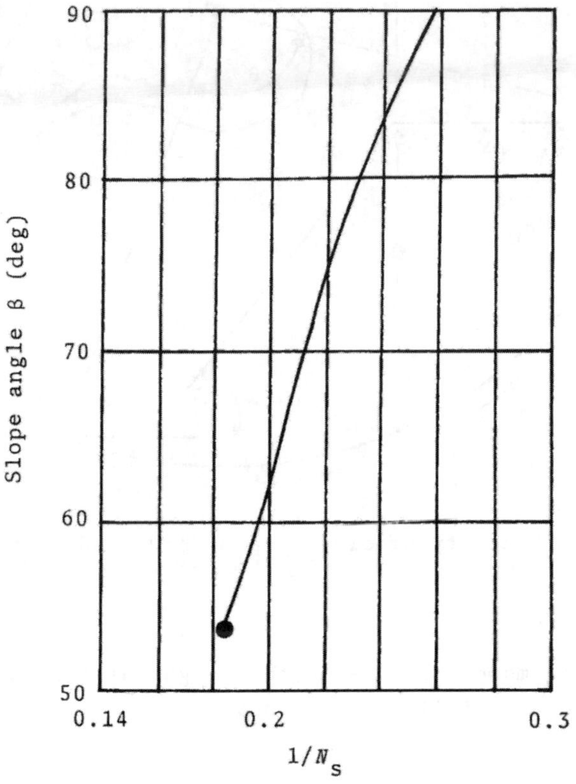

(a)

Figure 5.7. Taylor's analysis of stability of slopes with $\phi=0$ concept--plot of β against $1/N_s$.

$$N_s = \text{Stability number} = \frac{\gamma H_{cr}}{c_u} \quad \frac{\gamma H}{c_{u(d)}} \tag{5.21}$$

where $c_{u(d)}$ developed undrained cohesion for an actual height H of the embankment

$$D = \text{Depth factor} = \frac{\text{Depth from the top of the slope to the firm layer (Fig. 5.2c)}}{\text{Height of slope } H} \tag{5.22}$$

From Fig. 5.7, it can be seen that when the slope angle β is less than 53°, the critical failure circle may be a slope circle,

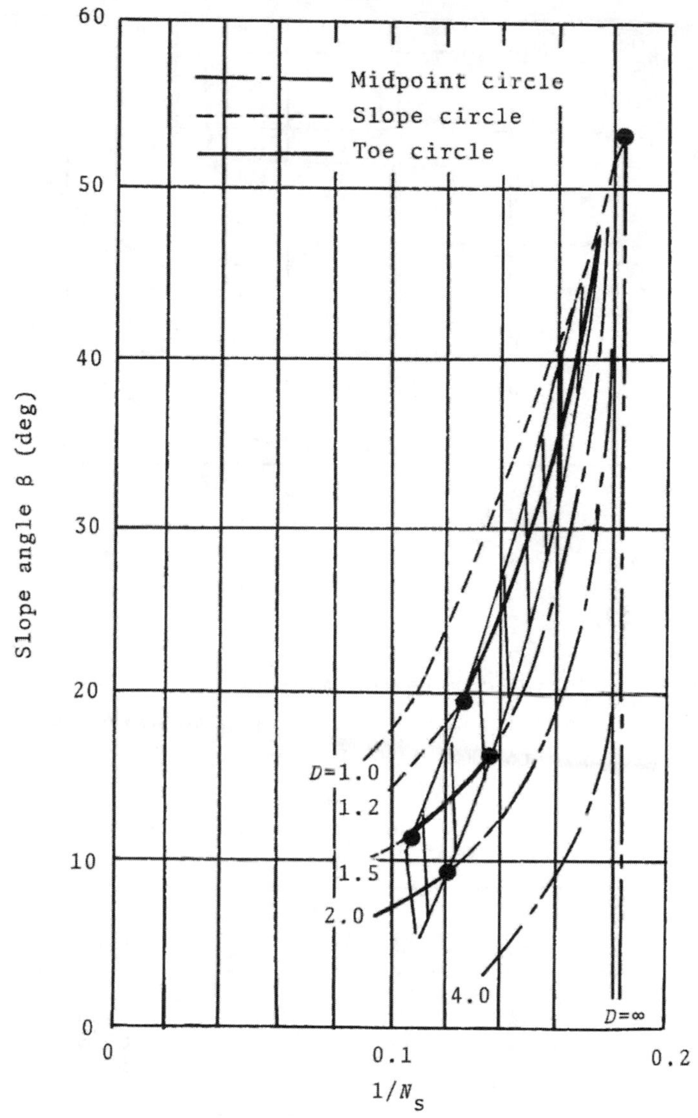

(b)

Figure 5.7. (Continued).

toe circle, or base failure circle (mid point circle) as shown in Fig. 5.2. The nature of the critical failure circle will depend on the depth factor. However when β>53°, the *critical failure circle is a toe circle.* Figure 5.8 can be used as an aid to locate

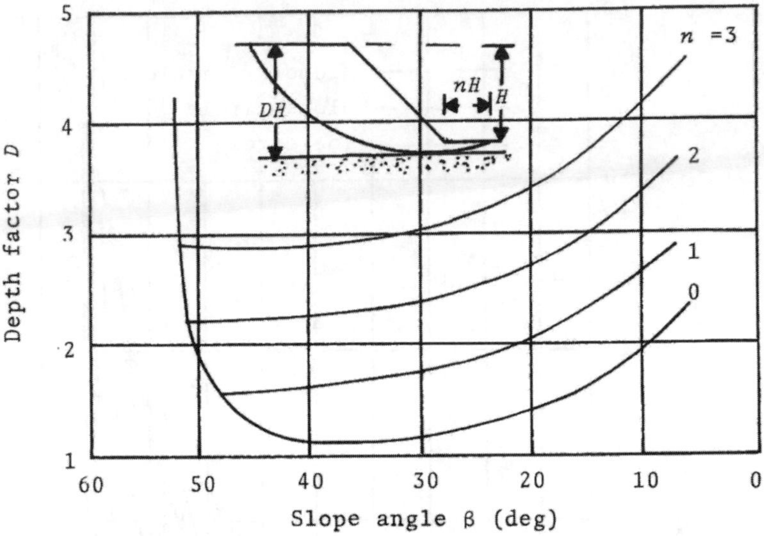

Figure 5.8. Location of critical midpoint circle (After Terzaghi and Peck, 1969).

the critical midpoint circle for a slope. The most critical toe circles for slopes with β<53° can be located with the aid of Table 5.2 with reference to Fig. 5.9 (Fellenius, 1927). However, it

Table 5.2. Location of Critical Toe Circles for β<53°[a]

β (deg)	η (deg)	ψ (deg)
45	28	82
33.68	26	68.68
26.57	25	61.57
18.43	25	53.43
11.32	25	48.32

[a]For definition of β, η, and ψ, refer to Fig. 5.9.

needs to be kept in mind that the critical toe circles are not necessarily the critical circles for a given slope. The centers of critical circles which are toe circles with β>53° can be located with the aid of Fig. 5.10.

Example 5.2. A saturated clay embankment has a height H=30 ft. A rock layer is located at a depth of 45 ft measured from the top of

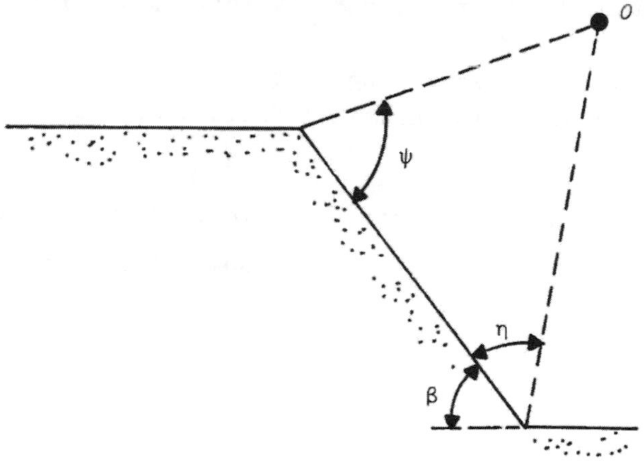

Figure 5.9. Location of most critical toe circles for β<53°.

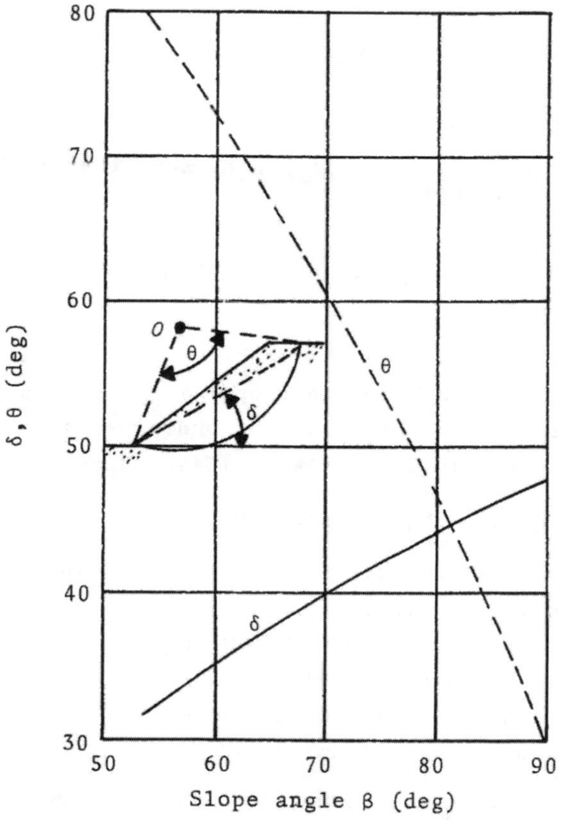

Figure 5.10. Location of critical circles for β>53°.

the embankment. Given β=35°, c_u=1000 lb/ft^2, and γ=120 lb/ft^3. Determine

a. the factor of safety against sliding F_s

b. the nature of the critical failure circle

Solution

Part a. Determination of factor of safety against sliding.

Given slope angle β=35°; D=45 ft/30 ft=1.5. From Fig. 5.7b, for β=35°

$$N_s \quad \frac{\gamma H}{c_{u(d)}} \simeq \frac{1}{0.169}$$

So

$$c_{u(d)} \quad (0.169)\gamma H \quad (0.169)(12)(30) \quad 608 \text{ lb/ft}^2$$

Hence

$$F_s = \frac{c_u}{c_{u(d)}} \quad \frac{1000}{608} = \underline{1.64}$$

Part b. Determination of nature of critical failure circle.

Referring to Fig. 5.7b, it can be seen that the critical circle is a ***midpoint circle***.

5.5 CLAY SLOPES WITH ANISOTROPIC STRENGTH PROPERTIES (ϕ=0 CONDITION)

Lo (1965) has analyzed the stability of saturated clay slopes exhibiting anisotropic strength properties. An empirical relation for the directional variation of the undrained shear strength of clay has been given in Chapter 4 as (Casagrande and Carrillo, 1944)

$$c_{u(i)} = c_{u(H)} + [c_{u(V)} - c_{u(H)}]\cos^2 i \tag{5.23}$$

where i = angle that the direction of major principal stress makes with the vertical

$c_{u(V)}$ undrained cohesion with the major principal stress in the vertical direction (i=0)

$c_{u(H)}$ undrained cohesion with the major principal stress in the horizontal direction (i=90°)

It was also pointed out in Chapter 4 that Lo (1965) conducted several undrained strength tests on specimens collected from the Welland, Ontario (Canada) area. According to his study, the angle

f between the failure plane and the major principal plane (Fig. 5.11) was about 55°. Using the results of this study, Lo has

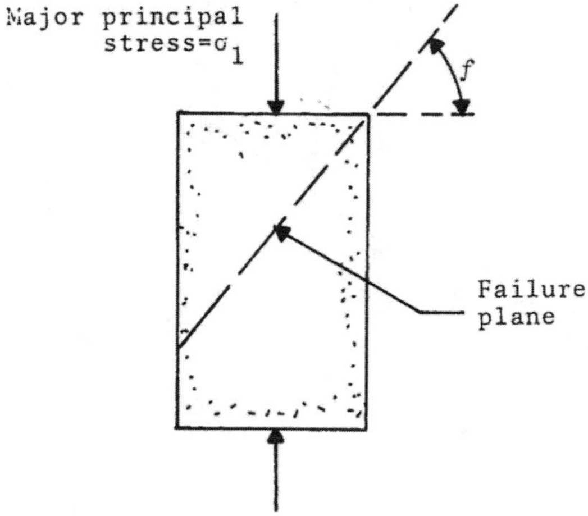

Figure 5.11. Definition of the angle f.

conducted an analysis for the stability of slopes (ϕ=0 concept) for which the depth of the hard stratum is located at a great depth. The results of this study can be explained with the aid of Fig. 5.12, in which BC is a trial failure with the center of the trial circle at O. The radius of the trial circle is equal to R.

From the geometry of Fig. 5.12, it is obvious that

$$i = f + \theta - \frac{\pi}{2} \tag{5.24}$$

According to Eq. (5.16)

$$F_s = \frac{M_r}{M_d} \tag{5.25}$$

The disturbing moment per unit length of the slope, M_d, can be given by the following equation

$$M_d = \frac{\gamma H^3}{12}[(1-2\cot^2\beta + 3\cot\lambda\cot\beta + 3\cot\alpha\cot\lambda \quad 3\cot\alpha\cot\beta) - 6n(n + \cot\beta - \cot\lambda + \cot\alpha)] \tag{5.26}$$

where γ unit weight of soil

or

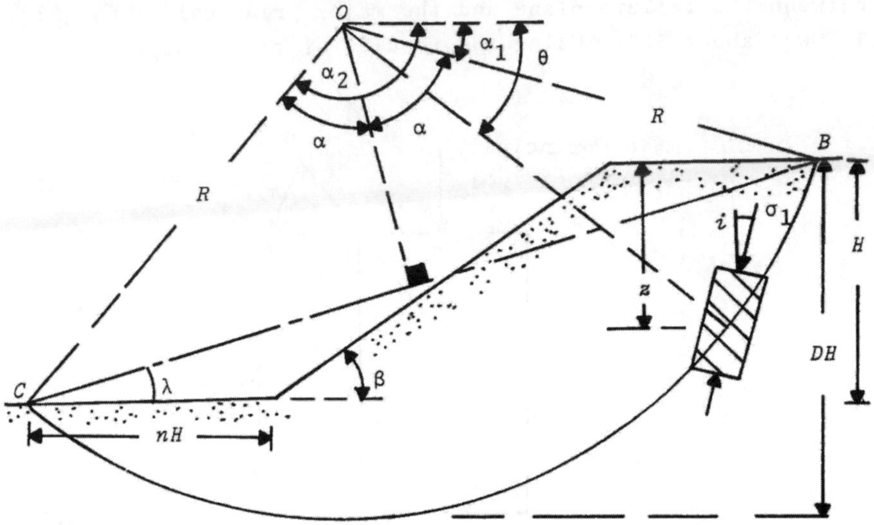

Figure 5.12. Stability of clay slope with anisotropic strength properties (ϕ=0 concept).

$$M_d = \frac{\gamma H^3}{12}(Y - Z) \tag{5.27}$$

where

$$Y = 1 \quad 2\cot^2\beta + 3\cot\lambda\cot\beta + 3\cot\alpha\cot\lambda - 3\cot\alpha\cot\beta \tag{5.28}$$

$$Z = 6n(n + \cot\beta \quad \cot\lambda + \cot\alpha) \tag{5.29}$$

If the values of $c_{u(V)}$ and $c_{u(H)}$ can be taken as constants with depth, then the resisting moment per unit length of the slope, M_r, can be given by the following equation

$$M_r = R\int_{\alpha_1}^{\alpha_2} c_u(\theta, Z) R d\theta \tag{5.30}$$

Combining Eqs. (5.23), (5.24), and (5.30)

$$M_r = R^2\int_{\alpha_1}^{\alpha_2} \{c_{u(H)} + [c_{u(V)} - c_{u(H)}]\cos^2(\alpha' + \theta)\} d\theta \tag{5.31}$$

where $\alpha' = f - \pi/2$

The preceding equation can be integrated to yield

$$M_r = R^2[(1+k)c_{u(V)}\alpha + \tfrac{1}{2}(1-k)c_{u(V)}\sin 2\alpha\cos(2f - 2\lambda)] \tag{5.32}$$

where $k \quad \dfrac{c_{u(H)}}{c_{u(V)}}$ (5.33)

The factor of safety can be obtained by combining Eqs. (5.25), (5.27), and (5.32), or

$$F_s = \frac{3}{\gamma H}\left\{\frac{c_{u(V)}\left[(1+k)\alpha+\frac{1}{2}(1-k)\sin 2\alpha\cos(2f-2\lambda)\right]}{\sin^2\alpha\sin^2\lambda(Y-Z)}\right\} \tag{5.34}$$

or

$$F_s = \frac{N_s}{\dfrac{\gamma H}{c_{u(V)}}} \tag{5.35}$$

where N_s = stability number

$$= 3\left[\frac{(1+k)\alpha + \frac{1}{2}(1-k)\sin 2\alpha\cos(2f-2\lambda)}{\sin^2\alpha\sin^2\lambda(Y-Z)}\right] \tag{5.36}$$

The minimum factor of safety can now be obtained by minimizing N_s with respect to α and λ, or

$$\frac{\partial N_s}{\partial \alpha} = 0 \tag{5.37}$$

and

$$\frac{\partial N_s}{\partial \lambda} = 0 \tag{5.38}$$

Assuming f=55°, the solution for the stability numbers evaluated in this manner are shown in Fig. 5.13. From this figure, it may be seen that for a given value of k there exists a critical slope angle $\beta \geq \beta_{cr}$ up to which the critical failure surfaces are toe circles. However for $\beta < \beta_{cr}$ the critical circles pass below the toe. Figure 5.14 shows the minimum values of stability numbers (N_s) for the *critical toe circles only*. Note that for $\beta < \beta_{cr}$ the stability numbers shown in Fig. 5.14 are not the absolute minimum values.

Figure 5.15 shows the variation of the nature of $N_{s(k)}$ to $N_{s(k=1)}$ (that is, isotropic case) for various values of k and slope angle β.

Example 5.3. Refer to Fig. 5.12. For the clay slope, the undrained shear strengths are as follows

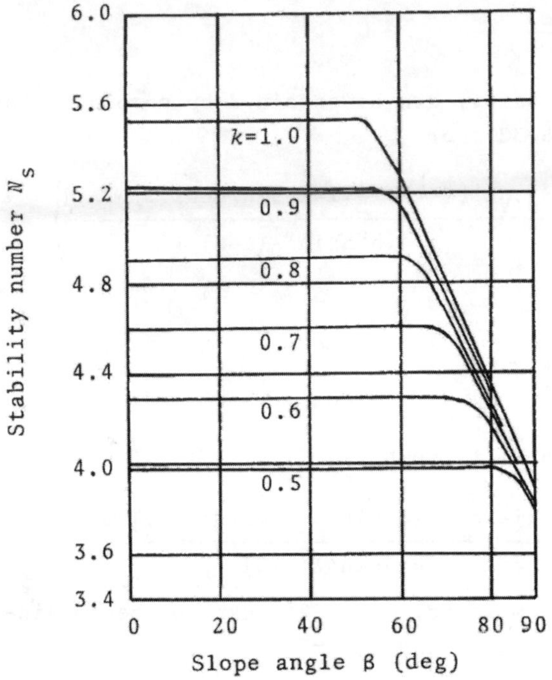

Figure 5.13. Variation of N_s with β--constant $c_{u(V)}$ with depth (After Lo, 1965).

$c_{u(V)}$ 750 lb/ft² (constant with depth)

$c_{u(H)}$ 525 lb/ft² (constant with depth)

If γ=115 lb/ft³, β=75°, and the height of the slope H=25 ft, determine the factor of safety (F_s).

Solution

From Eq. (5.33)

$$k = \frac{c_{u(H)}}{c_{u(V)}} = \frac{525}{750} = 0.7$$

From Fig. 5.13, for β=75° and k=0.7, the value of $N_s \simeq 4.4$. The factor of safety can be given as [Eq. (5.35)]

$$F_s \quad \frac{N_s}{\frac{\gamma H}{c_{u(V)}}} \quad \frac{(4.4)(750)}{(115)(25)} = \underline{1.15}$$

Example 5.4. For the slope described in Example 5.3, determine

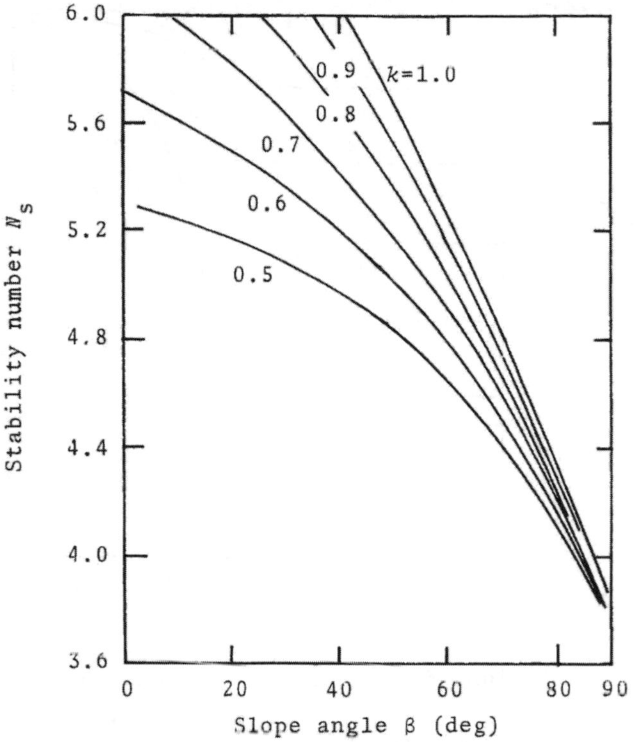

Figure 5.14. Minimum values of N_s for *critical* toe circles only (after Lo, 1965).

the critical height of the slope (H_{cr}).

Solution

For critical height $H=H_{cr}$, $F_s=1$. So

$$\frac{\gamma H_{cr}}{c_{u(V)}} = N_s$$

or

$$H_{cr} = \frac{N_s c_{u(V)}}{\gamma} = \frac{(4.4)(750)}{115} \quad \underline{28.7\ \text{ft}}$$

5.6 STABILITY OF SLOPES IN CLAY ($\phi=0$ CONCEPT) WITH c_u INCREASING WITH DEPTH

The analysis for stability of slopes given in Section 5.4 is

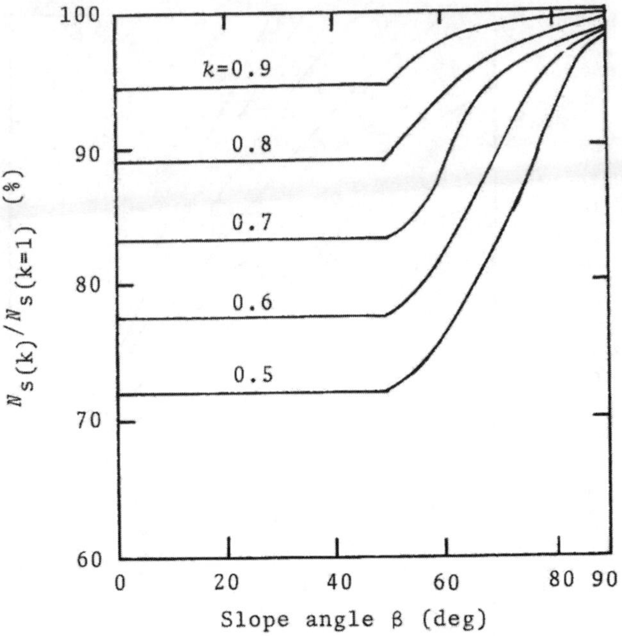

Figure 5.15. Variation of $N_{s(k)}/N_{s(k=1)}$ with β--constant $c_{u(V)}$ with depth (After Lo, 1968).

for the condition of constant value of c_u with depth. However, in many normally consolidated clays, the undrained shear strength increases with depth (Fig. 5.16); that is,

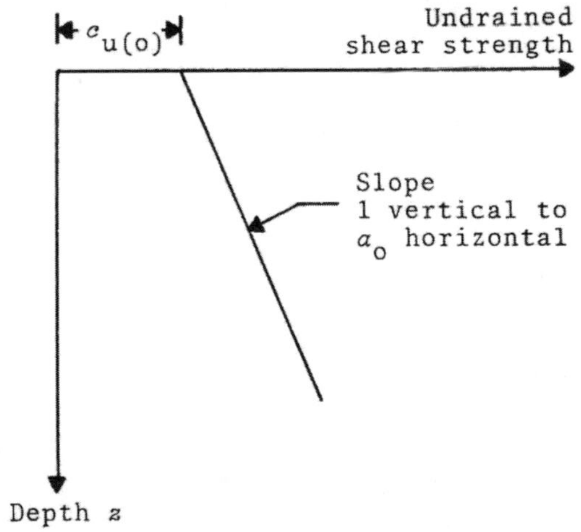

Figure 5.16. Variation of undrained shear strength in normally consolidated isotropic clay soil.

$$c_u = c_{u(o)} + a_o z \tag{5.39}$$

In most cases, the magnitudes of $c_{u(o)}$ can vary from zero to 200 lb/ft^2, and a_o may vary up to about 75 lb/ft^2. This type of condition will invalidate calculations made by using the graphs given in Fig. 5.7. Koppula (1984) has made an analysis for determining the factor of safety F_s for slopes with increasing undrained shear strength with depth. This analysis will be discussed in this section.

Figure 5.17 shows a slope of height H with a trial failure surface assumed to be an arc of a circle. Note that this is a *toe circle*, because the optimum critical slip surface will never be a midpoint circle since the strength of clay increases with depth. The disturbing moment M_d can be given as

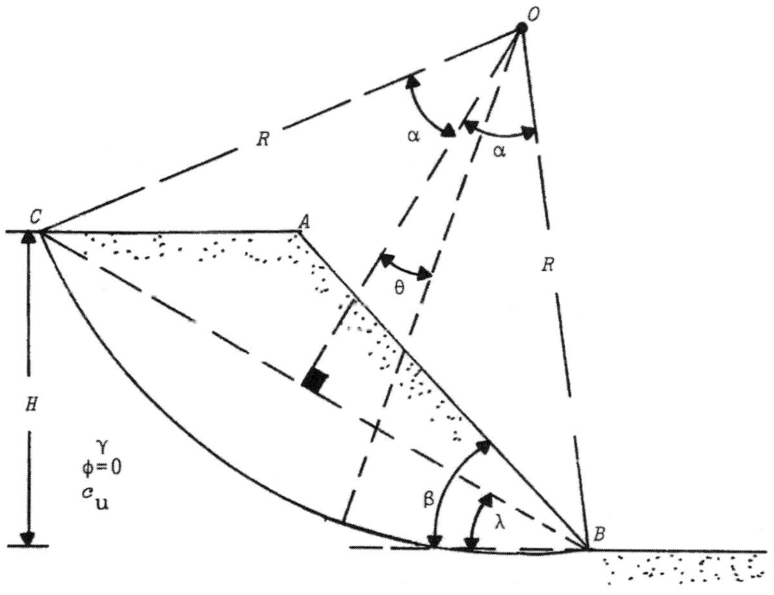

Figure 5.17. Analysis of slope in clay soil (φ=0 concept) with increasing undrained shear strength.

$$M_d = \frac{\gamma H^3}{12}(1 - 2\cot^2\beta - 3\cot\alpha\cot\beta + 3\cot\beta\cot\lambda + 3\cot\lambda\cot\alpha) \tag{5.40}$$

In a similar manner, the resisting moment

$$M_r = R \int_{-\alpha}^{+\alpha} c_u R d\theta \tag{5.41}$$

where $$R = \frac{H}{2\sin\alpha\sin\lambda} \tag{5.42}$$

Also

$$c_u = c_{u(o)} + a_o[R\cos(\lambda+\theta) - R\cos(\alpha-\lambda) + H] \tag{5.43}$$

Combining Eqs. (5.41), (5.42), and (5.43)

$$M_r \quad \frac{c_{u(o)} H^2 \alpha}{2\sin^2\alpha\sin^2\lambda} + \frac{a_o H^3}{4\sin^2\alpha\sin^2\lambda}(\alpha + \cot\lambda \quad \alpha\cot\alpha\cot\lambda) \tag{5.44}$$

However, the factor of safety

$$F_s = \frac{M_r}{M_d} \tag{5.45}$$

Combining Eqs. (5.40), (5.44), and (5.45), and simplifying

$$F_s = \left[\frac{c_{u(o)}}{\gamma H}\right] N_s \tag{5.46}$$

where N_s = stability number

$$= \frac{[6\alpha + 3c_R(\alpha + \cot\alpha - \alpha\cot\alpha\cot\lambda)]}{\sin^2\alpha\sin^2\lambda(1 - 2\cot^2\beta - 3\cot\alpha\cot\beta + 3\cot\beta\cot\lambda + 3\cot\lambda\cot\alpha)} \tag{5.47a}$$

$$c_R = \frac{c_{u(o)} + a_o H}{c_{u(o)}} \quad 1 \quad \frac{a_o H}{c_{u(o)}} \tag{5.47b}$$

The minimum value of N_s can be determined as

$$\frac{\partial N_s}{\partial \alpha} = 0$$

and

$$\frac{\partial N_s}{\partial \lambda} = 0$$

The absolute minimum values of N_s thus determined are given in Table 5.3.

Example 5.5. Refer to Fig. 5.17. Given H=10 m, $\tan\beta$=1/3; $c_{u(o)}$=

Table 5.3. Minimum Values of N_s [Eq. (5.46)][a]

c_R	N_s						
	1H:1V	1.5H:1V	2H:1V	3H:1V	4H:1V	5H:1V	10H:1V
0.1	6.32	6.86	7.22	7.69	8.03	8.30	9.38
0.2	6.75	7.41	7.86	8.54	9.05	9.50	11.37
0.3	7.18	7.95	8.50	9.36	10.05	10.67	11.27
0.4	7.61	8.48	9.13	10.17	11.03	11.79	15.09
0.5	8.04	9.01	9.76	10.97	11.99	12.90	16.90
1.0	10.16	11.63	12.85	14.88	16.66	18.31	25.63
2.0	14.34	16.77	18.89	22.55	25.77	28.84	42.61
3.0	18.49	21.87	24.88	30.19	34.77	39.26	59.38
4.0	22.62	26.95	30.80	37.62	43.74	49.60	76.01
5.0	26.74	32.01	36.72	45.13	52.66	59.91	92.63
10.0	47.32	57.29	66.31	82.51	97.24	111.32	175.49

[a]After Koppula (1984).

10 kN/m^2, a_o=2 kN/m^2, and γ=18.5 kN/m^3. Determine the factor of safety F_s for the slope.

Solution

From Eq. (5.47b)

$$c_R = \frac{a_o H}{c_{u(o)}} = \frac{(2)(10)}{10} = 2$$

Referring to Table 5.3, for a slope 1V:3H and c_R=2, the value of N_s is equal to 22.55. From Eq. (5.46)

$$F_s = \left[\frac{c_{u(o)}}{\gamma H}\right] N_s = \left[\frac{10}{(18)(10)}\right](22.55) \quad \underline{1.25}$$

5.7 STABILITY OF FINITE SLOPES WITH c-ϕ SOILS--MASS PROCEDURE

Stability analysis of finite slopes made of c-ϕ soils can be primarily divided into two major categories: (a) *mass procedure* and (b) *method of slices*. In the *mass procedure*, the mass of the soil located above the probable sliding surface is considered as a single unit. This procedure is very attractive when the slope is made of homogeneous soil. In the *method of slices*, the soil located above the probable sliding surface is divided into several vertical slices and the stability of each slice is separately calculated. In the following sections, several available methods for the calculation of the factor of safety (F_s) of slopes with mass procedure will be discussed.

5.7.1 Taylor's Friction Circle Method

Taylor (1937) developed a method usually referred to as the *friction circle method* for the analysis of stability of homogeneous slopes. This type of analysis can be classified under mass procedure. It can be explained by referring to Fig. 5.18, in which *AB*

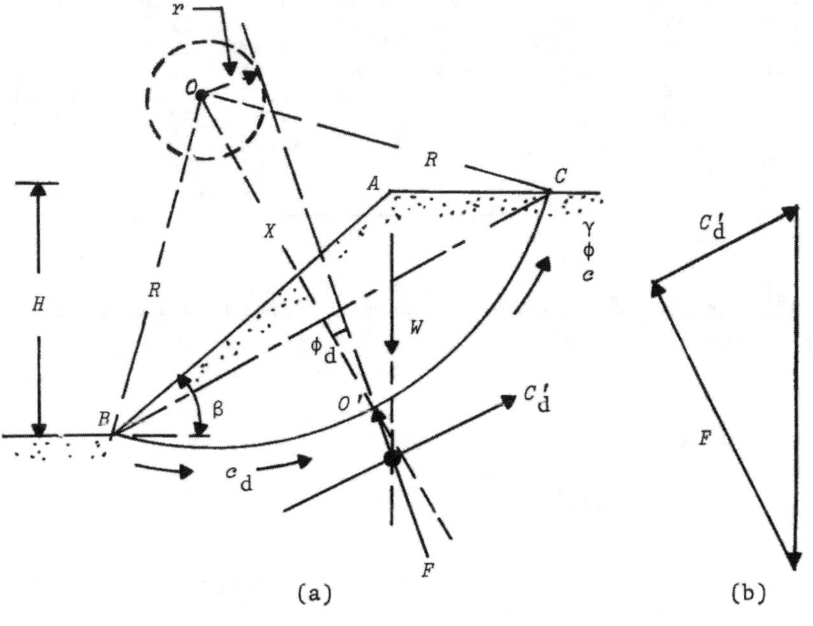

Figure 5.18. Stability analysis of a slope with c-ϕ soil--Taylor's friction circle method.

is a slope making an angle β with the horizontal. The soil constituting the slope is homogeneous with a unit weight γ, and its shear strength can be defined by the equation

$$s = c + \sigma' \tan\phi$$

In Fig. 5.18a, *BC* is an arc of a circle with its center at *O*. The trial circle has a radius *R*. For stability analysis, following are the forces per unit length of the slope on the trial failure mass *ABC*:

1. The weight of the wedge *ABC*--*W*
2. The force due to the developed cohesion c_d along the arc *BC*, or

$$C_d = c_d \widehat{BC} \tag{5.48}$$

where c_d = developed cohesion

The resultant cohesive force along the slip surface $\widehat{BC}$ is equal to a force having a magnitude of $C_d'=c_d\overline{BC}$ ($\overline{BC}$=length of the arc BC), and it acts in a direction parallel to the cord $\overline{BC}$ at a distance X from O. So

$$C_d'X = c_d\overline{BC}X = c_d\widehat{BC}R \quad C_dR$$

or

$$X = \frac{\widehat{BC}}{\overline{BC}} R \tag{5.49}$$

3. The resultant of the normal and frictional force acting along the potential sliding surface--F

Since the forces C_d', F, and W are in equilibrium, the line of action of F must pass through the intersection of W and C_d'. The force F will make an angle of ϕ_d (developed friction angle) with the radial line O-O' which is normal to the slip surface. The line of action of F will be tangent to a circle having a radius

$$r = k'R\sin\phi_d \tag{5.50}$$

The value of k' is always greater than one. The lower limit of $k'=1$ corresponds to a case in which the line of action of F is tangent to the circle having a radius of

$$r = R\sin\phi_d$$

This circle is referred to as the *friction circle*.

The value of k' is statically indeterminate and depends on (a) the central angle $BOC=\theta$ and (b) the distribution of intergranular stress along the slip surface AB. Taylor (1937) has determined the value of k' for two cases: (a) uniform distribution of intergranular stress along the slip surface, and (b) sinusoidal distribution of intergranular stress along the slip surface with zero values at the ends. This analysis shows that for $\theta\leq 50°$ the value of k' varies within 5-10% and, hence, the assumption of the normal stress distribution does not change the factor of safety calculation.

Since the magnitude of the force W is known and the directions of W, C_d', and F are known, we can construct a force polygon as shown in Fig. 5.18b from which the magnitude of C_d' can be obtained. Several trials can be made in this manner by taking different trial surfaces to obtain the mimimum value of C_d' (for a given value of

ϕ_d). Using the friction circle method, Taylor (1937) has obtained the critical values of $N_s = \gamma H / c_d$ for various values of ϕ_d and β. Figure 5.19 shows the plot of $1/N_s$ as obtained by Taylor (1937) using the friction circle method.

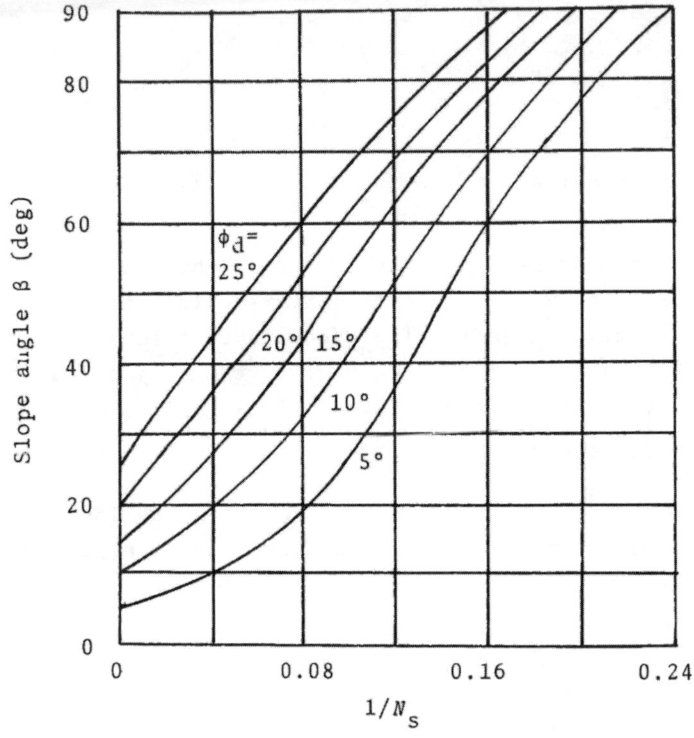

Figure 5.19. Variation of $1/N_s$ with slope angle β and ϕ_d--friction circle method (After Taylor, 1937).

Taylor's chart for stability numbers can be used in several ways, and they are as follows:

1. If β, ϕ, and c are known, the critical height (H_{cr}) of a slope can be determined. Note that when $H = H_{cr}$ the value of $F_s = 1$. For such a problem, $c = c_d$ and $\phi = \phi_d$. For known values of ϕ and β, the value of N_s can be obtained from Fig. 5.19. Now

$$N_s \quad \frac{\gamma H}{c_d} \quad \frac{\gamma H_{cr}}{c}$$

Hence

$$H_{cr} = \frac{N_s c}{\gamma} \tag{5.51}$$

2. If c, ϕ, and H are known, the steepest possible slope (at which $F_s=1$) can also be obtained. For this case, $H=H_{cr}$, $c=c_d$, and $\phi=\phi_d$. We can calculate the value of N_s as

$$N_s \quad \frac{\gamma H}{c}$$

With known values of N_s and $\phi=\phi_d$, the maximum value of β can be easily obtained from Fig. 5.19.

3. If c, ϕ, β, and H are known, the value of F_s can be calculated by the following step-by-step procedure.

a. Assume several values of $\phi_d \leq \phi$.

b. For each value of ϕ_d assumed in Step a, calculate the value of F_ϕ, or

$$F_\phi = \frac{\tan\phi}{\tan\phi_d}$$

c. For each value of ϕ_d assumed in Step a and with the given value of β, obtain N_s from Fig. 5.19. So

$$N_s \quad \frac{\gamma H}{c_d}$$

or

$$c_d = \frac{\gamma H}{N_s}$$

d. For each value of c_d obtained in Step c, calculate F_c, or

$$F_c = \frac{c}{c_d}$$

e. For any given value of ϕ_d, Steps b and d provide us with values of F_ϕ and F_c. Plot these values in a graph as shown in Fig. 5.20. The value of F_s can be obtained for the case where $F_c=F_\phi$.

Using the above procedure, Singh (1970) has obtained the values of F_s for various slopes. These results are shown in a non-dimensional form in Fig. 5.21.

Example 5.6. For a slope with $\beta=30°$, $H=45$ ft, $\phi=25°$, $c=600$ lb/ft^2, and $\gamma=110$ lb/ft^3, determine the factor of safety F_s using the procedure outlined in Section 5.7.1.

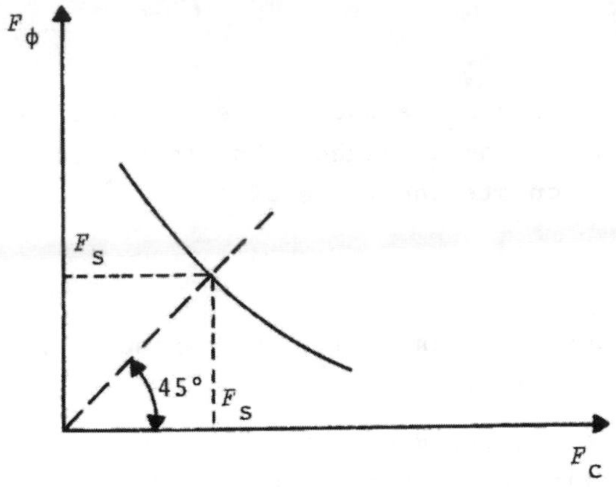

Figure 5.20. Determination of F_s from the F_ϕ vs. F_c plot.

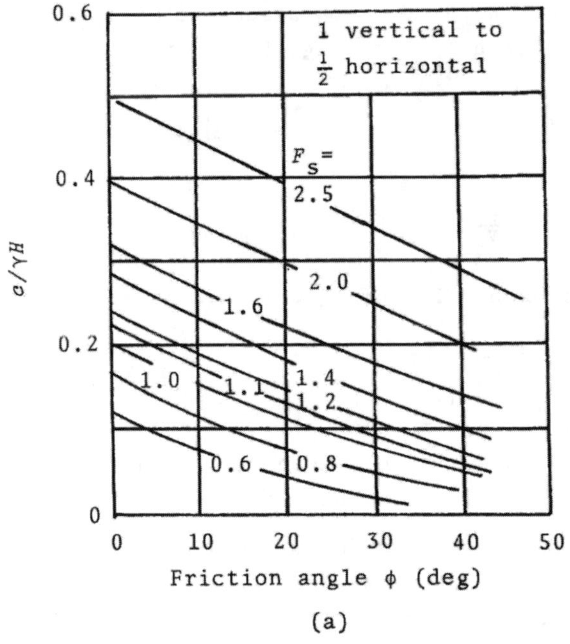

Figure 5.21. Factor of safety contours based on friction circle method (After Singh, 1970).

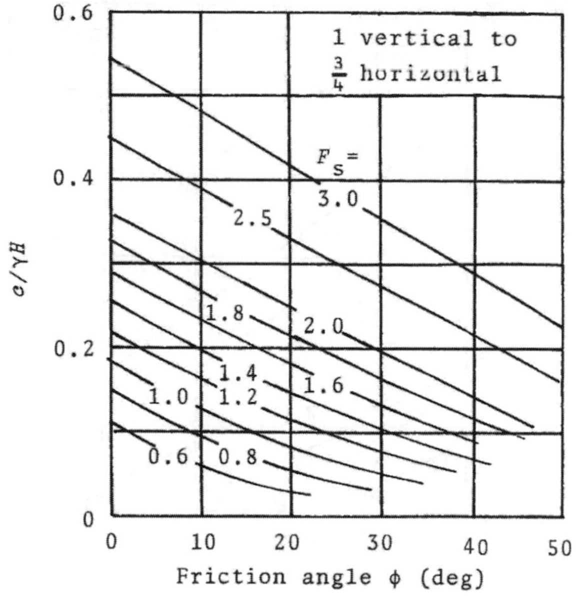

(b)

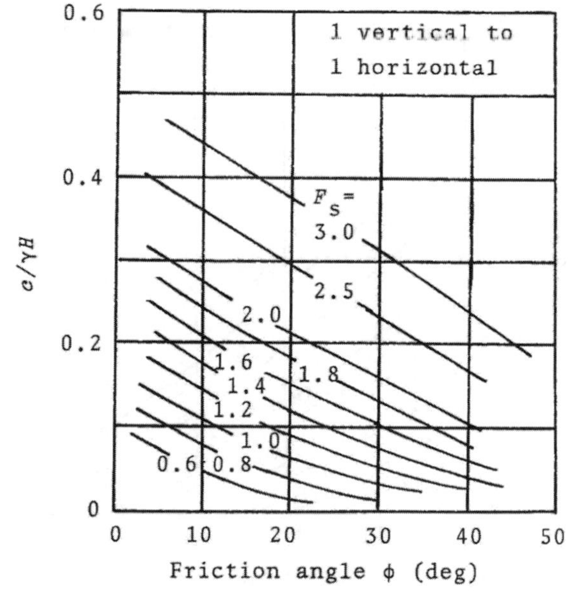

(c)

Figure 5.21. (Continued).

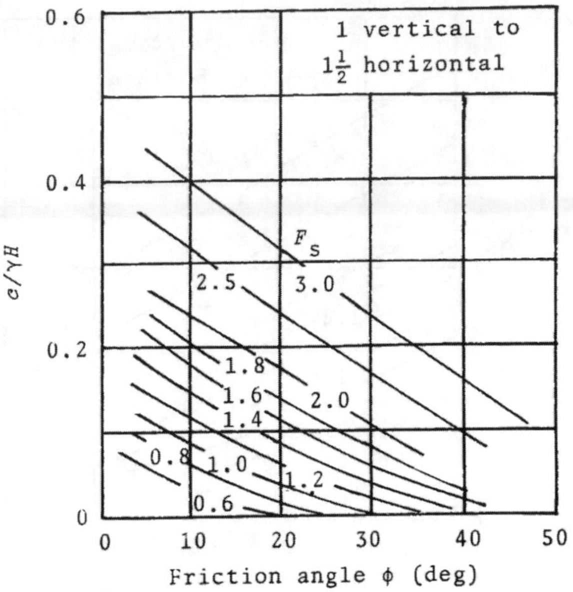

(d)

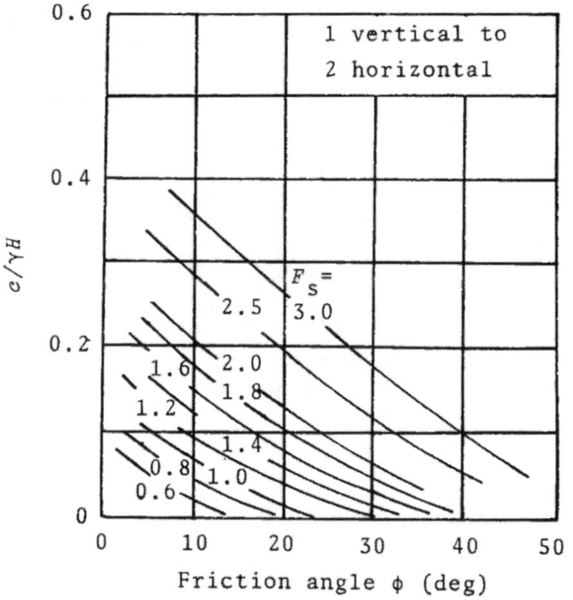

(e)

Figure 5.21. (Continued).

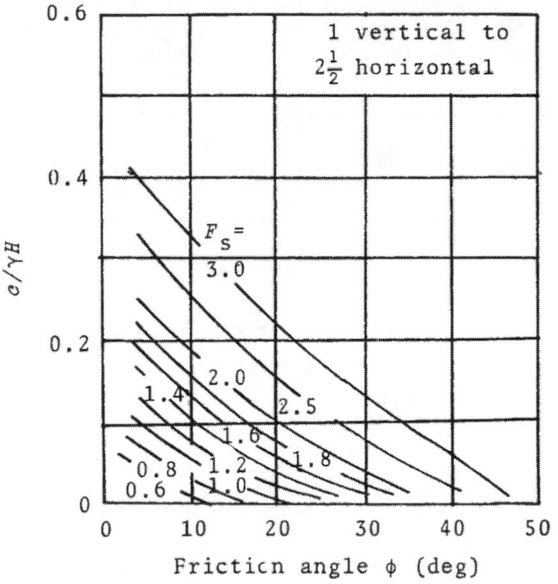

(f)

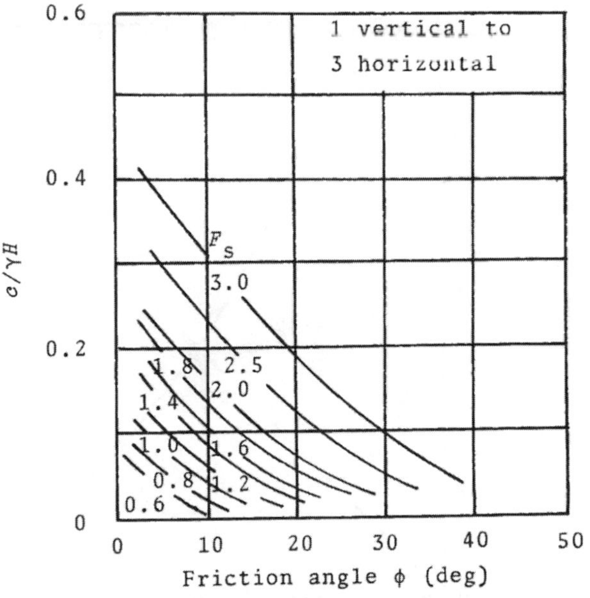

(g)

Figure 5.21. (Continued).

Solution

The following table can now be prepared.

Assumed ϕ_d (deg)	$F_s = \frac{\tan\phi}{\tan\phi_d}$	N_s (Fig. 5.19)	$c_d = \frac{\gamma H}{N_s}$ (lb/ft^2)	$F_c = \frac{c}{c_d}$
25	1	$\frac{1}{0.026} = 38.5$	128.6	4.67
15	1.74	$\frac{1}{0.05} = 20.0$	247.5	2.42
10	2.64	$\frac{1}{0.07} = 14.3$	346.2	1.73
5	5.33	$\frac{1}{0.107} = 9.3$	532.2	1.13

Figure 5.22 shows a plot of F_ϕ and F_c as obtained from the above table. From the figure

$$F_c \quad F_\phi = F_s = \underline{2.05}$$

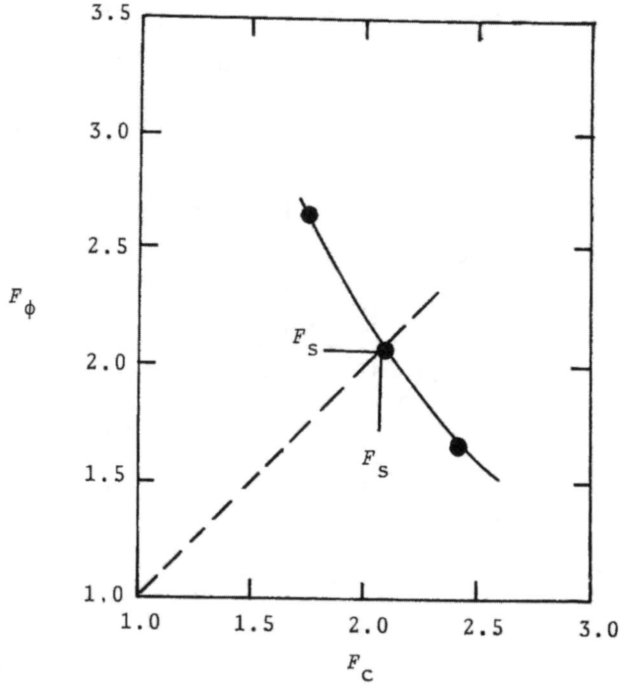

Figure 5.22.

5.7.2 Cousins' Stability Analysis (c-ϕ SOIL)

Cousins (1978) has used a procedure similar to that adopted by Taylor (1937) to determine the stability of simple homogeneous slopes taking into consideration the effect of pore water pressure. Figure 5.23 shows a slope with a trial failure surface which is an

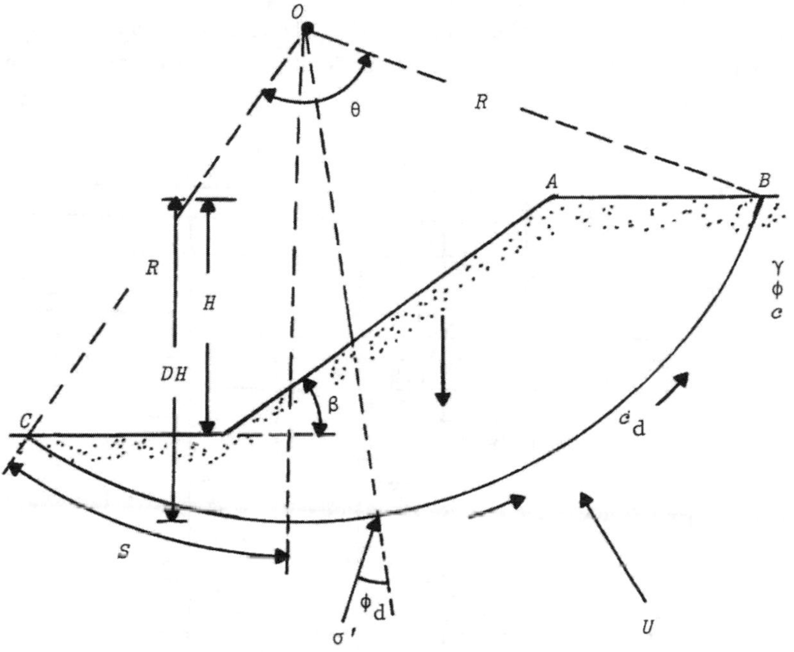

Figure 5.23. Cousins' analysis of slopes with c-ϕ soil.

arc of a circle with its center at O. The radius of the trial circle is equal to R. The forces on the soil located above the trial failure surface are:

1. The weight of the soil--W
2. The force due to pore water pressure along the trial failure surface--U
3. The developed cohesion per unit area along the trial failure surface--c_d
4. The intergranular effective stress per unit area along the trial failure surface--σ'. Note that, at any point, σ' will be inclined at an angle ϕ_d (developed friction angle) to the normal drawn to the failure surface.

The distribution of σ′ along the failure surface was assumed by Cousins to be of the following nature

$$\sigma' = p_1 \sin\left(\frac{\pi S}{R\theta}\right) + p_2 \sin\left(\frac{2\pi S}{R\theta}\right) \qquad (5.52)$$

where p_1, p_2 = constants
The terms S (distance along the failure arc) and θ in Eq. (5.52) have been defined in Fig. 5.23.

The solutions for the factor of safety of slopes have been presented by Cousins (1978) in graphical form, and these are given in Figs. 5.24 and 5.25. In preparing the graphs, the following

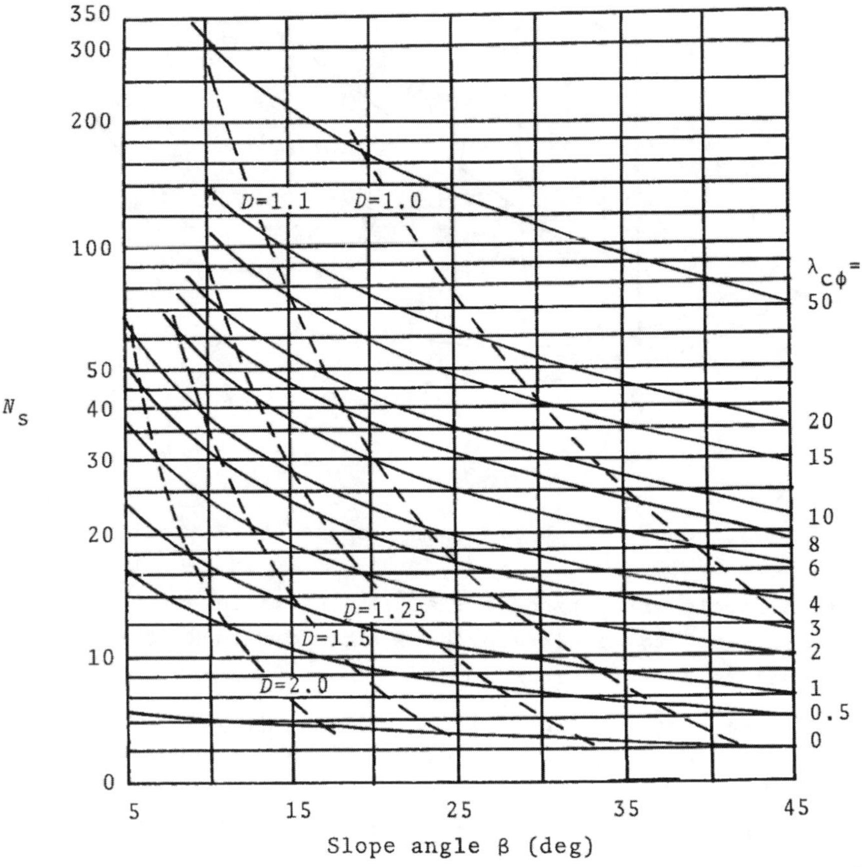

(a)

Figure 5.24. Variation of N_s with slope angle β for *critical toe circles*: (a) r_u=0; (b) r_u=0.25; and (c) r_u=0.5 (after Cousins, 1978). For definition of D, refer to Fig. 5.23.

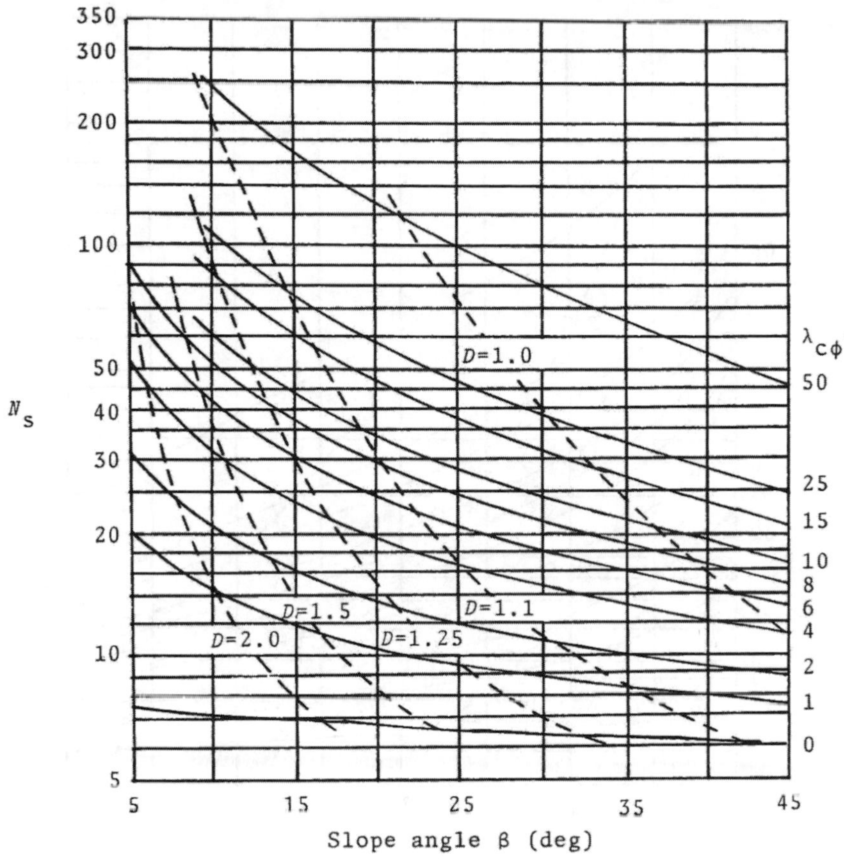

(b)

Figure 5.24. (Continued).

parameters have been used.

1. $\lambda_{c\phi} \quad \frac{\gamma H \tan\phi}{c}$ (5.53)

where H = height of the slope

γ = bulk unit weight of the soil

c = cohesion

ϕ = soil friction angle

2. $N_s \quad \frac{\gamma H F_s}{c}$ (5.54)

where N_s = stability number

F_s = factor of safety with respect to strength

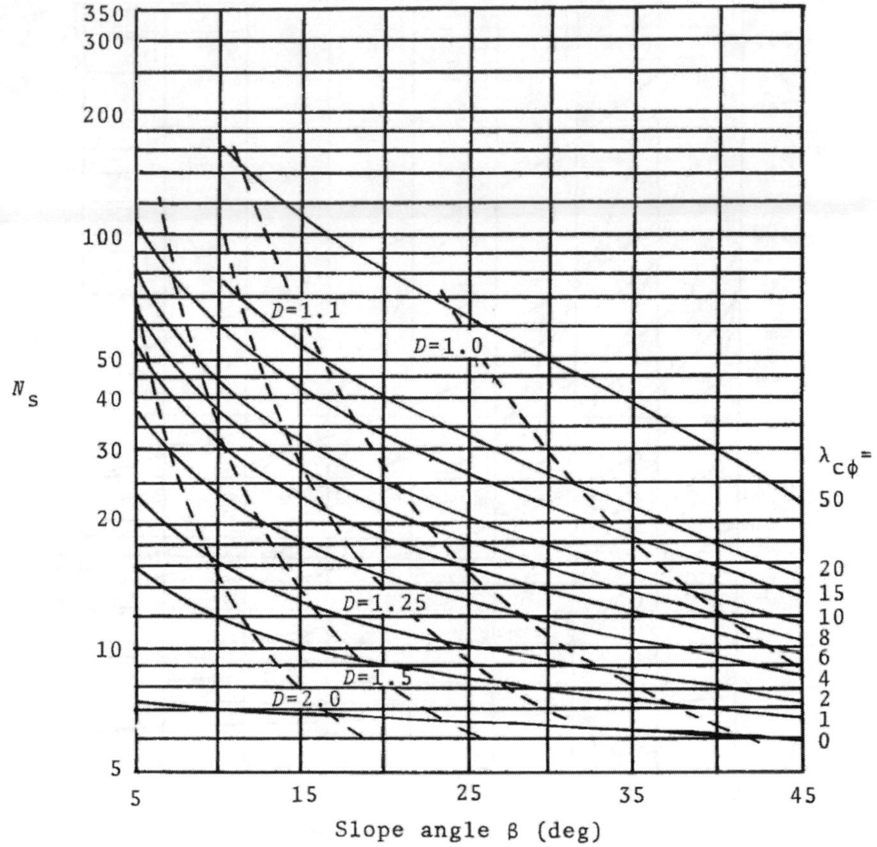

(c)

Figure 5.24. (Continued).

3. Depth factor D (as defined in Fig. 5.23)

4. *Pore pressure ratio* r_u. The pore pressure ratio at any point along a trial failure surface can be explained by referring to Fig. 5.26. At a point P on the trial failure surface, the pore water pressure can be given as

$$u = h\gamma_w \tag{5.55}$$

The total vertical stress at point P is equal to

$$\sigma_v = \gamma z \tag{5.56}$$

where γ = bulk unit weight of soil

z - depth of soil above point P

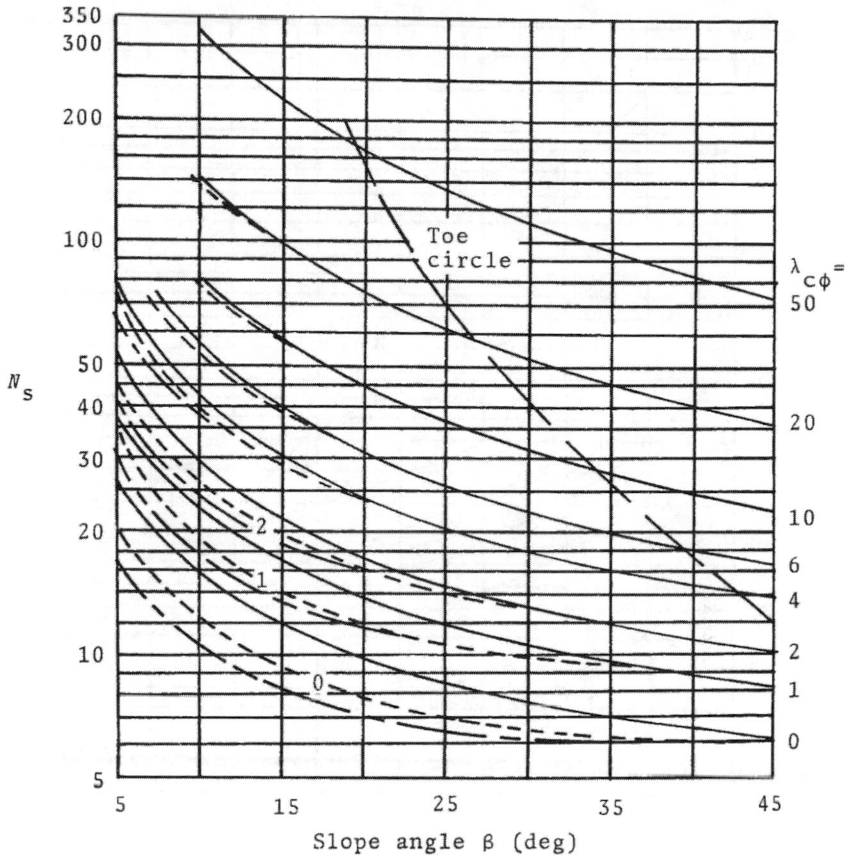

(a)

Figure 5.25. Variation of N_s with slope angle β for *critical circle* with specified depth factors: (a) r_u=0; (b) r_u=0.25; and (c) r_u=0.5 (after Cousins, 1978). *Note:* ——— D=1; – – – – – D=1.25; ——-——— D=1.5.

So, the *pore pressure ratio* at point P can be expressed as

$$r_u \quad \frac{u}{\sigma_v} = \frac{h\gamma_w}{\gamma z} \tag{5.57}$$

It needs to be pointed out that the pore pressure ratios shown in Figs. 5.24 and 5.25 are the *weighted average values* for the entire failure surface. Figures 5.27 and 5.28 are for the determination of the coordinates of *critical slip circles*. Figure 5.29 shows the origin for X and Y coordinates for use in Figs. 5.27 and 5.28 . The use of the charts are demonstrated in Example 5.7.

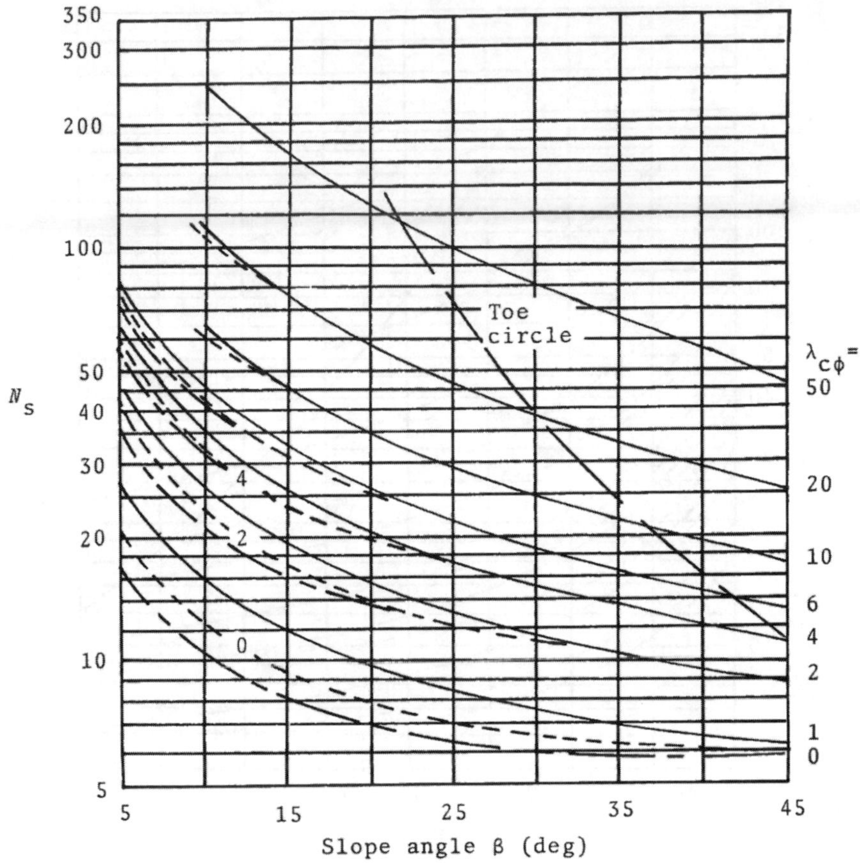

(b)

Figure 5.25. (Continued).

Example 5.7. A slope is 30 ft in height. For the slope, given β (slope angle)=30°, ϕ=25°, c=175 lb/ft^2, γ=120 lb/ft^3, and r_u (weighted average)=0.27. Determine:

a. The factor of safety F_s

b. The location of the critical circle (that is, X and Y)

Solution

Part a. Determination of factor of safety

According to Eq. (5.53)

$$\lambda_{c\phi} \quad \frac{\gamma H \tan\phi}{c} = \frac{(120)(30)(\tan 25)}{175} \quad 9.59$$

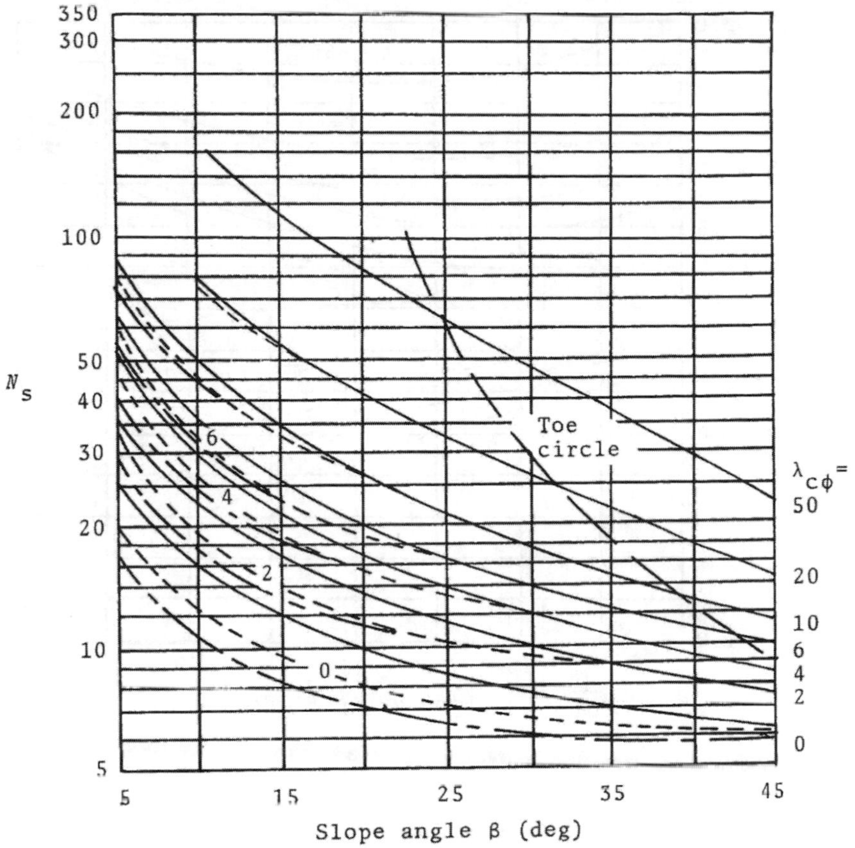

(c)

Figure 5.25. (Continued).

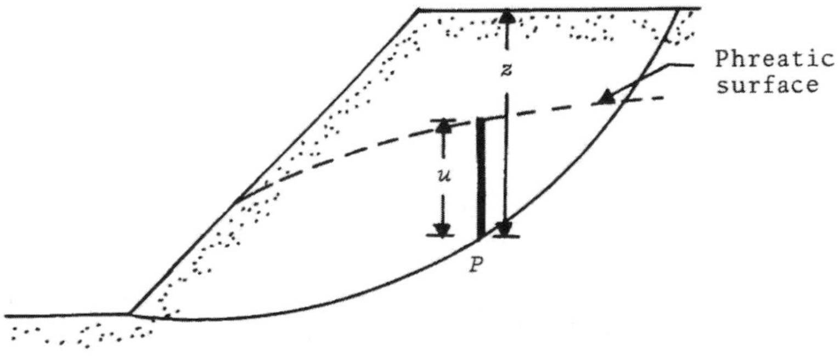

Figure 5.26. Definition of r_u [Eq. (5.57)].

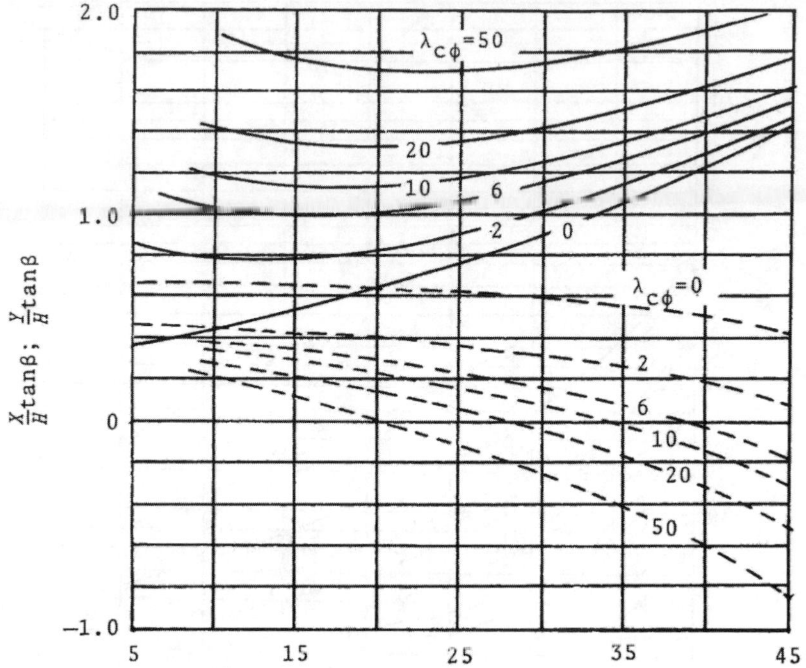

(a)

Figure 5.27. Location of the center of *critical toe circles*: (a) r_u=0; (b) r_u=0.25; and (c) r_u=0.5 (after Cousins, 1978). *Note*: -----$(X/H)\tan\beta$ and——— $(Y/H)\tan\beta$.

Check for the possibility of *toe* failure:

From Fig. 5.24b and 5.24c, for β=30° and $\lambda_{c\phi}$=9.59, the stability numbers

$$N_s = \frac{\gamma H F_s}{c} \simeq 23.5 \text{ (with } D \simeq 1.04) \text{ for } r_u = 0.25$$

and

$$N_s \simeq 17.5 \text{ (with } D \simeq 1.04) \text{ for } r_u = 0.5$$

So, by interpolation

$$N_s \simeq 23.02 \text{ for } r_u = 0.27$$

Check for the possibility of *base* failure:

From Fig. 5.25b and 5.25c, for β=30° and $\lambda_{c\phi}$=9.59

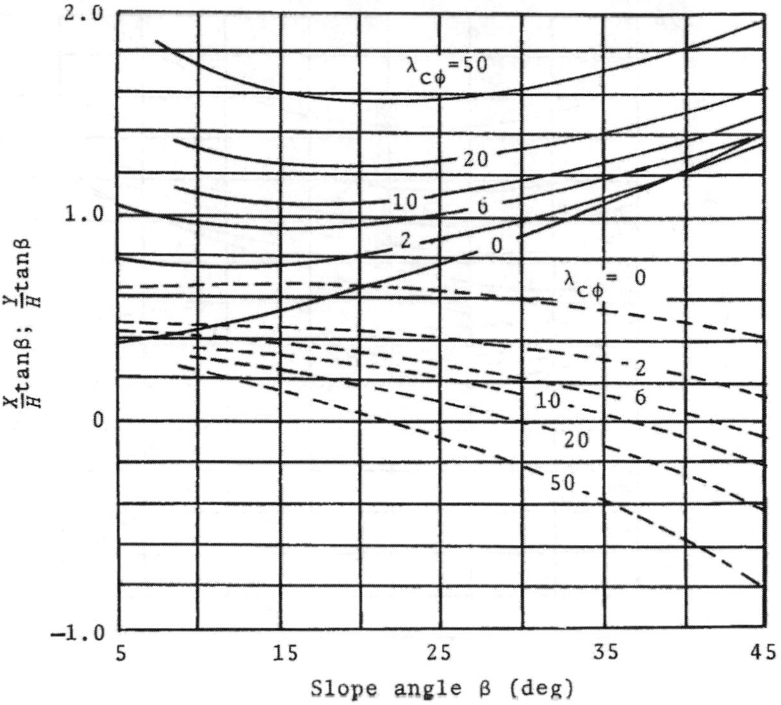

(b)

Figure 5.27. (Continued).

$$N_s \simeq 24.5 \text{ (with } D \simeq 1.0) \text{ for } r_u = 0.25$$

and

$$N_s = 17.5 \text{ (with } D \simeq 1,0) \text{ for } r_u = 0.5$$

By interpolation

$$N_s \sim 23.94 \text{ for } r_u = 0.27$$

The smallest value of N_s for r_u=0.27 is about 23.02 (toe failure). So

$$F_s \quad \frac{N_s c}{\gamma H} = \frac{(23.02)(175)}{(120)(30)} = \underline{1.119}$$

Part b. Determination of location of critical circle.

Referring to Fig. 5.27b and 5.27c, for $\lambda_{c\phi}$=9.59

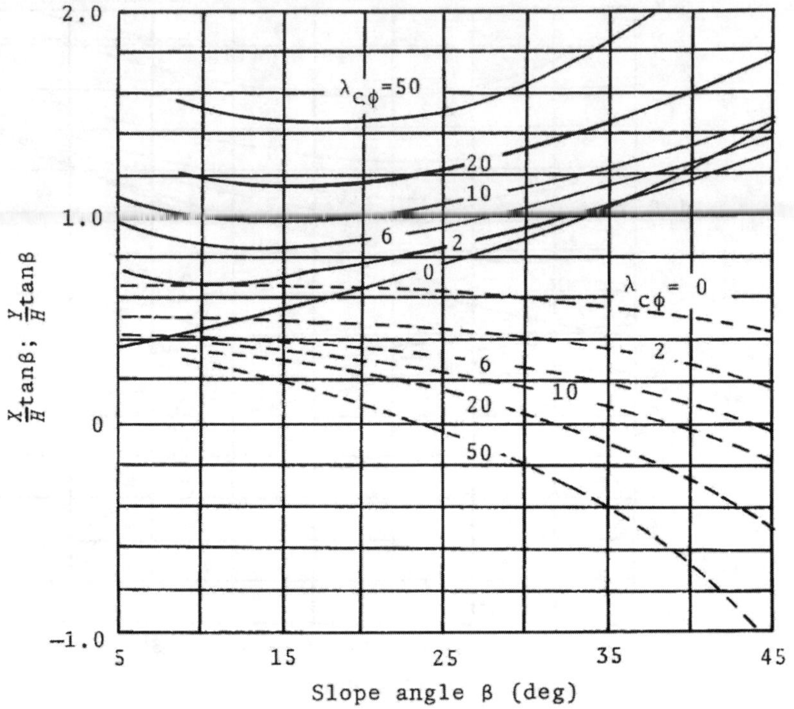

(c)

Figure 5.27. (Continued).

$$\left.\begin{aligned}&\frac{X}{H}\tan\beta = 0.12\\ \text{and}&\\ &\frac{Y}{H}\tan\beta \quad 1.15\end{aligned}\right\} \text{ for toe circle and } r_u=0.25$$

$$\left.\begin{aligned}&\frac{X}{H}\tan\beta = 0.15\\ \text{and}&\\ &\frac{Y}{H}\tan\beta = 1.1\end{aligned}\right\} \text{ for toe circle and } r_u=0.5$$

Hence, for toe circle and r_u=0.27

$$\frac{X}{H}\tan\beta = 0.122 \tag{a}$$

$$\frac{Y}{H}\tan\beta = 1.144 \tag{b}$$

So

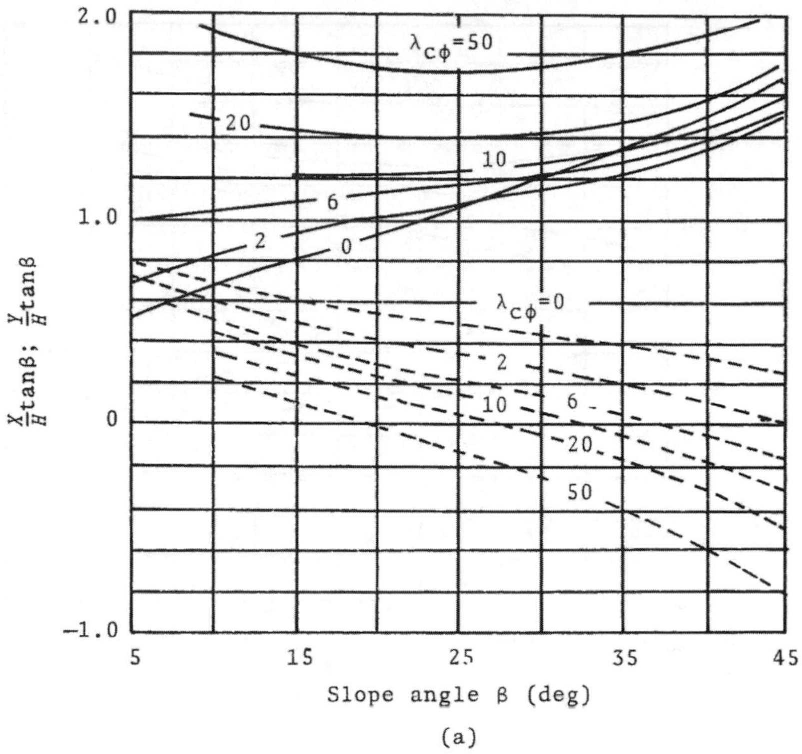

Figure 5.28. Location of *critical circles* with D=1: (a) r_u=0, (b) r_u=0.25; (c) r_u=0.5.(after Cousins, 1978). *Note*:---- $(X/H)\tan\beta$ and ——— $(Y/H)\tan\beta$.

$$X = \frac{(0.122)H}{\tan\beta} = \frac{(0.122)(30)}{\tan 30} = \underline{6.34\text{ ft}}$$

$$Y = \frac{(1.144)(30)}{\tan 30} = \underline{59.4\text{ ft}}$$

5.8 METHOD OF SLICES

As mentioned before, in the method of slices the soil located above the probable surface of sliding is divided into several vertical slices, and the stability of each slice is calculated separately. In the following sections, several available methods for calculation of the factor of safety by the slice procedure will be discussed.

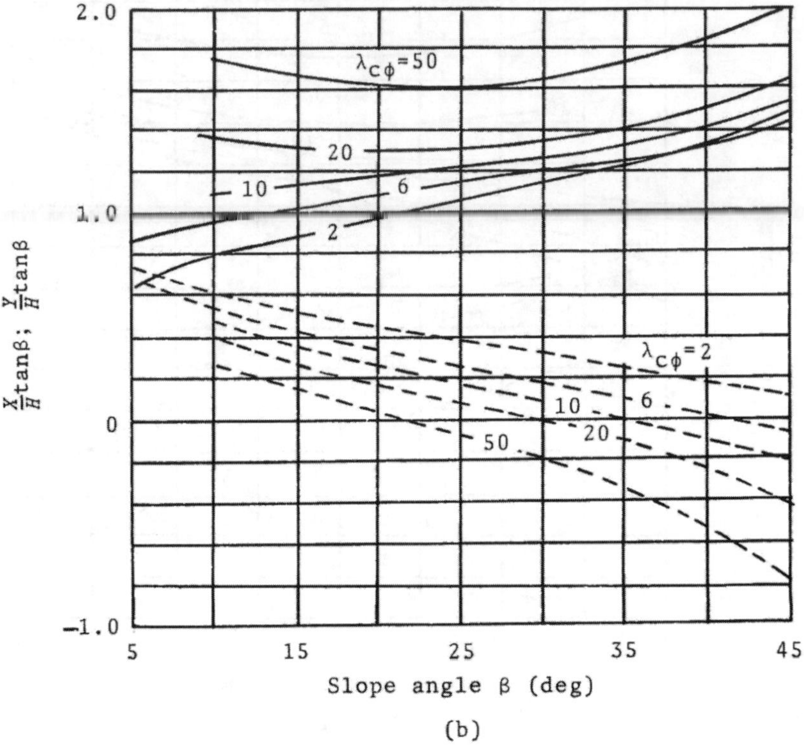

Figure 5.28. (Continued).

5.8.1 Ordinary Method of Slices

This method is more useful when a slope consists of several types of soil with different shear strength parameters, and also when the pore water pressure variation in the slope is known or can be estimated. The factor of safety against slope failure is determined by assuming the critical failure surface to be the *arc of a circle*. In order to explain this method, let us refer to Fig. 5.30. Figure 5.30a shows a slope *ABC*. The circular arc *AB'D* with its center at *O* is a trial failure surface. The soil mass *ABCDB'A* which is likely to fail is divided into several slices. These slices are not necessarily of equal width. Now, let us consider the forces per unit length of the slope (at right angles to the cross section) acting on the *n*th slice which is shown in Fig. 5.30b. Note that, in this figure

W_n = *total* weight of the slice

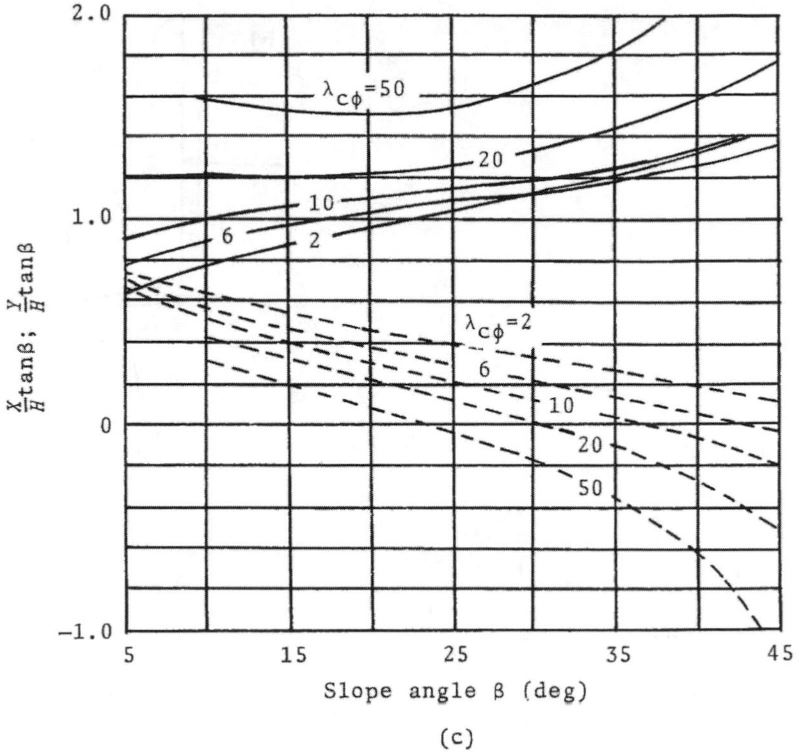

Figure 5.28. (Continued).

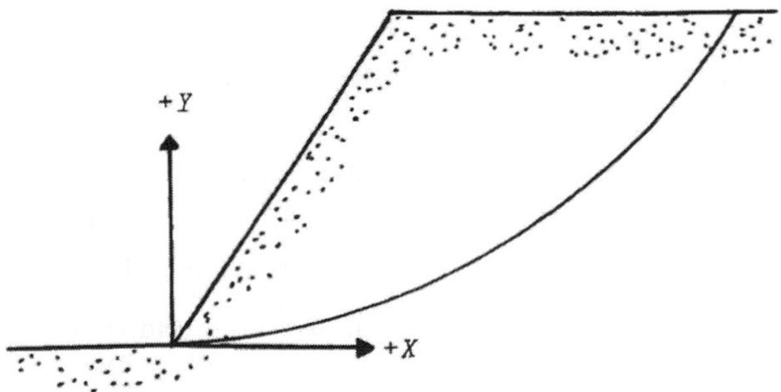

Figure 5.29. Definition of X and Y for use in Figs. 5.27 and 5.28 for location of critical failure surface.

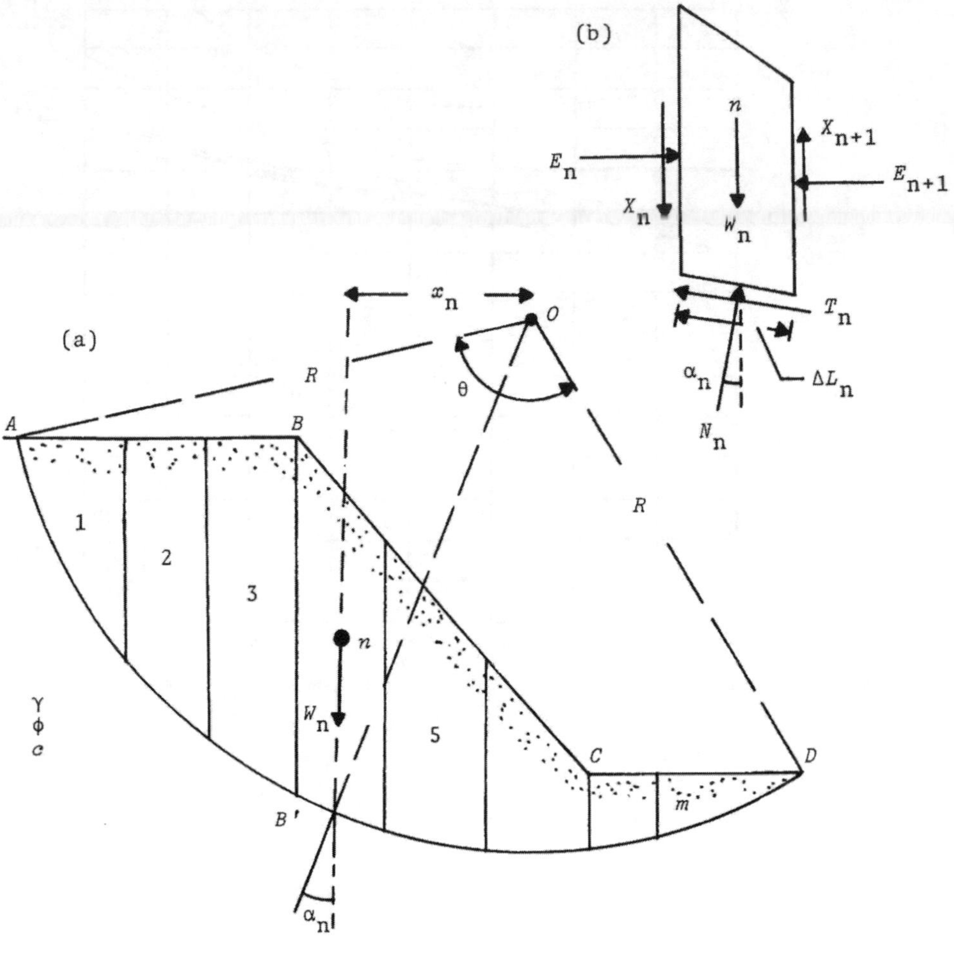

Figure 5.30. Ordinary method of slices.

E_n, E_{n+1} = total horizontal forces along the sides of the slices

X_n, X_{n+1} = vertical shear forces along the sides of the slices

N_n = *total* normal force at the base of the slice

T_n shear force at the base of the slice

The shear force T_n can also be given by the equation

$$T_n = s_m \Delta L_n = \left(\frac{c + \sigma_n' \tan\phi}{F_s} \right) \Delta L_n \tag{5.58}$$

where s_m = mobilized shear strength

ΔL_n = length at the base of the nth slice
c, ϕ = effective shear strength parameters of the soil
F_s = factor of safety
σ'_n effective normal stress at the base of the slice

Note that

$$\sigma'_n = \left(\frac{N_n}{\Delta L_n}\right) - u_n \tag{5.59}$$

where u_n = pore water pressure at the base of the slice

If the forces E_n, E_{n+1}, X_n, and X_{n+1} are neglected, for equilibrium

$$\Sigma \text{ Moment of all forces about } O \quad 0$$

or

$$W_n x_n \quad T_n R \tag{5.60}$$

where R radius of the circular arc

Combining Eqs. (5.58), (5.59), and (5.60)

$$W_n x_n = \left(\frac{c+\sigma'_n \tan\phi}{F_s}\right)\Delta L_n R = \frac{\{c+[(N_n/\Delta L_n)-u_n]\tan\phi\}}{F_s}(\Delta L_n R) \tag{5.61}$$

However, when E_n, E_{n+1}, X_n, and X_{n+1} are neglected

$$N = W_n \cos\alpha_n$$

So

$$F_s \quad R\left[\frac{c\Delta L_n + (W_n \cos\alpha_n - u_n \Delta L_n)\tan\phi}{W_n x_n}\right] \tag{5.62}$$

However, it can be seen from Fig. 5.30a that

$$x_n \quad R\sin\alpha_n$$

Combining the preceding two equations

$$F_s = \frac{c\Delta L_n + (W_n \cos\alpha_n - u_n \Delta L_n)\tan\phi}{W_n \cos\alpha_n} \tag{5.63}$$

The factor of safety for the entire sliding wedge can thus be given as

$$F_s = \sum_{n=1}^{m} \frac{c\Delta L_n + (W_n \cos\alpha_n - u_n \Delta L_n)\tan\phi}{W_n \cos\alpha_n} \tag{5.64}$$

Note that $\Sigma c\Delta L_n = cR\theta$. The minimum factor of safety against sliding for a slope can be obtained by using Eq. (5.64) for several trial failure surfaces and drawing contours of equal factors of safety.

5.8.2 Bishop's Simplified Method of Slices

The value of F_s as obtained by using Eq. (5.64) is referred to as the conventional, or ordinary, method of slices. Bishop (1955) has shown that the error introduced by neglecting the forces E_n, E_{n+1}, X_n, and X_{n+1} is about 15%. The accuracy may be improved in the following manner.

When the forces E_n, E_{n+1}, X_n, and X_{n+1} are not neglected, the summation of the vertical forces on the slice as shown in Fig. 5.30b can be given as

$$N_n\cos\alpha_n + T_n\sin\alpha_n = W_n + (X_n - X_{n+1}) \tag{5.65}$$

However

$$N_n = N_n' + u_n\Delta L_n$$

where N_n' effective normal forces at the base of the slice

Combining the preceding two equations

$$N_n'\cos\alpha_n \quad [W_n + (X_n - X_{n+1}) - T_n\sin\alpha_n - u_n\Delta L_n\cos\alpha_n] \tag{5.66}$$

Also, the tangential force in Eq. (5.66) may be given as

$$T_n \quad \frac{c\Delta L_n + N_n'\tan\phi}{F_s} \tag{5.67}$$

Combining Eqs. (5.66) and (5.67) yields

$$N_n' = \frac{[W_n + (X_n - X_{n+1}) - \frac{c}{F_s}\Delta L_n\sin\alpha_n - u_n\Delta L_n\cos\alpha_n]}{\cos\alpha_n + \frac{\tan\phi\sin\alpha_n}{F_s}} \tag{5.68}$$

Referring to Eq. (5.62)

$$F_s = \frac{R[c\Delta L_n + (N_n' - u_n\Delta L_n)\tan\phi]}{W_n x_n} = \frac{R[c\Delta L_n + N_n'\tan\phi]}{W_n R\sin\alpha_n}$$

$$= \frac{c\Delta L_n + N_n'\tan\phi}{W_n\sin\alpha_n}$$

Substitution of the relations for N_n' given in Eq. (5.68) into Eq. (5.69) yields

$$F_s = \frac{cb_n + (W_n - u_n b_n)\tan\phi + (X_n - X_{n+1})\tan\phi}{m_{\alpha(n)} W_n \sin\alpha_n} \tag{5.70}$$

where b_n = width of the nth slice $\Delta L_n \cos\alpha_n$

$$m_{\alpha(n)} = \cos\alpha_n + \frac{\tan\phi \sin\alpha_n}{F_s} \tag{5.71}$$

Bishop (1955) has shown that, if $X_n - X_{n+1}$ is neglected, Eq. (5.70) will give about 1% error. Hence, neglecting the $X_n - X_{n+1}$ term, the factor of safety for all slices combined can be given as

$$F_s = \sum_{n=1}^{n=m} \frac{cb_n + (W_n - u_n b_n)\tan\phi}{m_{\alpha(n)} W_n \sin\alpha_n} \tag{5.72}$$

Determination of the factor of safety by using Eq. (5.72) is generally referred to as the *Bishop's Simplified Method of Slices*. In using this procedure, the following general guidelines should be used:

1. Select a trial failure surface.
2. Divide the soil mass above the trial failure surface into several slices.
3. Determine b_n, α_n, W_n, and u_n for each slice.
4. Assume a factor of safety F_s to calculate $m_{\alpha(n)}$ [Eq. (5.71)].
5. Use Eq. (5.72) to calculate F_s.
6. If the F_s calculated in Step 5 is not the same as that assumed in Step 4, assume another F_s and repeat Steps 4 and 5 until $F_{s(\text{assumed})}$ is equal to $F_{s(\text{calculated})}$.
6. Repeat Steps 1 through 6 for other trial surfaces from which the minimum F_s is determined.

5.8.3 Bishop and Morgenstern's Method of Slices

Bishop and Morgenstern (1960) used the equation for obtaining the factor of safety of slopes (F_s) by Bishop's Simplified Method [Eq. (5.72)] to provide a number of useful charts in a nondimensional form. In order to understand these charts, we refer to Fig. 5.31 for the definitions of the parameters. Note that the depth factor D and the pore water pressure ratio r_u is defined in a similar manner as in Section 5.7.2.

According to Eq. (5.72)

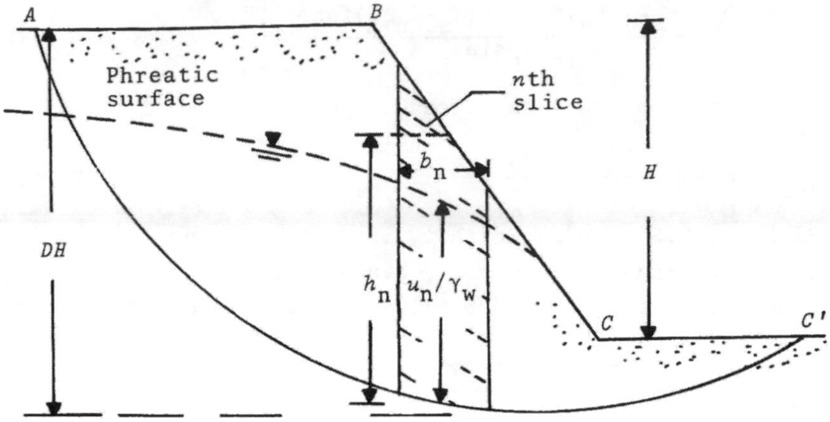

Figure 5.31. Bishop and Morgenstern's method of slices.

$$F_s \quad \sum_{n=1}^{n=m} \frac{cb_n + (W_n - u_n b_n)\tan\phi}{m_{\alpha(n)} W_n \sin\alpha_n} \tag{5.73}$$

However

$$W_n = \gamma b_n h_n \tag{5.74}$$

and

$$r_u = \frac{u_n}{\gamma h_n} \tag{5.75}$$

Substituting Eqs. (5.74) and (5.75) into Eq. (5.73), the factor of safety can be given by the relation

$$F_s = \left[\frac{1}{\sum_{n=1}^{n=m} \frac{b_n}{H}\frac{h_n}{H}\sin\alpha_n}\right] \sum_{n=1}^{n=m} \left\{\left[\frac{c}{\gamma H}\frac{b_n}{H} + \frac{b_n}{H}\frac{h_n}{H}(1-r_u)\tan\phi\right]\frac{1}{m_{\alpha(n)}}\right\} \tag{5.76}$$

It can be seen from the preceding equation that, for given values of $c/\gamma H$, ϕ, and r_u, the value of F_s depends on the geometry of the sliding body if the pore water pressure is *assumed constant throughout*. For steady state seepage condition, a *weighted average* value of r_u may be taken which may range up to a maximum of about 0.5. The averaging technique will at best overestimate the factor of safety by about 7%.

The factor of safety F_s can then be given by an expression

$$F_s = m' - n' r_u \quad (5.77)$$

where m', n' = stability coefficients

The values of m' and n' for various values of $c/\gamma H$, D, ϕ, and slope angles (β) are shown in Table 5.4.

Table 5.4. Values of m' and n' [Eq. (5.77)]

A. Stability coefficients m' and n' for $c/\gamma H$=0

	Stability coefficients for earth slopes							
	Slope 2:1		Slope 3:1		Slope 4:1		Slope 5:1	
ϕ	m'	n'	m'	n'	m'	n'	m'	n'
10.0	0.353	0.441	0.529	0.588	0.705	0.749	0.882	0.917
12.5	0.443	0.554	0.665	0.739	0.887	0.943	1.109	1.153
15.0	0.536	0.670	0.804	0.893	1.072	1.139	1.340	1.393
17.5	0.631	0.789	0.946	1.051	1.261	1.340	1.577	1.639
20.0	0.728	0.910	1.092	1.213	1.456	1.547	1.820	1.892
22.5	0.828	1.035	1.243	1.381	1.657	1.761	2.071	2.153
25.0	0.933	1.166	1.399	1.554	1.865	1.982	2.332	2.424
27.5	1.041	1.301	1.562	1.736	2.082	2.213	2.603	2.706
30.0	1.155	1.444	1.732	1.924	2.309	2.454	2.887	3.001
32.5	1.274	1.593	1.911	2.123	2.548	2.708	3.185	3.311
35.0	1.400	1.750	2.101	2.334	2.801	2.977	3.501	3.639
37.5	1.535	1.919	2.302	2.558	3.069	3.261	3.837	3.989
40.0	1.678	2.098	2.517	2.797	3.356	3.566	4.196	4.362

B. Stability coefficients m' and n' for $c/\gamma H$=0.025 and D=1.00

	Slope 2:1		Slope 3:1		Slope 4:1		Slope 5:1	
ϕ	m'	n'	m'	n'	m'	n'	m'	n'
10.0	0.678	0.534	0.906	0.683	1.130	0.846	1.365	1.031
12.5	0.790	0.655	1.066	0.849	1.337	1.061	1.620	1.282
15.0	0.901	0.776	1.224	1.014	1.544	1.273	1.868	1.534
17.5	1.012	0.898	1.380	1.179	1.751	1.485	2.121	1.789
20.0	1.124	1.022	1.542	1.347	1.962	1.698	2.380	2.050
22.5	1.239	1.150	1.705	1.518	2.177	1.916	2.646	2.317
25.0	1.356	1.282	1.875	1.696	2.400	2.141	2.921	2.596
27.5	1.478	1.421	2.050	1.882	2.631	2.375	3.207	2.886
30.0	1.606	1.567	2.235	2.078	2.873	2.622	3.508	3.191
32.5	1.739	1.721	2.431	2.285	3.127	2.883	3.823	3.511
35.0	1.880	1.885	2.635	2.505	3.396	3.160	4.156	3.849
37.5	2.030	2.060	2.855	2.741	3.681	3.458	4.510	4.209
40.0	2.190	2.247	3.090	2.993	3.984	3.778	4.885	4.592

Table 5.4. (Continued).

C. Stability coefficients m' and n' for $c/\gamma H$=0.025 and D=1.25

ϕ	Slope 2:1		Slope 3:1		Slope 4:1		Slope 5:1	
	m'	n'	m'	n'	m'	n'	m'	n'
10.0	0.737	0.614	0.901	0.726	1.085	0.867	1.285	1.014
12.5	0.878	0.759	1.076	0.908	1.299	1.089	1.543	1.278
15.0	1.019	0.907	1.253	1.093	1.515	1.311	1.803	1.545
17.5	1.162	1.059	1.433	1.282	1.736	1.541	2.065	1.814
20.0	1.309	1.216	1.618	1.478	1.961	1.775	2.334	2.090
22.5	1.461	1.379	1.808	1.680	2.194	2.017	2.610	2.373
25.0	1.619	1.547	2.007	1.891	2.437	2.269	2.897	2.669
27.5	1.783	1.728	2.213	2.111	2.689	2.531	3.196	2.976
30.0	1.956	1.915	2.431	2.342	2.953	2.806	3.511	3.299
32.5	2.139	2.112	2.659	2.686	3.231	3.095	3.841	3.638
35.0	2.331	2.321	2.901	2.841	3.524	3.400	4.191	3.998
37.5	2.536	2.541	3.158	3.112	3.835	3.723	4.563	4.379
40.0	2.753	2.775	3.431	3.399	4.164	4.064	4.958	4.784

D. Stability coefficients m' and n' for $c/\gamma H$=0.05 and D=1.00

ϕ	Slope 2:1		Slope 3:1		Slope 4:1		Slope 5:1	
	m'	n'	m'	n'	m'	n'	m'	n'
10.0	0.913	0.563	1.181	0.717	1.469	0.910	1.733	1.069
12.5	1.030	0.690	1.343	0.878	1.688	1.136	1.995	1.316
15.0	1.145	0.816	1.506	1.043	1.904	1.353	2.256	1.567
17.5	1.262	0.942	1.671	1.212	2.117	1.565	2.517	1.825
20.0	1.380	1.071	1.840	1.387	2.333	1.776	2.783	2.091
22.5	1.500	1.202	2.014	1.568	2.551	1.989	3.055	2.365
25.0	1.624	1.338	2.193	1.757	2.778	2.211	3.336	2.651
27.5	1.753	1.480	1.380	1.952	3.013	2.444	3.628	2.948
30.0	1.888	1.630	2.574	2.157	3.261	2.693	3.934	3.259
32.5	2.029	1.789	2.777	2.370	3.523	2.961	4.256	3.585
35.0	2.178	1.958	2.990	2.592	3.803	3.253	4.597	3.927
37.5	2.336	2.138	3.215	2.826	4.103	3.574	4.959	4.288
40.0	2.505	2.332	3.451	3.071	4.425	3.926	5.344	4.668

Bishop and Morgenstern (1960) published the factor of safety tables for various slopes for $c/\gamma H$=0, 0.025, and 0.05. O'Connor and Mitchell (1977) have given similar tables for $c/\gamma H$=0.075 and 0.1.

In order to obtain the minimum value of F_s for sections *not located directly on a hard stratum*, the following procedure may be used.

Table 5.4. (Continued).

E. Stability coefficients m' and n' for $c/\gamma H$=0.05 and D=1.25

ϕ	Slope 2:1		Slope 3:1		Slope 4:1		Slope 5:1	
	m'	n'	m'	n'	m'	n'	m'	n'
10.0	0.919	0.633	1.119	0.766	1.344	0.886	1.594	1.042
12.5	1.065	0.792	1.294	0.941	1.563	1.112	1.850	1.300
15.0	1.211	0.950	1.471	1.119	1.782	1.338	2.109	1.562
17.5	1.359	1.108	1.650	1.303	2.004	1.567	2.373	1.831
20.0	1.509	1.266	1.834	1.493	2.230	1.799	2 643	2.107
22.5	1.663	1.428	2.024	1.690	2.463	2.038	2.921	2.392
25.0	1.822	1.595	2.222	1.897	2.705	2.287	3.211	2.690
27.5	1.988	1.769	2.428	2.113	2.957	2.546	3.513	2.999
30.0	2.161	1.950	2.645	2.342	3.221	2.819	3.829	3.324
32.5	2.343	2.141	2.873	2.583	3.500	3.107	4.161	3.665
35.0	2.535	2.344	3.114	2.839	3.795	3.413	4.511	4.025
37.5	2.738	2.560	3.370	3.111	4.109	3.740	4.881	4.405
40.0	2.953	2.791	3.642	3.400	4.442	4.090	5.273	4.806

F. Stability coefficients m' and n' for $c/\gamma H$=0.05 and D=1.50

ϕ	Slope 2:1		Slope 3:1		Slope 4:1		Slope 5:1	
	m'	n'	m'	n'	m'	n'	m'	n'
10.0	1.022	0.751	1.170	0.828	1.343	0.974	1.547	1.108
12.5	1.202	0.936	1.376	1.043	1.589	1.227	1.829	1.399
15.0	1.383	1.122	1.583	1.260	1.835	1.480	2.112	1.690
17.5	1.565	1.309	1.795	1.480	2.084	1.734	2.398	1.983
20.0	1.752	1.501	2.011	1.705	2.337	1.993	2.690	2.280
22.5	1.943	1.698	2.234	1.937	2.597	2.258	2.990	2.585
25.0	2.143	1.903	2.467	2.179	2.867	2.534	3.302	2.902
27.5	2.350	2.117	2.709	2.431	3.148	2.820	3.626	3.231
30.0	2.568	2.342	2.964	2.696	3.443	3.120	3.967	3.577
32.5	2.798	2.580	3.232	2.975	3.753	3.436	4.326	3.940
35.0	3.041	2.832	3.515	3.269	4.082	3.771	4.707	4.325
37.5	3.299	3.102	3.817	3.583	4.431	4.128	5.112	4.735
40.0	3.574	3.389	4.136	3.915	4.803	4.507	5.543	5.171

1. Determine the slope angle β, ϕ, and $c/\gamma H$ for the slope.

2. Estimate r_u.

3. Enter charts for D=1, 1.25, and 1.5 (and also for toe circle if given) for proper values of $c/\gamma H$ and obtain m' and n'.

4. Calculate the values of F_s using Eq. (5.77) and the values of m' and n' determined in Step 3 for various depth factors.

Table 5.4. (Continued).

G. Stability coefficients m' and n' for $c/\gamma H$=0.075 and toe circles

ϕ	Slope 2:1		Slope 3:1		Slope 4:1		Slope 5:1	
	m'	n'	m'	n'	m'	n'	m'	n'
20	1.593	1.158	2.055	1.516	2.498	1.903	2.934	2.301
25	1.853	1.430	2.426	1.888	2.980	2.361	3.520	2.861
30	2.133	1.730	2.826	2.288	3.496	2.888	4.150	3.461
35	2.433	2.058	3.253	2.730	4.055	3.445	4.846	4.159
40	2.773	2.430	3.737	3.231	4.680	4.061	5.609	4.918

H. Stability coefficients $m'$and n' for $c/\gamma H$=0.075 and D=1.00

ϕ	Slope 2:1		Slope 3:1		Slope 4:1		Slope 5:1	
	$m'm'$	n'	m'	n'	m'	n'	m'	n'
20	1.610	1.100	2.141	1.443	2.664	1.801	3.173	2.130
25	1.872	1.386	2.502	1.815	3.126	2.259	3.742	2.715
30	2.142	1.686	2.884	2.201	3.623	2.758	4.357	3.331
35	2.443	2.030	3.306	2.659	4.177	3.331	5.024	4.001
40	2.772	2.386	3.775	3.145	4.785	3.945	5.776	4.759

I. Stability coefficients m' and n' for $c/\gamma H$=0.075 and D=1.25

ϕ	Slope 2:1		Slope 3:1		Slope 4:1		Slope 5:1	
	m'	n'	m'	n'	m'	n'	m'	n'
20	1.688	1.285	2.071	1.543	2.492	1.815	2.954	2.173
25	2.004	1.641	2.469	1.957	2.972	2.315	3.523	2.730
30	2.352	2.015	2.888	2.385	3.499	2.857	4.149	3.357
35	2.728	2.385	3.357	2.870	4.079	3.457	4.831	4.043
40	3.154	2.841	3.889	3.428	4.729	4.128	5.603	4.830

J. Stability coefficients m' and n' for $c/\gamma H$=0.075 and D=1.50

ϕ	Slope 2:1		Slope 3:1		Slope 4:1		Slope 5:1	
	m'	n'	m'	n'	m'	n'	m'	n'
20	1.918	1.514	2.199	1.728	2.548	1.985	2.931	2.272
25	2.308	1.914	2.660	2.200	3.083	2.530	3.552	2.915
30	2.735	2.355	3.158	2.714	3.659	3.128	4.128	3.585
35	3.211	2.854	3.708	3.285	4.302	3.786	4.961	4.343
40	3.742	3.397	4.332	3.926	5.026	4.527	5.788	5.185

Table 5.4. (Continued)

K. Stability coefficients m' and n' for $c/\gamma H$=0.100 and toe circles

ϕ	Slope 2:1		Slope 3:1		Slope 4:1		Slope 5:1	
	m'	n'	m'	n'	m'	n'	m'	n'
20	1.804	2.101	2.286	1.588	2.748	1.974	3.190	2.361
25	2.076	1.488	2.665	1.945	3.246	2.459	3.796	2.959
30	2.362	1.786	3.076	2.359	3.770	2.961	4.442	3.576
35	2.673	2.130	3.518	2.803	4.339	3.518	5.146	4.249
40	3.012	2.486	4.008	3.303	4.984	4.173	5.923	5.019

L. Stability coefficients m' and n' for $c/\gamma H$=0.100 and D=1.00

ϕ	Slope 2:1		Slope 3:1		Slope 4:1		Slope 5:1	
	m'	n'	m'	n'	m'	n'	m'	n'
20	1.841	1.143	2.421	1.472	2.982	1.815	3.549	2.157
25	2.102	1.430	2.785	1.845	3.458	2.303	4.131	2.743
30	2.378	1.714	3.183	2.258	3.973	2.830	4.751	3.372
35	2.692	2.086	3.612	2.715	4.516	3.359	5.426	4.059
40	3.025	2.445	4.103	3.230	5.144	4.001	6.187	4.831

M. Stability coefficients m' and n' for $c/\gamma H$=0.100 and D=1.25

ϕ	Slope 2:1		Slope 3:1		Slope 4:1		Slope 5:1	
	m'	n'	m'	n'	m'	n'	m'	n'
20	1.874	1.301	2.283	1.558	2.751	1.843	3.253	2.158
25	2.197	1.642	2.681	1.972	3.233	2.330	3.833	2.758
30	2.540	2.000	3.112	2.415	3.753	2.858	4.451	3.372
35	2.922	2.415	3.588	2.914	4.333	3.458	5.141	4.072
40	3.345	2.855	4.119	3.457	4.987	4.142	5.921	4.872

N. Stability coefficients m' and n' for $c/\gamma H$=0.100 and D=1.50

ϕ	Slope 2:1		Slope 3:1		Slope 4:1		Slope 5:1	
	m'	n'	m'	n'	m'	n'	m'	n'
20	2.079	1.528	2.387	1.742	2.768	2.014	3.158	2.285
25	2.477	1.942	2.852	2.215	3.297	2.542	3.796	2.927
30	2.908	2.385	3.349	2.728	3.881	3.143	4.468	3.614
35	3.385	2.884	3.900	3.300	4.520	3.800	5.211	4.372
40	3.924	3.441	4.524	3.941	5.247	4.542	6.040	5.200

Note: Tables A through F after Bishop and Morgenstern (1960) and Tables G through N after O'Connor and Mitchell (1977)

5. The smallest value of F_s obtained in Step 4 is the critical factor of safety.

Example 5.8. A slope is 46.5 ft in height. Given slope 2H:1V, ϕ= 20°, c=400 lb/ft², γ=115 lb/ft³, and r_u=0.28. Determine the factor of safety F_s by using Table 5.4.

Solution

$$\frac{c}{\gamma H} = \frac{400}{(115)(46.5)} = 0.075$$

Referring to Tables 5.4G, H, I, and J, the following table may be prepared.

Toe circle or D	m'	n'	$F_s=m'-n'r_u$
Toe circle	1.593	1.159	1.268
1.0	1.610	1.100	1.302
1.25	1.688	1.285	1.328
1.5	1.918	1.514	1.494

So

F_s $\underline{1.268}$

5.8.4 Morgenstern's Method of Slices for Rapid Drawdown Condition

Morgenstern (1963) used the Bishop and Morgenstern method of slices (1960) to determine the factor of safety F_s during rapid drawdown. The factor of safety charts given by Morgenstern (1963) are shown in Figs. 5.32, 5.33, and 5.34. In these graphs, the following notations have been used

a. L=height of drawdown

b. H=height of the embankment

c. β=slope angle

In preparing Figs. 5.32, 5.33, and 5.34, it has been assumed that the embankment is made of homogeneous material and rests on an impervious base. It is also assumed that initially the water level coincides with the top of the embankment, and during the drawdown period dissipation of pore water pressure does not occur (Fig. 5.35). Also it is assumed that the unit weight of soil (saturated) $\gamma=2\gamma_w$ (γ_w=unit weight of water). Morgenstern's charts cover a range of $c/\gamma H$=0.0125 to 0.05.

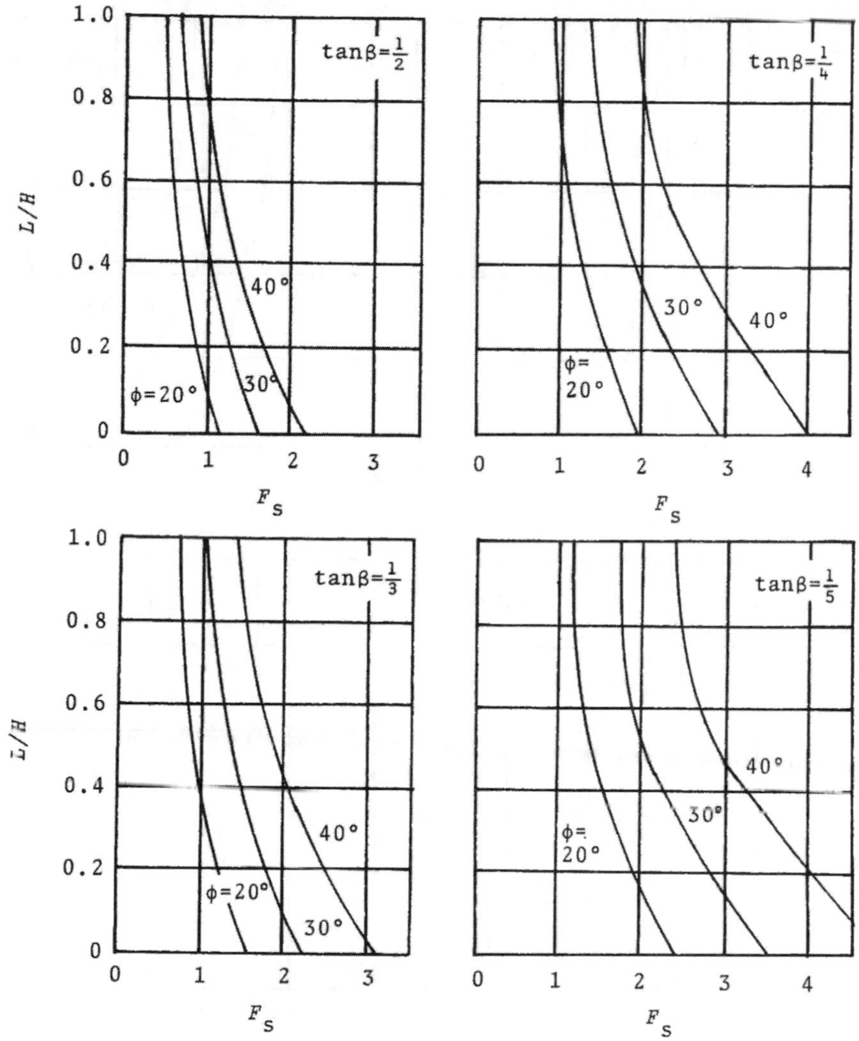

Figure 5.32. Drawdown stability chart for $c/\gamma H$=0.0125 (after Morgenstern, 1963).

5.8.5 Spencer's Method of Slices

Bishop's simplified method of slices as described in Section 5.8.2 satisfies the equations of equilibrium with respect to moment, but not with respect to the forces. Spencer (1967) has provided a method to determine the factor of safety (F_s) by taking into account the *interslice forces*, which does satisfy the *equations of equilibrium with respect to moment and forces*. It is also

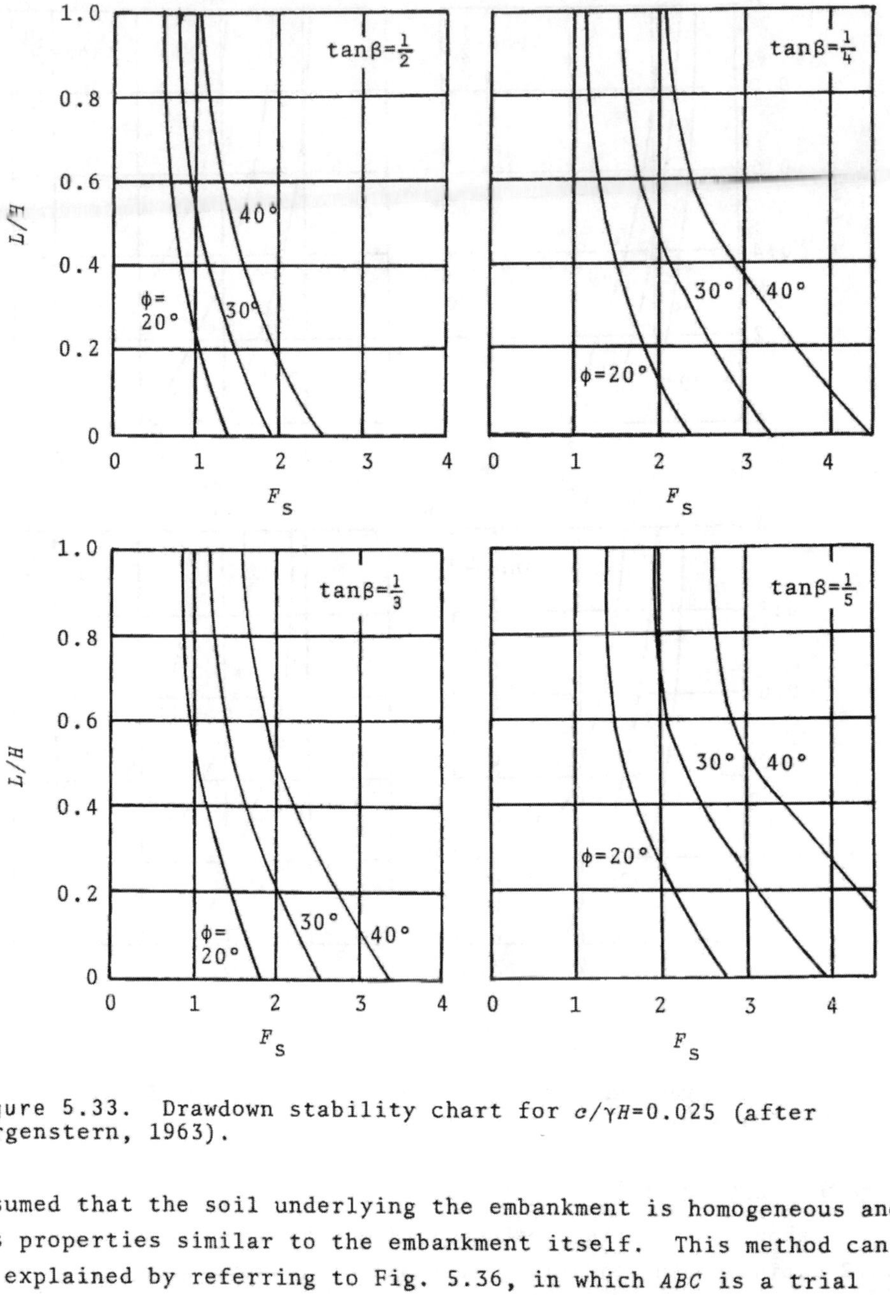

Figure 5.33. Drawdown stability chart for $c/\gamma H$=0.025 (after Morgenstern, 1963).

assumed that the soil underlying the embankment is homogeneous and has properties similar to the embankment itself. This method can be explained by referring to Fig. 5.36, in which *ABC* is a trial failure arc of a circle (Fig. 5.36a). The center of the circle is located at *O*, and the radius of the circle is *R*. Figure 5.36b shows the *n*th slice along with the forces acting on it. Note that the width of the slice is b_n, and it has an average height of h_n.

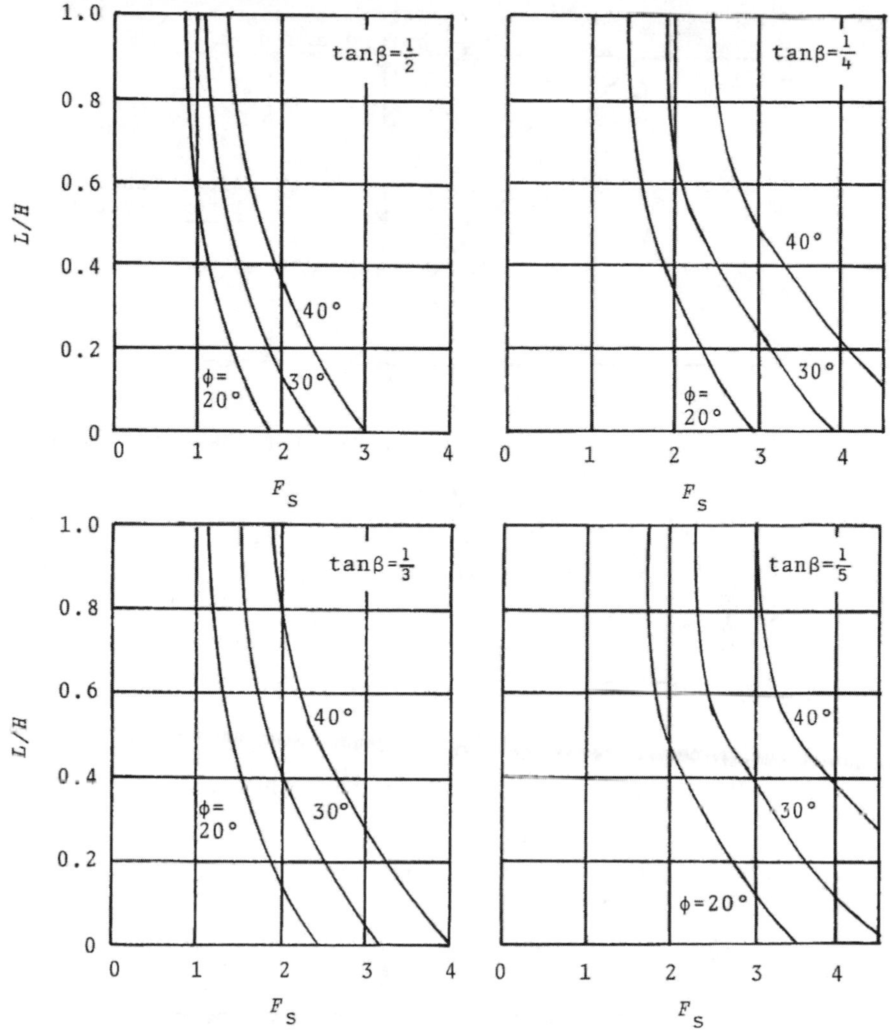

Figure 5.34. Drawdown stability chart for $c/\gamma H$=0.05 (after Morgenstern, 1963).

The forces on the nth slice are

1. Total weight of the slice--W_n
2. Total normal reaction at the base--N_n. The force N_n consists of two parts:
   a. Effective force N'_n
   b. Force due to pore water pressure, the magnitude of which is equal to $u_n b_n \sec\alpha_n$ (u=pore water pressure). Hence

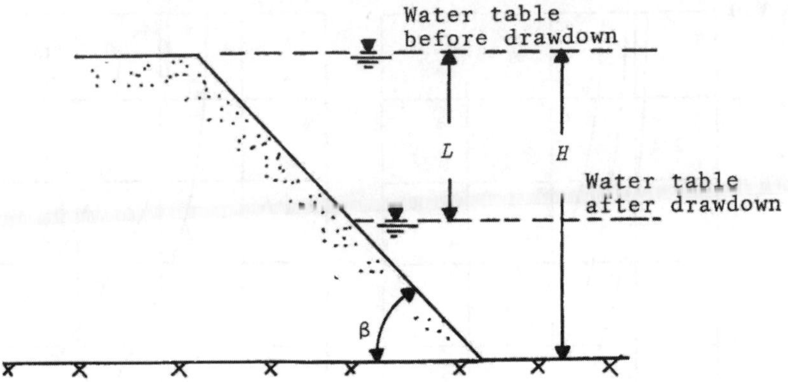

Figure 5.35. Stability analysis for rapid drawdown condition.

$$N_n = N'_n + u_n b_n \sec\alpha_n \tag{5.78}$$

3. The mobilized shear force at the base of the slice--T_n, which can be expressed as

$$T_n = \frac{c b_n \sec\alpha_n + N'_n \tan\phi}{F_s} \tag{5.79}$$

4. The interslice forces--Z_n and Z_{n+1}. Figure 5.36c shows the force polygon for the slice. Note that in Fig. 5.36c, Q_n is the resultant of Z_n and Z_{n+1}.

By resolving the above-mentioned forces in the horizontal and vertical directions

$$Q_n = \frac{\dfrac{c b_n \sec\alpha_n}{F_s} + \dfrac{\tan\phi}{F_s}(W_n \cos\alpha_n - u_n b_n \sec\alpha_n) - W_n \sin\alpha_n}{\cos(\alpha_n - \theta)\left[1 + \left(\dfrac{\tan\phi}{F_s}\right)\tan(\alpha_n - \theta)\right]} \tag{5.80}$$

However

$$W_n = \gamma b_n h_n \tag{5.81}$$

and

$$u_n = r_u \gamma h_n \tag{5.82}$$

where γ = bulk unit weight of soil

r_u = pore pressure ratio

Substituting Eqs. (5.81) and (5.82) into Eq. (5.80), we obtain

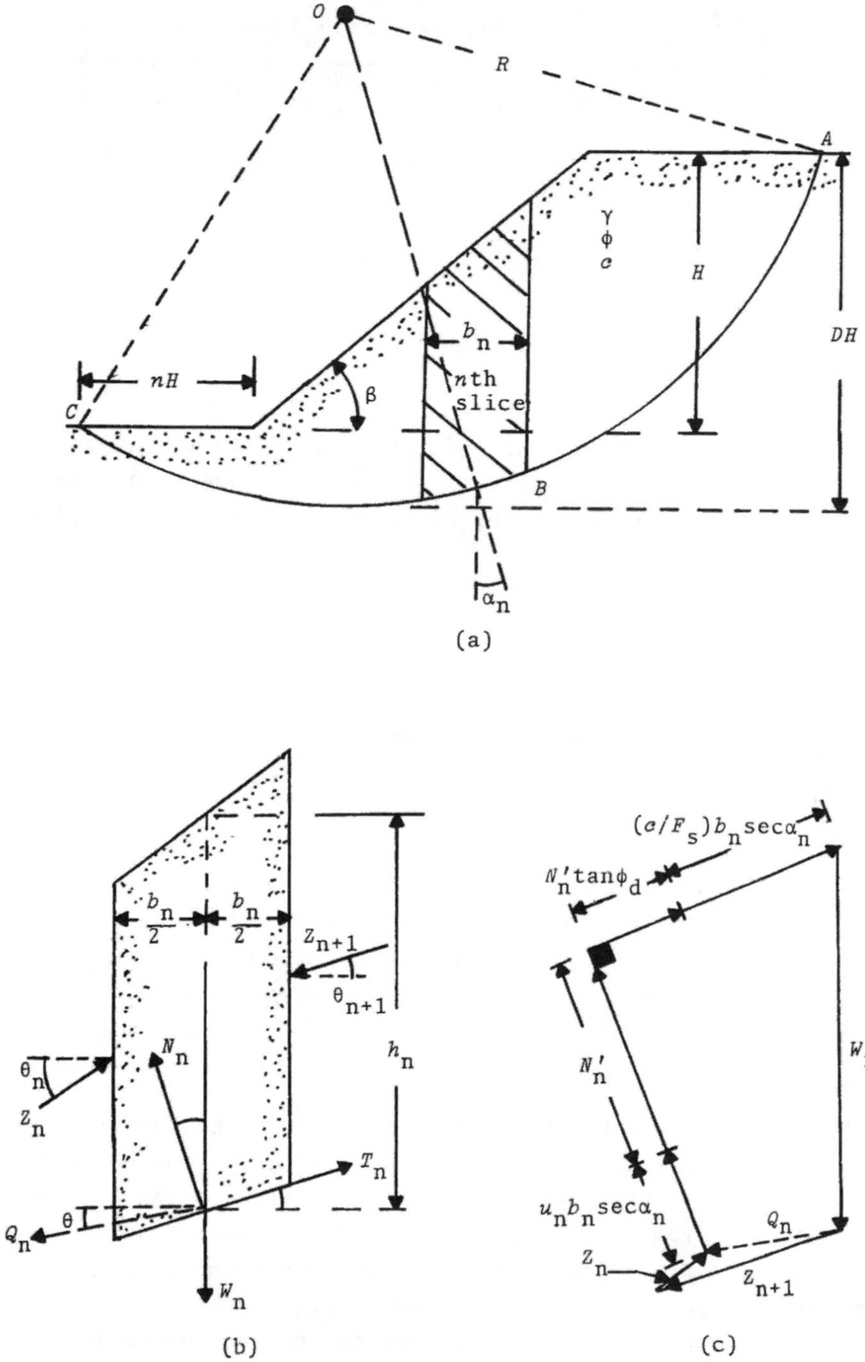

Figure 5.36. Spencer's method of slices.

$$Q_n \quad \gamma H b_n \left\{ \frac{\frac{c}{F_s \gamma H} + \frac{1}{2}\,\frac{h_n}{H}\,\frac{\tan\phi}{F_s}(1-2r_u+\cos 2\alpha_n) - \frac{1}{2}\,\frac{h_n}{H}\sin 2\alpha_n}{\cos\alpha_n \cos(\alpha_n-\theta)\left[1+\left(\frac{\tan\phi}{F_s}\right)\tan(\alpha_n-\theta)\right]} \right\} \qquad (5.83)$$

If the external forces on the embankments are in equilibrium

$$\sum_{n=1}^{n=m} Q_n \cos\theta \quad 0 \qquad (5.84)$$

$$\sum_{n=1}^{n=m} Q_n \sin\theta \quad 0 \qquad (5.85)$$

Also, if the sum of the moments of the external forces about the center of rotation O is zero, the sum of the moments of the interslice forces about the center of rotation must be equal to zero, or

$$\sum_{n=1}^{n=m} Q_n [R\cos(\alpha_n-\theta)] = 0 \qquad (5.86)$$

However, since R is a constant

$$\sum_{n=1}^{n=m} Q_n \cos(\alpha_n-\theta) = 0 \qquad (5.87)$$

Spencer has shown that, if the value of θ is assumed constant for all slices (that is, $n=1, \ldots, n=m$), it does not result in a large error. So, Eqs. (5.84) and (5.85) transform to

$$\sum_{n=1}^{n=m} Q_n \quad 0 \qquad (5.88)$$

Now, several values of θ can be chosen and the values of F_s can be determined from Eqs. (5.87) and (5.88) and plotted in the manner as shown in Fig. 5.37. The point of intersection of the results obtained from Eqs. (5.87) and (5.88) as shown in Fig. 5.37 is the *actual factor of safety* for the trial wedge which satisfies the *moment and force equilibrium equations*.

Several trial wedges of this type can be considered and the factors of safety determined. Based on the contours of equal factors of safety, the minimum value of F_s for the slope can be obtained.

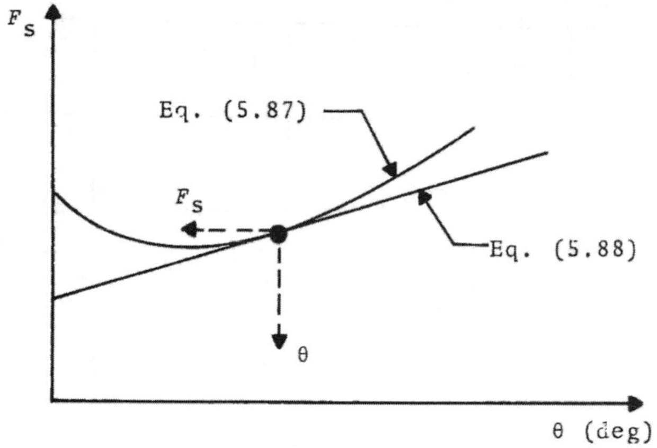

Figure 5.37. Solution for F_s for Eqs. (5.87) and (5.88).

Figure 5.38a, b, and c show the plots of $c/F_s\gamma H$ vs. slope angle β as obtained by Spencer's method of slices. Note that, for

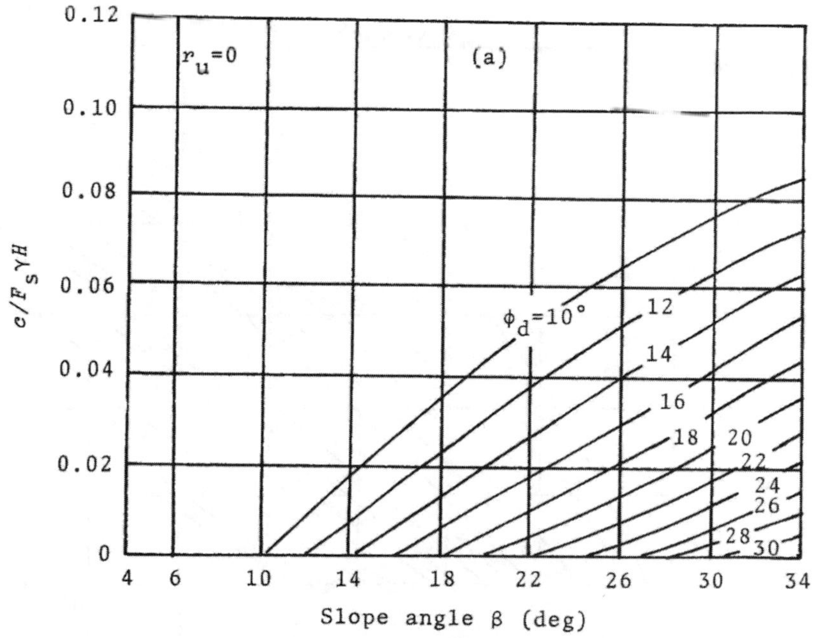

Figure 5.38. Plot of $c/F_s\gamma H$ vs. β for various values of ϕ_d (after Spencer, 1967).

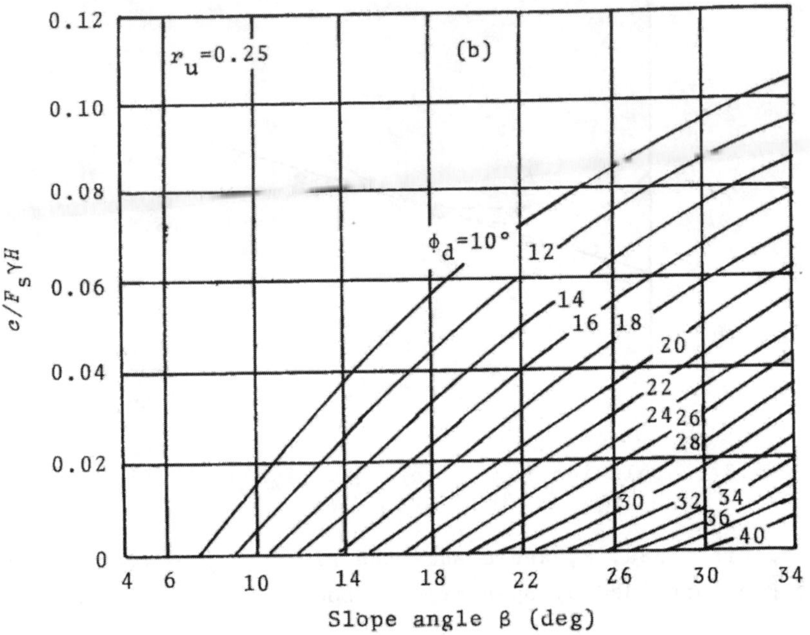

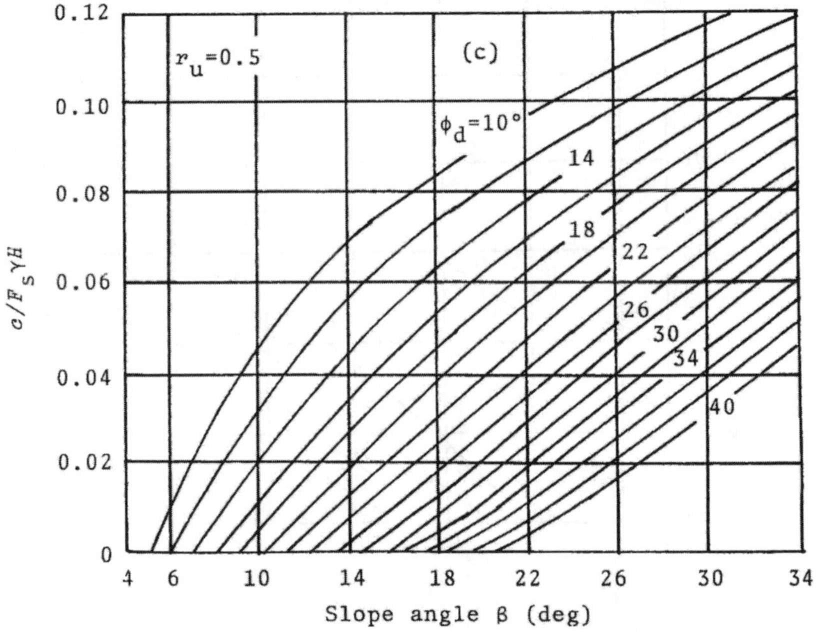

Figure 5.38. (Continued).

these calculations, an *average value of* r_u has been taken (similar to Bishop and Morgenstern's method). In these figure, ϕ_d is the developed angle of friction, or

$$\phi_d = \tan^{-1}\left(\frac{\tan\phi}{F_s}\right)$$

Figures 5.39, 5.40, and 5.41 can be used to locate the critical failure circles for a given slope. The notations for the

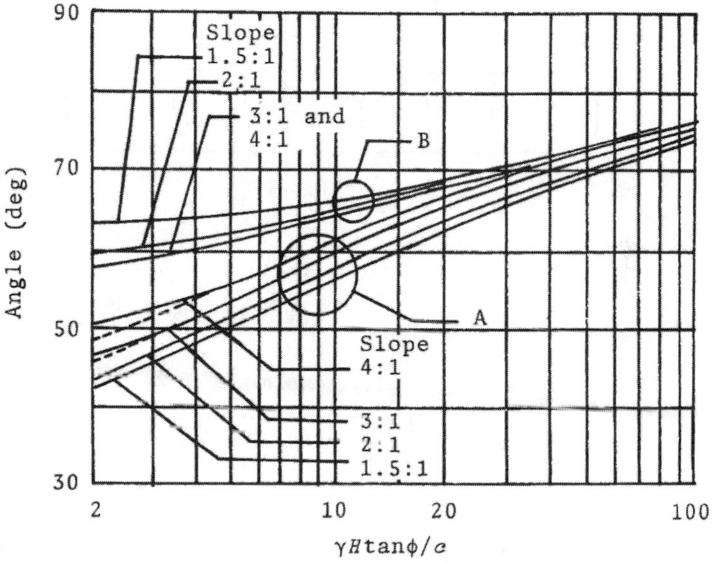

(a)

Figure 5.39. Relationship between angles *A*, *B*, and $(\gamma H\tan\phi)/c$: (a) r_u=0; (b) r_u=0.25; and (c) r_u=0.5 (after Spencer, 1967).

A and *B* and *n* and *D* have been defined in Fig. 5.42. It needs to be pointed out that the broken lines in Figs. 5.39 and 5.41 are for *critical toe circles which are not necessarily the most critical circles*. The solid lines refer to the parameters relating to the most critical circles.

Example 5.9. A slope is shown in Fig. 5.43a. Determine the factor of safety of the slope (F_s). Use Spencer's solution.

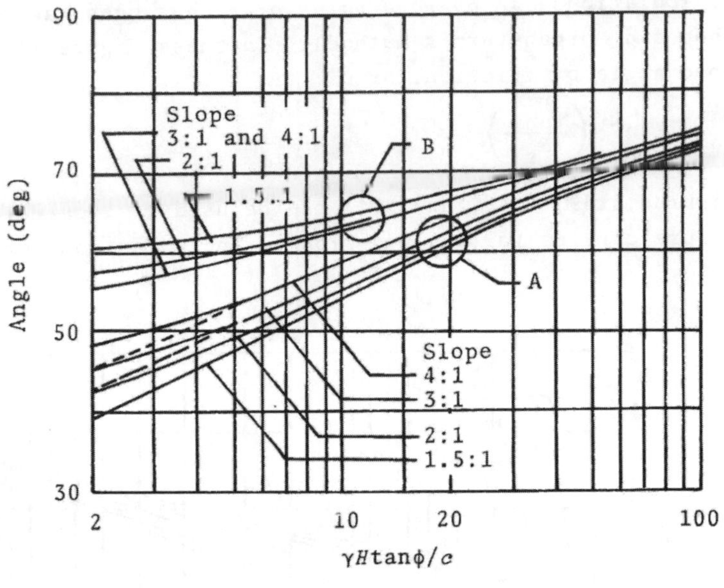

(b)

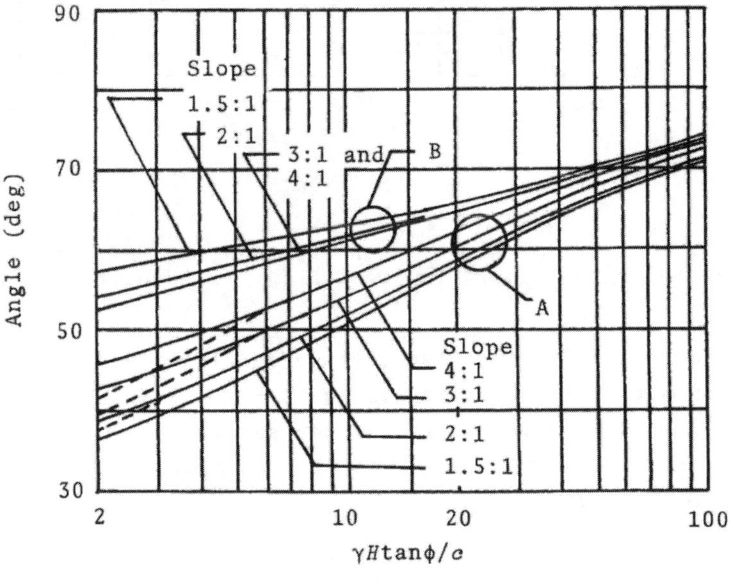

(c)

Figure 5.39. (Continued).

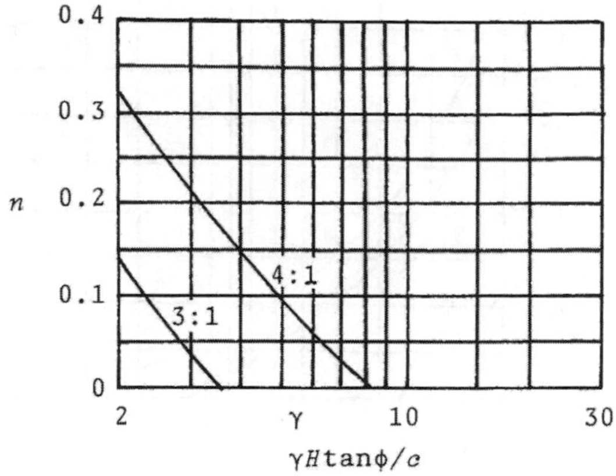

(a)

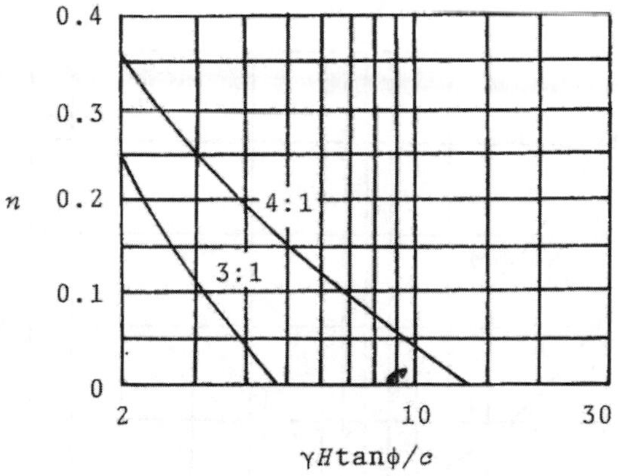

(b)

Figure 5.40. Relationship between n and $(\gamma H \tan\phi)/c$: (a) r_u=0, (b) r_u=0.25, and (c) r_u=0.5 (after Spencer, 1967).

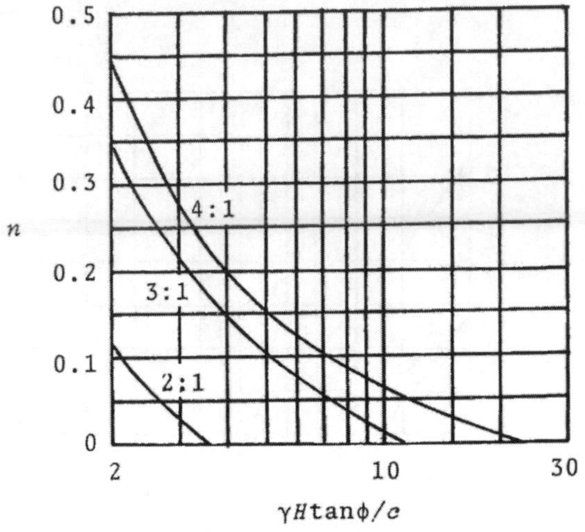

(c)

Figure 5.40. (Continued).

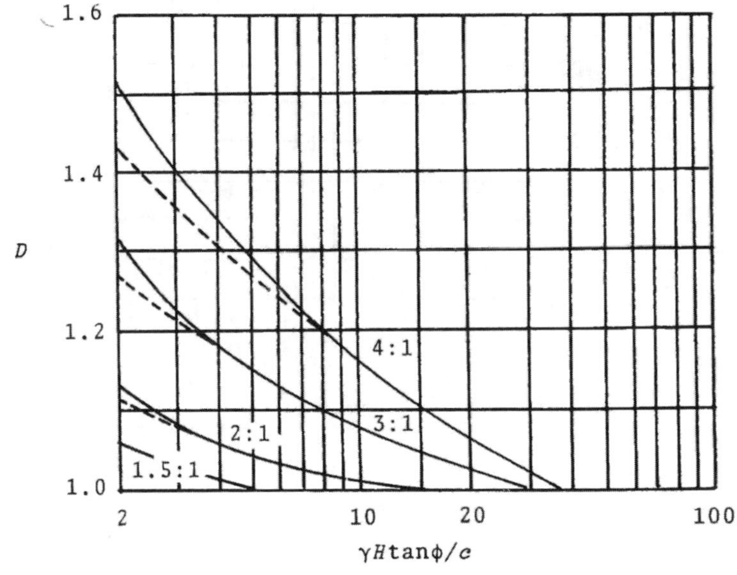

(a)

Figure 5.41. Relationship between parameters D and $(\gamma H \tan\phi)/c$: (a) r_u=0; (b) r_u=0.25, and (c) r_u=0.5 (after Spencer, 1967).

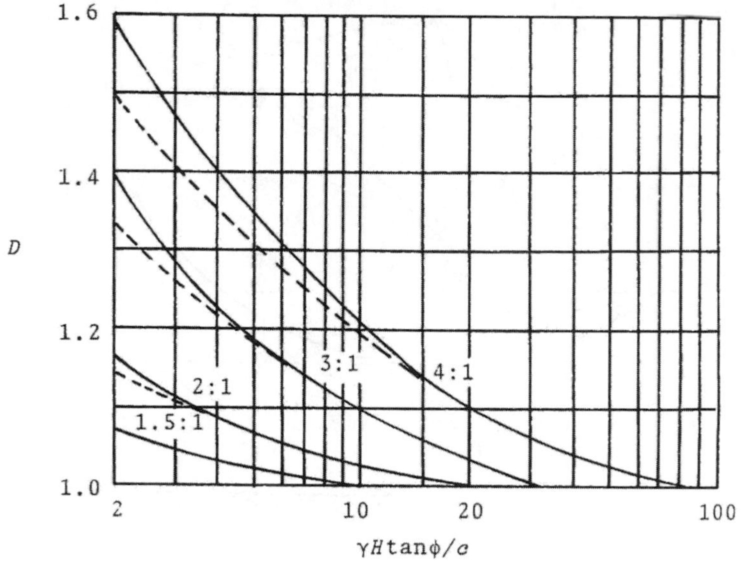

(b)

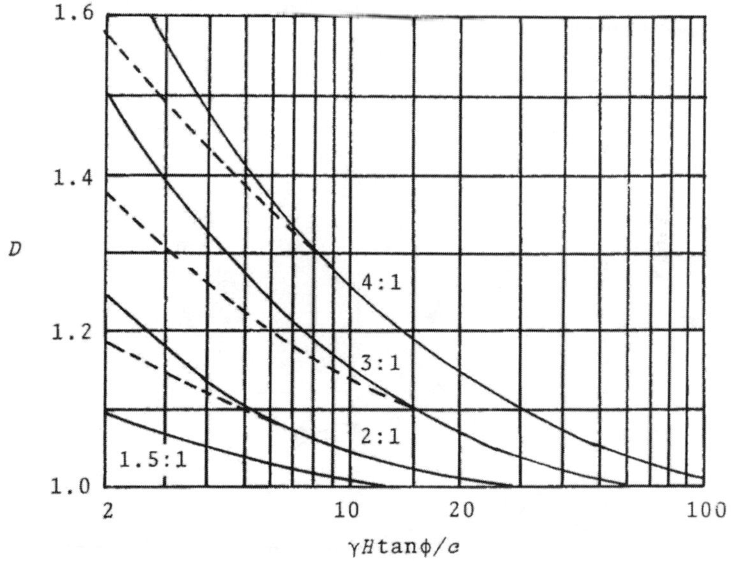

(c)

Figure 5.41. (Continued).

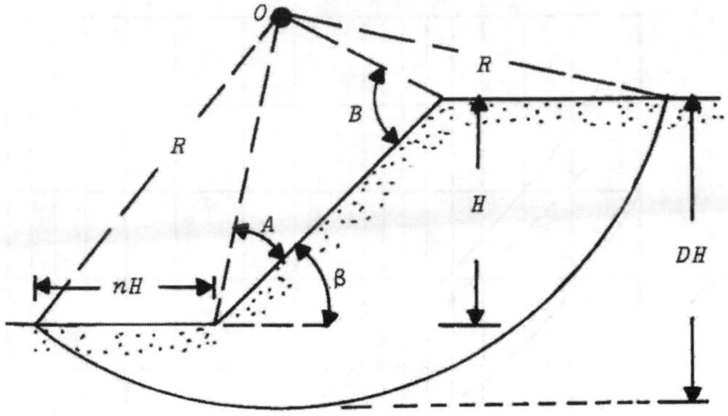

Figure 5.42. Definitions for parameters A, B, n, and D used in Figs. 5.39, 5.40, and 5.41.

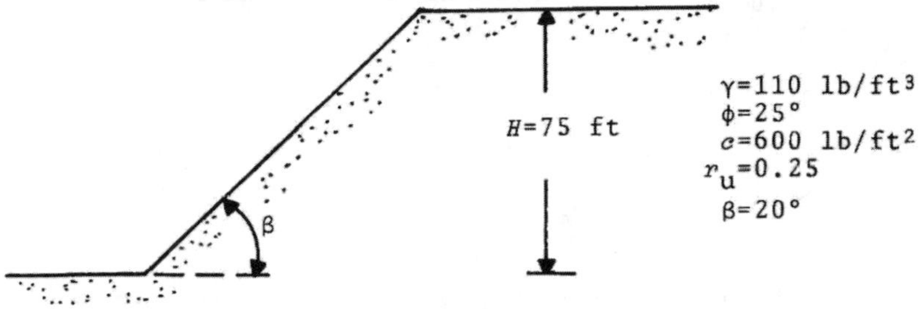

(a)

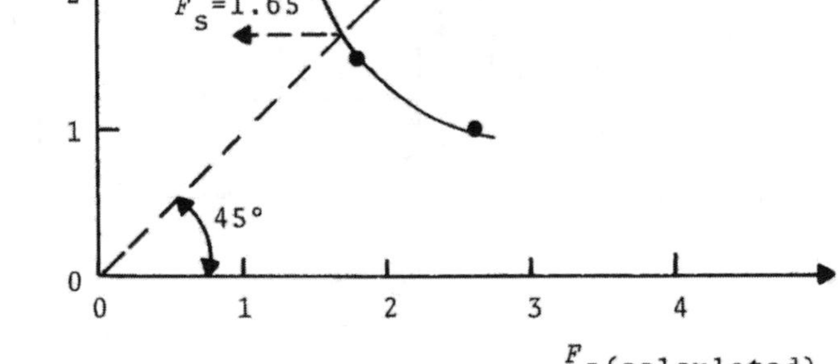

(b)

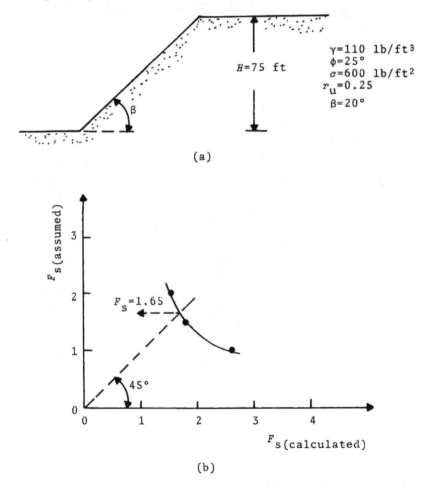

Figure 5.43.

Solution

A table can be prepared with several values of F_s as follows.

$F_{s(\text{assumed})}$	$\frac{c}{F_{s(\text{assumed})}\gamma H}$ [a]	ϕ_d [b]	$F_{s(\text{calculated})}=\frac{\tan\phi}{\tan\phi_d}$
1.0	0.0727	10	2.645
1.5	0.0485	14.5	1.803
2.0	0.0364	17	1.525

[a] c=600 lb/ft^2, γ=110 lb/ft^3, H=75 ft
[b] From Fig. 5.38b (for β=20°)

A graph can now be plotted with $F_{s(\text{assumed})}$ and $F_{s(\text{calculated})}$ (as shown in Fig. 5.43b). From this figure, when $F_s \approx 1.65$, the value of $F_{s(\text{assumed})}$ is equal to $F_{s(\text{calculated})}$. So, for this slope

$$F_s = \underline{1.65}$$

5.9 LIMIT ANALYSIS SOLUTIONS FOR SLOPES

The theories in the preceding sections that have been used to determine the stability of slopes are based on the *limit equilibrium* method. *Limit analysis solutions* (Drucker and Prager, 1952) can also conveniently be used for stability analysis. The upper bound theorem of generalized theory of perfect plasticity has been used by Chen, Giger, and Fang (1969) and Chen and Giger (1971) to analyze the stability of slopes. According to the analysis of Chen and Giger (1971), the failure surface is assumed to be the *arc of a logarithmic spiral* as shown in Fig. 5.44. The logarithmic spiral follows the relationship $r=r_o e^{\theta\tan\phi}$. According to the *upper bound* theorem, the slope would collapse if the external work done by soil exceeds the rate of dissipation of internal energy. The rate of external work done by the region *ABDCA*

$$\dot{w} = \underset{\substack{\uparrow\\ \text{Region}\\ OBDO}}{\dot{w}_1} \quad \underset{\substack{\uparrow\\ \text{Region}\\ OABO}}{\dot{w}_2} \quad - \quad \underset{\substack{\uparrow\\ \text{Region}\\ OADO}}{\dot{w}_3} \quad - \quad \underset{\substack{\uparrow\\ \text{Region}\\ ACDA}}{\dot{w}_4}$$

$$= \gamma\Omega r_o^3(f_1-f_2-f_3-f_4) \tag{5.89}$$

where f_1,f_2,f_3,f_4 = function of H, r_o, β, β', α, θ_o, θ_h, and L

Ω = angular velocity of the region

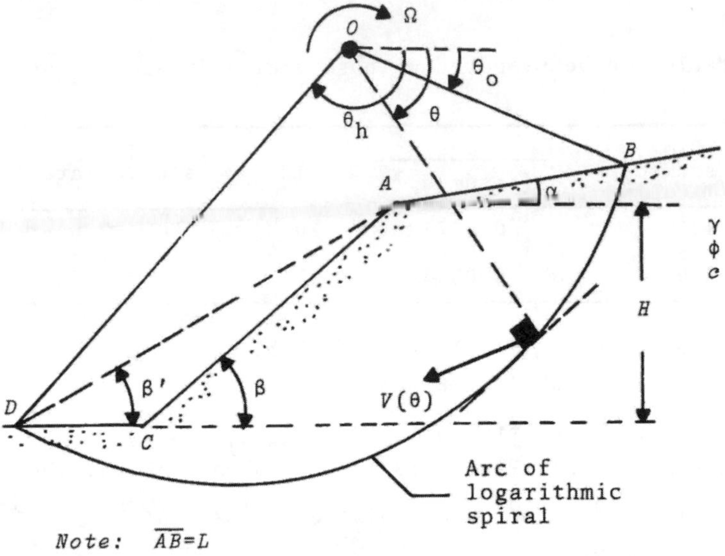

Figure 5.44. Limit analysis solution of slopes.

(Note that $OB = r_o$ and $AB = L$.) For example

$$f_4 = \left(\frac{H}{r_o}\right)^2 \frac{\sin(\beta-\beta')}{2\sin\beta\sin\beta'}\left[\cos\theta_o - \left(\frac{L}{r_o}\right)\cos\alpha - \frac{1}{3}\left(\frac{H}{r_o}\right)(\cot\beta' + \cot\beta)\right] \quad (5.90)$$

The internal dissipation of energy occurs along the surface of discontinuity BD. Equating $\dot{w}$ with the rate of dissipation of energy, it can be shown that

$$H \quad \frac{c}{\gamma} f(\theta_h, \theta_o, \beta') \quad (5.91)$$

where

$$f(\theta_h, \theta_o, \beta') = \left[\frac{\sin\beta'\{\exp[2(\theta_h - \theta_o)\tan\phi] - 1\}}{2\sin(\beta'-\alpha)\tan\phi(f_1 - f_2 - f_3 - f_4)}\right]\{\sin(\theta_h + \alpha)$$

$$\exp[(\theta_h - \theta_o)\tan\phi] - \sin(\theta_o + \alpha)\} \quad (5.92)$$

The function $f(\theta_h, \theta_o, \beta')$ has a minimum and, hence, it indicates a least upper bound when

$$\frac{\partial f}{\partial \theta_o} = 0 \quad (5.93)$$

$$\frac{\partial f}{\partial \theta_h} = 0 \quad (5.94)$$

$$\frac{\partial f}{\partial \beta'} = 0 \tag{5.95}$$

The values of θ_o; θ_h, and β' which will satisfy Eqs. (5.93), (5.94) and (5.95) will given the minimum value of $f(\theta_h,\theta_o,\beta')$. Hence, from Eq. (5.91)

$$\frac{\gamma H}{c} = N_s \tag{5.96}$$

where N_s min $f(\theta_h,\theta_o,\beta')$ (5.97)

The variation of $1/N_s$ for various values of ϕ and β as obtained by Chen and Giger (1971) for α=0 are given in Fig. 5.45. It needs to be pointed out that, for small values of soil friction angle ϕ, the most critical failure surface will pass below the toe of the slope as shown in Fig. 5.44. For larger friction angles,

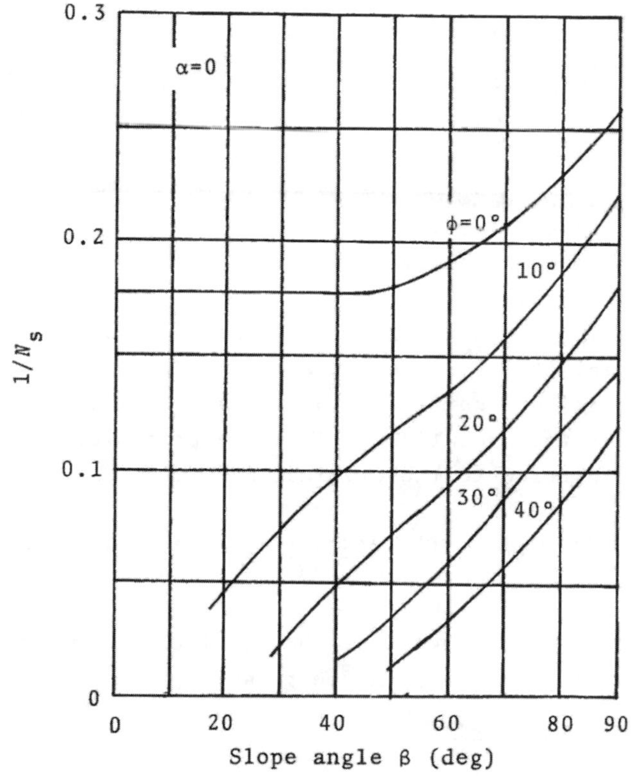

Figure 5.45. Variation of $1/N_s$ vs. β for various values of soil friction angle ϕ--based on the limit analysis solution.

the failure surface will pass through the toe.

Table 5.5 gives a comparison of $1/N_s$ as obtained from the limit analysis results with those obtained from the friction circle method (Section 5.7.1--Limit Equilibrium Solution). It can be seen that the results obtained from these two methods compare well.

Table 5.5. Comparion of $1/N_s$ (α=0)

Friction angle ϕ (deg)	Slope angle β (deg)	$1/N_s$	
		Limit analysis solution[b]	Friction circle method[a]
0	15	0.181	0.145
5	15	0.0695	0.068
15	15	0.023	0.022
0	30	0.181	0.156
5	30	0.1095	0.110
15	30	0.046	0.046
0	45	0.181	0.170
5	45	0.136	0.136
15	45	0.083	0.083
0	60	0.1905	0.1908
5	60	0.1623	0.1618
15	60	0.116	0.116

[a]Based on Section 5.7.1

[b]Based on Chen and Giger's solution (1971)

Chen, Snitbhan, and Fang (1975) used the limit analysis to study the stability of slopes in anisotropic but homogeneous soil assuming the failure surface to be the arc of a logarithmic spiral. The results of this solution will not be presented here, and the readers are referred to the original paper.

5.10 ESTIMATION OF DITCH SAFETY BY USING A CYCLOIDAL FAILURE SURFACE

Equation (5.14) shows that the critical height, H_{cr}, of a slope with a slope angle β can be given as

$$H_{cr} \quad \frac{4c}{\gamma}\,\frac{\sin\beta\cos\phi}{1-\sin(\beta-\phi)} \tag{5.14}$$

The preceding equation was derived on the assumption that the *potential failure surface is a plane* and the soil mass above the failure surface acts as a unit. For vertical cuts (β=90°), the above equation simplifies to the form

$$H_{cr} = \left(\frac{4c}{\gamma}\right) \tan(45+\phi/2) \tag{5.98}$$

Keeping in mind several collapses of open unsupported ditches and the casualities associated with them, Ellis (1973) proposed a conservative approach to evaluate the safety of ditch cuts. This approach assumes that the critical failure surface is the *arc of a cycloid*. Figure 5.46 shows the basic parameters for the arc *OA* of a cycloid which is generated by the rotation of a circle of radius *R*. The equations for the arc of the cycloid can be given as

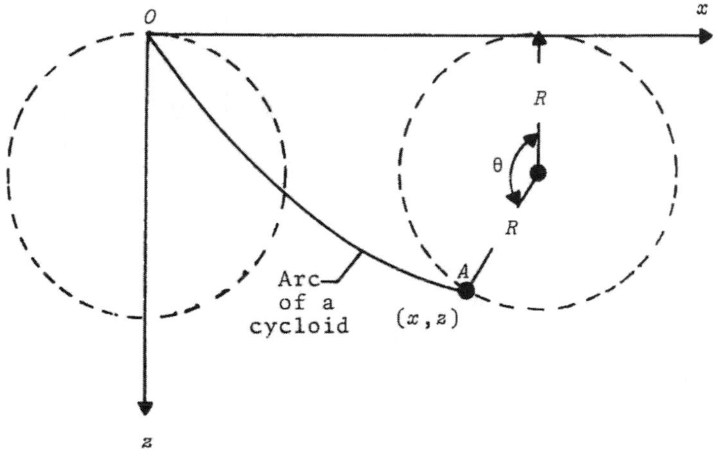

Figure 5.46. Parameters of a cycloid.

$$x = R(\theta - \sin\theta) \tag{5.99a}$$

and

$$z = R(1-\cos\theta) \tag{5.99b}$$

where θ angle of rotation of the circle between points *O* and *A*

Now, consider a vertical unsupported cut as shown in Fig. 5.47. The height of the cut is *H*, and *BC* is a potential failure surface, which is an arc of a cycloid. Let us also assume that the soil above the failure surface will not necessarily act as a solid mass, and also any excess of shear strength at some point along the failure surface *will not balance* the deficiencies of shear strength at some other point below that. In Fig. 5.47, *EFGJ*

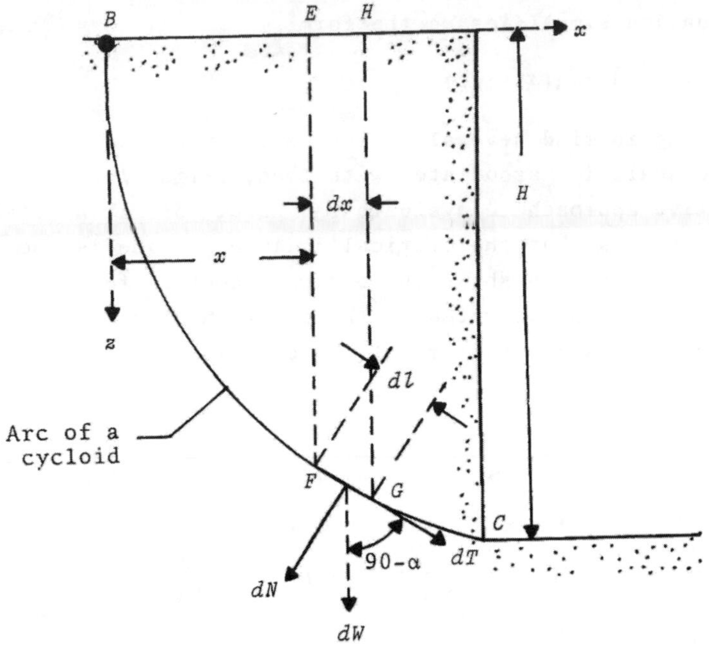

Figure 5.47. Vertical unsupported cut.

is an elementary strip of soil mass of width dx. The weight of the strip per unit length at right angles to the cross section can be given as

$$dW \quad \gamma z dx \tag{5.100}$$

where dW = weight of the strip

γ = unit weight of soil

Combining Eqs. (5.99a), (5.99b), and (5.100) gives

$$dW = \gamma[R(1-\cos\theta)]d[R(\theta-\sin\theta)] \quad \gamma R^2(1-\cos\theta)^2 d\theta \tag{5.101}$$

The normal and tangential components of dW can be expressed as

$$\text{Normal component} = dN = dW\cos\alpha \tag{5.102}$$

and

$$\text{Tangential component} = dT = dW\sin\alpha \tag{5.103}$$

The angle α is defined in Fig. 5.47 and can be given as

$$\alpha = \tan^{-1}\left(\frac{dz}{dx}\right)$$

Combining Eq. (5.99) and the preceding equations for α

$$\tan\alpha = \frac{\cos\theta/2}{\sin\theta/2} \qquad (5.104)$$

or

$$\sin\alpha = \cos\theta/2 \qquad (5.105a)$$

and

$$\cos\alpha = \sin\theta/2 \qquad (5.105b)$$

The effective normal stress (σ') and induced shear stress (τ) along the elementary length $FG=dl$ can now be determined from Eqs. (5.101), (5.102), (5.103), and (5.105)

$$\sigma' = \frac{dN}{dl} = \frac{[\gamma R^2(1-\cos\theta)^2 d\theta]\sin\theta/2}{dl} \qquad (5.106)$$

$$\tau \quad \frac{dT}{dl} \quad \frac{[\gamma R^2(1-\cos\theta)^2 d\theta]\cos\theta/2}{dl} \qquad (5.107)$$

However

$$dl = \sqrt{(dx)^2+(dz)^2}$$

From Eq. (5.99), $dx=R(1-\cos\theta)d\theta$ and $dz=R\sin\theta d\theta$. So

$$dl = \sqrt{(dx)^2+(dz)^2} = \sqrt{[R(1-\cos\theta)d\theta]^2 + R^2\sin^2\theta(d\theta)^2}$$

$$= 4R(\sin\theta/2)d(\theta/2) \qquad (5.108)$$

Substitution of Eq..(5.108) into Eqs. (5.106) and (5.107) yields

$$\sigma = 2R\gamma\sin^4(\theta/2) \qquad (5.109)$$

$$\tau = 2R\gamma\sin^3(\theta/2)\cos(\theta/2) \qquad (5.110)$$

The maximum shear strength that can be mobilized along the elementary length FG in Fig. 5.47 is

$$s = c+\sigma'\tan\phi$$

Substituting the relation for normal stress [Eq. (5.109)] into the preceding equation, we get

$$s \quad c+2R\gamma\sin^4(\theta/2)\tan\phi \qquad (5.111)$$

If the critical height $H=H_{cr}$ of a vertical cut is defined as the height for which *no point along the critical failure surface has a factor of safety of less than one*, then

$$s = \tau \qquad (5.112)$$

Substitution of the relations for s and τ [Eqs. (5.110) and (5.111)] into Eq. (5.112) yields

$$c+2R\gamma\sin^4(\theta/2)\tan\phi = 2R\gamma\sin^3(\theta/2)\cos(\theta/2)$$

or

$$R = \frac{c}{2\gamma[\sin^3(\theta/2)\cos(\theta/2) - \sin^4(\theta/2)\tan\phi]} \tag{5.113}$$

From Eq. (5.99b)

$$R \quad \frac{z}{1-\cos\theta} \tag{5.114}$$

Combining Eqs. (5.113) and (5.114)

$$z \quad \frac{c}{\gamma}\left[\frac{1}{\sin(\theta/2)\cos(\theta/2) - \sin^2(\theta/2)\tan\phi}\right] \tag{5.115}$$

For the critical value of θ where z is a minimum

$$\frac{\partial z}{\partial \theta} \quad 0 \tag{5.116}$$

From Eqs. (5.115) and (5.116)

$$\theta = \theta_c = 90-\phi \tag{5.117}$$

So

$$z = H_{cr} = \frac{c}{\gamma}\left[\frac{1}{\sin(\theta_c/2)\cos(\theta_c/2) - \sin^2(\theta_c/2)\tan\phi}\right] \tag{5.118}$$

The radius of the generating circle R for which $z=H_{cr}$ and $\theta=\theta_c$ can be obtained from Eqs. (5.99b), (5.117), and (5.118) as

$$R \quad \frac{H_c}{1-\cos\theta_c} = \frac{H_c}{1-\sin\phi} \quad \frac{2c}{\gamma}\left[\frac{\tan(45+\phi/2)}{(1-\sin\phi)}\right] \tag{5.119}$$

Therefore, the induced shear stress at any point along the critical slip surface can be obtained by combining Eqs. (5.110) and (5.119), or

$$\begin{aligned}\tau &= 2R\gamma\sin^3(\theta/2)\cos(\theta/2)\\ &= 2\left(\frac{2c}{\gamma}\right)\gamma\left[\frac{\tan(45+\phi/2)}{1-\sin\phi)}\sin^3(\theta/2)\cos(\theta/2)\right]\\ &= 4c\left[\frac{(1+\sin\phi)}{\cos\phi(1-\sin\phi)}\right]\sin^3(\theta/2)\cos(\theta/2)\end{aligned} \tag{5.120}$$

In a similar manner, the mobilizable shear strength at any point along the critical slip surface can be obtained by combining Eqs. (5.111) and (5.119), or

$$s \quad c+2R\gamma\sin^4(\theta/2)\tan\phi$$

$$c+2\left(\frac{2c}{\gamma}\right)\gamma\left[\frac{\tan(45+\phi/2)}{(1-\sin\phi)}\right]\sin^4(\theta/2)\tan\phi$$

$$= c\left[1+(4)\frac{\sin\phi}{(1-\sin\phi)^2}\sin^4(\phi/2)\right] \tag{5.121}$$

Now, the factor of safety against sliding for the cut can be given as

$$F_s = \frac{s}{\tau} = \frac{c\left[1+\dfrac{4\sin\phi}{(1-\sin\phi)^2}\sin^4(\theta/2)\right]}{4c\left[\dfrac{(1+\sin\phi)}{\cos\phi(1-\sin\phi)}\right]\sin^3(\phi/2)\cos(\phi/2)}$$

$$= \frac{\cos\phi(1-\sin\phi)}{4(1+\sin\phi)\sin^3(\theta/2)\cos(\theta/2)} + \tan\phi\tan(\theta/2) \tag{5.122}$$

Following is a step-by-step procedure to obtain the factor of safety against sliding for vertical cuts in soil.

1. Determine γ, c, ϕ, and H

2. Determine θ generated along the critical slip surface. This can be done by using Eq. (5.99b) which gives

$$z = H \quad R(1-\cos\theta)$$

or

$$\cos^{-1}\left(1-\frac{H}{R}\right) = \theta \tag{5.123}$$

For the critical slip surface [Eq. (5.119)]

$$R = \frac{2c}{\gamma}\left[\frac{\tan(45+\phi/2)}{(1-\sin\phi)}\right]$$

Hence, from Eqs. (5.119) and (5.123)

$$\theta = \cos^{-1}\left[1-\left(\frac{H\gamma}{c}\right)\left(\frac{(1-\sin\phi)}{2\tan^2(45+\phi/2)}\right)\right] \tag{5.124}$$

3. With the value of θ calculated from Eq. (5.124), determine the value of F_s by using Eq. (5.122).

If the value of θ calculated by using Eq. (5.124) is equal to or less than $\theta_c=90-\phi$, the factor of safety will be equal to or less than one. If $\theta>\theta_c$, then the value of F_s will be less than one.

It should be kept in mind that the factor of safety determined by the above procedure is conservative because of the fact that it gives the minimum factor of safety that occurs at any given point along the critical slip surface. (In Fig. 5.47, this point is *C*.)

In most stability analysis problems, the factor of safety would have been given as (refer to Fig. 5.47)

$$F_s = \frac{\int s\, dl}{\int \tau\, dl} \tag{5.125}$$

Example 5.10. Consider a vertical cut with the following: H=2 m, γ=17.5 kN/m^3, c=16 kN/m^2, and ϕ=12°. Determine the factor of safety of the cut by using the procedure outlined in Section 5.10.

Solution

From Eq. (5.124)

$$\theta = \cos^{-1}\left[1-\left(\frac{H\gamma}{c}\right)\left(\frac{(1-\sin\phi)}{2\tan^2(45+\phi/2)}\right)\right]$$

or

$$\theta = \cos^{-1}\left[1-\frac{(2)(17.5)}{16}\;\frac{(1-\sin 12°)}{2\tan^2(45+\phi/2)}\right]$$

$$= \cos^{-1}\left[1-\frac{(2.188)(0.792)}{3.05}\right] = \cos^{-1}(0.4318) = 64.4°$$

Note that θ=64.4°<θ_c=90°-ϕ=90°-12°=78°. Now, substituting θ=64.4° and ϕ=12° in Eq. (5.122) yields F_s=<u>1.4</u>.

5.10.1 Calculation of the Side Slopes For Ditch With Deeper Cuts

As pointed out in the preceding section (Section 5.10), if the value of θ calculated by using Eq. (5.124) becomes more than θ_c= 90-ϕ, the factor of safety calculated by using Eq. (5.122) will be less than one. For the safety of the ditch against sliding, the sides of the cuts should be inclined rather than vertical. This subsection discusses the procedure for determination of the critical value of the side slope for which the value of F_s will be equal to or more than one at any given point along the critical surface of sliding as recommended by Ellis (1973).

Consider the side of a ditch of height H as shown in Fig. 5.48. BCD is the critical sliding surface. Let the height H be greater than the critical height H_{cr} [Eq. (5.118)], and also let the vertical distance between points B and C on the critical sliding surface be equal to H_{cr}. Hence, at point C the value of F_s=1. Now, if the slope is inclined, the factor of safety will increase from one at C to infinity at point D. The critical slope angle, β_{cr}, can be determined in the following manner.

From Fig. 5.48, it can be seen that

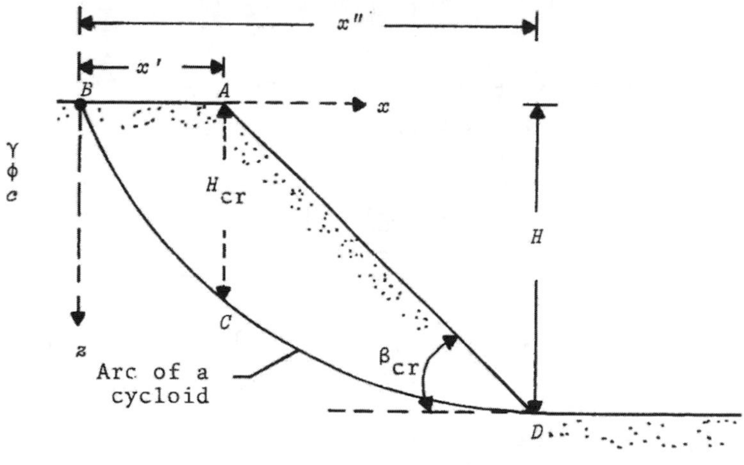

Figure 5.48. Side slope for ditch for deeper cut.

$$\beta_{cr} = \tan^{-1}\left(\frac{H}{x''-x'}\right) \tag{5.126}$$

where $x' = R(\theta'-\sin\theta')$ (5.127)

$$x'' = R(\theta''-\sin\theta'') \tag{5.128}$$

θ', θ'' = final angle which the cycloid generating circle has rotated to reach points C and D, respectively

Substitution of Eqs. (5.127) and (5.128) into Eq. (5.126) yields

$$\beta_{cr} = \tan^{-1}\left[\frac{H}{R(\theta''-\theta'-\sin\theta''+\sin\theta')}\right] \tag{5.129}$$

However, $\theta'=\theta_c=(\pi/2)-\phi$. Also, for critical failure surface [Eq. (5.119)]

$$R = \frac{2c}{\gamma}\,\frac{\tan(45+\phi/2)}{(1-\sin\phi)} \tag{5.119}$$

Substitution of the above relations for θ_c and R into Eq. (5.129) and simplification gives

$$\beta_{cr} = \tan^{-1}\left[\frac{H\gamma}{c}\,\frac{\cos\phi(1-\sin\phi)}{2(1+\sin\phi)(\theta''-\sin\theta''-\pi/2+\phi+\cos\phi)}\right] \tag{5.130}$$

Following is a step-by-step procedure to calculate β_{cr}:

1. Determine γ, c, ϕ, and H.
2. Determine θ'' by using Eq..(5.124).
3. If $\theta''>90-\phi$, then $H>H_{cr}$. So calculate β_{cr} by using Eq.

(5.129).

Example 5.11. Refer to Example 5.10. Other quantities remaining the same, if H=3 m determine β_{cr}.

Solution.

$$\theta = \cos^{-1}\left[1-\left(\frac{H\gamma}{c}\right)\ \frac{(1-\sin\phi)}{2\tan^2(45+\phi/2)}\right]$$

So

$$\theta = \cos^{-1}\left[1-\ \frac{(3)(17.5)}{16}\ \frac{1-\sin 12°}{2\tan^2(45+12/2)}\right]$$

$$\cos^{-1}\left[1-\ \frac{(3.281)(0.792)}{3.05}\right] \quad \cos^{-1}(0.148) = 81.49°$$

However

$$\theta_c = 90-\phi = 90-12 \quad 68°$$

So, $\theta > \theta_c$. Now, θ=81.49°=θ''. Substituting ϕ=12° and θ''=81.49° in Eq. (5.130) gives

$$\beta_{cr} = \tan^{-1}\left\{\frac{(3)(17.5)}{16} \times \frac{\cos 12(1-\sin 12)}{2(1+\sin 12)\left[(\frac{\pi}{180}\ 81.49)-\sin 81.49-\pi/2+(\frac{\pi}{180}\ 12)+\cos 12°\right]}\right\}$$

$$= \underline{87°}$$

References

Bishop, A.W., 1955. The use of the slip circle in the stability analysis of slopes. Geotechnique, 5(1):7-17.

Bishop, A.W. and Morgenstern, N.R., 1960. Stability coefficient of earth slopes. Geotechnique, 10(4):129-150.

Casagrande, A. and Carrillo, N. 1944. Shear failure of anisotropic soils. J. Boston Soc. Civ. Eng. Contribution to Soil Mechanics 1941-1953.

Chen, W.F., 1970. Discussion. J. Soil Mech. Found. Div., ASCE, 96(SM1):324-326.

Chen, W.F. and Giger, M.W., 1971. Limit analysis of stability of slopes. J. Soil Mech. Found. Div., ASCE, 97(SM1):19-26.

Chen, W.F., Giger, M.W., and Fang, H.Y., 1969. On the limit analysis of stability of slopes. Soils and Foundations, 9(4):23-32.

Chen, W.F., Snitbhan, N., and Fang, H.Y., 1975. Stability of slopes in aniostropic,nonhomogeneous soils. Canadian Geotech. J., 12:150.

Cousins, B.F., 1978. Stability charts for simple earth slopes. J. Geotech. Eng. Div., ASCE, 104(GT2):267-279.

Culmann, K., 1866. Die graphische statik. Meyer and Zeller,

Zurich, Switzerland.
Drucker, D.C. and Prager, W., 1952. Soil mechanics and plastic analysis of limit design. Q. Appl. Math. 10:157-165.
Ellis, H.B., 1973. Use of cycloidal arcs for estimating ditch safety. J. Soil Mech. Found. Div., ASCE, 99(SM2):165-179.
Fellenius, W., 1927. Erdstatische berechnungen. W. Ernst U. Sohn, Berlin.
Koppula, S.D., 1984. On stability of slopes on clays with linearly increasing strength. Canadian Geotech. J., 21(3):577-581.
Lo, K.Y., 1965. Stability of slopes in anisotropic soils. J. Soil Mech. Found. Div., ASCE, 91(SM4):85-106.
Morgenstern, N., 1963. Stability charts for earth slopes during rapid drawdown. Geotechnique, 13(2):121-131.
O'Connor, M.J. and Mitchell, R.J., 1977. An extension of the Bishop and Morgenstern slope stability charts. Canadian Geotech. J., 14(1):144-151.
Singh, A., 1970. Shear strength and stability of manmade slopes. J. Soil Mech. Found. Div., ASCE, 96(SM6):1879-1892.
Spencer, E., 1969. Circular and logarithmic spiral slip surfaces. J. Soil Mech. Found. Div., ASCE, 95(SM1):227-234.
Spencer, E., 1967. A method of analysis of the stability of embankments assuming parallel inter-slice forces. Geotechnique, 17(1):11-26.
Taylor, D.W., 1937. Stability of earth slopes. J. Boston Soc. Civ. Eng., 24:197-246.
Terzaghi, K., and Peck, R.B., 1967. Soil mechanics in engineering practice, 2nd ed. John Wiley and Sons, New York.

APPENDIX A
The Pole Method for Finding Stresses from Mohr's Circle

In geotechnical engineering, Mohr's circle is an extremely useful tool for determination of stresses on a given plane in a soil mass. The principles of Mohr's circle will be demonstrated in this appendix.

Let *ABCD* be a soil element as shown in Fig. A.1. The stresses on the horizontal faces (*AB* and *CD*) of the element are:

Normal stress	$+\sigma_x$
Shear stress	$+\tau$

Similarly, the stresses on the vertical faces (*AD* and *BC*) are:

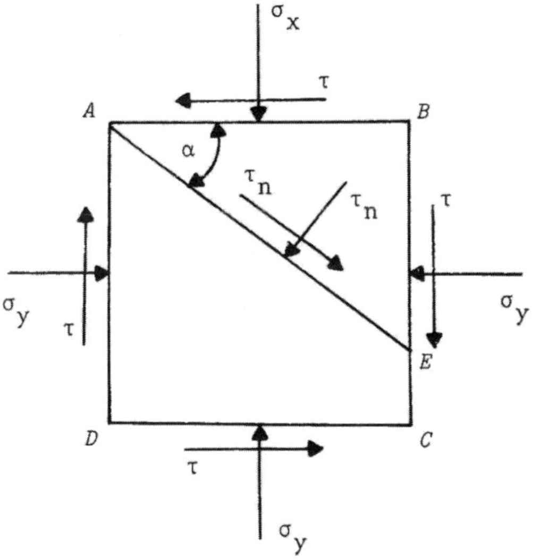

(a)

Figure A.1. The pole method for finding stresses along a plane.

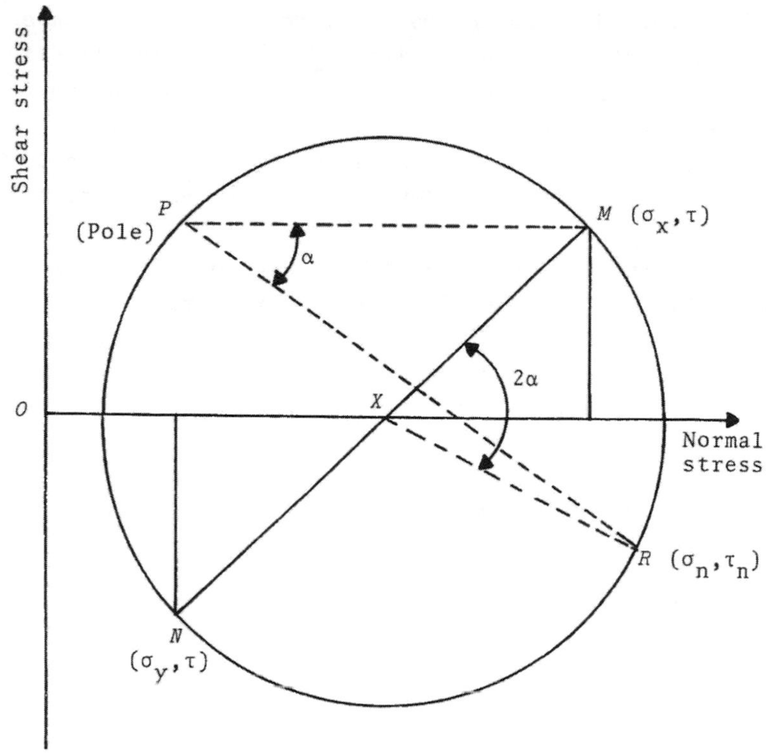

(b)

Figure A.1. (Continued).

Normal stress	$+\sigma_y$
Shear stress	$-\tau$

The sign conventions used for the stresses stated above are:

Normal stress	*+ve* for compression
Shear stress	*+ve* if they act on opposite faces in such as way as to create counterclockwise rotation

The above-mentioned stresses on the vertical and horizontal planes of the soil element are presented in a graphical form as points *M* and *N*, respectively, in Fig. A.1b. In this figure

$$OX \quad \frac{\sigma_x+\sigma_y}{2} \tag{A.1}$$

If a circle is drawn with *X* as the center and *XM* as the radius it will be referred to as the Mohr's circle. Point *M* represents the stresses on the horizontal plane. So, if a line is drawn from this point which is parallel to the plane on which the corresponding stresses act (in this case, horizontal plane), it will intersect the Mohr's circle at point *P*, which is called the *pole*. It is a unique point for the stress conditions shown on the element *ABCD* (Fig. A.1a).

If we are required to find the stresses on a plane such as *AE* as shown in Fig. A.1a, we need to draw a line from the pole which will be parallel to *AE*. This line is *PR* in Fig. A.1b. The coordinates of the point of intersection of this line with the Mohr's circle (point *R* in this case) will give the normal and shear stresses on the required plane.

APPENDIX B
Properties of Logarithmic Spirals and Logarithmic Spiral Sectors

In the analysis of lateral earth pressure, bearing capacity of shallow foundations, and stability of slopes, the entire or a part of the failure surface in the soil is assumed to be the arc of a logarithmic spiral. Following are some useful properties of logarithmic spirals and logarithmic spiral sectors.

The general equation of a logarithmic spiral (Fig. B.1) can be given as

$$r = r_o e^{\theta \tan\phi} \tag{B.1}$$

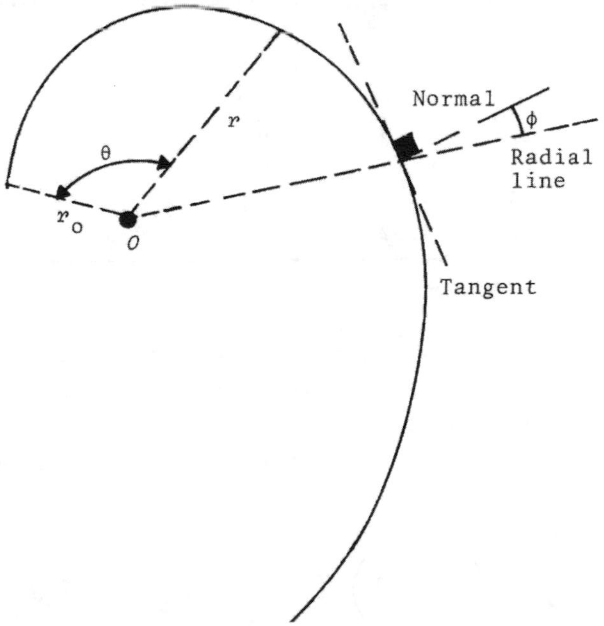

Figure B.1. Logarithmic spiral.

where θ = angle, in radians

ϕ = soil friction angle

Or

$$\frac{r}{r_o} = e^{\theta \tan\phi} \qquad \text{(B.2)}$$

Table B.1 gives the variations of r/r_o with θ and ϕ.

Table B.1. Variation of r/r_o with θ and ϕ

θ (deg)	r/r_o, ϕ (deg) 20	25	30	35	40
0	1	1	1	1	1
10	1.066	1.085	1.106	1.130	1.158
20	1.135	1.127	1.223	1.277	1.340
30	1.210	1.277	1.353	1.443	1.552
40	1.289	1.385	1.496	1.63	1.796
50	1.375	1.502	1.655	1.842	2.080
60	1.463	1.63	1.831	2.082	2.408
70	1.560	1.768	2.025	2.352	2.786
80	1.662	1.917	2.239	2.658	3.227
90	1.771	2.080	2.477	3.004	3.736
100	1.887	2.257	2.739	3.394	4.325
110	2.011	2.448	3.029	3.836	5.008
120	2.143	2.655	3.351	4.334	5.797
130	2.284	2.881	3.706	4.897	6.712
140	2.434	3.125	4.099	5.534	7.770
150	2.593	3.390	4.534	6.253	8.996
160	2.763	3.677	5.014	7.066	10.415
170	2.944	3.989	5.546	7.985	12.057
180	3.138	4.327	6.134	9.023	13.959

Another property of a log spiral is that, at any given point, the radial line makes an angle ϕ with the normal drawn to the spiral. This is shown in Fig. B.1.

The area of a sector of a logarithmic spiral can be given as (Fig. B.2)

$$A = \int dA = \int \tfrac{1}{2} r_o \left(r_o e^{\theta \tan\phi} \right)^2 d\theta \quad \int_0^{\theta_1} \tfrac{1}{2} r_o^2 e^{2\theta \tan\phi}\, d\theta$$

$$= \frac{r_1^2 - r_o^2}{4\tan\phi} \qquad \text{(B.3)}$$

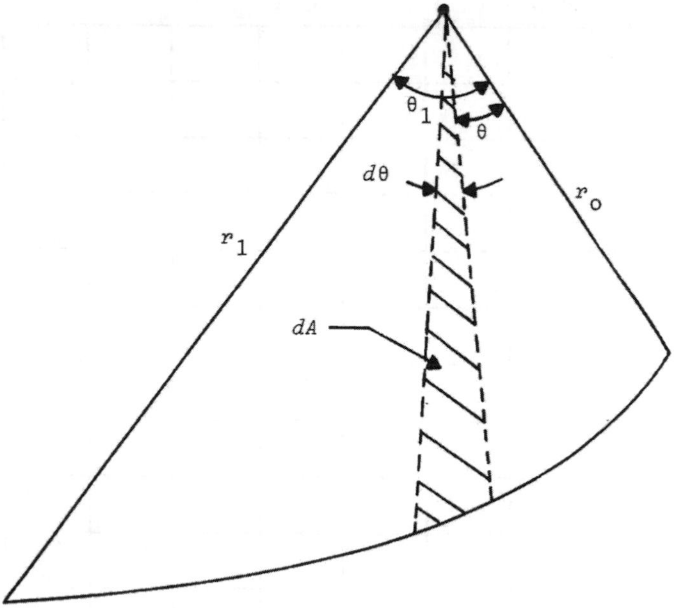

Figure B.2. Area of the sector of a logarithmic spiral.

$$A \quad \frac{r_o^2[(r_1/r_o)^2-1]}{4\tan\phi}$$

or

$$\frac{A}{r_o^2} = \frac{[(r_1/r_o)^2-1]}{4\tan\phi} \tag{B.4}$$

Figure B.3 shows the plot of A/r_o^2 with r_1/r_o and ϕ.

The location of the centroid (Fig. B.4) of a sector of a logarithmic spiral can be given as (Hijab, 1956)

$$a = \left[\frac{4r_o\tan\theta_1}{3(9\tan^2\phi+1)}\right]\left[\frac{(r_1/r_o)^3(3\tan\phi\sin\theta_1-\cos\theta_1)+1}{(r_1/r_o)^2-1}\right] \tag{B.5}$$

$$b = \left[\frac{4r_o\tan\phi}{3(9\tan^2\phi+1)}\right]\left[\frac{(r_1/r_o)^2-3\tan\phi\sin\theta_1-\cos\theta_1}{(r_1/r_o)^2-1}\right]$$

References

Hijab, W.A., 1956. A note on the centroid of a logarithmic spiral sector. Geotechnique, 6(2):96-99.

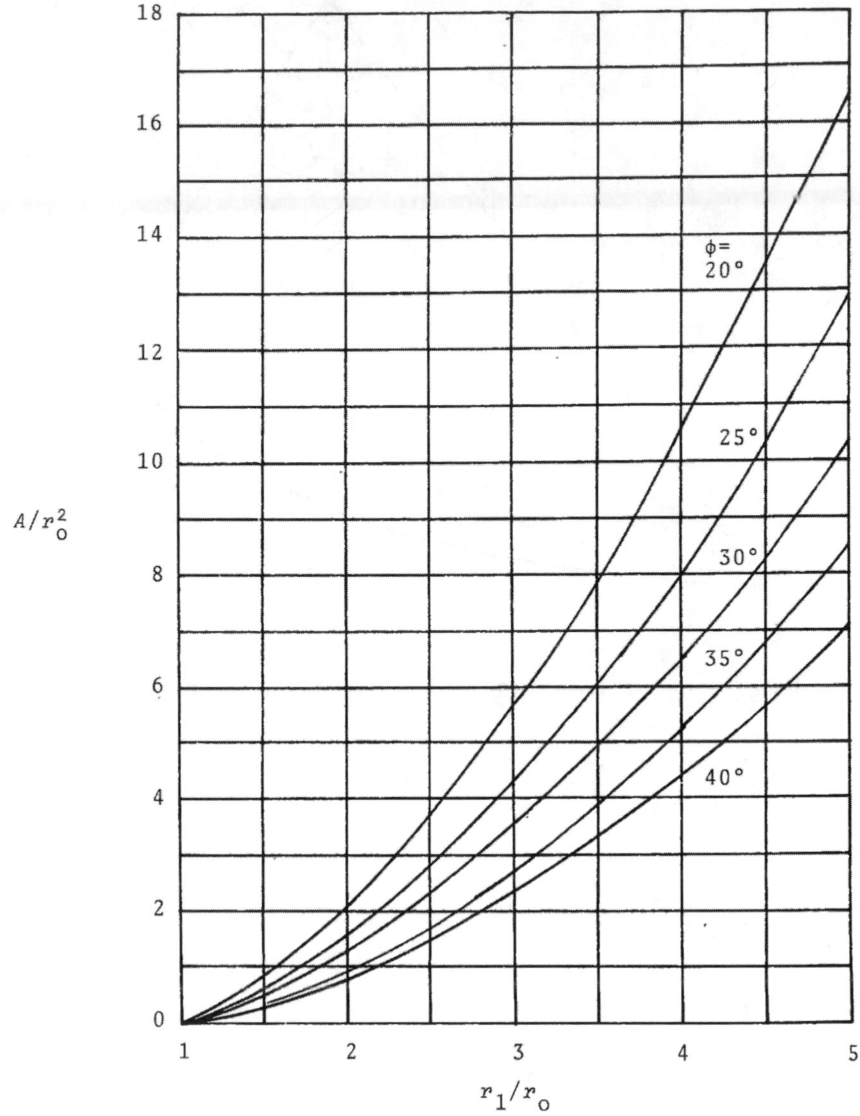

Figure B.3. Plot of A/r_o^2 with r_1/r_o and ϕ [Eq. (B.4)].

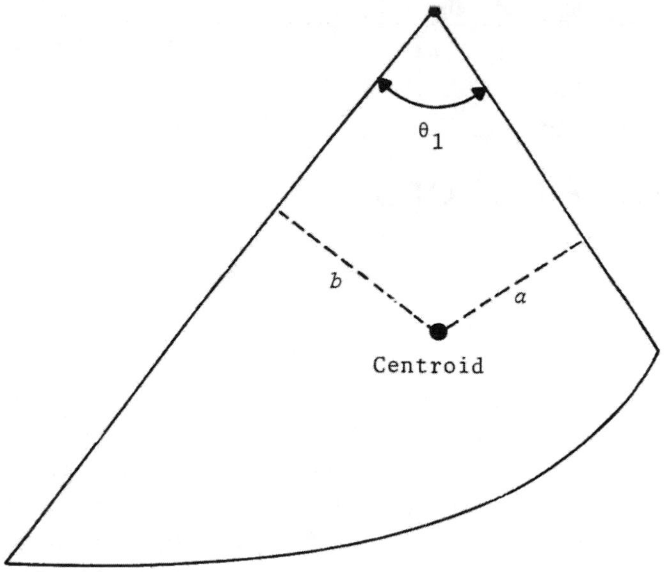

Figure B.4. Centroid of the sector of a logarithmic spiral.

APPENDIX C
Helical Anchors

In Chapter 3, the ultimate uplift capacity of helical anchors has been discussed. The helical anchors provide resistance to the uplifting loads on foundations. However, they can also increase the compressive load bearing capacity of shallow foundations by transferring load to the soil at greater depths at the soil-helix interface. Figures C.1 and C.2 show photographs of helical anchors. Figure C.3 shows the installation procedure for a helical anchor.

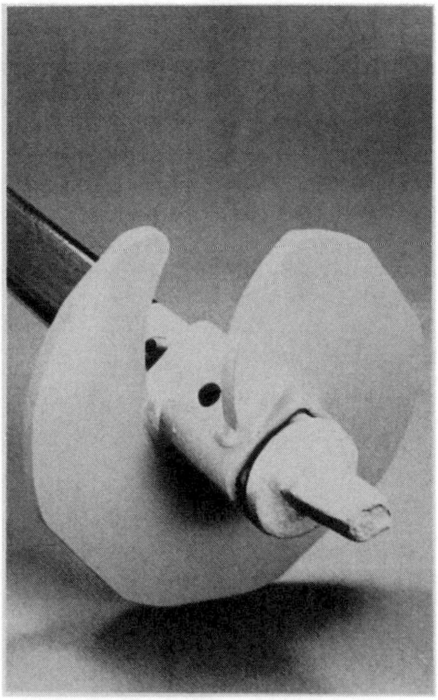

Figure C.1. Helical anchor with one helix. (Courtesy of A. B. Chance Company, Centralia, Missouri, U.S.A.)

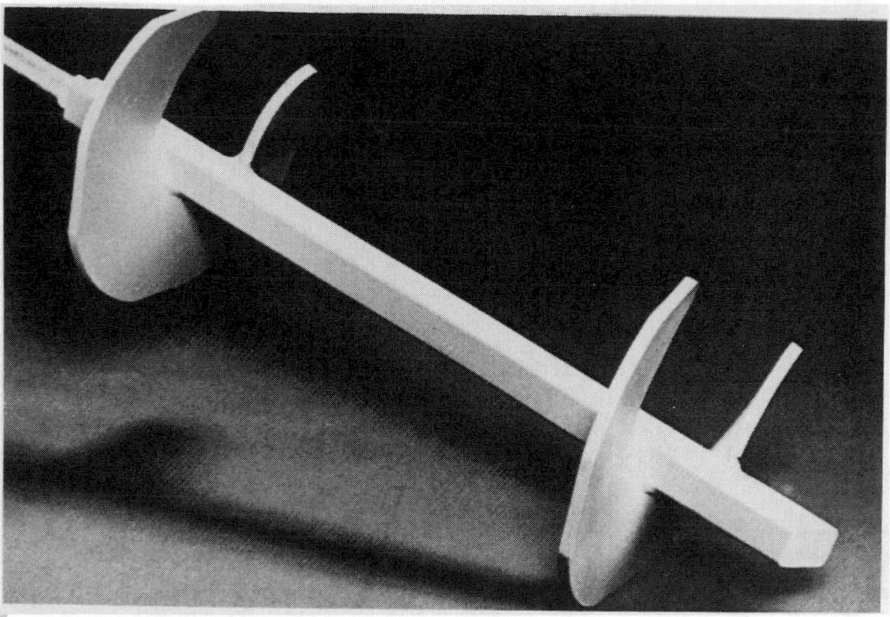

Figure C.2. Helical anchor with two helices. (Courtesy of A. B. Chance Company, Centralia, Missouri, U.S.A.)

Figure C.3. Installation of helical anchor. (Courtey of A. B. Chance Company, Centralia, Missouri, U.S.A.)

INDEX

C

D

E

Q

R

S

T

U

V

W